Experimental Embryogenesis
in Vascular Plants

EXPERIMENTAL BOTANY
An International Series of Monographs

CONSULTING EDITOR

J. F. Sutcliffe

School of Biological Sciences, University of Sussex, England

Forthcoming Titles

Experimental Embryogenesis in Vascular Plants

V. RAGHAVAN

Department of Botany
The Ohio State University

1976
ACADEMIC PRESS
London New York San Francisco

A Subsidiary of Harcourt Brace Jovanovich, Publishers

ACADEMIC PRESS INC. (LONDON) LTD.
24/28 Oval Road,
London NW1 7DX

United States Edition published by
ACADEMIC PRESS INC.
111 Fifth Avenue
New York, New York 10003

Library of Congress Catalog Number: 76-1096
ISBN: 0 12 575450 7

PRINTED IN GREAT BRITAIN BY
BUTLER AND TANNER LTD.
FROME AND LONDON

Preface

The study of embryogenesis in plants is of far more than academic interest, since most of the food products used by man are derived from processes related to embryogenesis and seed formation in higher plants. The survival of man and his civilization depends largely upon his ability to circumvent the normal reproductive processes in plants and thereby produce an abundance of food, efficiently, dependably and cheaply. To a considerable extent, practical methods to improve the quality of our agriculture have their origins in fundamental principles discovered in the laboratory, and therefore an understanding of those principles is likely to enrich our attempts in the field. These considerations, as well as the fact of my botanical interest being largely confined to the reproductive processes in vascular plants, prompted me to bring together in a comprehensive form an account of the experimental investigations on embryogenesis to reflect the great expansion in research that has taken place in this area during the last decade.

The growth and development of embryos is an aspect of plant science of considerable contemporary interest and constitutes one of the main focal subjects within the whole area of plant development. Historically, as was the case in many areas of botany, the study of embryos began as a descriptive field of study. Gradually, interest in embryogenesis has evolved in a direction utilizing modern techniques of tissue culture, cytology, physiology and biochemistry and unified by the thread of relationship between structure and function. This functional approach is concerned with an understanding of how embryos grow, how growth mechanisms determine their structure, and how these mechanisms can be modified. In more recent years, the strategy in the study of embryos has been completely changed by discoveries relating to the concept of totipotency and the production of somatic embryos from plant cells. The major emphasis in this book, which simply reflects an exciting era of research in developmental botany, is centered about these dynamic activities of embryos as they develop from their single-celled origin and interact with their milieu. In writing this book, I have in mind

advanced students, professional workers and teachers including agro-
nomists, foresters, horticulturists and plant physiologists interested
in the study of embryogenesis in plants. I believe that this book can
help students and research workers in these fields by providing them
with a ready source of information on experimental studies on plant
embryos. Some of the concepts described herein have practical over-
tones which will be of use to those wishing to apply them.

The sources of illustrations used in this book are credited in the
captions that accompany each. However, here I wish to express my
appreciation to my professional colleagues who kindly provided photo-
graphic prints from their works for inclusion in the book and to their
publishers, too many to list here, who generously gave me reproduction
rights for illustrations and tables used here. Although a succession of
secretaries have typed drafts of this book, the final version was typed
by Ms. Mary Malone and Ms. Pat Walker, to whom I express my
thanks for their careful and conscientious effort. I am also indebted
to my wife, Lakshmi, who gave up other interests to translate articles
from French for my use. This book has been in the making for a long
time. During this period, when I was spending untold hours in my
office, I was supported by my family, without whose understanding
and fortitude it would not have been possible to complete this work.

June, 1976   V. RAGHAVAN

Contents

Section II Adventive Embryogenesis

Section III From Seed to Seedling

Abbreviations

The following abbreviations are used in this book:

AMO-1618 = 4-hydroxy-5-isopropyl-2-methylphenyl trimethylammonium chloride, 1-piperidine carboxylate

AMP = 3′,5′-adenosine monophosphate

ATP = adenosine triphosphate

2,4-D = 2,4-dichlorophenoxyacetic acid

DNA = deoxyribonucleic acid

GA = gibberellic acid

GTP = guanosine triphosphate

IAA = indoleacetic acid

IBA = indolebutyric acid

NAA = naphthaleneacetic acid

NAD = nicotinamide adenine dinucleotide

NADPH = nicotinamide adenine dinucleotide phosphate

RNA = ribonucleic acid

mRNA = messenger RNA

rRNA = ribosomal RNA

tRNA = transfer RNA

TIBA = triiodobenzoic acid

For Lakshmi and Anita

1. Experimental Plant Embryogenesis: Problems and Prospects

The seed germinates heralding the active life of the plant. Given an adequate supply of water and mineral salts in the soil, the normal composition of the atmosphere, a favorable temperature and the energy of sunlight, the seed will form the seedling, the seedling will mature into the adult which will eventually flower and set seeds for the next generation—all on a predictable schedule. If we trace the history of the adult plant backwards through time we see that the master unit in the seed from which it has evolved is the germ or the embryo. The embryo thus represents the beginning of the new sporophytic generation. Even in an immature seed the embryo is sufficiently well developed with morphologically differentiated primordia of the future vegetative organs of the plant, namely, the radicle or the embryonic root, and the epicotyl or the embryonic stem which in many cases may already have the rudiments of the first pair of leaves. The process of germination is the outward and visible evidence of the innate capacity for growth of the different parts of the dormant embryo which depends upon subtle alterations in cell control mechanisms. Most of us are aware of, or have seen, the dramatic sequence of events from the time a seed is planted in the nursery bed to the point when the seedling appears above ground. Not visible to the naked eye, but nonetheless important, are several hormonal substances whose marvelous interplay controls and directs the organized growth of the plant from the embryo. Evidence for this generalization is overwhelming and some of the critical proofs are landmarks in plant physiology literature.

While plant physiologists and developmental botanists have succeeded to a remarkable degree in unraveling the causative factors that regulate growth of the component parts of the plant, surprisingly meagre and primitive is our knowledge of the determining factors in the organized growth of the embryo in undisturbed normal development. In both plants and animals, the fertilized egg or the zygote is the fundamental structural unit which, by a series of subtle and complex influences, becomes progressively expressed to give rise to the fully developed embryo. There is an extreme contrast between the zygote and the embryo that develops from it. The zygote is best appraised by stating that it is an inherently unstable entity: it maintains its seemingly simple organization only for a brief period of time before it subdivides into a large number of smaller cells. These cells group into self-contained pockets of tissues which in turn differentiate organs. Thus, out of a simple-looking cell emerges, by a series of developmental events closely coordinated in time and space, the complex embryo, endowed with an array of distinct parts, each destined for a specific function in the adult. Perhaps in no other organ in the plant do we find such dramatic changes in growth, differentiation and tissue formation crowded into a relatively brief span of time as in the embryo.

Biologists have often sought to answer the question: What causes a single-celled zygote to reproduce meticulously, generation after generation, a complicated pattern of many cells which are biochemically and structurally so unlike itself? It has proved no easy matter to answer this question without indulging in philosophical generalities, in spite of the fact that a considerable part of the contemporary effort in biology is dedicated to questions of this nature. It is reasonable, although a gross oversimplification, to assume that the genetic information built up in the zygote determines the specific pattern of cells that subsequently arise from it. As all the cells formed by the division of the single-celled zygote receive the same genetic blue-print and thus conserve the characters and potentialities of the parent cells, we are further provoked to ask, without being able to give the answer: What factors order or stimulate one type of growth in certain cells, and another type of growth in certain other cells of the same organ?

This book does not claim to be able to answer these questions and perhaps many others of equal significance which have aroused widespread and general concern among biologists for some time now. What I intend to do is to bring together some of the findings from the literature in the experimental embryogenesis of vascular plants (pteridophytes, gymnosperms and angiosperms), both past and present, in the hope of stimulating the kind of research that will answer the questions in future.

What is experimental plant embryogenesis? So far the greatest effort in plant embryology has been directed towards descriptive accounts of sporogenesis and gametogenesis and enumeration of the steps in the transformation of the zygote into a fully fledged embryo, by precise observations of histological preparations of the male and female gametophytes and embryos of different ages. This work, described in some classical books (Coulter and Chamberlain, 1917; Campbell, 1930; Schnarf, 1929, 1931; Bower, 1935; Eames, 1936; Johansen, 1950; Maheshwari, P., 1950) and scores of scientific papers, has been carried forward by many investigators in many laboratories around the world; it has as its major goal an understanding of the architectural principles on which gametes are formed and embryo types are constructed. These studies provided a useful framework for an appreciation of the range of variation in the development and organization of gametophytes and embryos of vascular plants and created a deeply entrenched idea that some of these events in each plant follow a typical pattern according to a blue-print characteristic of each species.

Evaluation of the accumulated data relating to sporogenesis, gametogenesis and the mode of embryo formation in a large number of species led to a diversification in the outlook in plant embryology. The utilization of embryological data in assigning phylogenetic affinities of certain families, genera and species of flowering plants was the immediate outcome of such diversification (Maheshwari, P., 1950, 1964; Johri, 1963b). Numerous and valuable indeed are the papers which describe gametophyte development and embryogenesis in the different species of plants and application of the data to solve phylogenetic problems and taxonomic relationships.

Since 1930, advances made in the fields of plant physiology, biochemistry and genetics, and refinements in the culture of plant organs and tissues under aseptic conditions, have had much influence in the orientation of modern order in plant embryology. This has given rise to the comparatively new discipline of experimental embryology, involving control of pollination and fertilization and manipulations of the anther, pollen grain, ovule, ovary, and embryo, by excision and culture, by chemical, hormonal and surgical treatments, and by exposure to selected day-length and temperature conditions to study the controlling mechanisms that affect their form and structure, in the hope that as we gain knowledge of the laws that control the unfolding of form and structure in the reproductive organs of plants, we may get new clues to control them to our advantage. Thus experimental embryology has changed an era of observations and inferences into an era of experiments and deductions, designed to discover what processes are involved in the evolution of embryonic form, how they are related and how they are controlled.

The boundary between embryology and embryogenesis seems to be somewhat vague. In contrast to embryology which includes all of the events connected with microsporogenesis, megasporogenesis, development of the male and female gametophytes, fertilization and endosperm and embryo formation, embryogenesis (embryogeny) is concerned with the whole constellation of post-fertilization events, and is regarded as the continuum of processes involved in the origin, growth, and orderly transformation of the zygote into a fully fledged embryo. With this limitation in mind, I have followed an orthodox approach in the text, beginning with a consideration of the formation of the zygote and ending with the embryo complete with all of its tissues and organs, emphasizing experimental studies which provide causal explanations of growth, development and morphogenesis during embryogenesis. Although experimental work on plant embryos is in its very embryonic stages, a survey of the work done so far, summarized in some recent reviews (Narayanaswami and Norstog, 1964; Maheshwari, P. and Rangaswamy, 1965; Wardlaw, 1965b; Degivry, 1966; Raghavan, 1966) will unfold the challenging future opportunities which will keep plant embryologists occupied for a long time.

In summing up, experimental study of the formation of the embryo, its nutrition, its responses to external stimuli and modification of its growth by hormonal and environmental factors, is the main theme of this book. The development of the text can be briefly summarized as follows. In the first three chapters, I have closely examined the present state of our knowledge of the structure, growth, and organization of the egg, zygote and embryo in representative species of vascular plants. The background of events leading to the formation of the embryo thereby established serves as a basis for the subsequent two chapters on the nutritional aspects of embryogenesis. The next six chapters are concerned with growth, organization, and differentiation of the embryo during progressive embryogenesis and its changing patterns of nutrition and metabolism. Although the framework of discussion is provided mainly by vascular plants, significant contributions from other taxonomic groups are stressed, especially when they illustrate principles that are more generally applicable. Included herein toward the end is also an account of the major advances made in the past few years in the applied aspects of embryo culture and in the experimental production of adventive embryos. Although the metabolism of the embryo during storage and germination of seeds is outside the scope of this book, some aspects of the physiology of the dormant embryo and the developmental and biochemical aspects of germination are discussed in the final two chapters, insofar as they seemed useful for a comprehension of the initiation of development. Because of the paucity of knowledge, I have not stressed in this book the many intriguing and fascinating problems in differentiation at the molecular level which must certainly be taking place during embryogenesis. Notwithstanding, it is hoped that this book will be of use in bringing together some of the scattered literature on the developmental physiology and morphogenesis of embryos and will give a general idea of the current and future perspectives in the embryogenesis of vascular plants.

Section I

From Egg to Embryo

2. Structure and Organization of the Egg, Zygote and Embryo

Embryogenesis involves extensive changes in form in defined and dramatic ways and a progressive change from the undifferentiated to the differentiated state. Beginning with the first division of the zygote, the plant remains in a state of continuing embryogenesis, producing new cells, tissues and organs throughout its life history. Common to the developmental changes during progressive remodeling of embryo structure are such processes as cell division, cell expansion, cell maturation, cell differentiation and formation of meristems, but the physiological and biochemical changes underlying the histological diversity of organs formed are probably different. Cells produced at each stage of development are arranged according to specific and predetermined patterns to give the embryo certain proportions that remain constant throughout its growth. Changes in the pattern of cell arrangement leading to changes in size and shape of the embryo are determined by the plane of cell division and by the relationship between the frequency of cell division and the rate and direction of cell elongation. The plane of cell division is, in turn, determined by the orientation of the mitotic spindle. In a very young embryo all cells are involved in producing a new generation of daughter cells; but as the embryo develops, through a period embracing differentiation and the onset of functional activities, cell division becomes segregated in certain areas predictable by their position in cell lineage to produce specialized cells, tissues and organs. It follows therefore that if differences in the cellular activities of embryos of different ages are of significance, there may be recognizable differences in the organization and composition of their cells.

The present chapter is devoted to an exploration of the structure of the egg, zygote and developing embryo and an analysis of the pattern of embryogenesis in representative species of angiosperms, gymnosperms and pteridophytes. The emphasis will be on the manner in which enhanced appreciation of the structural organization of cells has contributed to the broad advancement of the understanding of their functional aspects. The fact that the center of interest has shifted from light microscopic studies to the level of the electron microscope is particularly indicative of the new tendencies in the study of embryogenesis. The range of studies in the development of embryos in plants is so wide that details of descriptive embryogenesis are kept to a minimum to serve as a background for experimental work to be considered in later chapters. For more comprehensive treatment, references cited in the respective sections should be consulted.

I. THE EGG AND ITS MILIEU

A. Angiosperms

In angiosperms, the egg is formed in a privileged location in the female gametophyte which itself is buried within the ovule with its multi-layered covering of integuments and nucellus. The gametophyte is derived from the product of a reduction division in the megaspore mother cell, followed by a series of mitotic divisions (Fig. 2.1). Depending upon the species, one, two or four megaspores may take part in the formation of the gametophyte. The mature female gametophyte or embryo sac has most commonly eight genetically identical haploid nuclei, although 4-nucleate, 16-nucleate and multinucleate embryo sacs have been described (Maheshwari, P., 1950; Johri, 1963a). These nuclei organize in the embryo sac as naked or partially covered cells according to a characteristic plan.

The complexity of organization of the embryo sac is such that a certain terminology is used to describe the individual nuclei or groups of nuclei. In a typical eight-nucleate embryo sac, the nuclei are arranged

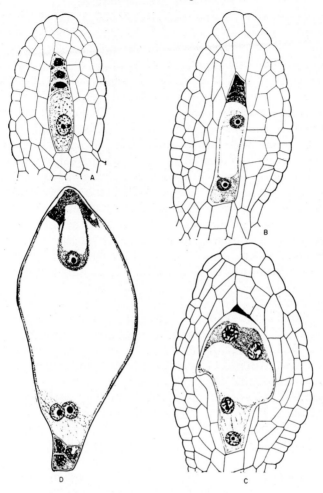

FIG. 2.1. *Hydrilla verticillata* illustrates the development of a typical eight-nucleate embryo sac. **A**, Tetrads of megaspores after reduction division; the gametophyte is formed from the large chalazal magaspore. **B**, The first mitotic division to form two haploid nuclei. **C**, Four-nucleate embryo sac after the second mitotic division. **D**, Eight-nucleate embryo sac after the third mitotic division. (After Maheshwari, P., 1950)

in two quartets at the micropylar and chalazal ends. One of the nuclei at the micropylar pole becomes delimited by a small amount of cytoplasm and organizes itself as the egg cell. It is flanked by two smaller cells called synergids. According to P. Maheshwari (1950), synergids are entirely absent in *Plumbago* and *Plumbagella* types of embryo sacs, while in several species of *Peperomia* there is but one synergid. The synergid is a pear-shaped cell, vacuolate at the chalazal end and lined by the nucleus at the micropylar end. At the micropylar end of the synergid there are also elaborate proliferations of the wall material known as the filiform apparatus which extends as small digitate projections into the cytoplasm. In embryology literature the egg cell and synergids are collectively known as the egg apparatus. At the chalazal end of the embryo sac three nuclei differentiate into the antipodal cells. During gametogenesis two polar nuclei move from the opposite quartets to a central position in the embryo sac and fuse to form a diploid polar fusion nucleus. The part of the embryo sac which remains after the egg apparatus and antipodals are formed is known as the central cell. This cell is uninucleate or binucleate depending upon whether the polar nuclei have fused together or are free. Despite the diversity of origin of the female gametophyte, in the majority of plants examined the organization of the egg apparatus presents a fairly uniform picture and only a few variations from the basic pattern have been reported.

The Egg

The potentiality of the zygote for initiating sporophytic growth is probably reflected in the cytoplasmic organization of the egg. Despite its importance, information on the structure of the egg is relatively fragmentary, and only in recent years have some details been forthcoming, thanks to the resolving power of the electron microscope. The main contributions that record the progress in the field are those of Jensen's group (Jensen, 1964, 1965c, 1968a,b; Schulz, R. and Jensen, 1968c; Cocucci and Jensen, 1969a), the Vazarts (Vazart, B. and Vazart, J., 1965b, 1966; Vazart, J., 1969) and Diboll (Diboll, 1968; Diboll and Larson, 1966). The import of these studies which differ only in minor details is summarized below. In view of the immense diversity in angiosperms, there is a great deal more to be seen and recorded from thin sections of the egg than has so far been attempted, to construct a coherent picture applicable generally.

As seen in cotton (*Gossypium hirsutum*) (Jensen, 1964, 1965c, 1968a,b) and *Capsella bursa-pastoris* (Schulz, R. and Jensen, 1968c), the egg is vacuolate with a thin layer of cytoplasm surrounding the vacuole. Plastids, mitochondria, ribosomes, dictyosomes and strands of endoplasmic reticulum (ER) are randomly and sparsely distributed in the cytoplasm while the nucleus is generally confined to the base or to one side of the egg (Fig. 2.2). In the egg cells of flax (*Linum usitatissimum*) (Vazart, B. and Vazart, J., 1965b, 1966) and *L. catharticum* (D'Alascio-Deschamps, 1973), plastids, ER, and droplets of lipid are concentrated

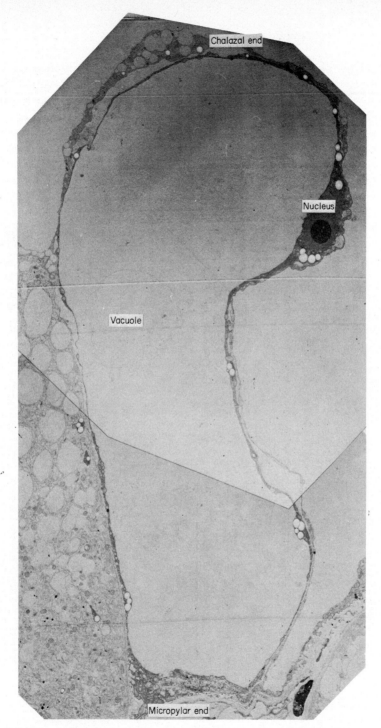

Fig. 2.2. The median section of the egg of *Gossypium hirsutum* shows a large central vacuole and a thin layer of cytoplasm surrounding the vacuole. The nucleus is towards the upper right-hand corner. (After Jensen, 1968b; photograph courtesy of W. A. Jensen)

around the nucleus with the ribosomes occupying a thin layer of cytoplasm surrounding the vacuole. The nucleus is devoid of typical chromosomes but contains variable amounts of a chromatin-like fibrillar
material. In sunflower (*Helianthus annuus*) (Newcomb, 1973a), *Crepis tectorum* (Godineau, 1973) and *Petunia hybrida* (van Went, 1970b), the eggs
show evidence of a polar axis by the eccentric position of the nucleus
and a major part of the cytoplasm towards the chalazal end. According
to Cocucci and Jensen (1969a), the egg of the orchid *Epidendrum scutella*
has a centrally located nucleus surrounded by numerous small vacuoles,
but there is an apparent lack of dictyosomes in the organelle comple-

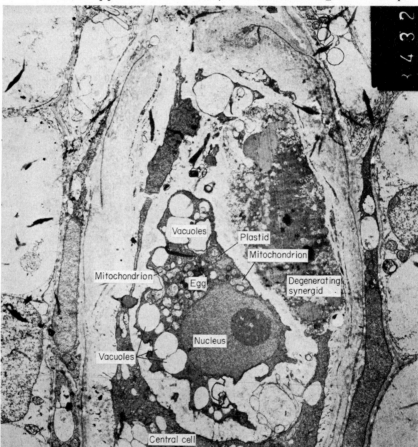

Fig. 2.3. The egg cell of *Epidendrum scutella* has a centrally located nucleus surrounded
by numerous vacuoles in the cytoplasm. Portions of the central cell and the degenerating synergid are also seen. (After Cocucci and Jensen, 1969a; photograph courtesy
of W. A. Jensen)

ment (Fig. 2.3). The egg of maize (*Zea mays*) is more like that of *E. scutella* in lacking a central vacuole, but unlike the orchid, it possesses many dictyosomes (Diboll and Larson, 1966). In the egg of *Myosurus minimus*, cup-shaped organelles resembling plastids, ribosomes and helical polysomes are present (Woodcock and Bell, 1968b). Although the picture is incomplete, the major attributes of an angiosperm egg as revealed in the electron microscope are sparsity of ER and dictyosomes, and the relatively small number of polysomes; these are clearly features of cells undergoing depressed metabolic and synthetic activities.

A very suggestive feature of mature eggs of cotton (Jensen, 1964, 1965c), *Torenia fournieri* (van der Pluijm, 1964), maize (Diboll and Larson, 1966) and *Petunia* (van Went, 1970b) is the disappearance of the cell wall towards the chalazal end. Both with regard to the maintenance of contact with the nutritive substances of the central cell and the entry of the sperm during fertilization, this cytological feature might appear to be involved, and it is therefore wise to keep an open mind about its ultimate advantage.

The Milieu of the Egg

Certain features of the embryo sac and its component cells need to be emphasized here, particularly features relating to their function. There is little doubt that the main function of the central cell is to sustain the biochemical growth of the egg to the minimum developmental stage necessary for fertilization, and all sorts of built-in devices and short-cuts may be expected to be present for an increased efficiency of nutrition. Tubular extensions and haustorial outgrowths are some of the more evident morphological features of embryo sacs seen in the light microscope which enable them to exploit food resources beyond their normal reach (Masand and Kapil, 1966). In the electron microscope the central cells of some plants have revealed massive membrane-lined wall projections (Vazart, B. and Vazart, J., 1966; Godineau, 1971; Newcomb, 1973a; Newcomb and Steeves, 1971; Newcomb and Fowke, 1973; Berger and Erdelská, 1973) which bear a striking resemblance to the "transfer cells" described in various parts of plants (Gunning and Pate, 1969). In series, these cellular adaptations afford such an efficient means of absorption and translocation of metabolites from the vicinity of the nucellus and integuments to the embryo sac (Fig. 2.4).

The central cell also contains reserve food materials in the form of starch grains, protein bodies and lipids which are stockpiled in the cytoplasm (Vazart, B. and Vazart, J., 1965b; Jensen, 1965c; Diboll, 1968; Diboll and Larson, 1966; Kuran and Marciniak, 1969; Newcomb, 1973a; Schulz, P. and Jensen, 1973). Electron microscopic observations

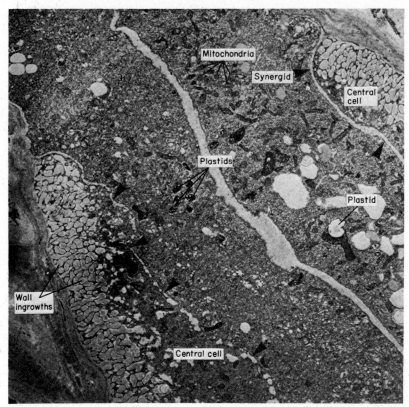

FIG. 2.4. Development of wall projections along the embryo sac wall of *Helianthus annuus*. The cytoplasm of the central cell separates the outer synergid wall and wall ingrowths. Arrows point to the synergid hooks formed by a curving of the outer synergid wall. (After Newcomb, 1973a; photograph courtesy of W. Newcomb)

indicate that the central cell is metabolically very active with an extensive network of ER, numerous well developed chloroplasts and mitochondria, dictyosomes and polysomes (Eymé, 1965; Jensen, 1965c; Diboll, 1968; Diboll and Larson, 1966; Mikulska and Rodkiewicz, 1967a, b; Cocucci and Jensen, 1969a; Vazart, J., 1969; Vazart, B. and Vazart, J., 1966; van Went, 1970b; Newcomb, 1973a; Schulz, P. and Jensen, 1973). Organelles known as glyoxysomes in which enzymes of the glyoxylic acid cycle are localized also occur in the central cell (Newcomb, 1973a; Schulz, P. and Jensen, 1973). It is logical to assume that the metabolic activity of the central cell contributes to the nutrition of the egg by supplying energy required for absorption of nutrients and conversion of stored reserves into simple precursors.

The synergids have a limited life span, and in many plants they do not survive beyond fertilization, while in others one or both of them persist as haustorial cells for a period of time after fertilization. The synergids have a peculiar and decisive significance for it is due to them that in some plants metabolites are presumably transported to the central cell. Since the filiform apparatus of the synergid is an extension of the wall material, it has kinship with the labyrinth-like wall outgrowths of the central cell described earlier. In cotton (Jensen, 1965b), *Capsella* (Schulz, R. and Jensen, 1968b), *Epidendrum* (Cocucci and Jensen, 1969a), *Petunia* (van Went, 1970a) and *Aquilegia formosa* (Vijayaraghavan *et al.*, 1972) the absorptive function of the filiform apparatus is accentuated by the lining of the plasma membrane around its lobes. The distribution of cytoplasmic organelles of the synergid in close proximity to the filiform apparatus also hints of a role for these cells in nutrient absorption and translocation (Vazart, B. and Vazart, J., 1966; Vazart, J., 1969; Jensen, 1965b; Godineau, 1966; van Went and Linskens, 1967; Diboll, 1967, 1968; Diboll and Larson, 1966; Rodkiewicz and Mikulska, 1967; Schulz, R. and Jensen, 1968b; Cocucci and Jensen, 1969a; Vijayaraghavan *et al.*, 1972; D'Alascio-Deschamps, 1973). In *Plumbago capensis* (Cass, 1972) and *P. zeylanica* (Cass and Karas, 1974) which lack synergids, the micropylar wall of the egg generates a facsimile of the filiform apparatus which, in all likelihood, performs the same functions as its counterpart in the synergid. Recent studies of fertilization in several angiosperms have shown that the pollen tube enters the embryo sac through a synergid which perhaps facilitates the process by secreting a chemotropic factor (van der Pluijm, 1964; van Went and Linskens, 1967; Diboll, 1968; Jensen and Fisher, 1968; Schulz, R. and Jensen, 1968b; Cocucci and Jensen, 1969b; Cass and Jensen, 1970).

Much less is known about the antipodals which are obliterated before fertilization and which rarely become persistent and haustorial (see Chapter 5). Members of the family Gramineae have antipodals that multiply and their sheer number itself suggests a role in the nutrition of the embryo sac. In maize, digitate wall projections are found on the inner face of the antipodal walls bordering the nucellus and along some of the cell walls within the proliferated antipodal mass (Diboll and Larson, 1966). The functional analogy of these wall outgrowths to similar features on the walls of the central cell and synergids is obvious. In sunflower, after the antipodals are formed, they embark on a synthetic phase characterized by an increase in nucleic acid and organelle complement of the cytoplasm (Newcomb, 1973a). These features, coupled with the presence of plasmodesmata in their end walls facing the central cell, suggest that materials synthesized in the antipodals are probably channeled to the central cell and egg apparatus in preparation for fertilization. At the present time, interpretations of the functions of the central cell, synergids and antipodals are somewhat speculative, but they are in keeping with the current ideas in plant and cell physiology.

B. Gymnosperms

The structural plan of the egg is more or less similar in the majority of gymnosperms examined, and as in angiosperms, it reaches a high level of differentiation and specialization. The gymnosperm egg is initiated in the archegonium which develops in the female gametophyte. The archegonium is generally reduced in nature and in its mature state consists of one or two neck cells, a ventral canal cell and an egg cell (Fig. 2.5). In the two highly evolved genera of Gnetales, *Gnetum* and

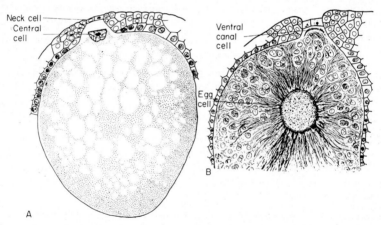

Fig. 2.5. Development of the archegonium in *Pinus wallichiana*. **A**, Archegonium with two neck cells and a central cell. **B**, Archegonium after division of the central cell to form the egg and the ventral canal cell. (After Konar and Ramchandani, 1958)

Welwitschia, archegonia are absent and certain specialized cells of the female gametophyte serve as functional eggs (Pearson, 1929; Vasil, 1959; Sanwal, 1962). The gymnosperm egg is relatively massive and at maturity consists of a huge nucleus surrounded by the greatest variety of visibly distinct cytoplasmic masses ever described in vascular plants.

Much of what we know of the structure of the gymnosperm egg at the fine level of scrutiny has come from the examination of selected genera belonging to the Coniferales and Ginkgoales. Chesnoy and Thomas (1971) have recently reviewed the structure of the gymnosperm egg with due consideration to the contributions from ultrastructural studies.

The basic model of the egg in the family Pinaceae is of a bag inflated with small cytoplasmic inclusions in the immediate vicinity and large inclusions towards the periphery of a centrally placed nucleus (Fig. 2.6). A widespread misconception is that these inclusions are proteid vacuoles; however, in the electron microscope, the small inclusions appear essentially as masses of cytoplasm rich in ER, mitochondria and dictyosomes, bound by a single membrane and entrapped in a small vacuole. The large inclusions are made up of several smaller cytoplasmic masses separated from each other by electron-dense material. Double membranes girdle the whole inclusion as well as the individual cytoplasmic masses (Camefort, 1959, 1960, 1962, 1963, 1965a, 1967b; Chesnoy and Thomas, 1969; Thomas and Chesnoy, 1969). In Cupressaceae (Chesnoy, 1967, 1969a, b, 1971) and Taxodiaceae (Camefort, 1970; Gianordoli, 1973) the egg cell is filled with a mass of dense vacuolate cytoplasmic nodules which are comparable to the small inclusions of Pinaceae. The egg cytoplasm of *Ginkgo biloba* is also separated into small organelle-rich inclusions which, however, lack a vacuole (Camefort, 1965b).

In *Pinus laricio* var. *austriaca*, plastids and mitochondria evince some degree of change in form and structure with maturity of the egg (Camefort, 1962). In a relatively young egg, plastids appear in the form of a ring surrounding a mass of cytoplasm containing various organelles, while in a mature egg they become engorged with cytoplasmic invaginations. In this condition the plastid retains only traces of its original structure such as the double membrane and its function is now distinctly different from that of a normal plastid. Early in its development the egg contains comparatively few mitochondria, but during further growth of the egg there is an increase in mitochondrial number. A good number of mitochondria become complex by increase in length and in the number of cristae which tenaciously bind the enzymes. In the mature egg ready for fertilization, the long mitochondria disappear while the remaining ones undergo structural regression by the loss of their internal membrane configuration. In the mature eggs of *Larix decidua* (Camefort, 1967b) and *Pseudotsuga menziesii* (Chesnoy and Thomas, 1969; Thomas and Chesnoy, 1969), these mitochondria which are now no more than simple double-membrane vesicles form a conspicuous ring around the nucleus. Since the changes observed in the plastids and mitochondria of mature eggs are those commonly associated with metabolically inactive cells, it seems that the gymnosperm egg, like its counterpart in angiosperms, maintains a low metabolic profile at the time of fertilization.

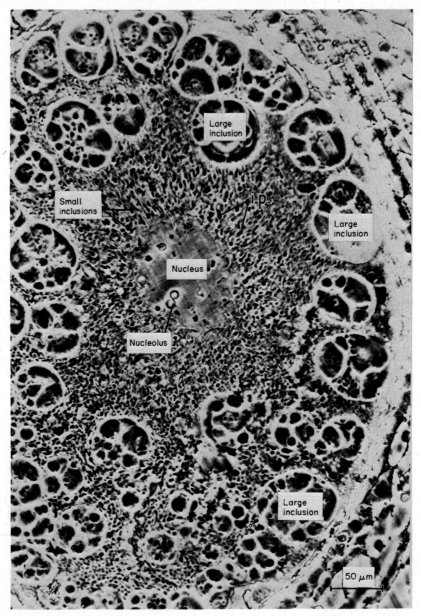

Fig. 2.6. The egg cell of *Pinus laricio* var. *austriaca* has small cytoplasmic inclusions in the immediate vicinity of the nucleus and large inclusions towards the periphery. Phase contrast picture. (After Camefort, 1962; photograph courtesy of H. Camefort)

C. Pteridophytes

Unlike gametophytes of angiosperms and gymnosperms, those of pteridophytes can be grown in sterile culture with relative ease and gametogenesis, fertilization and subsequent growth of the embryo followed under controlled conditions. For this reason, they offer great scope as experimental systems to study embryogenesis and related problems. The egg is generally sheltered in the archegonium which shows a remarkable uniformity in structure throughout the group. The archegonium is typically flask-shaped and consists of a neck which projects above the surface of the gametophyte and a basal region called the venter which is sunk in the gametophyte. The neck contains a row of neck canal cells protected by multiple tiers of neck cells. The structural subdivisions of the venter are the egg situated at the base of the archegonium and the ventral canal cell lying in the axial row adjacent to the egg (Fig. 2.7). The parent cell which gives rise to the latter cells

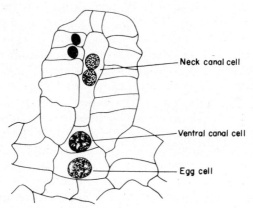

FIG. 2.7. Median section of a mature archegonium of *Todea barbara*, showing a binucleate neck canal cell, the ventral canal cell and the egg cell. (Redrawn from DeMaggio, 1961a)

is usually designated as the primary ventral cell. The progenitor of the cells of the entire axial row is known as the central cell. The cytoplasm of the egg is enriched with plastids, mitochondria, ribosomes, ER and dictyosomes. Vacuoles are rare in the cytoplasm and, if present, are of small dimensions. An unusual feature of the egg nucleus of some ferns is the absence of individual chromosomes, which in the more complex situations appear as sheets of nuclear material (Tourte, 1968; Bell, 1970, 1974).

On the de novo Origin of Plastids and Mitochondria

Recent electron microscopic investigations have disclosed some interesting changes in the organelles of fern eggs during oogenesis and have led to a lively controversy on the origin of plastids and mitochondria in cells. In the immature bracken fern (*Pteridium aquilinum*) egg, the cytoplasm contains vacuolate bodies identified as plastids which disappear as the egg matures. However, the fully mature egg regenerates new plastids which seemingly appear as amoeboid bodies possessing scattered lamella. Similar changes take place in the mitochondrial population of the egg during maturation. In the fairly immature egg, mitochondria appear at best as distended vacuolate bodies with fragments of cristae. A few days before maturity and particularly in the mature egg there is an abundance of mitochondria of various forms freely scattered in the cytoplasm. At this time there is also a substantial increase in the amount of ER in the periphery of the egg where it tends to form concentric layers (Bell and Mühlethaler, 1962b, 1964a; Bell et al.,·1966). What these profound changes mean is that the mature fern egg is in a high metabolic state.

The tendency for structural elaboration seen in the mature egg of *Pteridium* is further accentuated by the active blebbing of the nuclear membrane; according to Bell and Mühlethaler (1964b; Bell, 1972) some of these blebs develop into proplastids and mitochondria. The evidence from this study appears to indicate that plastids and mitochondria degenerate during early oogenesis, but are formed *de novo* in the mature egg from the nuclear membrane that peels off into the cytoplasm. Tourte's (1968) study of this species has also demonstrated peeling off of the nuclear membrane, but he contends that the peels are progenitors of the small vesicles found abundantly in the cytoplasm. In a subsequent work, Tourte (1970) has disputed the observation that plastids and mitochondria disappear or degenerate at specific stages of oogenesis. It is difficult to give a convincing explanation for the differences noted on the same material in two laboratories.

Menke and Fricke (1964) have also observed detached nuclear evaginations in preparations of the egg of the male fern (*Dryopteris filix-mas*). Although these authors did not gather evidence for the formation of proplastids and mitochondria from the nuclear blebs, Bell (1966) claims to have observed in this species as well the degeneration of organelles and their subsequent regeneration by evagination of the nuclear envelope. Studies of Camefort (1962, 1963, 1965a) on the development of the egg cell of *Pinus* described in the previous section indicate degeneration and elimination of plastids and mitochondria, but the evidence in support of their *de novo* formation is equivocal. Ultrastructural investigation of the developing egg cell of the liverwort *Sphaerocarpus donnellii* (Diers, 1966) has provided no indication of the degeneration and elimination of plastids and mitochondria during oogenesis. Although nuclear evaginations of the type described in *Pteridium* occurred during oogenesis, they were not involved in the formation of cytoplasmic organelles(Diers, 1964, 1965). Megagametogenesis in angiosperms thus far examined in the electron microscope is also not accompanied by any unusual nuclear activity (van der Pluijm, 1964; Diboll and Larson, 1966; Schulz, R. and Jensen, 1968c). In an ultrastructural study of wheat (*Triticum vulgare*) endosperm tissue, Buttrose (1963a) has indicated the possibility of *de novo* origin of organelles. So far as this author is aware, this is the only support which Bell and Mühlethaler's hypothesis has received from another laboratory. Since much skepticism surrounds these interpretations, it seems unlikely that the view on the *de novo* origin of plastids and mitochondria described above will gain general acceptance at the present time, but it will be unfair to refute it or confirm it from evidence based on a limited number of electron micrographs alone. The difficulties involved in the reconciliation of static but high resolution electron micrographs with the actual changes occurring in the cytoplasm are apparent here.

In the light of a relationship between cytoplasmic vesicles, plastids and mitochondria of the egg cell referred to above, it was natural that the vesicles should excite interest in their possible intercellular origin (Tourte, 1968; Bell, 1969). According to Bell (1969), cytoplasmic vesicles in the egg cell of *Pteridium* are either empty with electron-transparent contents or opaque with conspicuous amounts of dense material. In the mature egg, the empty vesicles are largely peripheral in distribution while the opaque vesicles occur throughout the cytoplasm frequently interspersed between the mitochondria and plastids. From the electron micrographs it is indisputable that vesicles are characteristically present during oogenesis of *Pteridium*; whether they are formed by the blebbing of the nuclear membrane (Tourte, 1968), or by the progressive subdivision of the large vacuoles or by the breakdown of the plastids and mitochondria (Bell, 1969) is still a moot question. Similar cytoplasmic vesicles have been described in the eggs of *Pinus* (Camefort, 1962), *Ginkgo* (Camefort, 1965b) and *Sphaerocarpus* (Diers, 1966).

Another special feature of the mature egg of *Pteridium* is the presence of a membrane lying between the plasmalemma and the cell wall (Bell and Mühlethaler, 1962a; Cave and Bell, 1974a). This osmiophilic membrane consists predominantly of lipids and is probably analogous to the membrane observed around the eggs of *Todea barbara*

(DeMaggio, 1961b) and *Dryopteris filix-mas* (Menke and Fricke, 1964). In appearance and staining behavior the membrane resembles the material of the vesicles; this observation, coupled with the disappearance of vesicles as the membrane is formed, assumes special significance suggestive of its formation from the remains of the old organelles. However, the presence of the membrane is yet to be demonstrated in other species, and even if the reality of its general existence is granted, it is hard to see precisely what function it serves. What is attractive about this membrane is that its presence ideally explains the reported failure of labeled nucleotides to diffuse freely into the embryo (Bell, 1961).

The organization of a mature egg represents the end of a phase of growth at the morphological, biochemical and ultrastructural levels to signify its "readiness for fertilization". These changes may be far more complex than the limited data presented here would indicate, and for this reason further studies along these lines might be of great interest.

II. THE ZYGOTE

In the majority of animals, and in virtually all algae, fungi and higher cryptogams, a full-grown embryo is initiated from the fertilized egg or zygote. In seed-bearing plants (gymnosperms and angiosperms) it is not unusual for embryos to be initiated from cells other than the zygote (see Chapters 14, 15). The zygote is the unicellular stage of a multicellular organism and the essential properties of interactions occurring in the adult are found in this cell. Generally, in angiosperms, one half of the zygote is anchored to the wall of the embryo sac at the micropylar end and the chalazal pole is free. Within the confines of the embryo sac the zygote is bathed in a fluid medium. The fluid constitutes the endosperm which possesses a chemical composition as yet dimly understood in a large majority of plants that makes it a rich source of nutrition for the zygote as well as for the embryo developing from it (see Chapters 5 and 6). Thus the zygote seems to be ideally packaged in anticipation of development with adequate safeguard for nurture of the young embryo. Before commencing growth, the zygote looks like any other undifferentiated cell and lacks the characteristics of a specialized type of cell. In cotton and other members of the Malvaceae which have a conspicuous micropylar vacuole in the egg, there is a dramatic decrease in size of the zygote immediately after fertilization to half its size before fertilization (Pollock and Jensen, 1964; Jensen, 1968b; Ashley, 1972). In contrast, in *Datura stramonium* the zygote actually enlarges in size while it prepares for the first division (Satina and Rietsema, 1959). An almost intermediate situation prevails in *Capsella bursa-pastoris* (Schulz, R. and Jensen, 1968c) where the micropylar vacuole undergoes a temporary decrease in volume followed by attainment of more than the original volume without any actual shrinkage in size of zygote. Whether post-fertilization change in size acts as a trigger for the zygote to divide, we do not know. One is not even sure that this is a universal phenomenon.

While changes in cell wall extensibility may conceivably lead to enlargement of the zygote, how can we explain its shrinkage prior to division? Does a collapse of the central vacuole result in shrinkage or must we consider that the cell wall itself is capable of extension and resorption? We must admit that answers to these questions cannot yet be given, but some ideas elaborated by Jensen (1968b) have helped to expose the issues and hence merit a brief summary. According to this investigator, a reduction in size of the zygote is probably due to loss of its water to the surrounding cells of the endosperm by simple osmotic diffusion. The accompanying volume change results in a decrease in size of the vacuole and in the area of the plasma membrane and vacuolar membrane. Electron microscopic studies have shown that a complex tube-containing ER with internal membranous components is formed in substantial quantity during this period. It is a fascinating possibility that the materials of the plasma membrane and vacuole membrane are utilized for elaboration of ER. Whether the mechanism by which membranes are transformed into tube-containing ER involves a configurational change of some membrane proteins, or a complete loss of membrane integrity or some other complex process, is unclear.

A. Ultrastructure of the Zygote

Our present knowledge of the ultrastructural changes in the fertilized egg stems from results of studies on a limited number of species and so there is an obvious need for more comparative work before valid generalizations on the true biological significance of the structural changes in the zygote can be attempted. Ultrastructural changes in the eggs of cotton, *Capsella* and *Epidendrum scutella* following fertilization have been studied in remarkable detail by Jensen and his coworkers (Jensen, 1968b; Schulz, R. and Jensen, 1968c; Cocucci and Jensen, 1969b). In summary, we possess the following picture. In cotton, the most impressive change observed in the zygote is a segregation of plastids, ER, mitochondria and ribosomes at the chalazal end, where they soon rearrange themselves in a ring around the nucleus. This is seen even as early as 4 h after fertilization (Fig. 2.8). What impels these organelles to begin movement after fertilization and what causes them to cease movement upon reaching the presumptive embryonal cell are unknown. In the egg of *Capsella* (Schulz, R. and Jensen, 1968c) where the micropylar vacuole is small and in the egg of *Epidendrum* (Cocucci and Jensen, 1969b) where there are numerous small, randomly distributed vacuoles, differences in the distribution of organelles before and after fertilization are not clear-cut.

Other changes in the egg upon fertilization are concerned with the increase in complexity of the cytoplasmic organelles. Transformation of ER in the fertilized egg was described before. As plastids are grouped around the nucleus the size and number of starch grains present in them increase markedly. It is in the zygote that mitochondria reach their highest degree of differentiation by elaboration of additional cristae. As is well known, a high degree of differentiation of the internal membrane system of mitochondria is suggestive of intense metabolic activity.

Changes that occur in the ribosome configuration of the egg upon fertilization are the very epitome of initiation of new developmental potencies. Ribosomes of mature eggs of cotton (Jensen, 1968b, c), *Capsella* (Schulz, R. and Jensen, 1968c), *Epidendrum* (Cocucci and Jensen, 1969b) and flax (Deschamps, 1969) are free in the cytoplasm, but after fertilization they strikingly group to form polysomes (mRNA–ribosome complexes). Additionally, in cotton, transformation of ribosomes originally present in the egg into polysomes is followed by the appearance of a new generation of ribosomes which aggregate into smaller polysomes. Less striking is the change observed in the zygote of maize; here helical polysomes present in the egg increase in length after fertilization (Diboll, 1968). A point of view which is useful to explain these observations in molecular terms might emphasize that fertilization signals the production of specific mRNAs and their engagement by ribosomes to form structural polysomes. However suggestive they are, electron micrographs of polysome formation do not constitute final proof of synthesis of mRNAs upon fertilization of the egg, but these studies bring us a step closer to the understanding of molecular events during this crucial period.

In the final aspect of completion of zygote formation, synthesis of additional cell wall material at the micropylar end and elaboration of a new cell wall at the chalazal end take place. Symptomatic of new wall synthesis is an increased activity of dictyosomes observed upon fertilization of eggs of cotton and *Capsella* (Jensen, 1968b; Schulz, R. and Jensen, 1968c).

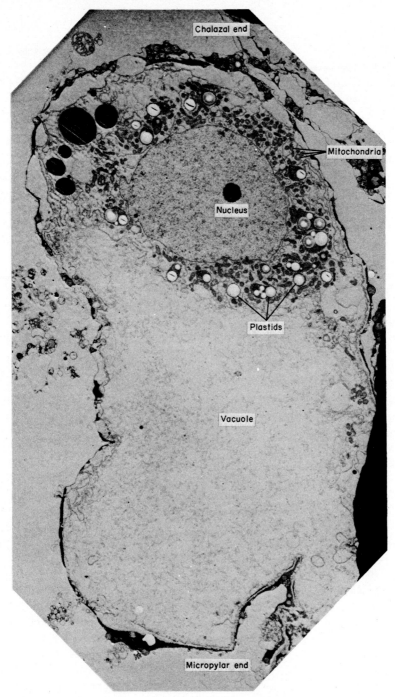

FIG. 2.8. The most impressive change observed in the egg of cotton, 4 h after fertilization, is the grouping of the cytoplasmic organelles around the nucleus at the chalazal end. (After Jensen, 1968b; photograph courtesy of W. A. Jensen)

Eggs of other plants examined more recently show variable cytological details, but little change after fertilization. In the eggs of *Petunia hybrida* (van Went, 1970c), flax (D'Alascio-Deschamps, 1972) and *Quercus gambelii* (Mogensen, 1972; Singh and Mogensen, 1975), an increase in lipid bodies after fertilization is a feature not observed in other plants.

Although it would be erroneous to emphasize a certain set of ultrastructural changes as being typical of the fertilized egg, we now have sufficient details which might serve as a basis for continuing investigations. Additionally, we need to know what factors in terms of enzymes, substrates or precursors in the fertilized egg cause mitochondria to divide and elaborate more cristae. How can we account for the storage of starch in the plastids of the egg following fertilization? Are microtubules involved in changes observed in the fertilized egg? What are the fine structural changes in parthenogenetically activated eggs? These questions are intriguing and require to be answered to meet the challenge of explaining fully how, following fertilization, the egg is functionally transformed into the zygote.

B. Polarity in the Zygote

Between fertilization and the first division, contents of the zygote undergo a nonuniform but orderly distribution to establish what is called polarity—a quality inherent in the cell which sets out a distinction between the apical and basal ends. In cotton, the nucleus moves to the chalazal end of the zygote. The cytoplasm is well concentrated at this end and steadily declines towards the opposite pole. Staining reactions have revealed a similar gradient in the distribution of RNA. We have already noted that cytoplasm of the zygote shows a gradient in the density of distribution of organelles seen in the electron microscope. The presence of a large vacuole at the proximal micropylar end, seen under both light and electron microscopes, completes the picture of polarity in the unsegmented zygote (Jensen, 1964, 1968b). Thus the future embryo seems to be fully predetermined in its unsegmented state by the distribution of cellular components, the potentialities for further development being concentrated at the upper half of the zygote. As we shall see later, the first division of the zygote separates the upper half which forms the embryo proper from the lower half which forms the suspensor.

How and when is polarity manifest in a fertilized egg? We know probably the least about polarity in eggs of vascular plants because they have not been amenable to experimentation without being damaged drastically. However, it is possible to extrapolate from results obtained with other forms that are convenient for such studies. The outstanding utility of eggs of the brown alga *Fucus*, and to a lesser extent other members of Fucales like *Pelvetia, Cystoseira, Hormosira, Ascophyllum* (see Jaffe, 1968, for references), *Sargassum* (Nakazawa, 1950) and *Coccophora* (Nakazawa, 1951), for the study of polarity has long been recognized. Like eggs of other plants, the egg of *Fucus* is an incredible unit of organization and action, with the advantage that it is easily available for experimentation. The egg has a central nucleus surrounded by a blend of complex biological molecules and cytoplasmic organelles. At the time it is shed into the sea, the egg is devoid of any visible axis, but it becomes polarized soon afterwards. Fertilization, which occurs sometime after the egg is released, is the signal for the initiation of polarity. The first sign of polarity occurs about 17–18 h after fertilization when a small pear-shaped protuberance appears on one side of the egg, causing its polarization into a rhizoid and an embryonal head. Next, the egg is cleaved in a plane at right angles to the emerging rhizoid, cutting off a rhizoidal cell at the base from an embryonal cell at the tip.

In the egg of *Fucus*, the problem of polarity is simplified to a consideration of factors that enable it to determine the best possible direction to send out a rhizoid. Studies in polarity of *Fucus* eggs have been made with great success by Whitaker in the 1940s and by Jaffe in recent years. For the interested reader, reviews by these workers (Whitaker, 1940; Jaffe, 1968) are recommended for additional information.

Of all ways that polarity is manifest in the egg, the point of sperm entrance is the most obvious. Yet, proceeding from the often observed fact that parthenogenetically activated eggs kept in the dark, as free as possible from external gradients of environmental factors, develop rhizoids, it can be argued that there is no built-in primary axis of polarity. The most compelling argument in favor of this idea comes from the work of Jaffe (1956), who has shown that when fertilized eggs are cultivated in plane polarized light, they are stimulated to germinate and form rhizoids at two opposite poles corresponding to the direction of vibration of the electric vector. This might happen only if the incipient polarity of the egg perhaps already established during oogenesis is labile and can be altered by effective external factors. Illumination of the environment in which eggs are allowed to grow is an example of an external vector which determines the axis of polarity. If eggs are illuminated with unilateral white light, the rhizoid develops from the shaded side. If eggs are subjected to a gradient of IAA they germinate towards the higher auxin concentration. In a small temperature gradient, the rhizoid is initiated from the warmer side. When measurements of electrical potentials were made in a normally germinating egg, the end towards which the rhizoid developed became increasingly negative. Fertilized eggs might also cluster in significant numbers at a given place in a culture dish after completely random movements; under such conditions they germinated towards the center of the group. We could multiply the examples (see Table 1 in Jaffe, 1968), but these are sufficient to show the diversity of external factors which can subtly influence the direction of axis in the cell. These factors, whatever their origin, might conceivably impose gradients in the cytoplasm as a result of which the rates of certain metabolic processes or the amounts of certain metabolites might become progressively more concentrated at one end of the egg than at the other.

This brings us to a consideration of polarity as a means by which the egg acquires both directional movement of metabolites and asymmetrization. Unfortunately, the specificity of the agents required to induce polarity leaves much to be desired, and so it is unlikely that these studies will give us useful leads to explore the basis of polarity. The problems of real importance are: How do environmental vectors leave an imprint in the egg to fix its axis irreversibly? What are the physical and metabolic processes involved in the initiation and fixation of polarity? Since, in the present state of our knowledge, polarity is not interpretable in metabolic terms, we are not in a position to answer these questions. However, as a working hypothesis it seems reasonable to conclude that the ultimate determination of polarity must reside in some cytoplasmic structures whose direction of movement is under control of an external vector. A fairly recent discovery of an unequal distribution of calcium ions in the prospective growth poles of polarizing *Pelvetia* eggs (Robinson and Jaffe, 1975) tends to suggest that an essential link in the polarization process may be a cellular component that strongly binds calcium. Since physiological activities are gene-controlled, one can imagine possible mechanisms by which a gradient of increased cytoplasmic structures could give directions to the nucleus for an increased synthesis of specific metabolites at specific sites in the fertilized egg. The idea that, in addition to chemical and metabolic gradients, polarity in the egg might actually be controlled by physical forces has largely been derived from studies on bryophytes and pteridophytes, where the egg is encapsulated in the gametophytic tissues and from angiosperms where it is bathed in a fluid of high osmotic pressure in the embryo sac. As far as the relationship of the physical forces surrounding the developing egg to its polarity is concerned, all we can say is that the physical environment presumably has an effect on the orientation of the cytoplasmic structures in such a way that the associated organization of the chemical gradient in a certain direction is favored.

To sum up, the concept of polarity presents a challenge to developmental biologists in terms of clarifying and describing its exact basic causation. In all of his writings, Wardlaw (1955, 1965c, 1968a, b), an outstanding student of plant morphogenesis, has given much thought to the problems inherent in polarity and his writings serve as a poignant reminder of the major gaps in our comprehension of this phenomenon in plant and animal systems. Despite the serious limitation which our present ignorance imposes, the review of the basis of polarity in this subsection might appear worth making if only because it may provide some perspective for dealing with this question.

III. EMBRYOGENESIS

The structure and organization of embryos in the different groups of vascular plants show the same variability exhibited by other plant characteristics. For this reason, a complete enumeration of the types of embryos in each taxon is considerably beyond the scope of this work, but brief summaries of the salient features in the embryogenesis of representative species of angiosperms, gymnosperms and pteridophytes seem permissible.

A. Angiosperms

Most of the steps in the intricate process of embryogenesis have been catalogued for a number of species of flowering plants (Johansen, 1950; Davis, 1966). At first glance, the task of following the formation of a three-dimensional embryo from a single-celled zygote might seem almost insurmountable. Yet, by painstakingly analyzing serial sections it is quite possible to assign the different parts of the final organization to definitive cells in the first few division phases of the embryo, and to map the origins of tissue layers. The embryo of each species has its own characteristic sequence or pattern of formation and every step in the division process occurs with the utmost precision. This latter statement, which finds support in contemporary works on angiosperm embryology (Johansen, 1950; Maheshwari, P., 1950; Souèges, 1951; Davis, 1966), has been questioned by recent cell lineage studies. By methods which were refreshing departures from earlier ones, Pollock and Jensen (1964) found that in the formation of the globular embryo of cotton, only the first division of the zygote followed the accepted pattern, while the later divisions were largely irregular. Differences from the classical pattern of cell disposition were also observed in the division of cells of the dermatogen of young embryos of *Capsella bursa-pastoris* (Pollock and Jensen, 1964). Similarly, in the proembryos of silver maple (*Acer saccharinum*), random growth and irregular cell division patterns were characteristically observed (Haskell and Postle-thwait, 1971). In other developmental studies (Vallade, 1973; Vallade and Bugnon, 1974) attention is directed to the quality and quantity of growth and histological differentiation of tissues in embryos to gain a true picture of the morphogenetic aspects of embryogenesis. The results of these several investigations tend to underscore the need to correlate cell lineage studies in embryogenesis with some knowledge of the nature of the influences in the immediate vicinity of the embryo which may be the key factors that determine the orientation of the mitotic spindle.

The Dicotyledonous Embryo

In *Capsella bursa-pastoris*, which has a long history of use as a classroom material to illustrate embryogenesis in a typical dicotyledonous plant, the zygote is about 30 μm long. This cell undergoes the first division by a transverse wall to form a large, highly vacuolate basal cell and a small densely cytoplasmic terminal cell (Schulz, R. and Jensen, 1968a, c). Each of these cells makes a crucial decision as soon as it is formed. The apical cell forms the embryo proper. The large cell near the base of the embryo sac does not pause after it is formed, but is partitioned immediately into two cells. These new daughter cells multiply by transverse divisions to form a 9–10-celled filament called a suspensor. We shall return to this structure later.

In the embryo of cotton, a number of differences are evident between the terminal and basal cells after the first division of the zygote (Jensen, 1964). There are a lot more plastids in the terminal embryonal cell than in the basal suspensor cell. Mitochondria appear large and variable in shape in the terminal cell. Vacuoles, most of them of the dense membranous type, are conspicuous in the basal cell. Associated with the vacuoles are small vesicles, apparently fusing with them to form large vacuoles. Thus, ultrastructurally the terminal and basal cells of the two-celled embryo amplify the structural features of the chalazal and micropylar ends, respectively, of the zygote described earlier.

In Fig. 2.9 are presented the easily recognizable stages in the transformation of the embryonal cell of *Capsella* into a full-grown embryo (Schaffner, 1906). The first division of this cell in a longitudinal plane is followed by a second division at right angles to the first to produce a quadrant. The cells of the quadrant in turn become partitioned to

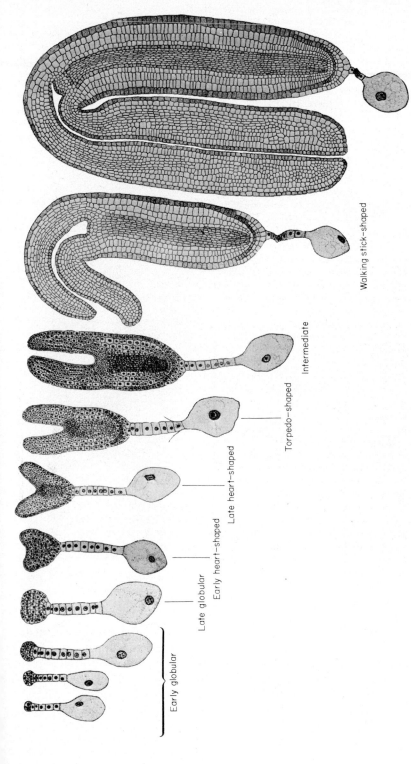

FIG. 2.9. Stages in the embryogenesis of *Capsella bursa-pastoris* in longitudinal sections. The lower end of the embryo is directed towards the micropyle. (After Schaffner, 1906)

Early globular

Late globular

Early heart-shaped

Late heart-shaped

Torpedo-shaped

Intermediate

Walking stick-shaped

Mature

form the eight-celled embryo or octant (Fig. 2.10). At this stage the differentiative potencies of the upper and lower tiers of cells of the octant are established. The former gives rise to the stem tip and cotyledons, while the hypocotyl is derived from the latter. The first recognizable

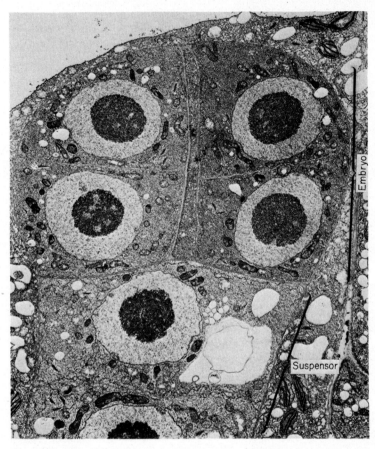

FIG. 2.10. Cells of the octant embryo of *Capsella bursa-pastoris* have a greater density of ribosomes and are less vacuolate than those of the suspensor. (After Schulz, R. and Jensen, 1968a; photograph courtesy of W. A. Jensen)

step in tissue differentiation in the embryo is a periclinal division in each of the octant cells to form a 16-celled sphere, the outer cells of which form the protoderm and the inner cells give rise to the procambium and ground meristem (Fig. 2.11). From now on, the progress of embryogenesis is discerned from the changing shape of the embryo. The developing embryo passes successively through "globular" (less

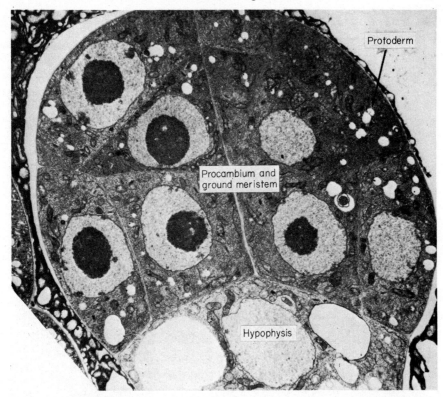

FIG. 2.11. In a 16-celled embryo of *Capsella bursa-pastoris* the outer cells form the protoderm, and the inner cells form the procambium and ground meristem. The hypophysis has a lower ribosome density and is more vacuolate than cells of the embryo. (After Schulz, R. and Jensen, 1968a; photograph courtesy of W. A. Jensen)

than 80 μm long), "heart-shaped" (81–250 μm long), "intermediate" (251–400 μm long), "torpedo-shaped" (401–700 μm long), "walking-stick-shaped" (701–1000 μm long), "inverted U-shaped" (1001–1700 μm long) and "mature" (more than 1701 μm long) stages.

The globular embryo, also known as proembryo,* has a radial symmetry which it inherits from the egg and which persists until the beginning of the heart-shaped stage. It continues to enlarge by divisions within the sphere. At this stage the suspensor cell nearest to the embryo

---

* According to Souèges (1936) the term proembryo refers to any developmental stage preceding cotyledon initiation. Since cotyledon initiation takes place at the heart-shaped stage, the terminology used here is consistent with that of Souèges.

designated as the hypophysis divides transversely to form two cells; the daughter cell from this division lying towards the embryo is partitioned again to form two cells. The cells of the root cortex, root cap and root epidermis of the mature embryo are contributed by descendants of these

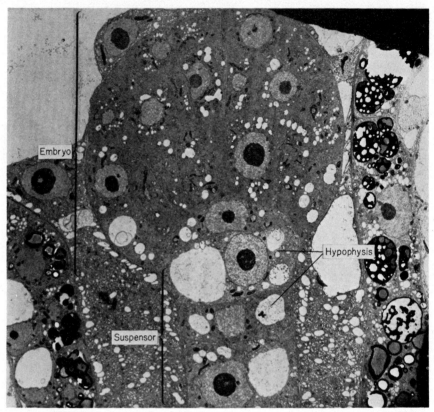

Fig. 2.12. Structural differences between cells of the embryo proper and those of the hypophysis observed in the 16-celled embryo of *Capsella bursa-pastoris* are accentuated in this photograph of the globular embryo. (After Schulz, R. and Jensen, 1968a; photograph courtesy of W. A. Jensen).

two cells. Ultrastructurally, cells of the hypophysis of the globular embryo are more vacuolate and have a lower ribosome density than cells of the embryo (Fig. 2.12).

At the end of the globular stage, the embryo more or less expands laterally as a result of periclinal divisions in the two opposite poles. This configuration is what constitutes the embryonic leaves or cotyledons. It is only at this stage of development that any ultrastructural differences are noticeable between cells of the different regions of the embryo. For example, cells of the procambium and ground meristem are more vacuolate than those of the protoderm. Plastids of the cells of the ground meristem and protoderm have lamellar systems while those of the procambium remain undifferentiated. When the embryo embarks upon the autotrophic phase of growth, as in the heart-shaped and older stages, the density of ribosomes in cells decreases.

During further development, cotyledons and the future stalk called the hypocotyl elongate (intermediate and torpedo-shaped stages) and a mound of rapidly dividing cells which constitute the apical meristem of the epicotyl is organized in the depression between cotyledons. Due to spatial restrictions inside the ovule, the embryo becomes curved at the tip (walking-stick-shaped) and finally assumes the shape of a horse-shoe (inverted U-shaped). The growth of the root system of the future plant is made possible by the organization of a prominent root apex at the proximal end of the embryo (Maheshwari, 1950; Rijven, 1952; Raghavan and Torrey, 1963). The basic plan of embryogenesis just described for *Capsella* shows the process in a relatively uncomplicated fashion and it is thus no accident that the species has long been used as an illustrative material.

By the time the embryo of *Capsella* has matured, its length has increased to more than 2 mm. This tremendous increase in size is due to the uptake of raw materials from the endosperm and their utilization in the synthesis of complex substrates. As shown by Pollock and Jensen (1964), as the embryo increases in size, the size of the individual cell is reduced and a single cell of the torpedo-shaped embryo is no bigger than a quarter of the zygote.

Mature embryos have a procambial system differentiated throughout the hypocotyl and cotyledons. Although mature xylem elements are found within the procambium (Miller and Wetmore, 1945), more commonly, embryos have only differentiating xylem or phloem elements (Bisalputra and Esau, 1964); rarely, neither type of conducting elements is present (Calvin, 1966).

The Suspensor. In the study of angiosperm embryogenesis, interest has been largely confined to the spectacular transformation of the apical cell, resulting from the first division of the zygote, into the embryo. Yet, from a comparative morphogenetic standpoint the lower cell which develops into the suspensor complex also deserves attention because the latter is a unique structure with specialized functions. In *Capsella* (Schulz, P. and Jensen, 1969) at the octant stage of the embryo the suspensor has six to seven cells which increase to about 10 cells in the globular embryo (Fig. 2.13). It is terminated by a large basal cell at the micropylar end and by the embryo proper at the chalazal end. As stated previously, a curious and interesting cell of the suspensor is the hypophysis which lies adjacent to the embryo and partly contributes to its formation. Compared to the cells of the embryo, those of the suspensor are more vacuolate and contain more ER but look impoverished of ribosomes; they also stain less intensely for proteins and nucleic acids. The suspensor cells are rich in dictyosomes, mitochondria, plastids and spherosomes. In the suspensor cells of the globular embryo, there is a decrease in the density of ribosomes and in the concentration of nucleic acids and proteins. Symptomatic of the short-lived nature of the suspensor, cytoplasmic degeneration of the cells followed by depletion of ribosomes and loss of nucleic acids and proteins begins at the heart-shaped stage of the embryo. Gradually, the middle lamella of cells begins to loosen and eventually the suspensor is obliterated by the growing embryo. The large vacuolate basal cell generally remains active for some time after the other cells have degenerated. On the basis of these changes which have also been described in sunflower (Newcomb, 1973b) and chickweed (*Stellaria media*) (Newcomb and Fowke, 1974) the suspensor could be characterized as a weak structure capable of abrupt physiological transition.

In the adventitiously formed embryos of carrot (*Daucus carota*) a suspensor as such is absent, but in a group of cells destined to produce adventive embryos, a distinction between a dividing embryonal cell and a nondividing suspensor-like cell is possible. The onset of functional differentiation between the cell types coincides with an increase in the ribosome content of the embryonal cell while that of the suspensor-like cell remains unchanged (Halperin and Jensen, 1967).

A nuclear phenomenon observed in the suspensor cells of some plants is endopolyploidy, that is, polyploidy resulting from repeated chromosome duplication which is not followed by cell or nuclear division. In *Phaseolus vulgaris* (French bean) and *P. coccineus*, one can recognize a progressive increase in the degree of endopolyploidy in the suspensor cells towards the micropylar region. Cells of the stalk of suspensor exhibit low to medium degree of endopolyploidy whereas the giant cells towards the micropylar end have polytene chromosomes showing "puffing" (Nagl, 1962a, b, 1970; Brady, 1973a, b). Cytochemical, cytophotometric and autoradiographic evidence shows that many heterochromatic regions of chromosomes of endopolyploid and polytene cells undergo extra synthesis of DNA, part of which is extruded into the cytoplasm as micronuclei. There is some evidence to show that cytoplasmic DNA so formed is capable of coding for RNA (Avanzi *et al.*, 1970, 1971).

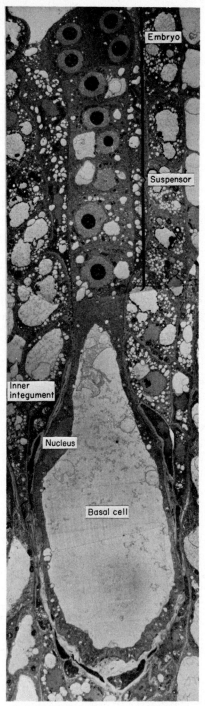

FIG. 2.13. The basal cell, a six-celled suspensor and an octant embryo of *Capsella bursa-pastoris*. The nucleus of the basal cell is in the peripheral cytoplasm. The cells of the inner integument are crushed by the expanding basal cell. (After Schulz, P. and Jensen, 1969; photograph courtesy of W. A. Jensen)

Persistent reports on the metabolic status of the suspensor have appeared in the literature. They are known to contain food materials, including various carbohydrates, oils, proteins and amino acids. The high metabolic activity of the suspensor cells of some species of *Crotalaria* is probably due to the presence of photosynthetically active pigments (Rau, 1950). According to Satina and Rietsema (1959), suspensor cells of several species of *Datura* serve as a direct source of oil to the proembryo. In careful cytochemical studies using fresh material, these workers detected oil first in the basal cell of the proembryo which was destined to form the suspensor. Other cells of the proembryo acquired oil only after its appearance in the presumptive suspensor cell. With some justification, this has led to the assumption that oil is channeled to the growing embryo through the suspensor.

If the suspensor is thought to function as a channel for the conduction of food materials, it would appear that there must be some structural modification in the cells for this purpose. What form might this modification take? In the suspensor cells of *Capsella* (Schulz, P. and Jensen, 1969, 1974) and *Pisum sativum* (pea) (Marinos, 1970) several finger-like projections from the lateral walls are observed to extend up into the endosperm (Fig. 2.14). The micropylar and lateral walls of the basal cells of suspensors of *Capsella* and chickweed (Newcomb and Fowke, 1974) embryos also contain elaborate networks of invaginated protoplasmic surfaces resembling the wall projections in the filiform apparatus of the synergid described earlier. In *Phaseolus vulgaris* modifications of the suspensor cells take the form of external protuberances which run helically round the cells (Schnepf and Nagl, 1970), while in *P. coccineus* (Clutter and Sussex, 1968) the wall projections are frequently branched. In addition, cytoplasmic connections in the form of plasmodesmata run between the individual cells of the suspensor and probably between the suspensor and embryo. It has been proposed that these structural modifications of the wall are the prime movers in the absorption of nutrients from the endosperm and their transport to the embryo (Schulz, P. and Jensen, 1969). Although the function of the suspensor has been debated for years, in my view it is Schulz and Jensen's analysis which has led us closest to what appears to be a correct concept of its function. Nagl (1973) has made a pointed comparison between suspensors of plant embryos and mammalian trophoblasts and suggests that like the latter the suspensor may function in the synthesis of specific hormone precursors for embryo growth. A recent report on the possibility of culturing isolated suspensor (Brady, 1970) offers much promise of gaining further insight into its function.

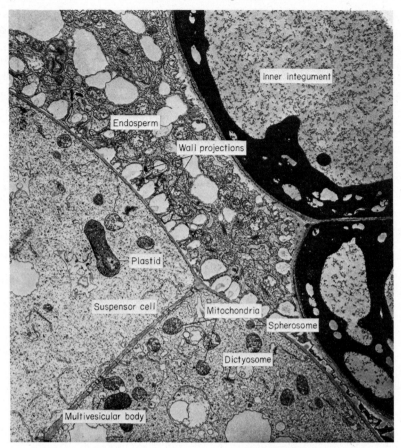

Fig. 2.14. From the lateral cell walls of the suspensor of *Capsella bursa-pastoris* several finger-like projections extend into the endosperm. (After Schulz, P. and Jensen, 1969; photograph courtesy of W. A. Jensen)

The Monocotyledonous Embryo

When viewed broadly, there are no discernible differences between embryos of dicotyledons and monocotyledons in the first few division cycles, although full-grown embryos of several monocotyledons, especially of the Gramineae, are so vastly different from those of dicotyledons. A useful example of a monocot to study the cleavage pattern of the embryo is *Luzula* (Juncaceae), which has a relatively simple embryogeny. In *L. forsteri*, the first division of the zygote is transverse (Souèges, 1923). The next round of cell divisions takes place longitudinally and transversely in the terminal and basal cells, respectively. The embryonal part is contributed by the terminal cell and by the upper cell products of the basal cell. As the embryo becomes spherical by further divisions, it is partitioned into two portions of which the terminal part gives rise to the lower part of the cotyledon, and the lower half contributes to the upper portion of the cotyledon and the hypocotyl. The shoot tip is initiated at the junction of the hypocotyl and cotyledon

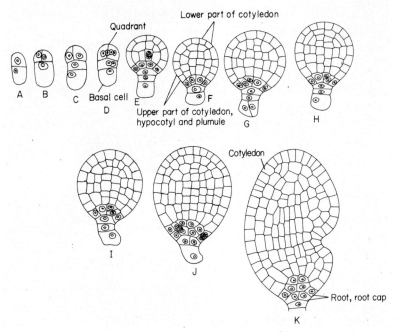

Fig. 2.15. Embryo development in *Luzula forsteri* in longitudinal sections. Lower end of the embryo in each drawing is directed towards the micropyle. **A–D**, Stages in the development of a quadrant. **E–H**, Later stages showing partitioning of the quadrant to form the lower part of the cotyledon and appear part of the cotyledon, hypocotyl and plumule. **I–K**, The root and root cap are formed from the entire one half of the original basal cell and partly from the other. (After Souèges, 1923).

as a result of suppression of lateral growth. The prospective fate of cells derived from one half of the original basal cell is to form the root and part of the root cap in the mature embryo; the other half of the original basal cell forms the remaining part of the root cap and a suspensor. Some stages in the embryogenesis of *Luzula* described above are shown in Fig. 2.15.

Embryos of the Gramineae have received much attention from morphologists and embryologists. It is therefore necessary to describe in some detail a typical graminean, embryo, examplified by maize. Because of the complexity of its structure, the full-grown maize embryo will be described first, and as far as possible, the different organs will be traced to definitive parts of the proembryo. The lower part of the embryo has a root and both root and root cap are covered by a sheath-like structure called the coleorhiza. The upper part of the axis of the embryo bears the epicotyl, which has leaf primordia and a shoot apex, all enclosed in a leaf-like sheath called coleoptile. Perhaps the most interesting part of the grass embryo is a shield-like structure called the scutellum to which the embryo is attached laterally (Fig. 2.16). The scutellum is of controversial morphology, but is considered equivalent to a cotyledon. Although absent in maize, most grass embryos have a flap-like structure known as the epiblast, usually situated above the coleorhiza and opposite the scutellum (Avery, 1930; Randolph, 1936; Kiesselbach, 1949; Abbe and Stein, 1954).

In maize, following the first division of the zygote, the terminal embryonal cell divides irregularly to form a mass of small cells. The basal suspensor cell divides only sparsely. In the embryo, which assumes a club-shaped form as a result of greater cell division in the upper region, a small lateral protuberance formed by a group of meristematic cells appears about 10–12 days after pollination. It is from this group of cells that primordia of the shoot and root are differentiated. The coleoptile appears as a sheathing structure on the protuberance. By the thirteenth day after pollination, the embryo becomes somewhat flattened by enlargement of its tip along the posterior side of the suspensorial region to form the scutellum. During the period of 14–20 days after pollination, the primordia of seedling leaves appear. A coleorhiza arises from the scutellar node and within the coleorhiza the primary root is recognized in about 20–22 days after pollination. The active life of the suspensor is terminated in about 20 days, after which it remains as a nonfunctional entity. A recent electron microscopic study of the early division phase of barley (*Hordeum vulgare*) embryos has revealed little, if any, evidence of cell specialization and polar organization (Norstog, 1972a). The only noticeable difference between the suspensor and embryonal cells appeared after two or three division cycles in the form of a progressive decrease in the number of vacuoles towards the suspensor region. The most irregular of the vacuoles were also concentrated towards the suspensor end (Fig. 2.17).

Based on extensive investigations on segmentation patterns in embryos of a large number of plants, Souèges (1951) has developed a system of classification of angiosperm embryos. It is impossible to discuss the details of this classification here; the interested reader will

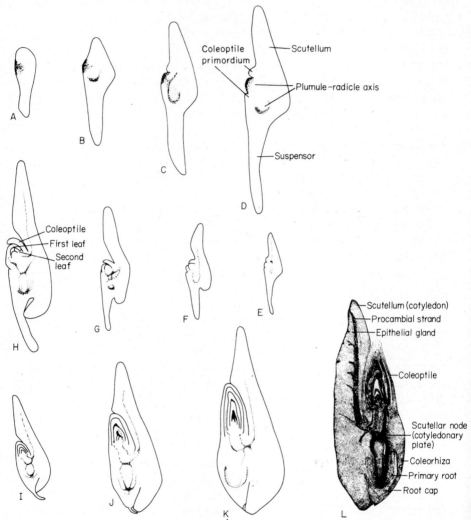

FIG. 2.16. Embryo development in *Zea mays* in outline drawings. **A–D**, Stages at 10, 12, 13 and 14 days after pollination showing differentiation of plumule-radicle axis, scutellum and coleoptile primordium. **E–H**, Stages at 14, 16, 18 and 20 days after pollination showing the appearance of primordia of first and second seedling leaves. **I–K**, Stages at 25, 35 and 45 days after pollination, showing formation of additional seedling leaves and obliteration of the suspensor. (After Randolph, 1936) **L**, Median longitudinal section of a mature embryo to show anatomical details. (After Avery, 1930)

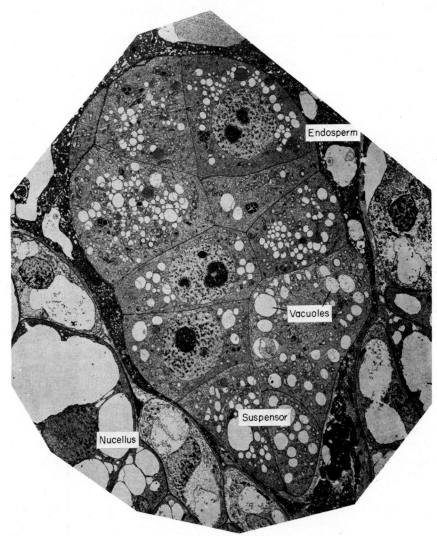

Fig. 2.17. Vacuoles tend to be irregular and fewer in number towards the suspensor end of a 20 day old barley embryo, which does not appear to have any intimate contact with the nucellus. Endosperm surrounds the embryo. (After Norstog, 1972a; photograph courtesy of K. J. Norstog)

find excellent summaries of this system in the books of P. Maheshwari (1950), Johansen (1950) and Wardlaw (1955), and in a review by Crété (1963). Briefly, in his proposed classification of angiosperm embryo types, Souèges has considered the mode of segmentation of the terminal cell of the two-celled proembryo, the mode of origin of the four-celled proembryo, the behavior of the basal cell and the relative contributions of each cell of the four-celled proembryo to the formation of the mature embryo. The wisdom of application of these laws to encompass embryos of the entire plant kingdom is questionable, particularly in the light of recent cell lineage studies in embryogenesis which have shown significant variations from the accepted patterns of division and disposition of cells of the proembryo.

B. Gymnosperms

The pattern of cell diversification during embryogenesis in gymnosperms bears a close resemblance to that found in angiosperms. The gymnosperm embryo is generally embedded in the ovule and obtains its nourishment from the tissues of the gametophyte. Indeed, during maturation of the egg, the metabolism of the surrounding tissues is actively pressed towards the synthesis of various reserve materials which are later used by the developing embryo. Thus, the gametophytic tissue in gymnosperms seems to exercise an overall control over the differentiation of the embryo.

In the gymnosperm egg, the period following fertilization is one during which a new submicroscopic structure characteristic of the future embryo is established in the cytoplasm, which thus becomes perfectly distinct from the egg cytoplasm with its remarkably complex storage inclusions. The first two divisions of the zygote nucleus take place in the anterior half of the cell. According to Camefort (1958, 1966b, 1967a, c, 1968b, 1969), during the first nuclear division the mitotic apparatus is formed in conjunction with the nucleoplasm and the cytoplasm immediately surrounding the nucleus. This cytoplasmic terrain, which is the progenitor of the cytoplasm of the embryo, is designated as "neocytoplasm". After the second division, the neocytoplasm and the engulfed free nuclei migrate towards the base of the egg where the proembryo is formed (Fig. 2.18). One of the more evident ultrastructural features of the neocytoplasm at this stage is the appearance of organelles including mitochondria, plastids and ribosomes; it is believed that, according to the species, some or all of the mitochondria and plastids found in the neocytoplasm have their origin in the male cytoplasm. In *Larix decidua*, the maternal mitochondria invade the neocytoplasm from the perinuclear zone of the egg, while the plastids eventually incorporated into the embryo appear to have come from the contents of the pollen tube. Some paternal mitochondria might have also been contributed to the neocytoplasm (Camefort, 1968a). In *Pinus*

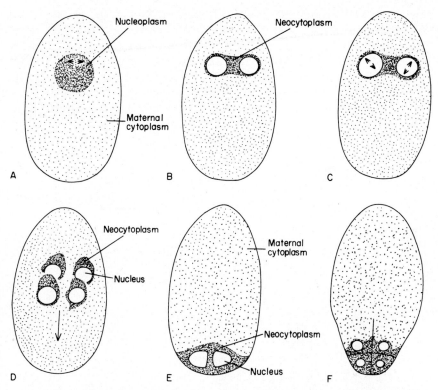

Fig. 2.18. Schematic representation of the formation of "neocytoplasm" during embryogeny of a gymnosperm. The nucleoplasm of the zygote and the neocytoplasm are heavily shaded, while the egg cytoplasm is lightly shaded. **A, B**, First mitosis. **C**, Second mitosis. **D**, Migration of free nuclei. **E**, Proembryo after migration of free nuclei. **F,** Cellular proembryo. (After Camefort, 1967a)

laricio, the situation is less clear cut, but at least the mitochondria appear to be of maternal origin (Camefort, 1966a, 1967a). In some members of the Cupressaceae most of the mitochondria in the neocytoplasm have their origin in the pollen tube (Chesnoy, 1969a, b, 1973). Because of the sheer number and complex ultrastructure of the storage inclusions of the egg cytoplasm, a full comprehension of their nutritional contributions to the developing proembryo has not been obtained; some evidence indicates that the storage inclusions simply begin to disintegrate when the division of the zygote nucleus is under way and thus flood the cell with simple metabolic precursors (Camefort, 1966c).

The later stages of embryogenesis of gymnosperms is best comprehended by a description of events in the embryogeny of *Pinus wallichiana*

(Fig. 2.19), which may be taken as a representative type (Konar and Ramchandani, 1958). The four free nuclei at the base of the egg divide simultaneously accompanied by wall formation to form eight nuclei arranged in two tiers of four cells each. Two additional divisions occur in the upper and lower tiers of cells to form a total of 16 cells in the proembryo. These cells are now arranged in four uniform tiers of four cells each and each tier of cells is earmarked for a particular

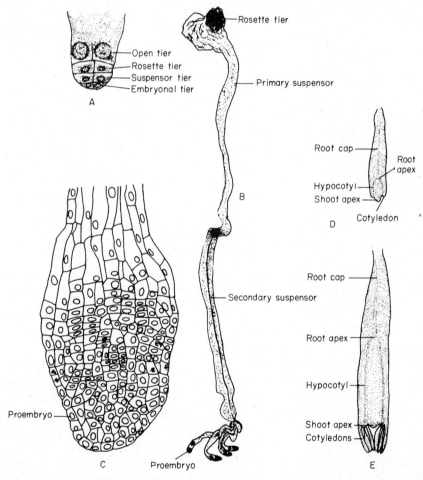

FIG. 2.19. Embryo development in *Pinus wallichiana* in longitudinal sections. **A**, Proembryo showing four tiers of cells. **B**, Proembryos with primary and secondary suspensors and rosette tier. **C**, Section of a proembryo. **D**, Embryo showing differentiation of root cap, root apex, shoot apex and cotyledons. **E**, A nearly mature embryo. (After Konar and Ramchandani, 1958)

function. The lowermost tier of cells in the proembryo contiguous with the base of the archegonium gives rise to the embryo proper. The cells of the second tier constitute the primary suspensors which elongate and push the terminal cells into the gametophyte. Cells of the third tier, called rosette cells, become meristematic and may form supernumerary embryos. The function of the uppermost tier is thought to be transmission of food materials to the growing embryos; this view is strengthened by the fact that these cells are only partially enclosed by walls and are thus in close contact with the nutritive resources of the gametophyte. In this context, one of the most explicit functional expressions of the upper tier of cells is seen in the differentiation of a plate-like fibrillar structure traversed by microvilli and cytoplasmic channels at its boundary with the lower tier of rosette cells (Thomas, 1973b).

Continued cell divisions in the upper embryonal cells eventually lead to the formation of pockets of meristematic tissues each of which differentiates into an embryo. This is, however, preceded by the formation of unusually long and coiled secondary suspensor cells which originate from the embryonal cells and thrust the growing embryo into a strategic location in the gametophytic tissue. The upper region of the embryonal tissue differentiates into the shoot apex, while the basal region forms the root. The mature embryo has a root apex, hypocotyl and shoot apex capped by several cotyledonary primordia. Well developed embryos of some cycads possess stomata and cuticle (Pant and Nautyal, 1962).

Embryogenesis in gymnosperms is complicated and variable owing to the occurrence of polyembryony (Coulter and Chamberlain, 1917; Johansen, 1950). In some genera (for example *Cedrus, Keteleeria, Tsuga, Larix, Pseudolarix*, etc.), each cell of the lowermost tier or of the rosette layer may form separate embryos. Polyembryony may also result from the simultaneous fertilization of several eggs as in *Picea, Larix, Pseudolarix*, etc. However, during further growth all embryos abort, or are absorbed into the gametophytic tissues except the one which outstrips the others and matures in the seed.

C. Pteridophytes

Depending upon the nature of modification for the elaboration of food materials for the embryo, three main types of embryogeny can be distinguished among pteridophytes. In Lycopodiales and Marattiales, embryos have suspensors which mediate in their interactions with the environment. In Psilotales, Equisetales and Isoetales a special absorbing organ (the foot), which transmits reserve materials from the gametophyte to the embryo, develops. In leptosporangiate ferns, embryos are lateral and lack a suspensor (Bower, 1926, 1928; Eames, 1936).

The pteridophyte egg is fertilized while it is still protected in the venter. During progressive embryogenesis, cells of the venter and those surrounding it form a hard protective covering for the embryo called the calyptra. For our purposes, we will confine our description to the embryogeny of the osmundaceous fern, *Todea barbara* (DeMaggio, 1961b).

After fertilization, there is an uneven distribution of the cytoplasmic contents of the zygote, as most of the cytoplasm containing chloroplasts and nucleus is accumulated in the portion of the zygote towards the base of the archegonium. The first division of the zygote 6 days after fertilization separating it into anterior and posterior hemispheres is characteristically in the plane of the archegonial axis and at right angles to the axis of the prothallus. This is followed by a division in the plane of the archegonium leading to the quadrant stage at 7 days after fertilization. An octant is formed at 9 days by transverse walls in each cell of the quadrant. The resulting spherical mass of cells can be arbitrarily divided into a superior and an inferior hemisphere, each composed of two anterior and two posterior octants. In the 11 day old embryo there are 16 cells. The next division to form a 32-celled embryo is complete by the thirteenth day. At this stage of development the embryo displays the first indication of the formation of a foot contributed by descendants of two anterior and two posterior octants comprised in the superior half. This organ is often conspicuous in the embryo about 17 days after fertilization because of its thin-walled, starch-filled cells. The prospective fate of the remaining cells, those of the inferior hemisphere, is to form the leaf, stem and root initials.

Leaf initials which can be traced to the anterior inferior octant are observed in 18 day old embryos and by 22 days a stage of leaf initiation suggestive of an early stage in crozier development is clearly seen. At this stage, a tetrahedral apical cell formed from the outer layers of the inferior half of the octant also takes form. In 20–22 day old embryos there are also conspicuous signs of stem and root initials emerging almost simultaneously from opposite regions of the embryo. The former is derived from the surface cells of the anterior portion of the embryo and the latter from the posterior inferior portion. Embryogenesis is essentially complete with the formation of the adult organ initials. In a full-grown embryo about 30 days after fertilization, the root is the first organ to rupture the calyptra and emerge. Some of the stages described above are illustrated in Fig. 2.20. This description of embryogenesis in *Todea barbara* is generally applicable to other ferns and fern allies of different systematic affinities, although formation of the various

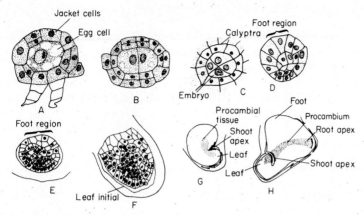

Fig. 2.20. Development of the embryo of *Todea barbara*. **A**, Five day old fertilized egg in longitudinal section. **B**, Transverse section of a two-celled embryo, 6 days old. **C**, Transverse section of a four-celled embryo, 7 days old. **D**, Median longitudinal section of a 17 day old embryo showing the formation of the foot. **E**, An embryo at 18 days showing well developed foot, an outer region of elongate cells and an inner region of isodiametric cells in the inferior hemisphere. **F**, Leaf initiation in a 20 day old embryo. **G**, Embryo showing the leaf curling over the stem initial; 28 days old. The procambial tissue extends from the central region of the embryo into the developing leaf and stem. **H**, Embryo showing the leaf still enclosed by the calyptra and the emerging root; 30 days after fertilization. (As illustrated by Wetmore and Steeves, 1971, after DeMaggio, 1961b)

organs may not be rigidly limited to a particular morphological region of the octant embryo. More obviously, differences in the timing of developmental sequences in embryogeny have also been observed.

IV. COMMENTS

The segmentation patterns in the early embryogeny of representative species of angiosperms, gymnosperms and pteridophytes recorded in this chapter appear to show many features in common. In attempting to analyze the factors that may be involved in the precise division pattern in fern embryos, Wardlaw (1955, 1965c) has marshaled impressive evidence to show that homologies in the basic pattern of embryonic organization may be largely determined by nutritional and metabolic gradients in the embryo itself or by physical or environmental forces in the surrounding milieu. An understanding of the nature of the morphogenetic influences and metabolic adjustments operating during embryogenesis would seem to be of much importance in elucidating the basis for the organizational differences in embryos of different groups of vascular plants.

3. Biochemical Embryogenesis

The process by which the fertilized egg is transformed into an embryo involves the production of stable and diverse cell types by the progressive structural and functional specialization of an increasing population of cells. Chemical transformation and build-up of complex molecules and differentiated structural substances are of significance at the molecular and macromolecular levels of specialization of cells in a developing embryo. A knowledge of the chemical composition of the embryo at defined stages of development and distribution of components in its cells and tissues are often useful attributes to interpret the causal relationships involved in cell specialization and embryonic growth. In animal embryology, extensive cytochemical, histochemical and biochemical studies of eggs and embryos of sea urchins and several amphibians have contributed to a general understanding of the relationship between cellular composition and morphological development. The simplest working hypothesis evolved from these studies is that metabolic changes and regional differences in biochemical properties are linked to morphogenesis.

The above type of relationship is not well understood in plant embryos, where no comparative studies at a level worthy of modern histochemistry and biochemistry have been feasible. In view of the achievements of animal embryologists, biochemical and histochemical approaches to the problem of embryogenesis in plants might appear highly rewarding as a means for a more penetrating analysis of the basis of cell activities. It is therefore essential to consider the limits of the present knowledge of the biochemical mechanisms in developing embryos, and perhaps thereby to indicate how much is yet to be explained before the relationship between cellular composition and morphogenesis can be understood. To this end, we shall examine results from studies on the distribution of chemical components in eggs and embryos of plants. The findings may also conceivably illuminate the dynamics of cellular metabolism in the growing embryo.

It is obviously impossible within the scope of this chapter to provide a survey of the methods that have contributed to our present understanding of the chemical composition of embryos. It is also unnecessary since studies to be discussed here involve procedures commonly and adequately discussed in standard books on cytochemical and biochemical methods.

I. NUCLEIC ACIDS OF THE EGG AND ZYGOTE

A. Deoxyribonucleic Acid

Recent innovations in cytochemical and histochemical methods for the microscopic localization of cell constituents and development of the method of autoradiography for visual localization of radioactive biological markers have prompted a number of workers to map the distribution of biochemical constituents in plant tissues that can be characterized by these methods. Components of the egg and zygote that have come under intensive study are DNA and RNA which have been shown to undergo impressive metabolic alterations during embryogenesis. Feulgen staining has revealed that nuclei of mature egg cells of some ferns and flowering plants lack DNA and behave aberrantly when they are ready for fertilization, although in the early stages of their formation they are strongly Feulgen-positive. Excellent summaries of the early work appear in the reviews by J. Vazart (1956) and B. Vazart (1958). A review of the more recent work is given by Raghavan (1975a).

Experiments by Bell (Bell, 1960, 1963; Sigee and Bell, 1968) have shown that DNA content of the egg nucleus of *Pteridium aquilinum* decreases progressively with maturity of the egg. Feulgen staining of the nucleus of the mature egg was exceedingly faint. Conventional autoradiographs of ^{3}H-thymidine incorporation showed that in the immature egg DNA synthesis was confined to the nucleus, but in the mature egg DNA was synthesized in large amounts in the cytoplasm, principally in the periphery of the cell. In later studies with the improved resolution offered by electron microscopic autoradiography, it was shown that extranuclear incorporation of ^{3}H-thymidine occurred partly into undifferentiated mitochondria and plastids and partly into the ground plasm of the egg, the latter probably representing degenerating organelles (Bell and Mühlethaler, 1964b; Sigee and Bell, 1971; Sigee, 1972b). Although the issue has been less systematically explored in the somatic cells of the archegonium, such data as are available suggest that there is little, if any, extraorganelle DNA synthesis in them (Sigee, 1972a).

The above is not a random observation, since a few gymnosperm and angiosperm eggs examined have also revealed the existence of unusual DNA metabolism. The nuclei of mature eggs of members of Cycadales are relatively huge in size, and in *Zamia* the egg nucleus may reach up to 1000 μm in length (Coulter and Chamberlain, 1903). In *Z. umbrosa* the chromatin contracts to a small globule in the mature egg (Bryan and Evans, 1956, 1957), which also shows a remarkable modification of the nuclear membrane into a reticulate pattern of ridges and depressions, simultaneously accompanied by the extrusion of Feulgen-negative vacuolate bodies. Shimamura (1935) has established that in *Ginkgo biloba* there is a complete absence of Feulgen staining in the nuclear reticulum of the egg, whereas nuclei of the prothallial cells are Feulgen-positive. The cytological structure of the egg also changes at maturity by the appearance of numerous granules of undetermined function in the cytoplasm. In *Cycas revoluta* (Shimamura, 1935) and *Cephalotaxus drupacea* (Singh, 1961), Feulgen staining is restricted to a small spot in the entire egg nucleus. The egg nucleus of *Pinus thunbergii* becomes Feulgen-negative at the same time as the number and stainability of storage inclusions of the cell decrease (Shimamura, 1956; Takao, 1959). Maturation of the egg of *Pseudotsuga menziesii* is associated with the depletion of DNA from the nucleus and its reappearance in the mitochondria which arrange themselves around the nucleus as a conspicuous Feulgen-positive ring (Chesnoy and Thomas, 1969; Thomas and Chesnoy, 1969). Complex as these various phenomena are, we shall not expect to be able to comprehend their significance until we gain some understanding of the mechanism by which chromatin is resorbed into the nucleus or released therefrom.

Microspectrophotometric measurements have revealed that the egg, synergids and antipodals of *Tradescantia paludosa* have the lowest amounts of DNA of any cells of the gametophytic or sporophytic tissues, but amounts are probably equal to those expected from their haploid nature (Woodard, 1956). Klyuchareva (1960) has followed cytochemically the disappearance of Feulgen-positive DNA from eggs of several cereal grains. In barley, wheat and rye (*Secale cereale*), this was accompanied by the disappearance of the chromatin network from eggs just before fertilization (Fig. 3.1). Immediately after fertilization, there was a dissolution of sperm DNA, and synthesis of new DNA in the form of granules inside the nucleolus. It has been suggested that chromosomes of the zygote are probably formed from these DNA granules,

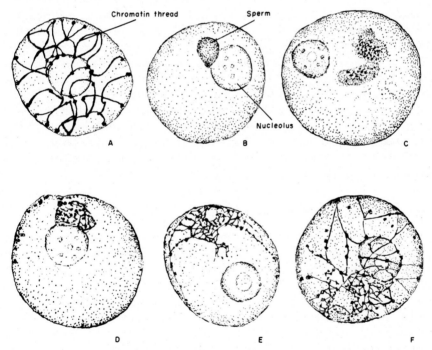

Fig. 3.1. Chromatin threads disappear in the egg nucleus of barley just before fertilization but are resynthesized later. **A**, Chromatin present before fertilization. **B**, Disappearance of chromatin after entry of sperm. **C**, Egg with two sperms. **D**, Fertilization and sperm disintegration. **E**, Appearance of male and female pronuclei. **F**, Zygote nucleus with new chromatin threads. (After Klyuchareva, 1960)

although actual evidence for this is not obvious. Feulgen-positive DNA was not present in the unfertilized egg of *Stellaria media*, although it was readily detected in antipodals and synergids (Pritchard, 1964a).

The thrust of this discussion is in the direction of emphasizing that we are dealing with a general phenomenon of unusual DNA metabolism occurring in the mature eggs of all three major groups of vascular plants. Indeed, there are sufficient similarities in the first few rounds of divisions of the zygote in members of these groups for such cytochemical likenesses to be not entirely surprising. Although we cannot fail to be impressed by the absence of Feulgen staining in the nucleus of the mature egg, the behavior of chromatin is interpretable in different ways. Some of the changes described above may be looked upon as a prelude to the participation of the nucleus in the impending act of fertilization. Since the egg is highly swollen at the time of fertilization, it is possible that DNA present in the cell has been diluted, and is thus beyond the resolution of Feulgen reaction. It is also conceivable that disappearance of Feulgen reaction is due to chemical transformation of chromosomal DNA. Woodcock and Bell (1968a) have demonstrated by fluorescent microscopy and ultraviolet microspectrography the presence of DNA in the egg of *Myosurus minimus*, although Feulgen reaction is negative. Therefore, a nucleus failing to show Feulgen reaction does not necessarily lack the nucleotides of DNA molecule. In any case, transformation of DNA is thoroughly interlocked with overall processes which regulate enzyme synthesis, precursor levels, and the physiological state of its primers, and resolution of these aspects of DNA metabolism of the egg, based on analytical evidence, is a necessary step in the elucidation of the changes described above.

B. Ribonucleic Acid and Proteins

In spite of their aberrant DNA metabolism, the egg and zygote do not reveal any unusual features in RNA and protein metabolism. It looks as though the metabolism of RNA and protein in the egg and zygote is not closely related to that of DNA, since if it were, it would change with changes in DNA metabolism. For example, in contrast to the absence of DNA, level of RNA in the egg of *Stellaria media* was very high, particularly in the nucleolus and cytoplasm (Pritchard, 1964a). The egg nucleus of the orchid,* *Vanda* (cultivar), was high in protein and

---

* With reference to orchids, the term embryo is used in this book to denote a multicellular globular structure enclosed in the dry, papery seed coat.

RNA (Alvarez and Sagawa, 1965a). The cytoplasm, although denser at the chalazal end, exhibited a low concentration of these substances except at the cell boundary in contact with the synergids. The association of RNA and protein in the chalazal region of the egg along with the nucleus reinforces the conclusion that polarity is already determined in the egg at this stage. The egg of *Capsella bursa-pastoris* (Schulz, R. and Jensen, 1968c) provides another example that strengthens the above point. Here also, RNA and proteins are localized in the nucleus, nucleolus and cytoplasm at the chalazal end of the egg, and this pattern does not change even after fertilization (Fig. 3.2).

Only in the egg of *Pteridium aquilinum* is the pattern of RNA synthesis well enough documented to give any prospect of relating it to nuclear activity. Jayasekera and Bell (1971) have shown that synthesis of autoradiographically detectable RNA during oogenesis is confined to two discrete intervals, the first during formation of the central cell of the archegonium, and the second during maturation of the egg. These "waves" of RNA synthesis, as the authors call them, are seen in Fig. 3.3 as a heavy labeling of ^{3}H-uridine in the nucleus and cytoplasm of the central cell of the archegonium, and as a nearly uniform incorporation of the label into the nucleus and cytoplasm of the maturing egg, compared to the patterns of incorporation observed in the newly formed and fully mature eggs. These temporal patterns of RNA synthesis presumably reflect nuclear activity in terms of gene activation responsible, respectively, for oogenesis, and for the differentiation of the egg and the initiation of sporophytic growth. In line with this interpretation, it was found that an unfertilized egg was especially vulnerable to inhibition by the pyrimidine analog, 2-thiouracil, which rendered it nonviable without modifying its ultrastructure, whereas application of the analog to a fertilized egg caused, during its subsequent growth in the absence of the analog, only a retardation of embryogenesis and, less frequently, transformation of the sporophyte into gametophytic entities (Bell, 1972; Jayasekera and Bell, 1972). The reasons for the developmental anomaly in the egg receiving 2-thiouracil treatment are not known in detail, but do not, apparently, differ from its often-postulated role as an inhibitor of RNA synthesis (Jayasekera *et al.*, 1972). The restriction of recovery of the fertilized egg exposed to the analog thus suggests that the fundamental process involved is a block in RNA synthesis concerned with sporophytic growth.

When ^{3}H-leucine and ^{14}C-arginine were used as markers for protein synthesis in *Pteridium* eggs, the principal protein synthesis was found to occur early during oogenesis (Fig. 3.4). Subsequently, incorporation of the isotopes diminished sharply with increasing maturity of the egg

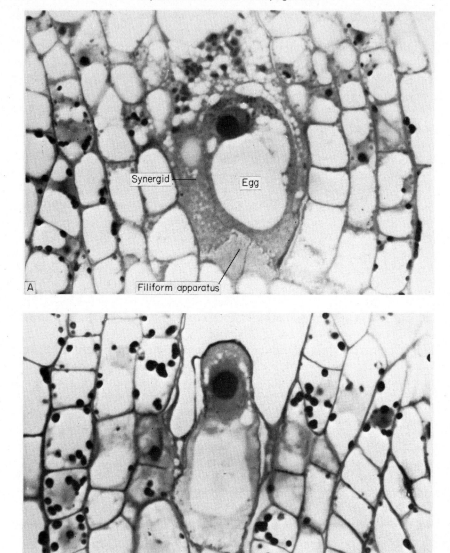

FIG. 3.2. Longitudinal sections through the egg (**A**) and zygote (**B**) of *Capsella bursa-pastoris* showing aniline blue black staining of nucleus, nucleolus and cytoplasm for proteins. The filiform apparatus of the synergid in **A**, and the chalazal wall in **B**, stain for insoluble carbohydrates with periodic-acid-Schiff reagent. (After Schulz, R. and Jensen, 1968c; photographs courtesy of W. A. Jensen.)

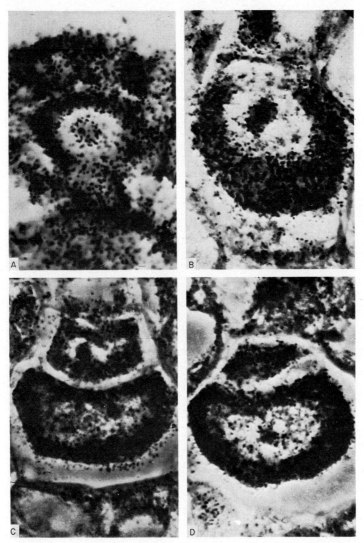

Fig. 3.3. Incorporation of ³H-uridine into developing archegonia of *Pteridium aquilinum*. Phase contrast pictures, with the dense cytoplasm and nucleolus appearing dark, and the nucleoplasm light. **A**, Central cell, about half developed, neck initials above. **B, C, D**, Primary ventral cell, maturing egg cell and fully mature egg cell, respectively. (After Jayasekera and Bell, 1971; photographs courtesy of P. R. Bell)

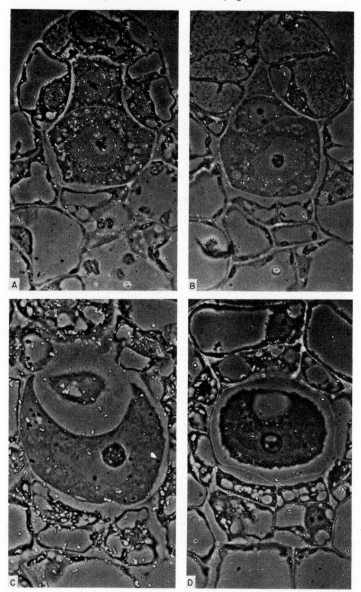

FIG. 3.4. Incorporation of ^{3}H-leucine into developing archegonia of *Pteridium aquilinum*. Phase contrast pictures. **A**, **B**, **C**, **D**, Primary ventral cell, young egg cell, maturing egg cell and fully mature egg cell, respectively. (After Cave and Bell, 1974b; photographs courtesy of P. R. Bell)

and became negligible in the mature egg (Cave and Bell, 1974b). Particularly interesting was the finding that the steadily diminishing incorporation of amino acids occurred during the period when RNA synthesis in the egg was high. Since RNA synthesized does not simultaneously encode the proteins, the force of these findings stems from their implication, now current from work with animal systems, that RNA is integrated in some fashion into the egg to be used as a template for initiating sporophytic growth.

It is unfortunate that no effort has gone into attempts to demonstrate by refined biochemical methods the connection between transcription and translation processes before, during or after fertilization of the egg in vascular plants. Moreover, up to the present time, only a few studies have been performed with the more favorable nonvascular systems such as fucoidian eggs. In *Fucus*, one consequence of fertilization of the egg is an increase in the capacity for protein synthesis and a conversion of inactive polysomes into functional units (Koehler and Linskens, 1967; Peterson and Torrey, 1968; Linskens, 1969). Since polysome content and protein synthetic activity of the egg register their peaks at about the same time when RNA synthetic activity is relatively low, it has been suggested that RNA formed in the immature egg persists in a functional state to direct the initial burst of protein synthesis upon fertilization (Linskens, 1969). The fertilized egg does, however, make new RNA, much of it probably due to activation from the cellular genome in the form of mRNA at some period following fertilization and this mRNA appears indispensable for differentiation processes such as rhizoid initiation and the first cell division (Quatrano, 1968). Thus, a gene-delegated control of protein synthesis, requiring both a stored template and continuous flow information type of genetic instruction, can be visualized to characterize early embryogenesis in *Fucus*. It will be interesting to see whether initiation of development following fertilization in higher plants is subject to this complex system of dual control.

II. NUCLEIC ACIDS IN DEVELOPING EMBRYOS

A. Deoxyribonucleic Acid

The pattern of nucleic acid changes during embryogenesis has been unraveled with some success in a few species, although in view of the variability of plant materials used, it is scarcely to be expected that any valid generalizations will emerge at this stage. According to Pritchard (1964b), DNA was present almost entirely in the nuclei of cells of early division phase embryos of *Stellaria media* without any evidence of cytoplasmic localization at this stage. In the 30–60-celled embryo, there was a decrease in the intensity of nuclear staining and the appearance of Feulgen-positive granules in the cytoplasm. In cotton embryos, the amount of DNA per cell remained fairly constant during embryogenesis (Fig. 3.5), with little, if any, extranuclear DNA (Fisher and Jensen, 1972). Against this finding must be set the fact that an earlier study (Yoo and Jensen, 1966) implied the synthesis of a considerable

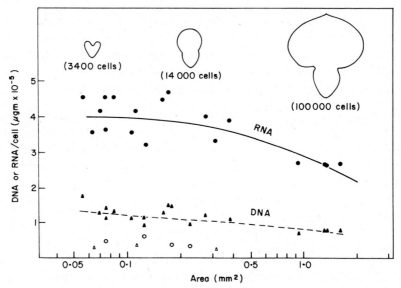

Fig. 3.5. In cotton, the amount of DNA per cell remains relatively constant during embryogenesis, but RNA content shows a decrease. Outlines of representative embryos with their respective cell number are also shown. (After Fisher and Jensen, 1972)

amount of cytoplasmic DNA in the young embryo, and a dramatic decrease in DNA content per cell as the embryo matured. The reasons for this difference remain enigmatic. In this context, references are appropriate here to two early studies on embryos of *Ginkgo biloba* (Shimamura, 1931) and *Pinus thunbergii* (Shimamura, 1956) where a decrease in size of the nucleus with successive divisions of the proembryo was noted. Cytological observations disclosed that this was due to the extrusion of chromatin from the nucleus at prophase and its later appearance in the cytoplasm as deeply staining basophilic granules.

One might suspect that the precise division of the zygote to produce a multicellular embryo might depend upon an efficient DNA duplicating mechanism, followed by cytokinesis. This picture is not immediately apparent in some studies because of complications introduced by the presence of cytoplasmic DNA as described above, and endopolyploidy. The amount of DNA in the nucleus of a haploid cell (gamete) is usually designated as 1C, and that in a diploid cell as 2C. Doubling of DNA content (1C to 2C or 2C to 4C) which occurs during a particular period (S phase) of the interphase is a prerequisite for cell division. Thus, one can associate the increase in DNA content of the nucleus with that part

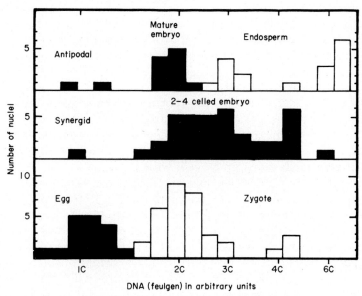

Fig. 3.6. Frequency of distribution of relative amounts of DNA in the nuclei of the mature egg, zygote and immature and mature embryos of *Tradescantia paludosa*. (After Woodard, 1956)

of the cell division cycle during which it has completed the S phase. In a study directed towards an analysis of the changing properties of chromosomal DNA during gametogenesis and embryogenesis in *Tradescantia paludosa*, Woodard (1956) has detected some conspicuous differences in DNA content of the cells of embryos of different ages. DNA was mostly in 2C or occasionally in 3C or 4C amounts in the fertilized egg, in 2C–4C amount in the two- to four-celled embryo and exclusively of 2C class in the mature embryo (Fig. 3.6). Like DNA, the nuclear histone content of the zygote and embryo showed a quantitative stability and occurred in multiples of the gametic quantity (Rasch and Woodard, 1959). A similar range of DNA values has been reported in the egg of *Petunia hybrida* at different periods after fertilization and before the occurrence of the first division wall (Vallade and Cornu, 1973). In the general context of embryogenesis, the results imply that DNA contents of the zygote and cells of early stage embryos faithfully follow expectations from their developmental history, involving simple mitotic divisions. However, as the seed matures, and the embryo becomes progressively dehydrated, other factors such as the available water supply might limit DNA synthesis (Brunori, 1967).

As a generalization, it was stated earlier that different parts of the embryo have different developmental potencies. During embryonic development, primary meristems arise by a series of ontogenetic changes which are probably different for each meristem. These considerations naturally lead us to inquire whether there are differences in the relative amounts of DNA in the cells of different parts of the embryo. There is some evidence that cells of the root apex, shoot apex, mesocotyl, coleoptile base and leaf primordia of maize embryo have 4C DNA, and those of the root cap, coleorhiza and scutellar node have both 2C and 4C DNA (Stein and Quastler, 1963). This is shown by the fact that no ^{3}H-thymidine is taken up by cells with 4C DNA for the first round of division which is otherwise normal. In contrast, the 2C cells of the root cap, coleorhiza and scutellar node take up ^{3}H-thymidine and resume DNA synthesis but do not proceed to divide. By a rather intricate experimental approach, in which autoradiography was combined with cytophotometry, Avanzi *et al.* (1969) have demonstrated that meristems of the shoot, and first and second leaf primordia of the mature embryo of *Triticum durum*, bear only 2C cells, while the primary root apex and the first pair of roots have both 2C and 4C cells. Similarly, the root meristem of the embryo of *Vicia faba* has a higher percentage of 4C cells than the epicotyl and leaf (Miah and Brunori, 1970). At the extreme end of the spectrum, in the embryos of lettuce (*Lactuca sativa*) and *Pinus pinea*, all cells are of 2C type (Brunori and D'Amato, 1967). The diversity in DNA values in embryos of different species and in different parts of the same embryo could result from conditions during embryonic maturation which control the interval between the inhibition of DNA synthesis and inhibition of mitosis. For example, in the meristems of the shoot apex, and first and second leaf primordia of *T. durum*, it is possible that mitotic activity is prolonged enough to deplete them completely of 4C cells, while in the root apex, this "depletion" phase is short-lived and less efficient (Avanzi *et al.*, 1969).

In the embryo of *Vanda* (Alvarez and Sagawa, 1965b), nuclear volume of the parenchymatous cells was nearly 40% higher than that of the meristematic cells. This was also correlated with an increased ploidy level of cells, suggestive of the absence of regular cytokinesis (Alvarez, 1968). Here, one can even construct a hierarchy in which cells towards the meristematic tip of the embryo have less and less volume and DNA content. On a more remarkable scale, analogous situations have been described in the cells of legume cotyledons which attain high degrees of polyploidy at the same time as the synthesis and accumulation of storage proteins are initiated in them. In *Pisum arvense*, DNA levels up to 32C–64C, representing 4–5 successive replications, have been observed when the cotyledonary cells embark upon their phase of enlargement, although during the desiccation phase of embryogenesis, DNA values remain at the 16C level (Smith, D. L., 1971, 1973). The quantity of nuclear DNA in the cells of the cotyledons of *P. sativum* also attained 32C–64C (Scharpé and van Parijs, 1971), while in *Vicia faba* it maintained an average value of 16C (Millerd and Whitfeld, 1973). With the information presently available, we cannot account for the nuclear behavior of the cells of legume cotyledons during their expansion phase of growth. The question to be answered is whether the extra DNA is used for the cell's current needs or is a convenient storage form of nucleosides and nucleotides for its future needs. An exciting thought that the increased DNA synthesis was an example of gene amplification concerned with the production of structural proteins did not stand the test of experimental verification (Millerd and Whitfeld, 1973).

B. Ribonucleic Acid

As a basis for discussion of the origin of functional differentiation in developing embryos, it is worthwhile to look at the changes in RNA content of their cells. In *Stellaria media* after the first division of the zygote, RNA concentration is more dense in the suspensor than in the embryonal cell, while the basic proteins appear to be uniformly distributed in both cells (Pritchard, 1964b). The attainment of mature size by the suspensor is accompanied by a substantial decrease in RNA content and reorganization of proteins into colorless proteinoplastids (Fig. 3.7). Histochemical observations have shown that in the two-celled embryos of cotton (Jensen, 1964) and *Vanda* (Alvarez and Sagawa, 1965b) a functional differentiation between the terminal embryonal cell and the basal suspensor cell is possible on the basis of a larger amount of RNA in the former (Jensen, 1964). The fate of the basal cell in embryos of cotton and *Vanda* on the one hand, and *Stellaria* on the other hand, might explain the obvious differences in their histochemistry. The basal cell in the embryos of cotton and *Vanda* divides to form a few-celled suspensor, while in *Stellaria* it never divides again, but functions directly as the primary suspensor cell, continuing to grow until cotyledonary initiation, and is completely resorbed in the mature embryo. As Fig. 3.8 shows, a higher RNA concentration in the organogenic apex of the embryo than in the suspensor is the rule in gymnosperms as well (Takao, 1960; Shafer and Kriebel, 1974).

In the two-celled through heart-shaped stages of embryos of several species examined, there are no generally characteristic histochemical changes in the distribution of RNA and protein that augur the onset

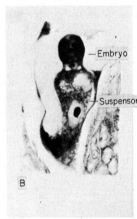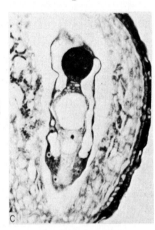

FIG. 3.7. In *Stellaria media*, azure B staining reveals changes in RNA content of cells of the suspensor and embryo proper. **A**, Three-celled embryo; note the high concentration of cytoplasmic RNA in the suspensor (lowermost cell). **B, C**, Progressive decrease in RNA content of the suspensor of 3 day and 4 day old embryos, respectively. (After Pritchard, 1964b; photographs courtesy of H. N. Pritchard)

of functional differentiation (Rondet, 1958, 1961, 1962; Schulz, R. and Jensen, 1968a; Norreel, 1972). Even in the protocorm stage of the embryo of *Vanda* where there is a clear distinction between the small cells of the meristematic region and the large cells of the parenchyma, there are hardly any noticeable differences in the distribution of RNA and protein (Alvarez and Sagawa, 1965b). We cannot say that differences do not exist, but they do not stand out with the dyes employed. At a later stage of development of embryos of *Capsella bursa-pastoris* (Schulz, R. and Jensen, 1968a), *Lens culinaris, Myosurus minimus, Alyssum maritimum* (Rondet, 1958, 1961, 1962), *Petunia hybrida* (Vallade, 1970) and tobacco (*Nicotiana tabacum*) (Norreel, 1972), cells of the presumptive cotyledons stain more intensely for RNA than the rest of the embryo. Evidence on which these statements are based is in Fig. 3.9. In *Stellaria*

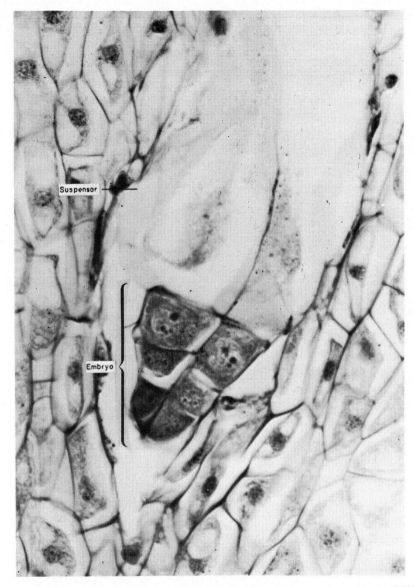

FIG. 3.8. The cells of the embryonic apex of *Pinus strobus* stain more intensely for RNA by methyl green-pyronin than do the cells of the suspensor. (After Shafer and Kriebel, 1974; photograph courtesy of T. H. Shafer)

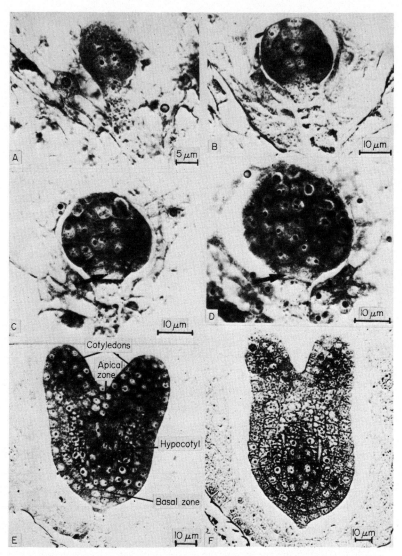

Fɪɢ. 3.9. RNA distribution during embryogenesis in *Nicotiana tabacum*. **A**, **B**, Globular embryos are homogeneously stained for RNA, but suspensor has little RNA. **C**, **D**, Globular embryos at a slightly advanced stage; arrows point to the basal cell of the suspensor which stains to the same extent as the rest of the embryo. **E**, The cells of the cotyledons and hypocotyl of the heart-shaped embryo stain more intensely than the region of the shoot apex and the basal zone. **F**, Torpedo-shaped embryo. (After Norreel, 1972; photographs courtesy of B. Norreel)

also there are indications that organ initiation in the young embryo is associated with an increased RNA localization (Pritchard, 1964b). Especially during cotyledonary initiation, high concentrations of cytoplasmic RNA are found in the cortical cells, procambial cells and in cells of the growing tips of the cotyledons. Once the cotyledons are mature, RNA content of the growing cells declines and assumes the proportion of other cells. However, quantitative changes in RNA content of ovules of poppy (*Papaver somniferum*) do not support claims about increased RNA localization associated with embryonic cotyledons since a correlation between peak in nucleic acid activity and cotyledon initiation did not materialize in this plant (Johri and Maheshwari, 1966a).

One may finally ask whether RNA accumulation in the individual cells keeps pace with the rapid cell multiplication that occurs during progressive embryogenesis. As Fig. 3.5 shows, although RNA content of cotton embryo increases with increase in embryo size, the amount per cell decreases with progressive embryogenesis (Yoo and Jensen, 1966; Fisher and Jensen, 1972). Results somewhat similar to, but more pronounced than, those described above have emerged from the work of Chang (1963a), who followed by autoradiography the rates of ^{32}P incorporation into nucleic acids of barley embryos over a period of time covering a range of physiological differentiation. The absolute amounts of incorporation of ^{32}P into RNA were found to increase continuously during embryogeny, but the relative amounts of incorporation of the isotope into RNA per cell per unit time decreased during progressive embryonic differentiation. It may be that the capacity of the nucleus to synthesize new RNA is not able to keep pace with the rapid cell multiplication that occurs during embryogenesis.

Some further insight regarding cellular rates of RNA synthesis during embryogenesis has come from studies on the embryos of *Phaseolus coccineus* in which, as pointed out in the previous chapter, the suspensor contains polytene cells while the cells of the embryo proper are diploid. One consequence of this distinct pattern of cell differentiation in the population is that the suspensor cells reach their peak synthetic activity long before the cells of the embryo proper. Moreover, on a per cell basis, synthesis of RNA and proteins in the cells of the suspensor is always hundreds-fold higher than in the embryo cells (Walbot, Brady, Clutter and Sussex, 1972; Sussex *et al.*, 1973). The implication of this finding is that the nondividing cells of the suspensor possess some selective advantage over the rapidly dividing cells of the embryo proper in being able to divert energy resources to processes connected with large-scale transcription rather than to those connected with cytokinesis. Some form of differential template activity has been suggested to account for the results (Clutter *et al.*, 1974), but direct evidence on this point has been hard to find.

To sum up all these observations, it should be emphasized that the basis for organ initiation in the developing embryo is undetermined. From the work reviewed above, it is difficult to defend the idea that an increased RNA accumulation is a prerequisite for initiation of cotyledons in the embryo, but these studies have given us some pointers to pursue further. However, bulk changes in RNA as observed histochemically or quantitatively are of little significance in interpreting possible mechanisms in morphogenesis, unless data are supplemented by studies of various RNA species or analysis of nucleotide composition. Although fractionation of RNA into easily identifiable components is now possible by acrylamide gel electrophoresis, embryos at the early stages of interest are too small to provide sufficient material for even the most sensitive physical and chemical techniques currently available.

III. ENZYMES IN EMBRYOGENESIS

Enzymes catalyze many, if not all, of the reactions which control functional activities in living systems. While progressive development of the embryo is manifest by form changes and associated alterations in structural elements, functional activity of newly formed organs becomes possible only through biochemical mechanisms. If this argument is valid, its relevance to the basis of enzyme studies in development is simple and obvious; enzymes involved in a function will attain maximum activity in appropriate spatial configurations prior to or synchronously with functional maturation. Since all enzymes are proteins, there is a strong temptation to assume that synthesis of enzymes might be linked to synthesis of specific proteins necessary for the initiation of functionally significant structures during morphogenesis. The genetic control of enzyme synthesis so well documented in microorganisms presupposes that the presence or absence of an enzyme at a given site might provide information on differential gene activity during progressive embryogenesis.

A word of caution is necessary here. Because of the extreme sensitivity of enzymes, the enzymatic approach to morphogenesis has its own pitfalls, and a negative result may more often be due to imperfect conditions for expression of enzyme activity rather than to lack of enzyme molecules.

A. Phosphatases

There is some evidence for the view that changes in the activity of certain enzymes are correlated with developmental processes in embryos. Among enzymes of importance are alkaline phosphatase, acid phosphatase, cytochrome oxidase and succinic dehydrogenase. Both alkaline and acid phosphatases are enzymes with low specificities. Chemically, phosphatases have been known to hydrolyze phosphate esters. For histochemical localization, free phosphate esters are precipitated as insoluble reaction products which are subsequently converted into colored or electron-dense derivatives. Critical evaluation of the original methods for detection and localization of phosphatases with special reference to plant materials is given by Jensen (1962).

Histochemical localization of phosphatases thus far has been confined to embryos of a few species of angiosperms. In coconut (*Cocos nucifera*) the earliest appearance of acid phosphatase activity was in cells at the free end of the embryo which eventually became the cotyledonary primordium (Wilson and Cutter, 1952, 1953). Subsequently, the epidermal layer and the vascular traces of the cotyledon exhibited greater enzyme activity than the rest of the cotyledon. In the mature embryo, primordia of the shoot and root, vascular bundle and epidermis were characterized by high acid phosphatase activity. The appearance of the enzyme first in the presumptive cells of the cotyledon has significance in that this organ is known to play a haustorial role in the digestion of the endosperm. The reason for the intense localization of the enzyme in the vascular tissue is certainly due to its role in the translocation of food materials throughout the embryo. The presumed role of the scutellum as an absorbing organ may similarly account for the first appearance of acid phosphatase in this organ during embryogenesis in rice (*Oryza sativa*) (Mizushima *et al.*, 1955).

Histochemical distribution pattern of acid phosphatase in embryos of *Capsella bursa-pastoris* has also revealed a relationship to morphogenesis (Raghavan, 1975d). It was not possible to reveal by specific histochemical tests the presence of enzyme in the unsegmented egg or in the first division phases of the embryo. One cannot deny their existence, but at most very few molecules can be present. The earliest evidence of enzyme was in the globular embryo (30–40 μm long) where it was uniformly distributed. The change from a globular to a heart-shaped embryo was accompanied by the appearance of the enzyme in the cotyledons and root apex. In the intermediate stage embryo, there was evidence of enzyme activity in the hypocotyl. In progressively older

embryos, the enzyme was mainly concentrated in the root apex, cotyledons and the shoot apex.

There is satisfactory evidence to show that in *Stellaria media* the
magnitude of acid phosphatase activity varies between the different
tissues of the heart-shaped embryo and between the embryo and the

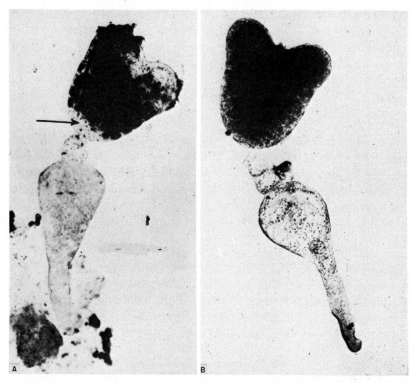

FIG. 3.10. In *Stellaria media*, (**A**) acid phosphatase activity is localized in the embryo,
and is absent in the suspensor, whereas (**B**) alkaline phosphatase is present in both
embryo and suspensor. The arrow points to the hypophysis. (After Pritchard and Bergstresser, 1969; photographs courtesy of H. N. Pritchard)

suspensor (Pritchard and Bergstresser, 1969). As seen in Fig. 3.10,
enzyme activity is confined to the cells of the embryo and is absent in
the suspensor cells. Within the cells of the embryo itself, the enzyme
occurs in the protoderm which differentiates the epidermis of the adult
plant, but the hypophysis which forms the root cap lacks the enzyme.
In contrast, alkaline phosphatase is present throughout the embryo and
suspensor. In the latter organ, there is a selective localization of the
enzyme in the cytoplasmic proteinoplastids. Since proteinoplastids are

rich in lipoproteins, alkaline phosphatase may be involved in the hydrolysis of the phosphate esters of lipoproteins.

B. Oxidative Enzymes

Succinic Dehydrogenase and Cytochrome Oxidase

These are typical mitochondrial enzymes associated with respiratory processes and one can obtain clues regarding the nature of the respiratory pathways in embryos of different ages from a distribution map of enzymes. In the mitochondria, both enzymes are found associated with the chain of other enzymes and cofactors in the electron transport pathway. Reduction of nitro-blue tetrazolium to insoluble formozans is a sensitive indicator of succinic dehydrogenase activity. Cytochrome oxidase is detected by the well known "Nadi" reaction in which a mixture of alpha-naphthol and dimethylparaphenylalanine is oxidized into indophenol blue by the enzyme. The reduction of vital dyes such as Janus green and methylene blue also gives some indication of the respiratory activity of cells since they owe their specificity to the presence of oxidative enzymes of mitochondria. Some authors have studied the conditions for optimum reduction of Janus green, methylene blue and tetrazolium chloride by embryo homogenates (Jensen et al., 1951; Smith, F. G., 1952; Sato, 1956a, b, c, 1962a, b) in the presence of different substrates and inhibitors. These papers should provide valuable background information to study changes in the oxidative enzymes of embryos on an accurate analytical scale.

The development of succinic dehydrogenase activity in cotton embryos of different ages furnishes a striking example of enzyme activity correlated in a predictable manner with functional needs of differentiating organs and tissues (Forman and Jensen, 1965). In the early stages of embryogeny when the suspensor was at its peak form, succinic dehydrogenase activity was more intense in this organ than in the embryo proper. In the early heart-shaped embryo, enzyme was confined to the cells of the developing cotyledons and the basal portion of the hypocotyl; activities so early established are carried over unchanged in the late heart-shaped stage to the torpedo-shaped stage. In post-torpedo-shaped embryo, accumulation of the enzyme was close to maximum, spreading to the radicle and the longitudinal axis. Within the cotyledons the developing procambial strands stood out clearly apart from the rest of cotyledonary tissue in their enzyme content. Finally, enzyme concentration in the relatively quiescent cells of the shoot apex was below normal (Fig. 3.11). Comparable observations have been made on the distribution of succinic dehydrogenase in the embryos

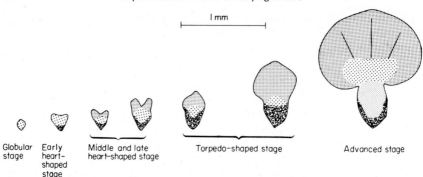

| 1 mm |

Globular stage | Early heart-shaped stage | Middle and late heart-shaped stage | Torpedo-shaped stage | Advanced stage

FIG. 3.11. Distribution of succinic dehydrogenase activity in cotton embryos of different ages is shown by the shaded areas. Lighter shading indicates lower activity. (After Forman and Jensen, 1965)

of other plants, notably of *Cypripedium* (Zinger and Poddubnaya-Arnoldi, 1966).

Since succinic dehydrogenase is bound to the mitochondrion, changes in the distribution of this organelle may reflect changes in enzyme activity. Can we attribute the higher enzyme activity in some cells of the embryo to an increase in mitochondrial population, or is it due to an increased development of enzyme molecules in the existing population? We will have no clear notion on the development of function in the mitochondria until much more is known about their ultrastructure and biochemistry in developing embryos.

As one of the enzymes of the Krebs cycle, an increase in the activity of succinic dehydrogenase during embryo development must obviously reflect an increase in respiratory rate. Data on oxygen uptake by developing cotton embryos are in good agreement with this expectation. Figure 3.12 represents a typical curve for oxygen consumption of developing embryos of cotton determined by Cartesian diver ultramicromethod (Forman and Jensen, 1965). It is seen that the net uptake of oxygen per hour per embryo increases with its size and morphological complexity; expressed on a per cell basis, the net uptake of oxygen is highest in relatively young embryos. Although an increase in the number of cells of the embryo without a corresponding rise in enzyme content will easily explain these results, other causes such as a greater amount of substrate availability or a greater affinity between the enzyme and substrate in the meristematic cells of the young embryo may provide an interpretation of the results at the biochemical level.

According to Cutter *et al.* (1952b), young coconut embryos in isolation consume more oxygen per mg dry weight than older ones. In the

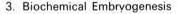

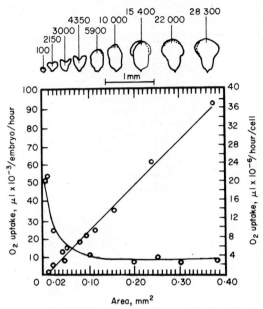

FIG. 3.12. Although net oxygen uptake by cotton embryo increases with age, expressed on a per cell basis, the activity is highest in relatively young embryos. On the top of the graph, developmental stages of the embryo with the approximate cell numbers are shown. (After Forman and Jensen, 1965)

embryos of *Papaver somniferum* and *Zephyranthes lancasteri*, peaks in respiratory intensity are probably related to elongation and growth of the embryonic cotyledons (Maheshwari, S. C. and Johri, 1963; Johri and Maheshwari, 1965, 1966b). Determination of succinic dehydrogenase activity in embryos of *Z. lancasteri* by following the rate of reduction of potassium ferricyanide spectrophotometrically gave a curve that resembled the one for oxygen uptake (Johri and Maheshwari, 1966b).

Reports on the development of cytochrome oxidase activity in embryos are limited. This is somewhat surprising since it is identified as one of the most important of the oxidative enzymes. In the heart-shaped embryo of *Stellaria* (Pritchard and Bergstresser, 1969) enzyme activity was especially intense in the meristematic tissues of the protoderm, ground meristem and procambium, and the nondividing cells of the suspensor (Fig. 3.13). In *Capsella* embryos also, one observes the familiar version of a progressive localization of cytochrome oxidase in the shoot apex, cotyledons and root apex (Raghavan, 1975d). If enzyme concentration is a factor that determines the rate of respiration, it is clear that cells of the meristematic regions of the embryo have a

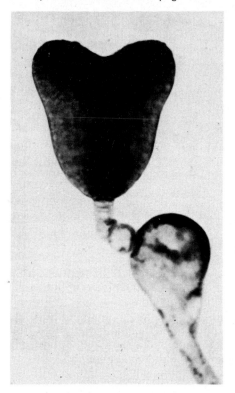

FIG. 3.13. Cytochrome oxidase activity is highest in the meristematic tissues of the protoderm, ground meristem and procambium, and in the suspensor cells of the heart-.shaped embryo of *Stellaria media*. (After Pritchard and Bergstresser, 1969; photograph courtesy of H. N. Pritchard)

more vigorously evolving metabolism than those of other regions. Although these studies indicate that enzyme development and functional activity go hand in hand, validity of the conclusion may be impaired on the ground that changes in enzyme activity might reflect an increased development of the organ, rather than the evolution of function. Appearance of different enzymes at different rates even in the same organ is a strong counter-argument against this.

Enzyme activity of isolated mitochondria. A more direct approach to the study of respiratory metabolism of developing embryos is to test the ability of isolated mitochondria to oxidize various intermediates of the Krebs cycle, and to demonstrate their relative efficiencies in coupling oxidation to phosphorylation. Such a study was undertaken by Prokof'ev and Rodinova (1966), who found an increase in the yield of mitochondrial proteins from sunflower embryos up to 20 days after pollination and a decrease thereafter. This indicates a possible shift in mitochondrial competence during embryo development, a conclusion further supported by other observations. Since oxygen consumption of mitochondria decreases with age of embryos, the change in their respiratory activity may have some relation to the shift in mitochondrial activity. During development of embryos, there was a pronouned decline in the ability of mitochondria to oxidize malate and α-glycerophosphate but coupling of electron transport to phosphorylation was enhanced. This may indicate a dominance of phosphorylation over oxidation as embryos mature, resulting in an increase in their high energy phosphate content. In seeking to explain the declining oxidative activity of mitochondria in the relatively older embryos, an interpretation based on a decrease in their enzyme content may appear attractive, but there is also the consideration that mitochondria of early embryos are more likely to release their enzymes and cofactors than those of the most mature ones.

Peroxidase

Peroxidase is of great biological interest because of its multiple isoenzymes and the large number of physiological, biochemical and oxidative processes in which it mediates. From the morphogenetic point of view, the enzyme appears to play a cardinal role in regulating the level of IAA in plant tissues by producing it, or by inactivating it (Galston and Davies, 1969). Needless to say, any information on the distribution of peroxidase activity in the embryos will be potentially useful in interpreting their morphogenetic mechanisms.

Alvarez and King (1969) studied the histochemical distribution of peroxidase in embryos of *Vanda* by following the oxidation of benzidine into a blue compound in the presence of hydrogen peroxide. In this plant, peroxidase activity was highest in the early stage of the embryo (protocorm), but it gradually declined with development of the protocorm (Fig. 3.14). In the protocorm, epidermal cells and their external surfaces showed an intense blue color, but as it developed meristematic and parenchymatous cells, root and vascular tissues acquired enzyme activity. When peroxidase isoenzymes were separated electrophoretically, a clear correlation was established between the stage of embryo development and the number of bands of isoenzymes; younger embryos which exhibited high enzyme activity had a larger number of bands while older ones which had the lowest activity had the smallest number of bands. If it can be shown that orchid embryos have low

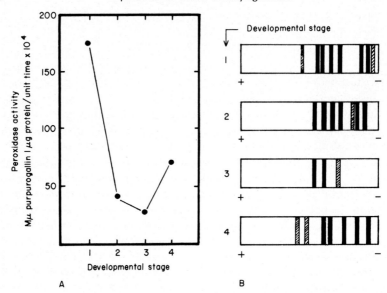

Fig. 3.14. **A**, Changes in the peroxidase activity of embryos of *Vanda* (cultivar) at different stages of development. **B**, Zymograms showing *iso*-peroxidase patterns in embryos corresponding to stages 1–4 in **A**. (After Alvarez and King, 1969)

concentrations of endogenous IAA at early stages of development and high concentrations at later stages, this would enhance the likelihood that peroxidase may control embryogenesis by inactivating IAA in the early stage embryos and by unblocking IAA synthesis in the late stage embryos. Although various investigators have attempted to study the effect of IAA on orchid embryos (Arditti, 1967a), the results are not clear-cut for any firm conclusions.

Adaptive Significance of Enzyme Activity

There are some reports which indicate that differences in the distribution of oxidative enzymes in embryos may have adaptive significance for their survival under unfavorable conditions. For example, acclimatization of wheat to extreme cold and unfavorable soil conditions is apparently associated with an increased activity of cytochrome oxidase and peroxidase and a declining activity of polyphenol oxidase in the embryos (Zenchenko, 1964). The capacity of embryos for nonenzymatic oxidation of substrates is also low in seeds of northern origin (Zenchenko, 1965). In the embryo of barley of northern origin, the trend in peroxide activity is similar to that in wheat (Zenchenko, 1954).

It will be recalled that peroxidase activity was intensely positive on the external surface of the epidermal cells of the embryo of *Vanda* (Alvarez and King, 1969). This is an interesting observation since it shows the existence of a mechanism for extracellular enzyme secretion by epidermal cells which have primarily a protective function. Since orchid seeds are shed when the embryo is still underdeveloped, this appears to have an ecological significance in the survival and continued growth of the embryo. In a recent study it has been shown that mitochondria of the dry peanut (*Arachis hypogaea*) embryo have a respiratory chain deficient in cytochrome *c* (Wilson and Bonner, 1971), conceivably as an adaption to prevent uncontrollable loss of substrates during the quiescent period of the embryo.

IV. METABOLIC INTERACTIONS DURING EMBRYOGENESIS

The notion of a developing organ implies accumulation of storage materials and utilization of products of their hydrolysis for synthetic processes. As a model system, the great diversity of embryos are no exceptions to this rule (Netolitzky, 1926; Tukey and Lee, 1937; Lee and Tukey, 1942; Devine, 1950; Raacke, 1957), and some offer exceptional opportunity for the study of storage and metabolism of ergastic substances, for they often combine relatively large size with highly complex metabolic reactions. As growth of the embryo takes place by repeated divisions of the fertilized egg, this is made possible by the simultaneous synthesis of new cell wall material. As the embryo increases in size, so does the need for various metabolites. In a growing embryo, an increase in dry mass is a good indication of the accumulation of storage materials. Much of this dry weight increase is associated with the accumulation of carbohydrates, proteins and fats, representing a growing reserve of potential sources of energy.

The gross chemical changes that occur during embryogenesis in plants are easily recognized by simple histochemical tests and are often accessible to quantitative treatment. Unfortunately, information on the metabolism of cellular constituents during embryogenesis is of a very incomplete nature. Although embryos have interested morphologists for a long time, biochemists have not been attracted to studies of their metabolism.

A. Histochemical Changes

Starch

Some interest has centered around the distribution, storage and utilization of starch during embryogenesis. Workers in descriptive embryogeny have frequently encountered this long-chain polysaccharide distributed ubiquitously in the embryo sac, zygote and early division phase embryos of numerous plants (Maheshwari, P., 1950; Subramanyam, 1960a). Starch is easily identified by physical and histochemical criteria and its distribution in the ovule has offered important clues regarding nutritional relationships of developing embryos. However, when we consider the variability of its occurrence and distribution, we are hardly encouraged to generalize. In the zygote of cotton, starch grains formed a striking mass surrounding the nucleus, but they were diluted during subsequent divisions and completely disappeared in the mature embryo (Reeves and Beasley, 1935; Jensen, 1964). Starch grains were also conspicuously present in the egg of *Capsella bursa-pastoris*, increasing upon fertilization and during the first division of the zygote. At the second division of the terminal embryonal cell, they completely disappeared and were not synthesized until the embryo attained maturity (Schulz, R. and Jensen, 1968a, c). Prior to the formation of adventive embryos in cultured nucellar tissues of *Citrus aurantifolia* there was a predominance of starch grains which decreased as embryos appeared (Sabharwal, 1962). The presence of starch grains was also a characteristic feature of adventitiously formed linear proembryos of carrot tissues and probably also of the parenchymatous cells from which potential embryoids arose (Halperin and Wetherell, 1964). The disappearance of starch from the subjacent cells prior to the formation of a whole embryo, or of a new cell in the existing embryo, might indicate that it is metabolized as a source of energy for initiating division processes. This interpretation lacks a biochemical basis but appears not wholly unreasonable in view of the well established evidence for starch–sugar transformation in cells.

In certain members of the Gramineae and Leguminosae, there is a correlation between the distribution of starch in the embryo and in the surrounding tissues (Mizushima *et al.*, 1955; Takao, 1962). The general picture emerging from these studies is that during early embryogenesis, starch granules are concentrated in the nucellus, integuments and endosperm with little evidence for their presence in the embryo itself. They appear in the embryo with the advent of cotyledonary initiation at about the same time as the nutritive tissues begin to disintegrate. A study of embryos of *Pinus thunbergii* (Takao, 1960) also disclosed a generally similar series of events. These observations imply that starch grains appearing in the embryo probably have their origin in the extraembryonal tissues.

According to Zinger and Poddubnaya-Arnoldi (1966), there are some differences in the timing of appearance and disappearance of starch grains during embryogenesis in certain primitive and advanced members of the Orchidaceae. In *Cypripedium insigne*, a relatively primitive species, starch grains appear in the chalazal region of the embryo sac and in the egg just before fertilization. At the time of fertilization, they are retained only in the egg, and later transmitted to the developing embryo. When the embryo attains half its normal size, starch grains begin to decrease until they disappear completely in the mature embryo. In contrast, in the advanced species like *Calanthe veitchii* and *Dendrobium nobile*, not only do starch grains appear before pollination, but they also form the main reserve material of the mature embryo. The evolutionary significance of the changes in the distribution of starch in the primitive and advanced species is not clearly established.

Proteins and Fats

Proteins, fats and oils are present in the cells of the embryo as protoplasmic inclusions which can be isolated by differential or gradient centrifugation (Yatsu and Altschul, 1963; Schnarrenberger *et al.*, 1972). According to Reeves and Beasley (1935), cotton embryos of different ages showed positive reactions for oil, protein, pentosans and sugars. In *Dianthus chinensis*, food reserves of fat and possibly protein, but no starch, are found in the cells of cotyledons and hypocotyl of torpedo-shaped and older embryos (Buell, 1952). Although oil was first observed early in the ontogeny of embryos of *Datura stramonium*, most of the oil deposition occurred during the sixth and seventeenth day after pollination, after which only small quantities were deposited (Satina and Rietsema, 1959). Periodic observations of living cells of embryos of certain orchids have revealed that the number of plastids and fat drops begins to increase right after fertilization to maturity (Poddubnaya-Arnoldi, 1960).

Mature embryos of *Cypripedium* do not contain reserve proteins detectable by biuret reaction, although biuret-reacting proteins are abundant in the young embryos. Proteins of embryos of *Dendrobium* and *Calanthe* lose their ability to react with biuret reagent earlier than those of *Cypripedium*. Presumably, lack of biuret reaction of the reserve proteins in the mature embryo is suggestive of changes in the structure of the protein molecules rendering them insensitive to the test. Cells of mature embryos of all three species are densely filled with fat which have their origin partly in the integument and partly in the pollen tube (Zinger and Poddubnaya-Arnoldi, 1966).

Embryos of *Taraxacum* (Poddubnaya-Arnoldi *et al.*, 1964) and *Cypripedium* (Zinger and Poddubnaya-Arnoldi, 1966) also showed distinct reactions to SH-groups. Embryos in various stages of development have chloroplasts containing photosynthetically active pigments (Meeuse and Ott, 1962).

The collective evidence from this survey seems to indicate that there is a wide open field for cytochemical and histochemical work on the cell constituents of embryos. The work described above, besides being very fragmentary, is also uncoordinated owing to the widely different choice of materials by different workers, and thus makes a very limited contribution to our understanding of this field.

B. Structural Changes

At the level of light and electron microscopes, storage proteins of embryos present an immense variety of size and shape which is probably associated with differences in the types of raw materials present and with the state of differentiation of the embryo. Bils and Howell (1963) have described, in soybean (*Glycine max*) embryo, cytoplasmic protein globules in association with a ribosome-rich ER, although no mention is made of the presence of membranes around the proteins. The classic concept of a storage protein granule is found in the embryo of *Tropaeolum majus*, where it occurs within a vacuole bounded by a single membrane (Nougarède, 1963a). In the cells of the mature embryo of *Pisum sativum*, storage proteins appear as granules within both vacuoles and ER cisternae and as needle-like crystalline structures in the cytoplasm (Bain and Mercer, 1966a; Yoo, 1970), while in cotton embryo, protein bodies are derived from a secretion of proteins into cisternae of the ER (Engleman, 1966). According to Öpik (1968), vacuoles of the young embryo of *Phaseolus vulgaris* are subdivided into small compartments which later become filled with proteins. The contents are initially flocculent, but as the embryo matures they become dense and membrane-bound (Fig. 3.15). Rarely as in the dormant embryo of *Fraxinus excelsior*, storage proteins may occur as intranuclear crystals, or as crystal-containing microbodies (Villiers, 1967b, 1968a).

The storage protein of the mature embryo of *Vicia faba* at its level of organization is a single unit, although ontogenetically it is a plurality of several individual protein bodies (Briarty *et al.*, 1969). In the developing embryo, the earliest rudiment of the protein body is deposited around margins of vacuoles of the outer cells of cotyledons. Once formed, the deposit increases in size, filling small vacuoles and spreading around the periphery of larger ones. Individual protein bodies fuse with one another to form larger masses. In this way, large proteins are formed until they become closely pressed together and are bounded by single membranes. A different situation prevails in the embryo of

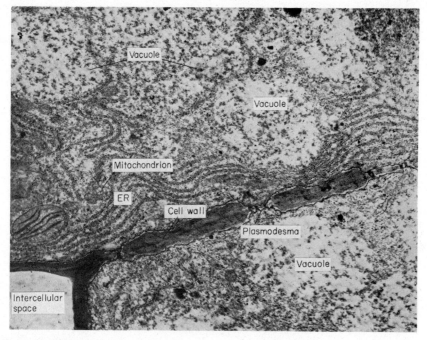

FIG. 3.15. Cells of 25 day old cotyledon of *Phaseolus vulgaris* showing ER lamellae near the wall, and vacuoles filled with flocculent protein. (After Öpik, 1968; photograph courtesy of H. Öpik)

Bidens cernua which has protein bodies containing globoids. In the early stage embryo, small vesicles which are initials of globoids are seen associated with protein masses at the edge of vacuoles and finally they are evenly distributed in the vacuole (Simola, 1971).

In the embryo of *Yucca schidigera* there are two morphological types, meshwork and core types, of membrane-bound protein bodies (Horner and Arnott, 1965, 1966). The meshwork type consists of electron-dense and electron-transparent regions in which are embedded birefringent bodies. The less frequent core type consists of a core surrounded by a matrix in which birefringent bodies are embedded (Fig. 3.16). Protein bodies closely similar to the meshwork and core types found in *Yucca* have also been described in the embryo of lettuce (Paulson and Srivastava, 1968). The embryo of *Sinapis alba* contains two types of proteins of vacuolar origin which are restricted to separate cells. One type designated as aleurone grain has inclusions of globoids while the other type designated as myosin grain is filled with homogeneous fibrillar material (Rest and Vaughan, 1972). Protein bodies found in the embryo of

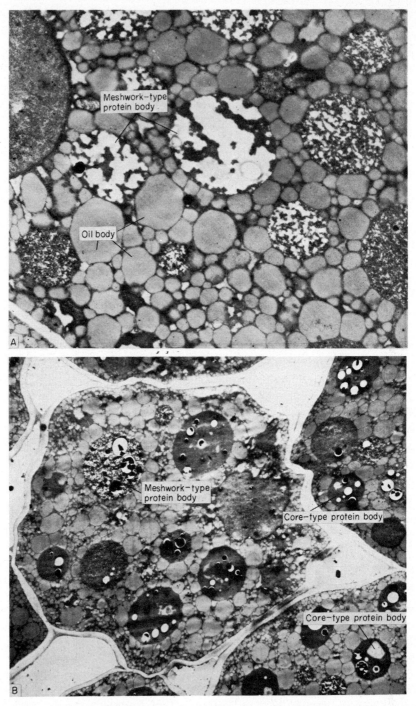

Fig. 3.16. Cells of the embryo of *Yucca schidigera* displaying protein bodies. **A**, Meshwork type protein bodies of different sizes, surrounded by oil bodies. **B**, Meshwork type and core type protein bodies. (After Horner and Arnott, 1965; photographs courtesy of H. T. Horner, Jr.)

Setaria lutescens are devoid of any inclusions whatsoever (Rost, 1972). Some reservations must, however, be made here, for the variability in the structure of protein bodies of embryos might be accounted for by the type or duration of fixation (Lott *et al.*, 1971). Not infrequently, protein bodies may serve as major sites of storage of specific proteins such as the globulin in legumes (Altschul *et al.*, 1961; Varner and Schidlovsky, 1963; Ericson and Chrispeels, 1973).

Thus, in the various species, the storage function of the embryo is taken up by vacuoles which operate as organizational centers for the formation of protein bodies. The synthesis of proteins in the storage cells is clearly related to the activity of stacks of ER with their attached ribosomes, which become prominent in cells at the same time as the appearance of vacuolar proteins, but how the physical movement of proteins into the growing granule is accomplished is unknown.

Much less is known about the development of other ergastic compounds in the embryos. An electron microscopic study of *Crambe abyssinica* embryos has shown that spherosomes have an origin distinct from that of oil bodies. The former, which appear before the onset of oil body formation, are formed from the dilated tip of the ER, while the oil bodies originate in small areas of the cytoplasm which may or may not be associated with ER (Smith, C. G., 1974).

C. Biochemical Trends

The overall biochemical changes of cell constituents during embryogenesis have been described only briefly and do not provide an adequate basis for discussion. Growth hormones of the type of auxins, gibberellins and cytokinins, and growth inhibitors such as abscisic acid, have been detected in embryos of diverse plants by chromatography and bioassay (see Chapter 4). In *Cassia fistula* most of the auxin becomes bound with the progress of maturity of the embryo, a fact which is registered as a decrease in auxin activity (Mukherjee *et al.*, 1966).

Rijven (1958) has shown that the insoluble nitrogen content of embryos of *Capsella* increases in direct proportion to their length. Growing embryos of *Datura* exhibit a dynamic protein metabolism; in young embryos, tissue proteins predominate, but in torpedo-shaped and older embryos, most of the proteins are stored as aleurones and serve as reserves on which embryos draw for the elaboration of raw materials during subsequent growth (Satina and Rietsema, 1959). There is an increase in the total nitrogen content of embryos until at least 6 weeks after pollination, although all through this period approximately 70% of nitrogen is in the form of proteins. It is possible that changes in insoluble nitrogen are obscured by the variability of the samples analyzed, and hence any decrease by combustion is not bound to be reflected in the data.

Free amino acids of embryos have a dual interest since they are both products of endosperm breakdown and the raw materials from which proteins are built. There is reason to believe that endosperm is digested at different rates and at different times at different regions. That there are regional differences in the concentration of free amino acids and in the synthesis of different proteins during embryogenesis is at present

TABLE 3.1. Free amino acids and amides in *Datura stramonium* embryos at various stages of development. (From Rietsema and Blondel, 1959; concentration given in mmol/g fresh weight)

Age (days)	15	19	23	29	33	40	50
Fresh weight of embryos (mg)	0·1	0·7	1·2	2·1	1·7	2·4	2·1
Aspartic acid	0·2	3·7	2·6	5·0	1·5	0·2	0·4
Glutamic acid	0·5	3·3	4·8	4·6	2·0	0·5	2·1
Serine	0·2	0·8	1·8	0·8	0·4	0·1	0·4
Asparagine	—	+	+	0·1	2·1	2·6	0·8
Threonine	+	1·1	1·9	+	—	+	+
Alanine	1·5	2·1	3·6	1·2	2·4	0·2	0·4
Glutamine	0·2	3·1	2·5	+	2·3	0·5	0·5
Lysine	0·1	2·1	7·0	1·5	0·2	0·1	—
Arginine	1·1	6·3	26·0	8·8	—	—	—

a likely possibility, although so far there are no data directly bearing on this.

Changes in free amino acids and amides of embryos of *Datura stramonium* at arbitrarily chosen stages of development from about 15 days after pollination to the maturity of the seed are presented in Table 3.1. Arginine, lysine, aspartic acid and glutamic acid occur in relative abundance, alanine and glutamine in moderate amounts, and serine, asparagine and threonine in trace amounts. Most of the amino acids show a peak concentration about 3–4 weeks after pollination. This period coincides with the termination of growth of the embryo, and formation of maximum reserve proteins. Thus the evidence argues against the formation of amino acids from the breakdown of proteins. The termination of active growth of the embryo is strikingly apparent in the appearance of asparagine and the disappearance of arginine from tissues (Rietsema and Blondel, 1959). The criticism can be raised against data presented that it is difficult to determine whether the amino acids appearing at any stage are left-overs from previous protein synthesis, or are assembled for the next round of protein synthesis.

In maize and barley embryos of different ages analyzed at different times after anthesis, there were rapid increases in the various chemical constituents such as proteins, fats, RNA, DNA, soluble nitrogen, total amino acids, sugar and soluble nucleotides (Ingle *et al.*, 1965; Duffus and Rosie, 1975). In maize, ribonuclease activity showed an exponential increase in embryos up to about 38 days after pollination (Dalby and Davies, 1967). The significance of enzyme activity in relation to changes in other cell constituents is uncertain. Although maize embryo is not primarily a protein storage tissue, a few noteworthy differences are present in the concentrations of different amino acids. For instance, in the intact mature embryo there are high concentrations of aspartic acid, glutamic acid, glycine, alanine, histidine, proline, valine and serine, but cystine, methionine, ornithine and tyrosine occur only in traces (Oaks and Beevers, 1964). Besides, although excised embryos have very low levels of tryptophan, they accumulate significant amounts of it upon culture in the dark. Utilizing exogenous supplies of tryptophan, the embryo can synthesize niacin (Nason, 1950).

Although a variety of changes occur during embryogenesis, those concerned with the synthesis of RNA and proteins appear to be crucial to signal the initiation of new developmental potencies. In legume seeds, RNA content exhibited a period of slow accumulation during early embryogenesis followed by a period of intense accumulation at the time of rapid growth of the embryo (Wollgiehn, 1960; Vecher and Matoshko, 1965; Galitz and Howell, 1965; Wheeler and Boulter, 1967). In the embryo axes of *Phaseolus vulgaris* which required 36 days from flowering to reach maturity, RNA was synthesized during the first 24 days of embryogeny. However, protein accumulation continued until the final stage of maturity of the embryo, long after RNA synthesis was terminated, probably by using as templates RNA synthesized earlier (Walbot, 1971). In line with this work, it has been observed that in pea cotyledons collected at different ages from fertilization of the ovule to seed maturity, there is a progressive decrease in the polysomal content and a loss in the capacity of ribosomes to incorporate amino acids into proteins (Beevers and Poulson, 1972). This draws our attention to the fact that synthesis of mRNA in the embryo becomes a limiting factor as embryogenesis is completed and seed maturation sets in. Inasmuch as these observations bear upon the concept of stored mRNA in the embryo of the mature seed, they will receive further treatment from the standpoint of templates for the first proteins of germination in Chapter 17. Cotyledons of developing embryos of leguminous plants have also been subjected to detailed studies relating to the characterization of storage proteins and fundamental information concerning the role of RNA, DNA and protein synthesis during cotyledon development has been obtained. This work has been reviewed very recently (Millerd, 1975).

Biochemical studies of gymnosperm embryos are rare. Some workers (Konar, 1958b; Durzan and Chalupa, 1968) have noted a close qualitative correspondence in the changes in free amino acids of gametophytes and embryos of pine. The chromatogram of the egg of *Pinus roxburghii* at the time of fertilization revealed the presence of aspartic acid, glutamic acid, serine, glycine, cysteine, histidine, threonine, β-alanine, tyrosine, proline, valine, arginine, leucine, asparagine and glutamine (Konar, 1958a). The formation of cotyledonary primordia and the subsequent attainment of maturity in the embryo were accompanied by the disappearance of all amino acids and glutamine; only asparagine was present in measurable quantities in the mature embryo. At the same time, there was an increase in the protein content of the embryo. Other striking features are the abundance of sucrose in the young embryos and a shift towards the production of starch and fat at later stages of embryogenesis.

V. COMMENTS

In this chapter an attempt has been made to bring together information from various sources and to integrate the biochemical and physiological aspects of embryogenesis with the structural features of development of the embryo. In summarizing our present knowledge on the biochemistry of plant embryos it becomes apparent that there are numerous gaps in our information. Details available are based on limited sampling and this has restricted generalizations applicable to a large number of representative plants. As biological systems, plant embryos, like animal embryos, are complex and dynamic, and an integrated approach involving the fostering of mutual interest between developmental anatomy, biochemistry and physiology may be expected to unearth the complexity and dynamism of this system. It seems to us that in such studies, to an important extent, the experimental embryologist needs to initiate investigations with very small embryos, for some of the secrets of functional differentiation are hidden in that period which precedes the manifestation of recognizable form.

4. Growth Correlations in the Embryo

Growth of the embryo is reckoned in terms of production of new cells, and organization of cells into tissues, tissues into organs, and organs into a unified whole organism. This progression of changes, which is the visible expression of underlying physiological events, creates structural diversity in what appears to be an initially homogeneous mass of cells. The early phase of embryogenesis is committed to the production of a population of deceptively simple and seemingly similar cells by rapid cell divisions. Multicellularity at this stage does not confer any advantages on the embryo owing to lack of coordination and specialization of cellular activities. Subsequently, however, the cells follow a course of cytodifferentiation resulting in the formation of groups of cells with specialized functions and establishment of specific mutual relationships among them. How do these seemingly identical cells mutually determine the location and type of growth and differentiation? An understanding of the factors that endow the embryo with the ability to undergo repeated and orderly divisions and eventual differentiation may be expected to begin with an understanding of the distribution and relative rates of cellular processes in the embryo and its interaction with the surrounding milieu. Before examining the determinative factors that are involved in achieving a full-grown embryo, it is therefore appropriate to examine some aspects of growth of the embryo and its surrounding tissues, to construct a coherent picture of growth correlations.

I. QUANTITATIVE ANALYSIS OF GROWTH OF THE EMBRYO

A. Angiosperms

In angiosperms, growth of the fruit is generally a consequence of pollination and fertilization, and frequently involves growth of the seed and enlargement of the ovary, or enlargement of the receptacle of the flower, or fusion of the floral parts with the ovary. Fruit growth is characterized by a rapid rate of cell division and cell enlargement in the ovary and by the growth of the embryo and endosperm from their single-celled beginnings. From the biochemical point of view, the problem in fruit growth becomes reduced to an analysis of the biosynthetic processes which are particularly intense during this period, resulting in the accumulation of reserve substances.

Growth of the Embryo, Ovule and Fruit

A broad array of studies performed on a number of species has sought to determine the quantitative relations between growth of the embryo, ovule and fruit. Since the embryo grows within the ovule, which itself may be encapsulated in the ovary, embryo growth must be regarded as part of a continuum of interdependent processes. As such, the nature of growth of the ovary and its supporting tissues is important for a proper appreciation of growth dependencies of the embryo. Embryos of most angiosperms remain microscopic until the fruit has developed considerably. Studies on the growth of several drupes, stone fruits and berries (Connors, 1919; Lilleland, 1930; Tukey, 1933b, c, 1934a, 1938; Tukey and Young, 1942; Winkler and Williams, 1935; Young, 1952; Nitsch et al., 1960) have indicated their development to fall into three distinct periods: (1) a period of rapid growth for about 6–7 weeks after bloom, (2) a period of delayed increase lasting about 1–6 weeks according to the species and variety and (3) a second period of rapid growth to fruit maturity. Taking peach (*Prunus persica*) as an example (Tukey, 1933b), growth curves of the pericarp, nucellus, integuments and embryo of four varieties from full bloom to maturity are plotted in Fig. 4.1. It is seen that during the period of rapid growth of the pericarp immediately after bloom, growth of the embryo is negligible, although considerable growth of the integument and nucellus occurs. Growth

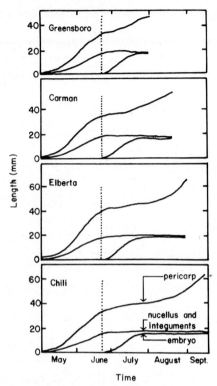

Fig. 4.1. Comparison of growth curves for pericarp, nucellus with integuments and embryo of four varieties of *Prunus persica* from flowering to fruit maturity. The dotted line indicates the time of resumption of rapid growth of the embryo. The order of curves for all varieties is the same as that given for 'Chili'. (After Tukey, 1933b)

increments in the latter tissues might have resulted mainly from expansion of cells already present. Growth of the embryo is resumed during the lag period of growth of the pericarp; as the embryo ceases to grow, the pericarp enters a second period of active growth. Thus, the great bulk of growth of the embryo is confined to a relatively brief period in the life of the fruit.

As the embryo begins to draw upon the nutrient supply of the endosperm, competitive limitations on the growth of the fruit begin to set in. This is clearly seen in studies where growth of the endosperm has been followed side by side with that of the embryo, ovule and ovary. In apple (*Malus sylvestris*) there is a notable periodicity in the growth of the fruit, endosperm and embryo (Luckwill, 1948; Murneek, 1954). During the first 3–5 weeks after fertilization, when the fruit develops rapidly, there is considerable build-up of a free nuclear endosperm, but the embryo remains in a state of arrested development. Rapid growth of the embryo occurs some time later, coincidentally with the transition of the endosperm from free nuclear to cellular state. Although the rate of fruit growth was negligible during the period of active growth of the embryo, there was yet another period of fruit growth which

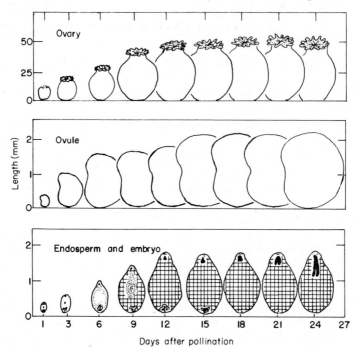

Fig. 4.2. Comparison of the rate of growth of the ovary, ovule, endosperm and embryo of *Papaver somniferum* from pollination of the flower to fruit maturity. (After Maheshwari, N. and Lal, 1961b)

occurred after the embryo attained its maximum size. Unlike the relationship between the endosperm and growth of the embryo, the form of the relationship between the endosperm and fruit growth is difficult to define.

In *Papaver somniferum* (Maheshwari, N. and Lal, 1961b; Johri and Maheshwari, 1966a), ovary and ovule reach their final size long before the embryo matures (Fig. 4.2). The endosperm, which is initially free-nuclear, follows the ovule in its trend of growth and attains its maximum growth at the same time as the latter. Growth of the embryo of *Zephyranthes lancasteri* (Johri and Maheshwari, 1966b) is marked by two periods of rapid increase in length from 8 to 11 days and 13 to

15 days after pollination, while the ovule and endosperm, which attain maximum growth by 8 days after pollination, grow very little thereafter (Fig. 4.3). A suggestion that is perhaps credible here is that in both cases the rapidity with which the embryo grows after the fruit has attained its maximum growth is related to the uninterrupted supply of nutrients from the endosperm.

Growth curves of dry fruits such as legumes have revealed two periods of active growth which are not, however, separated by a lag period

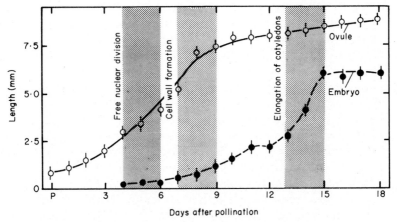

Fig. 4.3. Comparison of the rate of growth of the ovule and embryo of *Zephyranthes lancasteri* from pollination of the flower (P) to maturity of the seed. (After Johri and Maheshwari, 1966)

(Woodroof and Woodroof, 1927; Bisson and Jones, 1932; Ulrich, 1942, 1957; Nitsch, 1953; McKee *et al.*, 1955). Again, as in the case of fleshy fruits, different organs show quite varied curves for volume increment during growth so that proportional growth contributed by individual organs to the total growth varies. The embryo proper begins to grow during the latter phase of growth of the seed, showing a relatively rapid progression from the zygote until a point arbitrarily designated as maturity of the pod. The basic contrasts in the growth curves of dry and fleshy fruits apparently suggest differences in the periodicity of cell elongation and cell division in the embryo and in the surrounding tissues. In leguminous plants the cotyledons, which constitute a major part of the embryo, attain their final cell number less than halfway through embryogenesis and subsequent increases in fresh and dry weight of these organs are the result of cell expansion and the deposition of storage proteins (McKee *et al.*, 1955; Millerd *et al.*, 1971).

Growth curves do not seem to be distinctive for the different morphological types of fruits, since typical diauxic growth curves similar to those of fleshy fruits have been obtained in dry fruits by some investigators (Randolph, 1936; Rijven, 1952; Carr and Skene, 1961; Flinn and Pate, 1968; Burrows and Carr, 1970; Walbot, Clutter and Sussex, 1972). According to Carr and Skene (1961) during the period of exponential growth of the seed of French bean lasting from 9 to 20 days after pollination, the embryo grows to fill the entire embryo sac and as a result a lag phase of about 3 days sets in. This is followed by a second period of growth which is initially as rapid as in the exponential phase, but afterwards declines. During the latter period, the embryo and seed grow in volume more or less at the same rate. Carr and Skene also found a lag period when they replotted the data of Woodroof and Woodroof (1927), Bisson and Jones (1932) and McKee *et al.* (1955). At the metabolic level, decrease in the production of cytokinins, amino acids and sugar, increase in the starch content and appearance of storage proteins are probably involved in the onset of lag phase in the growth of the fruit (Flinn and Pate, 1968; Burrows and Carr, 1970).

The typical diauxic growth curves apparent in some fruits is not universal, however. Quantitative analyses of growth of the fruit, seed and embryo of the Para rubber tree (*Hevea brasiliensis*) have revealed that the fruit attains its maximum length in about 6 weeks after fertilization, but maximum seed size is not attained until 10–14 days after attainment of maximum fruit size. The embryo remains small and microscopic until about 7 weeks after fertilization, but it develops rapidly soon afterwards to reach the maximum size 10–14 days after termination of seed growth (Muzik, 1954). Here growth constitutes a series of successive stages in time which begin with the fruit and terminate with the embryo. In *Withania somnifera* (Solanaceae), the potentially linear curve for fresh weight increments in the fruit is interrupted by two periods of low growth rates, roughly corresponding to periods of initiation and elongation of the cotyledons (Mohan Ram and Kamini, 1964). As in all cases of organized growth, it is reasonable to assume that physiological signals exist in the developing fruit for communication between the different organs for a sequential expression of their growth potencies. For a meaningful interpretation of the results, the empirical correlations based on histological observations should be supplemented with physiological data on changes in growth hormone contents of the fruit.

Relative Growth of Embryos

Inherent in any attempt to relate embryonic growth with time is the difficulty that the embryo cannot be measured without being excised, and hence data of the nature discussed above presumably reflect only a part of the true aspect of growth. Although growth of the embryo follows the pattern of a typical sigmoid curve, slight variations have been found in the final form of the curves depending upon whether growth is recorded as length increments or in terms of fresh or dry weights. This is clearly illustrated by growth curves obtained for embryos of cotton by Reeves and Beasley (1935). When length measurements were made, it was found that the embryo entered its grand period of growth in length a week before it entered the grand period of growth in weight. Curves for fresh and dry weights of the embryos were similar and showed a formative period of rapid growth and a period of slow growth preparatory to termination of growth (Fig. 4.4). Rijven (1952) showed that in *Capsella bursa-pastoris*, maximum relative rate of growth during a 24 h period occurred in embryos 100–150 μm long (Fig. 4.5). Embryos about 100 μm long were initially

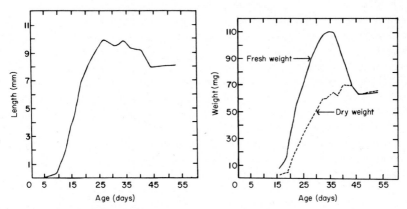

Fig. 4.4. Variations in the growth curves of embryos of cotton, expressed in terms of length or on fresh and dry weight basis. (After Reeves and Beasley, 1935)

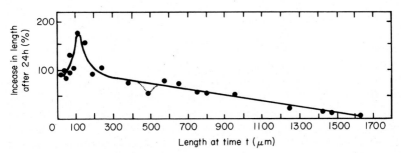

Fig. 4.5. Relative growth rate during a 24 h period in embryos of *Capsella bursa-pastoris*. (After Rijven, 1952)

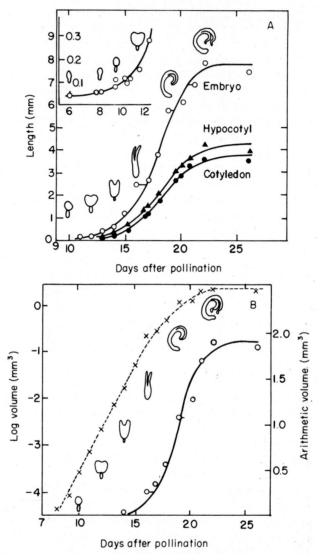

Fig. 4.6. **A**, Linear measurements of whole embryos, hypocotyls and cotyledons of *Datura stramonium* from pollination of the flower to maturity of the fruit provide conventional S-shaped curves. Inset is a growth curve of embryos in early division stages. **B**, Embryo volumes also yield an S-shaped curve when plotted on an arithmetic scale (continuous line), but a linear relationship is obtained when they are plotted on a logarithmic scale (broken line). (After Rietsema *et al.*, 1955)

heart-shaped, and exhibited mainly longitudinal growth until they became torpedo-shaped; minimum relative growth occurred in mature embryos. Linear measurements of whole embryos, hypocotyl and coty-ledons of *Datura stramonium* provided typical S-shaped curves when plotted (Fig. 4.6); most rapid growth of the embryo began when it turned heart-shaped at about 10 days after pollination, and continued until the twenty-second day after pollination (Rietsema *et al.*, 1955; Rietsema and Blondel, 1959). When volume measurements of embryos were considered, growth was exponential as one would expect from a system undergoing cell division at a constant rate, although about 16 days after pollination there was a lag phase. The onset of the lag phase was the result of the embryo filling the entire embryo sac, and con-sequently its further growth being limited by the ability of the integu-ments to expand. Fruit growth followed immediately after pollination and was complete at about the same time the exponential phase of growth of the embryo was complete. In this respect, growth relation-ships between the embryo and fruit of *D. stramonium* differ from that of other angiosperms studied.

B. Gymnosperms

In contrast to angiosperms, in several gymnosperms, including nearly all conifers, there is an extensive development of the ovule before fertil-ization. This endows a dual effect on embryo–ovule relationships. On the one hand, limitations in growth of the embryo at early stages of development imposed by the inability of integuments to expand are eliminated. On the other hand, there are negligible side effects of the contained embryo on the enlargement of the seed, while in angio-sperms, a possible effect of the embryo in stimulating growth of the seed is admittedly present.

From volumetric measurements of the rate of growth of the seed, megagametophyte and embryo, Buchholz (Buchholz, 1946; Buchholz and Stiemert, 1945) has divided the embryogeny of ponderosa pine (*Pinus ponderosa*) into two distinct periods. During the period lasting about 5 weeks from fertilization designated as the period of embryonic selection, there is little differentiation of the embryo, but a single suc-cessful embryo outstrips the others and becomes dominant. The second period of development, less than half as long as the first, is the period when the successful embryo embarks upon organ differentiation. This is followed in a few weeks by ripening of the seed. A graphic record of development of the embryo of *P. ponderosa* from fertilization to maturity described above is depicted in Fig. 4.7.

In another line of investigation, Buchholz (1945) has shown that linear measurements of embryos can serve as a reliable guide to determining possible causes of hybrid vigor in conifers. An obvious question posed has been whether hybrid embryos are initially larger, thus enabling seedlings to show hybrid vigor. This does not seem to be the

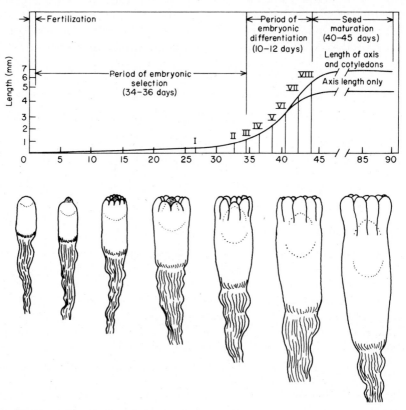

Fig. 4.7. Changes in length and morphological complexity of embryos of *Pinus ponderosa* from fertilization to maturity of seeds. (After Buchholz, 1946; Buchholz and Stiemert, 1945)

case since measurements of the embryos of *P. murryana* and *P. banksiana* and of the cross *P. murryana* × *P. banksiana* have shown that in their length and shape index the hybrid stands between the two parents. To put it another way, the presence of a large embryo does not result in a large seedling in the crossed varieties; possibly, the advantages of hybrid vigor are expressed during germination of the seed and subsequent growth of the seedling.

C. Pteridophytes

An approximation of the relationship outlined earlier for angiosperms is obtained in the growth of embryos of the fern, *Todea barbara* (DeMaggio, 1963). Uniformly growing embryos raised on gametophytes grown aseptically under controlled conditions of nutrition (mineral solution

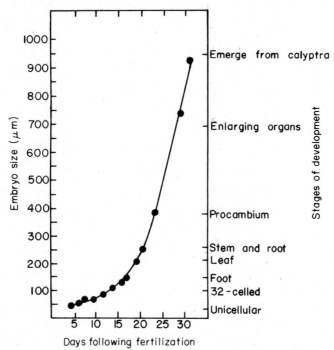

FIG. 4.8. Changes in embryo size and morphological differentiation following fertilization in *Todea barbara*. (After DeMaggio, 1963)

supplemented with 1% sucrose) and light intensity were used for computing growth data. In this species, although the duration of development of the embryo from the time of fertilization to emergence from the calyptra lasted about 30 days, during the first half characterized mostly by cell divisions, there was only a three-fold increase in size. Rapid increase in size associated with cell division, cell elongation, cellular differentiation and organ formation occurred during the later stages of development (Fig. 4.8).

II. GROWTH HORMONES AND GROWTH OF THE EMBRYO

Fruit growth is generally marked by two sequential stages, first, when cell division predominates, and second, when cell enlargement predominates; both these processes are subject to different kinds of subtle regulatory control. Such a distinction is less conspicuous in the growth of the encapsulated embryo where cell divisions account for most of the growth. Growth-promoting substances of the type of auxins, gibberellins and cytokinins are important in fruit development on account of their effect on cell division and cell enlargement. It is now known that the hormonal stimulus initiating fruit growth has its origin in the act of pollination, which sets in motion those physiological processes necessary to enable the fruit to grow right up to maturity. Whatever may be the mechanism by which pollination channels hormones to the developing ovule, studies on the role of growth hormones in mimicking the effects of pollination have established the involvement of auxins and gibberellins. The process of fertilization following pollination also has a major impact on fruit and seed growth by unleashing additional or perhaps new growth hormones, which emanate from the developing embryo and endosperm. In certain fruits, such as apples, cytokinins are present in relatively high concentrations shortly after fertilization, and it is hard to visualize their appearance without ascribing to them a significant role in the growth of the fruit. Thus, one can recognize stages of development in a fruit which are characterized by recognizable changes in a prescribed balance of growth substances. The problem here is to relate growth of embryos to changes in the quality and quantity of growth hormones in the ovule, to gain clues regarding the nature of the stimuli we may seek to emulate and the nature of the processes we may seek to control.

A. Auxins

It is very suggestive that a few studies have attempted to relate growth of the embryo to changes in the concentration of auxins, gibberellins and cytokinins. Wright (1956) has shown that in blackcurrant (*Ribes nigrum*) optimal periods of production of IAA and of an unidentified neutral auxin appear to coincide with the period of maximum growth of the embryo and endosperm (Fig. 4.9), but whether the auxin is produced in the embryo or elsewhere in the ovule is not determined. Later, as the auxin content decreases, growth of the embryo also subsides. If auxin is a specific requirement for growth of the embryo it seems more probable that the auxin dependence will be confined to the heterotrophic phase and will disappear as the embryo enters upon the auto-trophic phase.

In apple and tomato (*Lycopersicon esculentum*), the peak in auxin activity seems to be associated with the formation of cellular endosperm and not with the period of maximum growth of the embryo. In fact, the hormone content of the endosperm recorded a low level during the period of rapid growth of the embryo, and increased when the embryo growth subsided. These results are in accord with the view that auxin

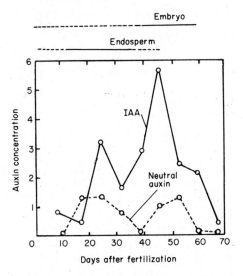

FIG. 4.9. Amounts of IAA and a neutral auxin extracted from fruits of *Ribes nigrum* from the time of fertilization to maturity coincide with phases of active growth of the embryo and of the endosperm. (From the data of Wright, 1956)

is consumed by the developing embryo more rapidly than it is produced in tne endosperm (Luckwill, 1948, 1953; Iwahori, 1967).

The role of growth hormones of the auxin type in the development of the embryo, endosperm and nucellus of 'Concord' grape (*Vitis labrusca*) was described by Nitsch *et al.* (1960). The data in Fig. 4.10 indicate that there are three bursts of auxin production from the time of fertilization to maturity of the berry. An interpretation of the changes in auxin level that occur during the ontogeny of the berry suggests that waves of auxin production correspond, respectively, to the development of the nucellus, endosperm and embryo. Changes in the auxin content of the ovules of cotton as they relate to the growth of the embryo have shown that differentiation of the embryonic cotyledons is accompanied by an increase in IAA levels (Maheshwari, S. C. *et al.* 1964). In all

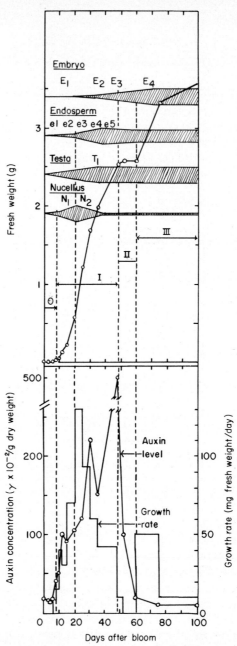

FIG. 4.10. The three waves of auxin production in 'Concord' grape correspond roughly to the development of the nucellus, endosperm and embryo. Period 0, 0–10 days after bloom; I, 10–50 days; II, 50–60 days; and III, 60 days to ripening. Nucellus: N_1, rapid development of nucellus; N_2, maximum volume of nucellus. Testa: T_1, seed full size, beginning of the hardening of seed coats. Endosperm: e_1, beginning of free nuclear endosperm; e_2, first cellular endosperm; e_3, beginning of rapid development; e_4, endosperm completely cellular; e_5, maximum development. Embryo: E_1, first division of zygote; E_2, 50 cells; E_3, cotyledons differentiated; E_4, full length embryo. (After Nitsch et al., 1960)

of these studies, auxin content of the tissues was determined by its activity in the oat first internode test and in Jerusalem artichoke tuber tissue test, or by color reactions. Since these tests are hardly specific for auxins and since the embryo forms only a small part of the seed, this creates considerable doubt about the role that auxins might play in the growth of the embryo. Moreover, the emphasis on auxin as the sole hormonal regulator of fruit growth might obscure the role of other hormones in the process.

B. Gibberellins

The presence of gibberellins may also be relevant for growth of the embryo. The existence of a close relationship between gibberellin content of the fruit or the seed and embryo growth is particularly impressive in some cases. The endosperm which provides the nurture for the embryo during its growth is extremely rich in gibberellin-like sub-

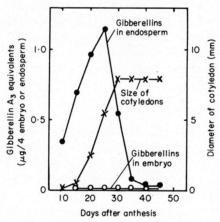

Fig. 4.11. Changes in growth of embryos of *Lupinus luteus* from the time of fertilization are paralleled by changes in the amount of "gibberellin-like substances" in the endosperm and not in the embryo. (After Ogawa, 1963a)

stances (Phinney *et al.*, 1957; Nitsch, 1958). Corcoran and Phinney (1962) determined the amounts of gibberellin-like substances in the endosperm-nucellus of *Echinocystis macrocarpa* and in the whole seed extracts of *Lupinus succulentus* and *Phaseolus vulgaris*, and found that the amount per seed increased during embryo growth and declined with its maturity. In *Lupinus luteus* (Fig. 4.11) no gibberellin-like substances were present in the embryo, but its growth increments were roughly paralleled by sustained supplies in the endosperm of a substance which showed some of the characteristics of GA_3 (Ogawa, 1963a). It might be logically expected that gibberellins are produced in the endosperm, and channeled to the embryo to accelerate its growth. Bhalla (1971) studied the changes in gibberellin-like substances in developing watermelon seeds (*Citrullus lanatus*). Since the peak gibberellin activity appeared in the seed after it attained maximum growth, it was argued that activity of the hormone might be correlated with the growth of the embryo and endosperm, and not of the seed *per se*. A convincing correlation between gibberellin content and embryo growth is seen in Japanese morning glory (*Pharbitis nil*) where gibberellins are present in both embryo and endosperm (Ogawa, 1963b).

Using chromatography and dwarf pea and dwarf maize bioassay, Skene (Skene, 1970; Skene and Carr, 1961; see also Radley, 1958; Hashimoto and Rappaport, 1966) recently assayed the fruits of *Phaseolus vulgaris* for gibberellins. On the chromatogram it was possible to identify a spot corresponding to GA_5 which appeared in the young fruit, and disappeared after cell division in the embryonic cotyledons had ceased, and another spot corresponding to GA_1 which appeared during the period when cotyledons grew mainly by cell elongation. This interesting observation suggests that waves of activity of different endogenous gibberellins are concerned with different aspects of embryo growth.

C. Cytokinins and Embryo Factors

The rich supplies of hormones in developing fruits and seeds provided the initial source from which endogenous cytokinins were first isolated from the plant. However, the possibility that cytokinins may contribute to the endogenous control of growth of the embryo remains to be firmly established. Burrows and Carr (1970) have noted a good synchronization between the cytokinin content of the seeds of *Pisum arvense* and rate of growth of the encapsulated embryo. Their data are given in Fig. 4.12, from which it appears that the two peaks in cytokinin content at 24 days and 30 days are coincident with the two maxima in growth of the embryo.

In seeds of *Datura tatula* and *Lupinus luteus*, control of growth of embryos has been attributed to the presence of unidentified growth-promoting and other substances referred to as "embryo factors". In both species, the relative embryo factor content per seed increased with growth of the embryo and decreased when its growth slowed down (Matsubara and Nakahira, 1965b). This raises the possibility, to be confirmed in other species, that growth of the embryo is due to an interaction of several substances acting at hormonal levels. For a more criti-

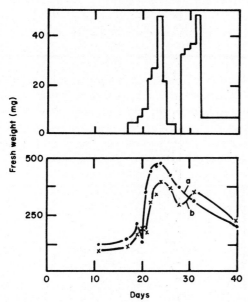

FIG. 4.12. The daily fresh weight growth increment of the embryo of *Pisum arvense* cv. 'New Zealand' (top) is synchronous with the cytokinin content of the seed (bottom). Cytokinin content is expressed in terms of milligrams callus fresh weight in soybean callus bioassay per gram fresh weight of seed (curve a) or per seed (curve b). (After Burrows and Carr, 1970)

cal evaluation, one should ideally begin by tracing the movement of metabolites from the endosperm into the embryo. Secondly, the increase or decrease in concentration of particular substances in the embryo as it develops and is nourished should also be determined.

III. CELLULAR ASPECTS OF EMBRYO GROWTH

The foregoing account has been entirely concerned with the growth of embryos manifest by an increase in volume, fresh or dry weight or length. However, as an indicator of growth which includes processes of cell division, cell elongation and cellular differentiation, volume, weight and length measurements leave much to be desired, since it is recognized that rates of more complex processes cannot be adequately translated by such methods. Some attempts have recently been made to relate growth of the embryo in cellular terms and these will be briefly described now. From analyses of histogenic patterns, cell number, cell size and mitotic frequencies, it has been shown that increases in cell number in the embryos of *Triticum vulgare* (Rédei and Rédei,

1955a) and *Oryza sativa* (Mizushima *et al.*, 1955), from soon after fertilization to full maturity, follow sigmoid patterns. On the other hand, in the embryos of *Capsella bursa-pastoris* and cotton (Pollock and Jensen, 1964; Yoo and Jensen, 1966) the form of the curve for increase in cell number is diauxic, taking an S-shaped form from the fertilized egg up to the globular stage and a different S-shaped form from the heart-shaped stage to the mature embryo (Fig. 4.13). The distribution of mitotic figures is uniform in very young embryos but is continually changed thereafter to reflect the onset of tissue and organ differentiation, assuming a U-shaped pattern in heart-shaped embryos and ending

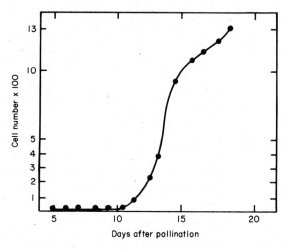

Fig. 4.13. Increase in cell number in embryos of cotton is represented by two S-shaped curves. (After Pollock and Jensen, 1964)

up in a Y-shaped pattern in older embryos (Fig. 4.14). Apparently, the onset of a specific histogenic pattern during progressive embryogenesis depends upon the attainment of a critical cell number before a change occurs, followed by unequal distribution of cell divisions. Thus, transition of the globular embryo to the heart-shaped form seems to result from increased cell divisions in the distal part of the embryo giving rise to cotyledons. A different pattern of distribution of cell divisions, accompanied by cell elongation, might account for the change in shape of the embryo from the heart-shaped to torpedo-shaped stage (Pollock and Jensen, 1964; Wochok, 1973b).

These studies tie in nicely with a familiar concept that recognizable differences in a system that prelude differentiation are signaled by an accelerated tempo in mitotic rhythm. As we have seen in the previous

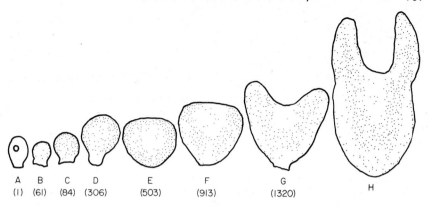

A B C D E F G H
(1) (61) (84) (306) (503) (913) (1320)

FIG. 4.14. Distribution of mitoses is represented in outline drawings of the embryos of cotton. Each dot represents one cell division. Numbers in parentheses below each drawing are the average total number of cells for that stage. (After Pollock and Jensen, 1964)

chapter, the increased mitotic activity is accompanied by corresponding increases in DNA and RNA synthesis. The key facts which may provide clues to a more explicit understanding of the factors that control differential cell divisions in a homogeneous tissue are, however, not available. One might suspect that some cytoplasmic gradient is being set up by cell division factors diffusing in the direction of the embryo where potential divisions occur, but as yet we have no evidence for this.

IV. COMMENTS

In summing up, it appears that subtle influences determine the fate of each cell as the embryo proceeds from stage to stage during embryogeny. However, individual cells of the embryo do not behave as isolated units of a static system, but function as parts of a developmental ensemble. As such they accomplish their function by mechanisms, still incompletely understood, which preserve the balance and integrity of the organ.

5. Nutritional Aspects of Embryogenesis— Morphological Considerations

Embryogenesis does not take place in a structural vacuum, although such an impression is conveyed by diagrams which emphasize only the different developmental stages of the embryo to the total obliteration of the infinite complexity of the surrounding environment. As a result, the relevant interactions between the embryo and its milieu have not been properly appreciated.

In plants and animals, early embryogenesis is of necessity linked to some form of energy needed for the synthetic activities of the embryo before it becomes self-supporting. The sources of energy are diverse and are provided in the egg or in the surrounding environment. In oviparous and ovoviviparous animals, production of yolk in the egg supplies the necessary fuel for the first few rounds of division of the zygote. The energy available from the yolk is limited, and the organism, no longer energetically self-sufficient, has to look to other sources of nutrition through the rest of the embryonic period. In invertebrates this latter phase of embryogenesis is accomplished by feeding the larva, while in placental mammals, essential nutrients from the mother are channeled to the embryo implanted in the uterus.

In plants, there is a good deal of evidence to show that the cytoplasm of the fertilized egg plays a vital role in providing the necessary energy requirements for the zygote nucleus to undergo the first few divisions. The visible inclusions in the egg have been described by histochemical methods and mostly consist of carbohydrates, proteins, nucleic acids and other energy-supplying materials. Sustained growth of the embryo is, however, possible as a result of absorption of food materials fabricated by the tissues of the gametophyte in which the embryo is constantly immersed or with which it maintains contact, although the nature of the materials absorbed is unknown. It is possible that the embryo can absorb only amino acids, mono- and disaccharides, inorganic ions and vitamins, and that the synthesis of more complex carbohydrates, proteins and fats occurs at the expense of this pool of low molecular weight substances. Indeed, the most striking feature in the embryogenesis of vascular plants is the elaboration of structures in the young embryo to absorb food materials from the gametophytic tissues. In the sections that follow, as a background for further discussion, vascular plants are considered from the point of view of nutrition of their embryos, particularly with regard to the organization and structure of the cells and tissues that nourish them.

I. ANGIOSPERMS

In contrast to the immense diversity in the pattern of embryo formation in angiosperms, similarity in the mode of their nutrition is impressive. At the time of fertilization, the embryo sac does not contain in sufficient quantity the type of nutrients necessary for continued growth of the embryo. The main line of nutrition of the embryo is derived from the large pool of reserve materials of a tissue known as the endosperm which is generated after fertilization and as the ovule is transformed into a seed. In exalbuminous seeds, the endosperm is short-lived and is used up during the growth of the embryo. In albuminous seeds, it remains as a permanent storage organ of the mature seed, and serves as a nurse tissue for the embryo during seed germination. With a few exceptions confined to certain members of the Orchidaceae (Swamy, 1949), Onagraceae (Johansen, 1931a, b, 1932), Podostemaceae (Razi, 1949) and Trapaceae (Ram, 1956), most angiosperms investigated contain an endosperm or its physiological equivalent on which the developing embryos are nourished. In some families, such as Piperaceae, Amaranthaceae, Portulacaceae, Capparidaceae, Zingiberaceae and Cannaceae (Masand and Kapil, 1966), in addition to the endosperm, food materials may be stored in a special tissue derived from the nucellus, called the perisperm. In several members of the Chenopodiaceae, Rosaceae, Polygonaceae and Caryophyllaceae, the nucellus may initially serve as an auxiliary source of nutrition for the embryo, but it is eventually replaced by the endosperm. So far as can be judged from ontogenetic studies it is difficult to draw a line between termination of the role of the nucellus and beginning of the function of the endosperm (White, 1950). The ease with which the inner epidermis of the integument lying closest to the embryo sac becomes differentiated in some members of the Sympetalae into a tissue appropriately known as the integumentary tapetum (endothelium) has been cited as an obvious feature of its nutritive function (Vazart, B. and Vazart, J., 1965a; Masand and Kapil, 1966). In the autonomous apomict, the dandelion (*Taraxacum officinale*), embryo and endosperm are formed without the intervention of nuclear fusion. However, the functional relationship between the two has been questioned since, often, the embryo may attain a relatively normal development while the endosperm remains undeveloped. Probably, development of an accessory storage tissue in the integument makes the endosperm superfluous (Cooper and Brink, 1949). In *Dianthus chinensis*, starch grains present in the embryo sac and placenta are utilized by the embryo without recourse to the storage products of the endosperm, whose major function appears to be the digestion and absorption of nucellar food materials (Buell, 1952). In any event, the association of the endosperm with the developing embryo is almost universal, indicating the significance of this tissue in the embryogenesis of angiosperms (Brink and Cooper, 1947). The convergence of sound morphological studies with histochemical and biochemical analyses of the cell constituents has shown that activities of the endosperm are due to an interplay of morphological development and biochemical complexity.

A. Endosperm Development

There are a number of descriptive studies on endosperm development in angiosperms, whose major objectives are related to classification of endosperm types and their phylogenetic origin and are useful in making comparative observations. A description of the modes of endosperm formation is beyond the scope of this book, but a brief orientation is necessary as a basis for the discussion to follow. During fertilization of the egg, one of the male gametes fuses with the diploid polar fusion nucleus in the central cell of the megagametophyte to form a triploid fusion nucleus (endosperm nucleus), which almost from its inception follows a developmental pattern sharply divergent from that of the zygote.

Endosperms of most angiosperms have a common origin from the triploid fusion nucleus, which in the course of differentiation divides by repeated mitoses to form a free nuclear or a cellular tissue charged with abundant food materials. Although normal development of the endosperm is deceptively simple, it presents a number of paradoxes that invite exploration. For example, we know practically nothing about the stimulus which the fusion of the male gamete imparts to the subsequent division of the endosperm nucleus, and the mechanism which results in the accumulation of metabolites in the formless mass of cells escapes us completely. It is to be stressed, however, that the endosperm in angiosperms is a new tissue with triploid number of chromosomes, although instances are known with diploid, tetraploid, pentaploid and higher numbers of chromosomes.

Free nuclear endosperm is characteristic of many more species of plants than was previously thought (Maheshwari, P., 1950; Chopra and Sachar, 1963). Here, the daughter nuclei formed by successive divisions of the endosperm nucleus may be randomly distributed in the embryo sac or confined to its periphery surrounding a large central vacuole, or may stream into incipient haustorial processes. In some species, after several divisions, the free nuclear endosperm embarks upon the cellular phase by the formation of walls around nuclei. The elaboration of highly differentiated organelle types in the cytoplasm immediately after nuclear division as seen in sunflower (Newcomb, 1973b) and *Capsella bursa-pastoris* (Schulz, P. and Jensen, 1974), or following wall formation as seen in wheat (Buttrose, 1963a), attests to the high metabolic activity of the endosperm. Free nuclear division is the primary mode of endosperm formation in plants belonging to Leguminosae, Cucurbitaceae and Gramineae, although some variations exist in the individual species with regard to the relative extent of free nuclear and cellular phases and the size and shape of the haustorial processes.

In certain families, including Scrophulariaceae, Acanthaceae, Lobeliaceae, Santalaceae and several others belonging to the Sympetalae, where the endosperm is of the cellular type, division of the endosperm nucleus is followed by cytokinesis and deposition of cell walls, resulting in the formation of a population of cells which eventually fill the entire embryo sac. New cells are repeatedly formed, expanding the area of the endosperm. An intermediate condition in which the endosperm nucleus divides into two nuclei having different developmental potencies has also been described. Here, the daughter nucleus formed towards the micropylar end of the embryo sac divides repeatedly to form a free nuclear endosperm while the one towards the chalazal end remains passive during most of the growth period of the embryo. This type of endosperm development found mainly in the order Helobiae of the monocotyledons is referred to as the "helobial" type (Swamy and Parameswaran, 1963).

There are a number of deviations from the ontogenetic patterns described above which defy accurate classification at present. A noteworthy one is found in coconut, the endosperm of which is also of special interest from the point of view of its use in tissue culture media. The endosperm of the mature drupe contains a clearly organized liquid portion and a hard solid portion enclosing the liquid. The complete endosperm is formed by the activity of both nucleus and cytoplasm; neither can be considered less important than the other. In the early stages of development of the endosperm, there are numerous amoeboid nuclei suspended in the organelle-rich cytoplasm to form a "liquid syncytium". The free nuclei appear in the embryo sac soon after fertilization and their appearance is presumably linked to the disintegration of the nucellus, or division of the primary endosperm nucleus (Quisumbing and Juliano, 1927). Some of the nuclei soon organize themselves as free cells. Later, the free nuclei and the individual cells migrate towards the periphery of the embryo sac where they initiate the solid cellular endosperm (Cutter et al., 1952a, b; Cutter and Freeman, 1954, 1955; Cutter et al., 1955). During the growth of the fruit there is a tremendous increase in size of the embryo sac cavity, which is also accompanied by an increase in volume of the sap and in the number of free nuclei. The latter are gradually incorporated into the bulk of the solid endosperm as the fruit matures. The difficulty in assigning endosperm development in coconut to one of the ontogenetic patterns described earlier is due to lack of information on the origin of the free nuclei (Quisumbing and Juliano, 1927; Maheshwari, P., 1950) and their subsequent division (Dutt, 1953; Cutter, et al., 1955; Datta, 1955; Henry, 1956). The vexing question is whether free nuclei can divide in the absence of cytoplasmic division. What results we have at present are controversial and do not lend themselves to equivocal interpretation.

B. Embryo–endosperm Dependencies

The concept that the embryo is dependent upon the endosperm for nourishment during a major part of its ontogeny is based principally on surmise from indirect evidence. Although fertilization of the egg by the male gamete and fusion of the second male gamete with the diploid polar nucleus are synchronous events, in the majority of angiosperms examined, the first few divisions of the endosperm nucleus and the zygote nucleus are asynchronous, with the former dividing earlier than the latter. The extent to which the endosperm develops prior to the division of the zygote varies, generally ranging from the formation of a few isolated nuclei to a fully developed tissue. In some plants, both endosperm nucleus and zygote nucleus divide concurrently, but the rate of mitosis in the endosperm exceeds that in the embryo (Brink and Cooper, 1947). All of this assures the availability of sufficient reserve food materials in the surrounding milieu at the time when the embryo is ready to embark upon its most rapid phase of growth.

Many studies have demonstrated that, in seeds of hybrid crosses in which there is a breakdown of endosperm formation, embryos usually fail to attain maturity (see Chapter 13). Of course, such observations do not establish a nutritional dependence of the embryo upon the endosperm, unless it can be shown that the embryo excised from the aborted seed grows to maturity in an artificial medium. Of the large number of investigations on the low fertility of hybrid crosses, in most cases there has been convincing success in rearing excised embryos in culture. Evidently, when the embryo is supplied with nutrients of the type presumably found in the endosperm, it reverts from an inhibited state to a normally growing state.

The manner in which the endosperm is used up by the embryo during its growth has also been cited in favor of its role in embryo nutrition. It is commonly observed that the endosperm is digested most at the site of the growing embryo. The essential events during this phase involve the individual cells of the endosperm, which simply disintegrate to give way to the embryo, until little, if any, of the former remains in the mature seed (List and Steward, 1965) (Fig. 5.1). Another line of supporting evidence comes from plants which do not normally have an endosperm or an equivalent tissue. These plants develop elaborate haustorial structures from the suspensor which serve what may be considered as the most obvious function, that is to say the transport of food substances from the accessory tissues to the growing embryo. Taken together, there is thus a body of evidence which is best interpreted in terms of a dependence of the embryo on the endosperm for nourishment during development. This evaluation has been strengthened in recent years by accumulation of information on the chemical composition of the endosperm of certain species, to be described in the next chapter which makes clear the potential of this tissue as a growth stimulant.

In spite of the built-in safeguard provided by the endosperm for embryo growth, it is important to emphasize that the precise stage in embryogenesis when the embryo begins to draw upon the food substances of the endosperm is not determined, and it is even doubtful

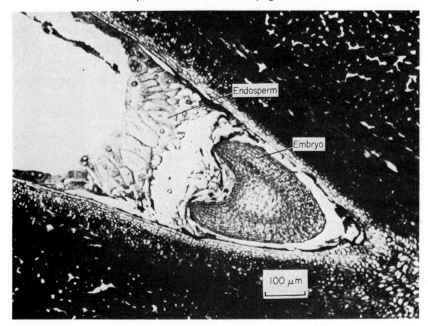

Fig. 5.1. In this section of the embryo and endosperm of *Aesculus woerlitzensis*, a progressive breakdown of the cellular endosperm in the vicinity of the cotyledons is seen. (After List and Steward, 1965)

whether there is any significant interaction between the endosperm and early division phase embryos. For example, during the early stages of embryogeny in some Compositae such as *Galinsoga ciliata* (Harris, 1935), *Podolepis jaceoides* (Davis, 1961a) and *Minuria denticulata* (Davis, 1963) there is no indication of digestion of endosperm such as occurs later. Supply of nutrients from a donor to a receptor depends upon the production in massive quantities of substances which the latter does not possess or has in minimal amounts. Since rapid endosperm multiplication which itself may require substantial quantities of cellular constituents and precursors to synthesize nucleic acids, proteins and carbohydrates takes place coincident with the early division of the zygote (Schulz, P. and Jensen, 1969, 1974; Newcomb, 1973b), it is unlikely that the limited resources of the endosperm will be used for fostering embryo growth at this stage. In all likelihood, early division phase embryos may use metabolites provided by the suspensor, persistent and degenerating synergids, antipodals and central cell, and a switchover to endosperm dependency occurs around the heart-shaped stage.

C. Endosperm Haustoria

A remarkable feature of the endosperm in some plants is the presence of haustoria, which have been the subject of considerable research at the morphological level. They attain complex morphology, appearing as branched, tubular processes which protrude into the nucellus or integuments. In recent years, organization of the haustoria has been most successfully studied by means of whole mounts of the endosperm, which facilitate the isolation of the various parts intact. A common grouping of the haustoria is based on their position in relation to the embryo sac. This classification divides haustoria into micropylar (that is, located at the micropylar end of the embryo sac) and chalazal (that is, originating from the chalazal end). More often, both micropylar and chalazal haustoria may be found on the same endosperm. References to a few carefully studied examples in each group will be sufficient to communicate the great variety and infinite complexity of the endosperm haustoria. The elaborate patterns of the haustoria raise many questions of importance concerning the nutritional mechanism of the embryo.

Micropylar Endosperm Haustoria

The best known example of a micropylar haustorium is that described in *Impatiens roylei* (Dahlgren, 1934). Here both micropylar and chalazal chambers may be distinguished after the first division of the primary endosperm nucleus. Subsequently, the micropylar cell divides into three small cells. The haustorium arises from the uppermost of these cells. For a variable length of time, the haustorium remains as a tubular structure until it ramifies extensively in the micropylar region of the embryo sac, even extending into the funiculus (Fig. 5.2 A). In *Thunbergia alata* (Mohan Ram and Wadhi, 1964), the primary micropylar haustorium is relatively undeveloped, and its place is taken up by six to seven tube-like secondary haustoria (Fig. 5.2 B). In *Blepharis maderaspatensis* (Phatak and Ambegaokar, 1963), the micropylar haustorium is extremely reduced and single-celled.

Chalazal Endosperm Haustoria

Since the last compilation by P. Maheshwari (1950), a number of plants have been shown to possess chalazal endosperm haustoria (Chopra and Sachar, 1963; Maheshwari, P. and Kapil, 1966; Masand and Kapil, 1966). The simplest forms of chalazal haustoria are found in Nymphaeaceae and Araceae: here the chalazal cell formed after the first division of the primary endosperm nucleus functions directly as a

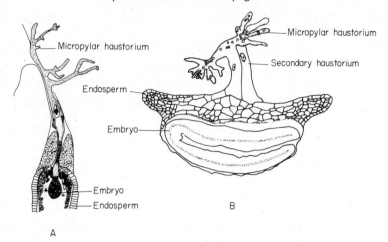

FIG. 5.2. Micropylar endosperm haustoria. **A**, In *Impatiens roylei*, the haustorium ramifies in the upper part of the ovule in the form of tubular structures. (After Dahlgren, 1934) **B**, In *Thunbergia alata*, tubular secondary haustoria arise from the micropylar haustorium. (After Mohan Ram and Wadhi, 1964)

haustorium (Maheshwari, P., 1950). In *Leptomeria acida* (Santalaceae), although the haustorium remains uninucleate, it becomes long enough to reach the base of the ovule and establish contact with the vascular supply of the placenta (Ram, 1959b) (Fig. 5.3A). In *Opilia amentacea* (Opiliaceae), the haustorium may even reach the extragynoecial tissue such as the pedicel (Swamy and Rao, 1963). In several members of the Leguminosae and Cucurbitaceae (Chopra and Agarwal, 1958; Johri and Garg, 1959; Chopra and Sachar, 1963; Chopra and Basu, 1965), the chalazal cell becomes coenocytic or infrequently cellular, and assumes a tubular or bulbous shape (Fig. 5.3 B). In this form it digests the persistent nucellus, and even encroaches upon the chalazal tissue. In *Macadamia ternifolia* (Proteaceae) (Kausik, 1938), the chalazal part of the endosperm containing free nuclei forms several lobes which appear as tubular sheaths in whole mounts (Fig. 5.3 C). These structures invade and maintain a haustorial relationship with the cha-lazal tissues of the ovule until food materials therein are exhausted. Less ramified, but more aggressive, is the haustorium observed in *Grevillea robusta* (Proteaceae) (Kausik, 1942). In the early stages of its growth, it appears as a coiled tubular structure containing many nuclei. Later, wall formation takes place around the nuclei and a new cellular tissue which constitutes a secondary endosperm is formed. In *Cansjera rheedii* (Swamy, 1960), the primary chalazal haustorium ramifies around the

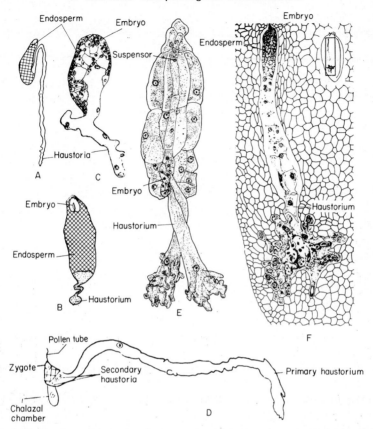

FIG. 5.3. Six representative species showing chalazal endosperm haustoria. **A**, The long, uninucleate haustorium of *Leptomeria acida*. (After Ram, 1959b) **B**, The haustorium of *Cassia sophera* is coenocytic and bulbous at the base. (After Johri and Garg, 1959) **C**, Tubular, sheath-like haustorium in *Macadamia ternifolia*. (After Kausik, 1938) **D**, The primary and secondary haustoria of *Comandra umbellata* originate from the caecum. (After Ram, 1957) **E**, The irregularly lobed bases of the two chalazal haustoria of *Exocarpus sparteus*. (After Ram, 1959a) **F**, The endosperm and its haustorium in *Nothapodytes foetida*. Note the unusual outline of the nucleus. Inset is a section of the entire ovule indicating the area enlarged. (After Swamy and Ganapathy, 1957)

vascular tissues of the ovule. Secondary haustoria differentiate from the cells of the endosperm that lie in close proximity to the primary haustorium. These cells elongate and form tubes toward the base which establish contact with the neighboring cells and with the primary haustorium. Endosperm haustoria of *Iodina rhambifolia* present a similar picture of complexity by coralloid branching of the primary chalazal haustorium (Bhatnagar and Sabharwal, 1966).

In *Comandra umbellata* (Santalaceae), the haustorium is long and invades the tissues of the chalazal region rapidly. The haustorium has its origin in a lateral outgrowth of the embryo sac called caecum (Ram, 1957). The first division of the endosperm nucleus is followed by a wall formed at the junction of the caecum and the embryo sac proper. The caecum divides again and the two cells formed lead to the formation of primary and secondary haustoria (Fig. 5.3 D).

The final form of haustorium in another member of the Santalaceae, *Exocarpus sparteus*, is shown in Fig. 5.3 E. Here the chalazal cell formed by the first division of the endosperm nucleus undergoes a longitudinal division. The two cells thus formed elongate enormously and grow into independent haustoria which produce numerous finger-like processes at the basal end (Ram, 1959a). In *Quinchamalium chilense* (Agarwal, 1962), the chalazal cell divides by two successive divisions to form four isobilaterally arranged cells which give rise to endosperm haustoria.

Swamy and Ganapathy (1957) have described a new type of endosperm haustorium in *Nothapodytes foetida* (Icacinaceae) and have contributed interesting observations on the interaction between the haustorium and the cells of the chalazal region of the ovule. The haustorium has its origin in the chalazal endosperm cell which enlarges to produce narrow branches. The continued growth of the haustorium is accomplished by contact between the haustorium initial and an adjacent cell of the chalazal tissue of the ovule. The outcome of this union is a transfer of the cytoplasm and nucleus of the chalazal cell into the haustorium. At the same time, the nucleus of the haustorium assumes a bizarre pattern as it streams into its branches (Fig. 5.3 F). These observations are of great importance from the point of view of function of the endosperm haustorium; they show that at some stage of its development the haustorium is enriched by the protoplasm from the adjacent chalazal cell, and by this means its function is deeply entwined with the nutrition of the embryo.

Endosperm with Micropylar and Chalazal Haustoria

Certain species belonging to Acanthaceae, Hydrophyllaceae, Lobeliaceae, Lentibulariaceae, Scrophulariaceae, Globulariaceae, Campanulaceae, Stylidiaceae, Verbenaceae and Loasaceae have endosperm tissues in which both chalazal and micropylar haustoria are present (Maheshwari, P., 1950; Subramanyam, 1960b; Masand and Kapil, 1966). Endosperms with fully developed chalazal and micropylar haustoria lose their original outline and look like grossly deformed structures which show great variations in their organization and behavior.

In Fig. 5.4 A is shown the endosperm of *Rhamphicarpa longiflora* (Scrophulariaceae) (Iyengar, 1942). Here both chalazal and micropylar haustoria are very aggressive structures which digest their way through the nucellar tissues and end up in close proximity to one another. A variation of this theme is found in *Melampyrum lineare* (Arekal, 1963) and *Alectra thomsoni* (Vijayaraghavan and Ratnaparkhi, 1972), both belonging to Scrophulariaceae, where the micropylar haustoria are extensively branched but the chalazal ones consist of single unbranched

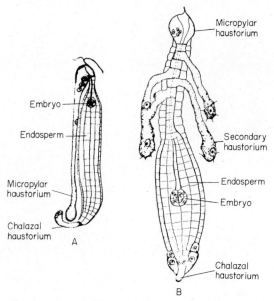

FIG. 5.4. **A**, Chalazal and micropylar endosperm haustoria of *Rhamphicarpa longiflora*. **B**, In *Centranthera hispida* both micropylar and chalazal haustoria are not well developed. However, conspicuous tubular secondary haustoria arise from the endosperm cells. (After Iyengar, 1942)

cells. Both establish contact with the vascular tissues of the ovule. In *Centranthera hispida*, also of Scrophulariaceae (Iyengar, 1942), both micropylar and chalazal haustoria are weakly developed single cells. Haustorial activity is due to the presence of several tubular, unbranched, uninucleate structures which arise from the endosperm cells just beneath the micropylar haustorium. They function as secondary haustoria and digest their way into the integuments (Fig. 5.4 B).

Subramanyam (1949, 1950, 1953) has investigated the development of the endosperm in some members of Campanulaceae, Lobeliaceae and Stylidiaceae. In two representatives of Stylidiaceae, *Levenhookia dubia*

(Subramanyam, 1950) and *Stylidium graminifolium* (Subramanyam, 1953), both micropylar and chalazal haustoria are very active and develop long tubular processes which grow in between the cells of the integument. Although well developed micropylar and chalazal haustoria are present in *Downingia* (Campanulaceae), they become inactive

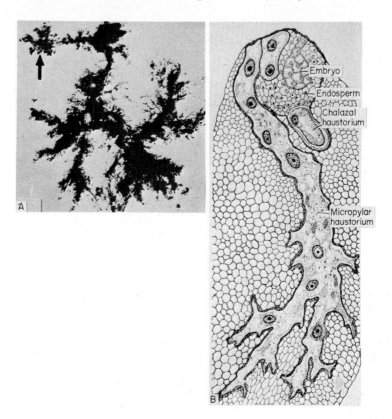

FIG. 5.5. **A**, Whole mount of a mature haustorium of *Avicennia officinalis*. Arrow indicates the micropylar end. **B**, Early stage in the development of the haustorium. (After Padmanabhan, 1964; photograph courtesy of D. Padmanabhan)

and hypertrophied early during embryogeny, raising some questions about the timing of haustorial activity in this plant (Kaplan, 1969).

The endosperm of *Avicennia officinalis* (Verbenaceae) is almost a classic example of a complex haustorium (Padmanabhan, 1964). This is due to the presence of an extensively branched micropylar haustorium which densely penetrates the tissues of the ovule, especially the placenta

(Fig. 5.5). The ultimate branches of the haustorium establish contact between themselves and with the cells of the endosperm through plasmodesmata-like connections. While all these processes are occurring the endosperm expands continuously and grows out of the ovule,

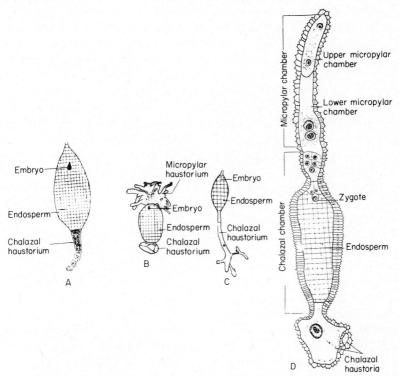

FIG. 5.6. Variations in the form and structure of the endosperm haustoria in members of the Loasaceae are shown in these figures. **A**, Unbranched chalazal haustorium of *Mentzelia laevicaulis*. **B**, The balloon-shaped chalazal haustorium, and the highly branched micropylar haustorium of *Loasa bergii*. **C**, The base of the chalazal haustorium of *Cajophora silvestris* is multicellular and narrow, whereas the terminal part is unbranched. (After Garcia, 1962b) **D**, Micropylar and chalazal haustoria of *Blumenbachia hieronymi*. Branches are seen on the latter. (After Garcia, 1962a)

carrying the embryo with it. In this species activity of the chalazal haustorium is confined to a short period before the micropylar one begins to ramify.

The endosperm haustoria of members of the Loasaceae appear to be variable in form and structure (Garcia, 1962a, b). The simplest form is found in *Mentzelia laevicaulis* where both micropylar and chalazal haustoria are unbranched (Fig. 5.6 A). In *Loasa bergii* there is a coenocytic and highly branched micropylar haustorium and a balloon-shaped uninucleate chalazal haustorium (Fig. 5.6 B). In *Cajophora silvestris* the base of the chalazal haustorium is multicellular and narrow, whereas the terminal portion is extremely branched; in comparison, the micropylar haustorium is coenocytic and unbranched (Fig. 5.6 C). Another variation is found in *Blumenbachia hieronymi* where the micropylar haustorium consists of two binucleate cells. The uninucleate chalazal haustorium is, however, more active and it invades the tissues of the integument by means of small branches (Fig. 5.6 D).

Mohan Ram and Wadhi (1964) have undertaken a systematic study of endosperm development in the Acanthaceae. No better example illustrates the range of variations in the structure of the micropylar and chalazal haustoria within members of a single family than the Acanthaceae, although rarely do the haustoria attain the complexity described earlier in some other species. In the majority of species investigated, the micropylar haustorium is binucleate, tubular and unbranched and persists in the mature seed. In contrast, the chalazal haustorium is short-lived, and usually degenerates before the embryo becomes self-supporting. In *Andrographis serpyllifolia*, *A. echioides* and *Haplanthus tentaculatus*, unbranched secondary haustoria arise from the peripheral cells of the cellular endosperm and digest their way into the integuments. In *Asystasia gangetica*, secondary haustoria originate as outgrowths of the free nuclear endosperm and become coenocytic.

Because of its intimate association with the ovular tissues on the one hand, and with the endosperm on the other, there is little question that the haustorium has a key role as an intermediary between the embryo and its nutrient supply. The question is how this happens. Does it play an active role by transporting soluble food from the ovular tissues to the endosperm? The presence of plasma membrane-lined wall ingrowths on the haustorium at the point of its contact with the ovule tissues as described briefly by Torosian (1971), seems to confirm its postulated role in the polarized transport of solutes. By what mechanism do cells disintegrate during ramification of the haustorium? Extraordinary though some of the haustoria may be, are they merely a natural consequence of a rapidly growing endosperm tissue with its initial endowment of nutrient substances? Finally, there is considerable uncertainty regarding the time at which the endosperm haustorium becomes active. Does it remain active throughout the life of the endosperm, or is its activity closely attuned to the physiological and developmental needs of the embryo at a particular stage, as, for example, when the basic nutritional support of the endosperm is exhausted? Clearly, such an important assumption as the role of the endosperm haustorium should be placed on a sound basis and it is probably within the range of current cytochemical and cytological techniques to establish these points.

D. Suspensor Haustoria

In an earlier chapter, the structure of the suspensor cells of some plants was described. Modifications of the walls of the suspensor cells as seen in the electron microscope have provided essential support for the belief that they function in the absorption and translocation of nutrients from the surrounding tissues to the embryo proper. Nutrient absorption is also thought to be accomplished by the embryo, which is brought by the suspensor into favorable position in the embryo sac with regard to food supply. The suspensor is seen in this role most dramatically during post-fertilization development in several members of the Loranthaceae (Maheshwari, P. and Johri, 1950; Maheshwari, P., and Singh, 1952; Maheshwari, P. et al., 1957; Singh, 1952; Narayana, 1954; Johri et al., 1957). Here, fertilization and polar fusion are accomplished in the stylar canal. Prior to division of the zygote, the primary endosperm nucleus travels down to the lower part of the embryo sac situated in the ovule and multiplies there to form a cellular endosperm. The first few divisions of the zygote take place in the stylar canal to form a long biseriate proembryo, the terminal cells of which function as the embryo proper. The rest of cells of the proembryo representing the suspensor elongate considerably to push the embryo into the vicinity of the endosperm. The distance to which the suspensor has to push the embryo through the stylar canal can be considerable—for example, according to Johri et al. (1957), in *Helicanthes elastica* the shortest organized embryo sac measured 4 mm while the longest measured 16 mm in a style 35 mm long.

This general outline of the role of the suspensor in pushing the embryo into the vicinity of the endosperm sets the stage for consideration of the remarkable haustorial modifications of suspensors. The suspensor haustoria in angiosperms exhibit a diversity in architectural design rarely encountered in any other part of the embryo. Detailed accounts of the numerous modifications of the suspensor are given by P. Maheshwari (1950), Subramanyam (1960b) and Masand and Kapil (1966), and therefore only a few examples have been selected for descriptive purposes.

Suspensor haustoria are present mainly in members of the Rubiaceae, Leguminosae, Haloragaceae, Tropaeolaceae, Fumariaceae, Crassulaceae, Orchidaceae and Trapaceae. Although several modifications of the suspensor have been described in the Leguminosae (Maheshwari, P., 1950), it is doubtful whether they are functionally effective. In some species of *Crotalaria* (Rau, 1950), however, the peripheral cells of the multicellular suspensor put out tubular processes which function as haustoria (Fig. 5.7 A). In *C. retusa* and *C. striata*, they extend as far as into the nucellus and integuments, respectively. In *Asperula azurea* (Lloyd, 1902) and *Rubia cordifolia* (Venkateswarlu and

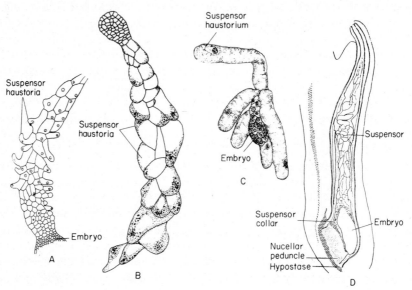

FIG. 5.7. Suspensor haustoria in angiosperms. **A**, In *Crotalaria retusa* haustoria are in the form of tubular structures from the suspensor cells. (After Rau, 1950) **B**, In *Rubia cordifolia* the suspensor haustoria are in the form of vesicular cells. (After Venkateswarlu and Rao, 1958) **C**, Long, tubular suspensor haustoria in *Cymbidium*. (After Swamy, 1942) **D**, The long, coiled, multicellular suspensor and suspensor collar of *Trapa bispinosa* in section. (After Ram, 1956)

Rao, 1958), members of the Rubiaceae, micropylar cells of the suspensor form several vesicles (Fig. 5.7 B). Unlike the haustorial processes of *C. retusa* and *C. striata*, the physiological activity of these structures is limited and is centered around the endosperm cells.

Haustorial activity is one of the outstanding physiological and developmental attributes of the suspensor in the Crassulaceae. In *Sedum ternatum* and *S. ochroleucum* (Subramanyam, 1960b), the basal cell of the suspensor enlarges and assumes a vesicular shape. Subsequently, thin-walled tubular extensions arise from the vesicle and invade the nucellus and integuments. In *S. ochroleucum*, these structures actively digest the cells of the inner integument and invade the raphe where they grow intracellularly. It may not be too soon to see in the suspensor haustoria considerable potential as organs for drawing nutrition from the somatic cells of the ovule and channeling them to the embryo.

In species belonging to the Orchidaceae, Trapaceae and Podostemaceae where the endosperm is completely suppressed right from the early stage, suspensor haustoria show a much greater range of growth and form than that exhibited by plants which possess a normal endosperm. In the orchid genera, *Eulophia* and *Cymbidium*, the zygote divides to form an irregular mass of six to ten cells. As seen in Fig. 5.7 C, some of these cells enlarge and form long fluffy structures which extend in various directions and come in contact with the integuments (Swamy, 1942, 1943).

In *Habenaria* (Orchidaceae), the suspensor grows out of the embryo sac and functions as an intraovular haustorium. Its free cell reaches the placenta at an early stage in development, and penetrates this tissue by the outgrowth of haustoria (Swamy, 1946). In *Trapa bispinosa* (Trapaceae) (Ram, 1956), the suspensor cell divides repeatedly by transverse and longitudinal walls to form a long, coiled, multiseriate haustorium which pushes the embryo deep into the embryo sac cavity. In addition, a collar-like layer of cells which has its origin in the suspensor covers the embryo on the side which faces the bulk of the nucellus (Fig. 5.7 D). The peripheral cells of the collar appear to be densely cytoplasmic and this observation has been cited in support of its haustorial role.

E. Synergids and Antipodals

Some questions have been raised with regard to the role of synergids and antipodals in the nutrition of the zygote and embryo in early division phases. A recurrent idea is that these specialized cells absorb nutrients from the nucellus and endosperm and channel them to the zygote or to the growing embryo. Indicative of an absorptive capacity of synergids and antipodals are findings by means of staining reactions of the similarity of compounds in these cells with those in the immediately surrounding endosperm cells (Perotti, 1913; Buell, 1952). The behavior of antipodals in the hybrid *Hordeum jubatum* × *Secale cereale* has led Brink and Cooper (1944) to suggest that they function in channeling nutrient materials for the initial multiplication of the endosperm nucleus. These findings have been supported in recent years by the discovery of several ultrastructural features in the synergids and antipodals designed to facilitate absorption and translocation of metabolites (see Chapter 2). Finally, although synergids and antipodals disappear soon after fertilization, in a number of species they persist until the development of the embryo is well under way and apparently function as haustoria. Examples of plants with persistent synergids are species of *Allium, Nothoscordum, Limnanthes, Albucus* and some members of the Cucurbitaceae and Compositae. Sometimes synergids increase in size and even extend beyond the limits of the embryo sac into the micropyle and funiculus as seen in some species of Compositae and Santalaceae (Maheshwari, P., 1950; Subramanyam, 1960a; Davis, 1961b, 1962; Johri, 1962b). A particularly striking example of synergid haustoria is that described in *Quinchamalium chilense* (Santalaceae) by Agarwal (1962). Here, even before fertilization, tips of synergids elon-

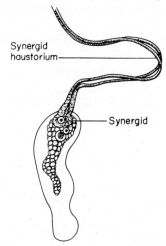

Synergid
haustorium

Synergid

Fig. 5.8. In *Quinchamalium chilense*, both synergid and antipodal haustoria reach large proportions. Only synergid haustoria are shown. (After Agarwal, 1962)

gate into tubular extensions which grow into the stylar tissues, reaching to about one-third the length of the style (Fig. 5.8). In this plant, the antipodal nuclei do not organize into individual cells, but form part of an "antipodal chamber", cut off from the embryo sac by a transverse wall. The tip of this cell elongates and traverses up to the placenta where it produces haustorial branches. Perhaps in no other angiosperms do the synergid and antipodal haustoria reach such proportions.

According to Venkateswarlu and Rao (1958), in *Rubia cordifolia* the basal antipodal cell elongates to form a tube which penetrates the chalazal part of the ovule. It is thought that the early growth of the embryo is due in part to the transmission of nutrients from the nonfunctional megaspore mother cells and arrested embryo sacs through the antipodal tube.

Persistent antipodals without any obvious haustorial structures are found in *Argemone mexicana* (Sachar, 1955), *Eschscholtzia californica* (Sachar and Mohan Ram, 1958), both belonging to the Papaveraceae, and *Caltha palustris* (Ranunculaceae) (Kapil and Jalan, 1962). In both *A. mexicana* and *E. californica*, antipodals are relatively large because of high metabolic activity. In the former, the depletion of the contents of the nucellus lying immediately beneath the antipodals is probably indicative of their haustorial role. In *C. palustris*, although uninucleate at first, the antipodals acquire a high degree of polyploidy by the division of the primary nucleus and subsequent fusion of the daughter nuclei. Polyploidization of synergids and antipodals recorded in other plants too (Johri, 1962b; Masand and Kapil, 1966) might also have significance in the nutrition of the young embryo.

F. Hypostase, Epitase and Chalazal Proliferating Tissue

Localized outgrowths in the ovule which do not differentiate into any distinct morphological structures may sometimes play a haustorial role in the nutrition of the early embryo in a way comparable to antipodals and synergids. Examples are proliferations of the nucellus at the chalazal and micropylar ends of the ovules of some species, designated as hypostase and epitase, respectively. By the evident position of these tissues, especially of the hypostase, it would seem natural to suggest their similarity to glandular tissues concerned with secretion of enzymes or growth hormones (Maheshwari, P., 1950). Curiously enough, the hypostase does not function in the transport of metabolites because in an autoradiographic analysis of ovules of *Zephyranthes drummondii* fed with $^{14}CO_2$, so little label appeared in this structure, although incorporation was high in the chalazal region (Coe, 1954).

A structure resembling hypostase has been described in the ovule of *Quercus gambelii* under the name "postament" (Brown and Mogensen, 1972). It is composed of a central column of elongate cells at the chalazal end of the embryo sac which extends into the funiculus, and maintains connection with the vascular supply of the ovule. It has been suggested that the postament has an important function in channeling food materials from the outer integument into the embryo sac (Mogensen, 1973).

Pollock and Jensen (1967) have described the ontogeny of a group of cells which proliferate from the nucellar cells at the chalazal end of the ovule of *Capsella bursa-pastoris*. At the fine·structural level (Schulz, P. and Jensen, 1971), these cells have revealed subcellular features in the cytoplasm designed for high metabolic activity, and in the cell wall for purpose of absorption of metabolites from neighboring cells. These observations have suggested a trend of thought, to be verified later, that the chalazal proliferating tissue may function in the absorption and storage of nutrient substances for the use of the developing embryo and endosperm.

In retrospect, we find that morphological adaptations in the ovule for the efficient nurture of the embryo are impressive. The array of structural modifications in the same plant for nutrition of the embryo may raise legitimate questions regarding the relative roles of the haustorial processes and the endosperm itself. Earlier in this chapter, we referred to the fact that the fertilized egg might contain enough stored metabolites for the first few cycles of cell division. Since synergids, antipodals and localized nucellar outgrowths do not survive beyond the heart-shaped stage of the embryo, it might be assumed that metabolites acquired by their respective haustoria may sustain the embryo until the endosperm is able to function as a nutrient source. It is generally agreed that during the major part of its ontogeny the embryo is nourished by food materials stored in the endosperm. By applying the above facts to embryogenesis, one can envision that progressive differentiation of the embryo in some plants depends upon a sequential availability of nutrients from different sources.

II. GYMNOSPERMS

In gymnosperms, the female gametophyte is formed from the functional megaspore which is generally retained on the sporophyte. Its nucleus undergoes repeated divisions to form a nutritive tissue. Perhaps in all species the megagametophyte goes through a free nuclear phase, before wall formation takes place to form the cellular tissue. The number of free nuclei formed and the extent of the free nuclear phase vary in different species. The cellular megagametophyte can be considered equivalent to the endosperm, but in contrast to the latter, the gymnosperm tissue is formed before fertilization and has a haploid chromosome number. The egg cell is deeply seated in the archegonium generally restricted to the micropylar end of the gametophyte, and during its weaning period the embryo is supplied with nutrients from the latter. For an account of the development of the female gametophyte in gymnosperms spanning the period from the uninucleate megaspore to the appearance of archegonia, the reader is referred to reviews by P. Maheshwari and Singh (1967) and Konar and Oberoi (1969).

Several of the early investigators in gymnosperm embryology were impressed by what seemed to be the great variability in the form of the suspensors. In many species the suspensor elongates rapidly and pushes the embryo deep into the gametophytic tissues. In *Pinus* and other conifers, embryology is complicated by the presence of secondary suspensors or embryonal tubes which serve a haustorial function (see Roy Chowdhury, 1962, for review). The appearance of the suspensor complex in a conifer genus is given in Fig. 5.9. Because of the pressure exerted by the megagametophyte to the rapid elongation of the suspensor, the latter is often seen as a coiled and sinuous structure. The embryo proper is generally located at the tip of the suspensor. In *Gnetum ula* (Vasil, 1959) and *G. gnemon* (Sanwal, 1962), the zygote divides into two cells by a transverse wall and each cell gives rise to long, branched

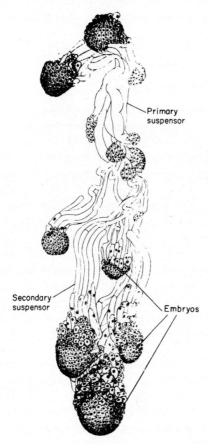

FIG. 5.9. The appearance of the embryo–suspensor complex in *Torreya californica*. Both primary and secondary suspensors produce embryos by budding. (After Buchholz, 1940)

tubes which ramify through the gametophyte and even extend into the nucellus. These tubes are comparable to the primary suspensor tubes in their function.

Since the egg and the innermost layer of cells of the megagameto-phyte share a common wall, soluble food materials can pass into the egg through the unthickened portions of the wall or through pit connections (Corti and Maugini, 1964; Maugini and Fiordi, 1970). The ability of the embryo to produce digestive enzymes has repeatedly been postulated in several species and does in fact play a cardinal role in the rapid digestion of nutrients. There is good reason to believe that absorption of the digested nutrients is facilitated by invaginations of the plasma membrane of the cells of the embryo (Maugini and Fiordi, 1970). It is thus clear that association of the embryo with the mega-gametophyte is a precise relationship involving a delicate mechanism for the nutrition of the embryo.

III. PTERIDOPHYTES

A brief consideration of the body plan of the gametophyte (prothallus) in pteridophytes which provides nourishment for the developing embryo obviously falls within the general framework of this discussion. It has its origin in a single cell, but the final form is somewhat unpredictable. In the simplest and most primitive pteridophytes represented by *Psilotum* and *Tmesipteris*, the prothalli are saprophytic, irregularly branched, subterranean bodies packed with mycorrhizal fungi and devoid of chlorophyll. Mycorrhizal association is also characteristic of the prothalli of several species of *Lycopodium*, and some members of the Ophioglossales, although surface-living prothalli are green and photosynthetic. In members of the family Marattiales, mycorrhizal growth occurs irrespective of whether the prothalli are green or not. The capacity for survival for a number of years is an important, but as yet unexplained, feature of the gametophytes of some species.

In several species of *Selaginella*, and in members of the Marsiliaceae and Salviniaceae, the prothalli are highly reduced, and limited to the confines of the megaspore or part of it. On the other hand, relatively large thalloid gametophytes are characteristic of Equisetales, and a majority of the leptosporangiate ferns. Although green and photosynthetic, some of them become mycorrhizal at later stages of development. While it is reasonable to assume that green and photosynthetic prothalli are auto-trophic in their nutrition, mycorrhizal association suggests the existence of mechanisms for the permeation of organic molecules into the pro-thallial cells, which may be utilized in significant amounts for growth.

In *Psilotum* and *Tmesipteris*, the zygote divides by a transverse wall to form an epibasal segment which gives rise wholly to the shoot, and a hypobasal segment which forms wholly a haustorial organ, the foot (Holloway, 1917, 1921, 1939). The foot, which often displays a remarkable ability for adaptation, is in contact with the tissues of the gametophyte through special haustorial processes (Fig. 5.10). These consist

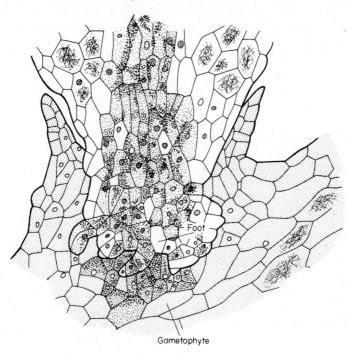

Gametophyte

Fig. 5.10. The foot region of an embryo of *Tmesipteris* at the point of attachment to the gametophyte shows elongate, round haustorial outgrowths. (As illustrated by Wardlaw, 1955, after Holloway, 1917)

of a single row of cells, which elongate and round off at the point of contact with the gametophyte, seemingly functioning as absorbing organs. In *Lycopodium* (Bruchmann, 1910), *Selaginella* (Bruchmann, 1912, 1913) and *Phylloglossum* (Thomas, 1901; Sampson, 1916), the path of nutrients from the prothallus is by way of the suspensor and foot to the apex of the embryo. The foot is also the characteristic organ mediating in the conduction of food materials from the prothallial cells in the majority of eusporangiate and leptosporangiate ferns (Bower, 1926, 1928).

The question of the precise role of the foot as the absorbing organ of the embryo has been controversial for some time. While it is possible that during the initial stages of development of the embryo the entire embryonic surface absorbs nutrients, the indications are that in comparatively older embryos the main absorptive function is delegated to the foot. To this general situation it may be added that in several species the presence of starch in the cells of the gametophyte, and its appearance in the embryonic foot after it has established contact with the former, may imply its transfer from the gametophyte to the sporophyte. It is reasonable to assume that the transfer is accomplished by secretion of enzymes which digest starch, and diffusion of the digested material

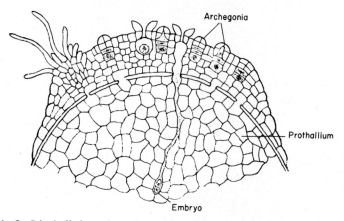

FIG. 5.11. In *Selaginella kraussiana*, the archegonial cavity is in the form of an elongate, tube-like structure which helps in the digestion of prothallial tissues. (After Bruchmann, 1912)

into the embryo where they are again resynthesized into complex molecules. In those species which have both suspensor and foot, it appears that the actual absorption is carried out by the foot while the suspensor helps to push the embryo deep into the prothallial tissues. In *Selaginella poulteri*, the suspensor not only thrusts the embryo into the prothallial tissues, but its distal end also secretes enzymes for the digestion of the nutritive cells (Bruchmann, 1912).

Bruchmann (1912, 1913) has shown that, in some other species of *Selaginella*, penetration of the developing embryo into the gametophytic tissues is accomplished in yet another way. In *S. kraussiana*, immediately after the division of the zygote, the archegonial cavity grows rapidly and penetrates deeply into the prothallus as an elongate tube-like structure (Fig. 5.11). No evident suspensor is present, but its function is taken up by the tubular structure referred to above. In *S. galeottei*, a similar sinuous tubular structure which helps in the digestion of the prothallial tissue during its passage is found.

Are differences in the developmental capacities of the egg and embryo paralleled by the distribution of metabolites? Wardlaw (1955, 1965a) has discussed at length the view that characteristic stages in the embryogenesis of pteridophytes are in some way attained under the guidance of metabolites that can be mobilized from the prothallial system. According to him, the typical alignment of the embryonic axis in the prothallus with the epibasal segment directed towards the prothallial apex may be related to the basipetal gradient of auxin along the prothallus. Perhaps the demonstration of the presence of auxin in a basipetal gradient in the prothallus (Albaum, 1938) is significant in this connection. In the post-octant stage, formation of the root, stem and foot is due in part, though not entirely, to differences in the nature of the morphogenetic stimuli in each of the presumptive regions. These interpretations still lack a biochemical foundation, but they are meaningful in the light of the role of the foot as a conduit for metabolites for the embryo.

IV. COMMENTS

In the foregoing sections only the more important facts concerning the morphology of tissues nurturing the developing embryos have been presented, and indeed, it would be presumptuous in a single chapter to try to give more than a few illustrative examples. The evidence, as we have seen, suggests that continued growth of the embryo is closely linked with the availability of a source of nutrient supply, and there is every indication that nutrients supplied by the female gametophyte or other accessory tissues exert a greater control on growth of the embryo than the materials manufactured within it. This might simply be a matter of supply of essential metabolites in the quantity required by the growing embryo, or alternately, the nutritive tissue may channel specific substances not manufactured by the embryo. However, for a proper understanding of the nutritional relationships of developing embryos, one needs to know the type of nutrients available to them, and other physiological aspects of their nutrition; these considerations form the guidelines for the next chapter.

6. Nutritional Aspects of Embryogenesis— Physiological Considerations

Chapter 5 was devoted to an account of the morphological features of nutritive tissues associated with developing embryos of vascular plants and their more efficient modifications for the accumulation, storage and utilization of food materials. It was implied therein that predictions based on morphological data have to some extent been realized by subsequent biochemical analyses of the constituents of a limited number of endosperm tissues of angiosperms and megagametophytes of gymnosperms. Although the picture is far from complete, it offers a promising approach to the understanding of the functional activities of extraembryonal tissues in the growth of embryos based on an integration of our knowledge of their morphology and chemical composition. No published investigation has, however, provided unequivocal evidence to show in a simple and direct fashion that nutrient materials of the surrounding milieu are absorbed and metabolized by the embryo.

In morphological and biochemical investigations of tissues from which developing embryos derive their nutrition, a problem arises from the fact that development of these tissues is inextricably tied up with the growth of embryos which are the prime targets for their various metabolites. It means that the primary analysis of these tissues concerns events taking place in a group of cells influenced in some way by a growing embryo. The need with regard to experimental analysis of the endosperm of angiosperms or its equivalent tissue in gymnosperms is to isolate them so that they can be studied under controlled conditions in the laboratory. Once we can obtain a continuously growing, free nuclear or cellular tissue of the endosperm or of the megagametophyte and subsequently achieve control of its inherent morphogenetic and biosynthetic potentialities, new insights will be possible into the nutritional needs of developing embryos, particularly with regard to the nature of the metabolites contributed by the extraembryonal tissues. Such a study will be especially meaningful if the cultured tissue is known to possess a characteristic chemical component which can be followed in an embryo implanted on it. Thus, against the background of the morphology of nutritive tissues of embryos presented in the previous chapter, it will be the purpose of this chapter to examine the various physiological aspects of embryo nutrition. In the main, chemical characterization of the endosperm of angiosperms and the megagametophyte of gymnosperms will form the substance of the first part of the chapter and will be followed by an account of their growth in culture, their conversion into continuously growing tissues, and control of their morphogenesis.

I. ANGIOSPERMS

A. Constituents of the Endosperm

The complexity of the nutritional resources of the endosperm is best comprehended from chemical analysis. Below, available information on the composition of the endosperms of coconut, maize, horse chestnut and others is presented. This account is limited to endosperm tissues which have produced growth stimulation in cultured embryos or in explanted tissues. Admittedly, much information on the chemical nature of other endosperm tissues, not included here, is available.

If we try to create a general picture of the contribution of the endosperm to the nutrition of the embryo beginning with the first division of the zygote, we should consider the food materials of both cellular and free nuclear endosperms. Unfortunately, barring a few exceptions, it is difficult to obtain large amounts of material of the free nuclear endosperm of plants for routine biochemical analysis, and so most of our information on the composition of the endosperm is derived from the solid cellular endosperm.

Coconut Milk

Much of our understanding of the biochemical composition of endosperm has come from a biological assay based on the ability of the endosperm of certain seeds to induce growth in explanted tissues such as slabs of the secondary phloem of carrot removed from a standard distance from the cambium. The favored material for this study has been the liquid endosperm of coconut, known in the botanical literature as "coconut milk" or "coconut water". The biochemical specialization of coconut milk as a growth induction stimulus in explanted tissues was first recognized by van Overbeek *et al.* (1941a, 1942) when they showed that immature embryos of *Datura stramonium*, which failed to grow even in the most complex media tried, could be cultured *in vitro* by the addition of coconut milk. Although indications are that the activity resides in the liquid portion of young and mature nuts, some workers have shown that a partially purified concentrate of the solid endosperm is active in inducing growth in carrot root tissue (Mauney *et al.*, 1952). However, after a certain stage loss of nutrients from the milk and a rise in oil content of the solid endosperm occur (Gonzalez y Sioco, 1914).

What are the factors inherent in coconut endosperm that dramatically induce growth of embryos and other organs? According to van Overbeek (1942) coconut milk contains at least three factors: (i) embryo factor, (ii) leaf growth factor and (iii) auxin. These early observations have been supplemented with results from chemical fractionation which have shown that coconut milk is a conglomeration of several inorganic ions (McCance and Widdowson, 1940; Tammes, 1959), amino acids (Pradera *et al.*, 1942; Paris and Duhamet, 1953; Baptist, 1956, 1963; Steward *et al.*, 1961; Tulecke *et al.*,, 1961), organic acids (Tulecke *et al.*, 1961), vitamins (Vandenbelt, 1945), growth hormones (Paris and Duhamet, 1953; Radley and Dear, 1958; Zwar *et al.*, 1963; Loeffler and van Overbeek, 1964; Letham, 1968), sugars (DeKruijff, 1906; Dunstan, 1906; LaHille, 1920; Tulecke *et al.*, 1961), sugar alcohols (Pollard *et al.*, 1961) and other substances. Indeed, some of the substances may not be of any real value with regard to growth promotion in cultured tissues; for example, coconut milk can be freed from mineral salts electrophoretically without serious impairment of its growth-promoting activity in carrot tissue bioassay (Smithers and Sutcliffe, 1967). The composition of coconut milk is summarized in Table 6.I.

TABLE 6.I. Composition of coconut milk

Substance	Amount	Reference
1. Inorganic ions	(mg/100 g)	
Potassium	312	
Chlorine	183	
Sodium	105	
Phosphorus	37	1
Magnesium	30	
Sulfur	24	
Iron	0·10	
Copper	0·04	
2. Amino acids	(μg/ml of coconut milk)	
Aspartic acid	65	
Threonine	44	
Serine	111	
Asparagine and glutamine *ca.*	60	
Proline	97	
Glutamic acid	240	
Alanine	312	
Valine	27	
Methionine	8	
Isoleucine	18	
Leucine	22	2
Tyrosine	16	
Phenylalanine	12	
β-alanine	12	
γ-Aminobutyric acid	820	
Lysine	150	
Tryptophan	39	
Arginine	133	
Ornithine	22	
Histidine	trace	
Pipecolic acid	+	
Hydroxyproline	trace	
Glycine	13·9	3
Homoserine	5·2	
Cystine (g/100 g) dried protein	0·97–1·17	4

Substance	Amount	Reference
3. Other nitrogen compounds (μmol/ml— analysis of Philippine coconuts)		
Dihydroxyphenylalanine	+	
Ethanolamine	0·01	5
Ammonia	+	
4. Organic acids (meq/ml)		
Malic acid	34·31	
Shikimic acid, quinic acid etc.	0·57	3
Citric acid	0·37	
Pyrrolidine carboxylic acid	0·39	
5. Vitamins (mg/ml)		
Nicotinic acid	0·64	
Pantothenic acid	0·52	
Biotin	0·02	
Riboflavin	0·01	6
Folic acid	0·003	
Thiamine	trace	
Pyridoxine	trace	
6. Sugars (mg/ml)		
Sucrose	9·18	
Glucose	7·25	3
Fructose	5·25	
Mannitol	0·8%	7
7. Growth substances (mg/ml)		
Auxin	0·07	8
Gibberellin	+	9
Cytokinins	+	10

Miscellaneous substances (mg/l)

1,3-Diphenylurea	5·8	11
Leucoanthocyanin	+	12
Phyllococosine	+	13
Sorbitol	15	⎫
myo-Inositol	0·1	⎬ 14
scyllo-Inositol	0·5	⎭
Acid phosphatase	+	15
Diastase	+	16
RNA polymerases	+	17
Dehydrogenase	+	⎫
Peroxidase	+	⎬ 18
Catalase	+	⎭

+ Indicates presence, quantity not determined.

References in Table I

1. McCance, R. A. and Widdowson E. M. (1940). *Medical Research Council* (Britain) Special Report Series No. 235.
2. Steward, F. C. with Shantz, E. M., Pollard, J. K., Mapes, M. O. and Mitra, J. (1961). *In* "Synthesis of Molecular and Cellular Structure" (D. Rudnick, ed.) pp. 193–246. New York.
3. Tulecke, W., Weinstein, L. H., Rutner, A. and Laurencot, H. J. Jr. (1961). *Contr. Boyce Thompson Inst.* **21**, 115–128.
4. Pradera, E. S., Fernandez, E. and Calderin, O. (1942). *Am. J. Dis. Child.* **64**, 977–995.
5. Steward, F. C. with Shantz, E. M., Mapes, M. O., Kent, A. E. and Holsten, R. D. (1964). *In* "Régulateurs Naturels de la Croissance Végétale", pp. 45–58. Paris.
6. Vandenbelt, J. M. (1945). *Nature, Lond.* **156**, 174–175.
7. Dunstan, W. R. (1906). *Trop. Agric.* **26**, 377–378.
8. Paris, D. and Duhamet, L. (1953). *C. r. hebd. Séanc. Acad. Sci. Paris* **236**, 1690–1692.
9. Radley, M. and Dear, E. (1958). *Nature, Lond.* **182**, 1098.
10. Zwar, J. A., Kefford, N. P., Bottomley, W. and Bruce, M. I. (1963). *Nature, Lond.* **200,** 679–680; Letham, D. S. (1968). *In* "Biochemistry and Physiology of Plant Growth Substances" (F. Wightman and G. Setterfield, eds) pp. 19–31. Ottawa.
11. Shantz, E. M. and Steward, F. C. (1955). *J. Am. chem. Soc.* **77**, 6351–6353.
12. Steward, F. C. and Mohan Ram, H. Y. (1961). *Advances in Morphogenesis* **1**, 189–265.
13. Kuraishi, S. and Okumura, F. S. (1961). *Nature, Lond.* **189**, 148–149.
14. Pollard, J. K., Shantz, E. M. and Steward, F. C. (1961). *Pl. Physiol. Lancaster* **36**, 492–501.
15. Wilson, K. S. and Cutter, V. M. Jr. (1955). *Am. J. Bot.* **42**, 116–119.
16. DeKruijff, E. (1906). *Bull. Dept. Agric. Indes Neerl.* No. **4**, 1–8.
17. Mondal, H., Mandal, R. K. and Biswas, B. B. (1972). *Eur. J. Biochem.* **25**, 463–470.
18. Sadasivan, V. (1951). *Archs Biochem.* **30**, 159–164.

A polysaccharide fraction obtained from coconut milk has been shown to support growth of carrot tissue explants even better than the original liquid itself, although the active ingredients of this fraction have not been characterized (Ramakrishnan et al., 1957). According to the analysis of Pollard et al. (1961), the main fraction of coconut milk responsible for growth induction in cultured carrot root tissue explants is somewhat arbitrarily divisible into a neutral fraction and an active fraction. The neutral fraction contains the hexitols, myo-inositol and scyllo-inositol. The components of the active fraction are far from being completely identified, but this part contains several potent substances such as diphenylurea and purines which contribute to its effectiveness at very low dilutions (Shantz and Steward, 1952, 1955b). A major purine component of coconut milk is a zeatin riboside which might account for its high cytokinin activity (Letham, 1968). Purine-like substances which resemble cytokinins in bioassays have also been isolated from the solid endosperm (Shaw and Srivastava, 1964). A fairly recent analysis of coconut milk (van Staden and Drewes, 1974) has assigned a major role to the amino acid, phenylalanine, in initiating cell divisions. However, the validity of this claim should be considered against the finding that the potent fractions were tested for activity in soybean callus bioassay which is specific for a cytokinin and that attempts to mimic the effect of active fractions by the synthetic amino acid were unsuccessful. Thus in summary, it seems fair to state that although several attempts have been made to isolate the essential growth factors from coconut milk, characterization of the nutritional components made little headway by analytical methods, and thus far, no single compound or group of compounds have come to be accepted as being able to duplicate the major activity of coconut milk in its entirety.

Endosperm of Maize and Other Cereals

Because of the obvious interest in human and animal nutrition, comprehensive studies on the chemical composition of the endosperm of cereals have been made. It is, however, beyond the scope of this book to discuss in any extensive manner these studies, and the interested reader will find ample information in standard books on cereal chemistry. A few words will be said in a general way about growth induction stimuli in extracts of maize kernels which have been the subject of considerable research. The growth stimulation of carrot tissue explants by extracts of maize endosperm is one of the same order of magnitude as coconut milk (Nétien et al., 1951; Steward and Caplin, 1952). Endosperm of developing maize contains IAA (Avery et al., 1941; Haagen-

Smit *et al.*, 1945, 1946; Guttenberg and Lehle-Joerges, 1947; Farrar *et al.*, 1958), indolepyruvic acid (Stowe and Thimann, 1953) and the indole-reacting amino acid, tryptophan (Stehsel and Wildman, 1950). There are some conflicting reports on the question whether IAA occurs as a free acid or bound to another molecule in an inactive form. Present evidence tends to indicate that the auxin may occur in one of the following forms: (i) an ethyl ester (Redemann *et al.*, 1951), (ii) a sugar conjugate (Steward and Shantz, 1959; Srivastava, 1963a), (iii) an inositol ester (Nicholls, 1967) and (iv) an inositol–sugar conjugate (Labarca *et al.*, 1965). Presumably, IAA is bound in such a manner that it can be easily hydrolyzed to the active form. In a few varieties of maize analyzed, the free auxin content of the kernels was extremely low at the time of pollination, but it rose rapidly thereafter, reaching a peak in the "milk" stage, followed by a decline in the mature endosperm (Avery *et al.*, 1942b). Among other cereals, endosperm tissues of wheat, rye, barley and rice are known to contain auxins (Avery *et al.*, 1941, 1942a; Hatcher and Gregory, 1941; Hatcher, 1943, 1945; Sircar and Das, 1951); barley (Ziebur and Brink, 1951) and wheat (Augusten, 1956) endosperms also contain "factors", probably hormonal in nature, which support growth of embryos. An unidentified substance, "syngamin", known to stimulate growth of excised embryos (McLane and Murneek, 1952), and the hexitol, *myo*-inositol (Pollard *et al.*, 1961), are also present in maize endosperm. Recent workers have identified several purines in maize endosperm (Miller, 1961; Srivastava, 1963b; Letham, 1964, 1968; Steward and Shantz, 1959) including zeatin, the naturally occurring cytokinin (Letham and Miller, 1965).

The metabolic interactions of the endosperm also depend upon the presence of amino acids and related substances. Alanine, glutamic acid, aspartic acid, serine, valine, leucine, tyrosine, glycine, proline, glutamine and asparagine were present in the endosperm of maize at nearly all stages from soon after pollination to maturity of the grain (Duvick, 1952). The proportion of amino acids and amides in equivalent wet weights of the endosperm declined with advancing age of the grain (Table 6.II). Both crude extract of the endosperm and amino acids separated from it promote growth and protein synthesis in cultured embryos (Györffy *et al.*, 1955; Oaks and Beevers, 1964). The fact that heat-denatured wheat endosperm prevented the growth of an embryo implanted on it led Kikuchi (1956) to consider proteins as its active ingredients.

TABLE 6.II. Free amino acids (μmol/0·5 g fresh weight) of the endosperm of maize (cv. 'Gourdseed Dent') of different ages. (From Duvick, 1952)

Days after pollination	Alanine	Glutamic acid	Aspartic acid	Serine	Valine	Leucine	Tyrosine	Glutamine
16	0·32	0·18	0·08	0·10	0·07	0·00	0·00	0·24
18	3·09	1·09	0·53	0·95	0·18	0·15	0·10	0·89
19	1·89	0·58	0·37	0·59	0·21	0·12	0·05	0·78
24	1·32	0·44	0·48	0·49	0·18	0·14	0·09	0·46
27	0·76	0·25	0·37	0·25	0·14	0·10	0·09	0·06
29	0·42	0·28	0·43	0·27	0·14	0·09	0·07	0·16
31	0·56	0·38	0·38	0·28	0·18	0·15	0·08	0·07
43	0·33	0·14	0·22	0·14	0·07	0·05	0·04	0·03

Endosperm of Horse Chestnut

By far the most potent of the extracts in carrot tissue bioassay is the liquid content of the vesicular embryo sac of horse chestnut (*Aesculus woerlitzensis*) which contains both free nuclear and cellular endosperms. Similar liquid obtained from the more common *A. hippocastanum* is much less effective. The vesicle is formed as an outgrowth of the embryo sac at its chalazal end, and apparently draws towards it various metabolites for storage and later utilization. As in the case of coconut milk, the growth-promoting activity of *A. woerlitzensis* endosperm is divisible into a neutral fraction and an active fraction, the former consisting of *myo*-inositol equivalent to 10% of its dry weight. This compound might serve the same function here as it does in coconut milk and maize endosperm in that it interacts with other fractions to promote growth. One other compound which has been identified with certainty from horse chestnut endosperm is the auxin synergist, chlorogenic acid. An unusual compound belonging to the class of leucoanthocyanins has been isolated from the endosperm extract and shown to promote growth of carrot tissue slabs as effectively as coconut milk. Endosperm extract also shows auxin activity in biossay (Shantz and Steward, 1955a, 1964; Steward and Shantz, 1959; List and Steward, 1965).

Recent fractionation of the active fraction and its characterization have shown that growth-promoting activity is distributed between IAA and adenyl compounds on the one hand, and between *myo*-inositol and glycosides on the other hand (Shantz and Steward, 1968).

Endosperm of Datura

Although the semisolid endosperm of *Datura stramonium* promotes growth of isolated embryos of the same species, except for a brief report on the presence of auxin, no evidence for the existence of other growth hormones in the endosperm has been obtained (Rietsema *et al.*, 1953a; Rappaport, 1954). Like the embryo, the endosperm contains large quantities of proteins and oils, and a great variety of free amino acids, and it is possible that the interaction of these compounds induces growth in cultured embryos. Changes in the concentration of free amino acids and amides in the endosperm at various stages of develop-

TABLE 6.III. Free amino acids and amides in the endosperm of *Datura stramonium* at various stages of development. (From Rietsema and Blondel, 1959; concentrations in mmol/g fresh weight)

Age in days	12·5	15	20	27	33	40	50
Fresh weight of endosperm (mg)	1·5	1·2	1·4	1·9	4·3	6·2	7·4
Aspartic acid	2·6	5·0	1·2	0·9	1·2	1·9	1·9
Glutamic acid	3·0	3·7	3·6	2·1	0·8	0·5	0·5
Serine	0·9	2·7	1·1	0·4	0·2	0·1	0·6
Asparagine	0·7	0·5	—	+	0·6	2·2	1·5
Threonine	1·2	1·3	1·5	0·7	0·1	—	+
Alanine	0·9	9·9	8·5	1·9	0·3	0·1	0·2
Glutamine	6·7	4·6	5·8	0·3	0·5	0·3	0·7
Lysine	0·6	1·1	0·9	0·2	—	+	+
Arginine	3·2	3·2	6·3	0·4	—	—	—

ment are given in Table 6.III. The peak concentrations of aspartic acid, glutamic acid, serine, alanine, arginine and glutamine were attained around the third week after pollination when embryos were already self-supporting in culture, and growth of the endosperm had terminated; threonine and lysine occurred in insignificant amounts, and completely disappeared in older seeds (Rietsema and Blondel, 1959). A comparison of the results presented in Table 6.III with those presented in Table 3.I shows that the general patterns of changes in activity of most of the amino acids in the endosperm and embryo are similar. The establishment of identical patterns in free amino acid changes in tissues of different genetic constitutions is a finding of some interest, and seems to indicate that the amino acid composition recorded in the endosperm is representative of the seed as a whole.

Other Sources

Steward and Caplin (1952) have shown that the gelatinous endosperm of *Juglans regia* can support the growth of carrot tissue explants to the same extent as coconut milk, although the chemical composition of the tissue is not determined. According to Nitsch (1953), a heat-stable complex capable of causing proliferation in cultured tissues is present in the liquid endosperm of *Allanblackia parviflora*. Nakajima (1962) has shown that the endosperm of *Cucumis sativus* contains substances which support growth of excised embryos of the same plant. Extracts of the endosperm of *Prunus persica, P. amygdalus* and *P. armeniaca* exhibited activity on a level comparable with coconut milk in the carrot bioassay system (Trione *et al.*, 1971a, b). In the case of *P. persica*, the activity appears to be due to the presence of a cytokinin (Powell and Pratt, 1964). According to Takeuchi (1956) the free nuclear endosperm of *Phaseolus vulgaris* has free amino acids and soluble sugars. The early stage endosperm of ivory nut (*Hyphaene natalensis*) is a liquid, whose growth-promoting activity is attributed to a cytokinin (van Staden, 1974).

Embryo factors from seeds of *Lupinus luteus* furnishing materials for the growth of isolated embryos of *Datura tatula* are distributed in both the endosperm and embryo (Matsubara and Nakahira, 1965b). Both cytokinins and gibberellin-like substances have been separated from crude preparations of the endosperm by methods too intricate to go into detail here, but it appears unlikely that either of these substances is the specific embryo factor (Matsubara and Ogawa, 1963; Matsubara and Koshimizu, 1966; Matsubara *et al.*, 1966).

There is considerable evidence to show that the site of production of auxin in the seed is generally the developing endosperm. For instance, in apple (Luckwill, 1948, 1953), a minor peak in the concentration of free auxin appears when the endosperm changes from a free nuclear to a cellular state. A major peak in auxin activity is observed when the endosperm replenishes itself by new divisions to form secondary endosperm tissue. In the seeds of *Cassia fistula*, IAA was the major auxin component and was present in greater amounts in the endosperm of seeds of all ages than in embryos (Mukherjee *et al.*, 1966). In cereal grains, disintegration of the endosperm might liberate auxins which influence nuclear divisions in the embryo (Nutman, 1939).

Endosperm tissues of several angiosperms have been assayed to determine the presence of substances with gibberellin-like properties in them. Endosperms of *Aesculus californica, Echinocystis macrocarpa, Zea mays, Persea americana, Prunus domestica, P. armeniaca, P. amygdalus, Juglans califor-*

nica, J. regia, Lupinus succulentus, L. luteus and apple (Phinney *et al.*, 1957; Nitsch, 1958; Corcoran and Phinney, 1962; Ogawa, 1963a) have given positive results in the well known dwarf corn or dwarf pea bioassay for gibberellins.

An unappreciated aspect of the composition of endosperm from the point of view of its nutritive potential is its sugar content. The knowledge that high concentrations of sucrose might serve as osmoticum in the induction of growth in proembryos of some species raises the possibility that changes in sugar concentration of the endosperm may have important bearing on the extracellular control of embryo growth (Ryczkowski, 1962).

B. Storage Forms of the Endosperm

The foregoing analysis shows that growth hormones, amino acids and other metabolites represent important constituents of the endosperms of diverse plants. Although these substances provide the zygote and early division phase embryos with the appropriate nutritional support to initiate cell divisions, it will not be unreasonable to argue that continued growth and morphogenesis of the embryo are under control of substances released by biochemical changes in the stored reserves of the endosperm. Before we consider the mobilization of the reserves, we shall turn first to gain a general picture of the food materials of the endosperm present in special storage forms. The increasing use of the electron microscope and of histochemical techniques has led to a clearer understanding of the nature of these substances.

While it may appear convenient to speak of the food substances of the endosperm as being present in special storage form, one should not forget that they represent a heterogeneous mixture of several compounds. Further, the hard core of information on the storage materials of the endosperm has come from analysis of mature, quiescent seeds where there is no conceivable interaction between the endosperm and the developing embryo (Crocker and Barton, 1957; Miller, 1958).

Carbohydrates

The important storage materials of the endosperm are carbohydrates, proteins, fats and oils. The principal storage carbohydrate is starch, which is present in the form of grains in the chloroplast in green tissues and in the leucoplast in storage tissues. Starch grains as they exist in the plastids of cereal endosperms differ not so much in their origin as in their number, size, shape and markings. According to Buttrose (1960, 1963a), who has studied starch formation in cereal endosperms in the electron microsope, the starch granule is initiated at any point in the plastid stroma. In barley and wheat endosperm, each proplastid develops only one starch granule, in contrast to that of oat (*Avena sativa*) and rice which has many. In cereal grains where starch is the main storage product of the endosperm, its accumulation begins immediately after fertilization and continues up to maturity of the grain. This is also accompanied by a rapid decline in the concentration of reducing sugars (Jennings and Morton, 1963).

The endosperms of some plants such as date (*Phoenix dactylifera*) and *Phytelephas macrocarpa* have carbohydrate reserves stored as hemicelluloses which yield mainly mannose and other monosaccharides upon hydrolysis (Meier, 1958; Miller, 1958). Unlike starch which is stored inside the cell, hemicelluloses form the principal component of cell walls where they are laid down as heavy wall thickenings. There is considerable variability in the chemical characteristics of storage hemicelluloses of endosperms of different species. One such variation is the hemicellulose called amyloid which, like starch, stains well with iodine–potassium iodide (Kooiman, 1960).

Another variant is presented by some members of the Palmae, the endosperm of which is of special interest because of the presence of tannins. As an example may be cited arecanut (*Areca catechu*) which is used as a masticatory in the Orient and Middle East (Raghavan and Baruah, 1958). The endosperm of the nut owes its marked degree of astringency to the presence of tannins which occur in the form of amorphous granules in the cytoplasm, or adsorbed to the cell wall, or as loose covering around carbohydrate granules. Compared to the immature endosperm, cells of the fully mature endosperm contain fewer tannins which, moreover, occur dissolved in the cytoplasm. There is some evidence to indicate that some of the tannins, which disappear during maturity of the endosperm, are transformed into carbohydrates.

Proteins

In cereal grains, the bulk of the storage proteins of the endosperm is confined to special cells called aleurone cells which surround the more bulky starch-containing endosperm. The aleurone cell derives its name from the protein-rich aleurone grains which fill the cell. As we shall see later, this spatial layout of aleurone cells and starchy cells has profound implications in the mobilization of the endosperm reserves. The starch-containing cells may also contain amorphous protein known as gluten.

In wheat, about 15% of dry weight of the endosperm is in the form of protein. Protein accumulation commences between 5 and 12 days after fertilization and is accompanied by a rapid decline in the concentration of free amino acids. In the electron microscope, isolated wheat protein bodies appear as evenly stained particles with no substructures (Graham *et al.*, 1963). On the other hand, protein bodies of the developing endosperm cells appear as free, osmiophilic, electron-dense bodies in the cytoplasm, or with a faint lamellar structure covered with a distinctive lipoprotein membrane (Jennings *et al.*, 1963). Buttrose (1963a) has observed small golgi vesicles associated with protein granules in the intact endosperm and believes that protein secretion may be condensed into granules by the action of the golgi apparatus. Although an earlier study of protein bodies of rice endosperm cells showed evenly stained particles (Mitsuda *et al.*, 1967), a later study of isolated protein bodies has revealed the existence of a substructure of concentric rings within the particle (Mitsuda *et al.*, 1969). Protein bodies of barley, like those of wheat and rice, appear to have a fine structure of electron-dense material in ordered concentric layers. The lamellar structure may appear entirely within the protein body or as an appendage to it (Ory and Henningsen, 1969). The bulk of the storage reserves of lettuce endosperm is also proteinaceous in nature and the granules contained within membrane-bound profiles are interconnected by tubular extensions (Jones, R. L., 1974).

Buttrose (1963b) has studied the fine structure of development of the aleurone grain in wheat. According to this author, the aleurone grain originates as a deposit within vacuoles of young aleurone cells about 2 weeks after anthesis. In the mature endosperm, the aleurone grain appears to consist of a peripheral bounding membrane within which is enclosed a matrix containing spherical inclusions. The cytoplasm of the aleurone cell also contains numerous spherical bodies of proteinaceous or carbohydrate nature. These bodies are organized as a layer on the surface of the aleurone grain as the seed matures. In barley, aleurone grains have two types of spherical inclusions embedded in a matrix, and limited by a unit membrane (Paleg and Hyde, 1964; van der Eb and Nieuwdorp, 1967; Jones, 1969). The inclusions are probably storage protein and phytin. Spherosomes, apparently similar to the unidentified bodies of Buttrose (1963b), surround the aleurone grains and are also arranged along the plasma membrane. Moreover, aleurone cells contain organelles typical of other plant cells, including ER with numerous polyribosomes.

Fats

The typical high fat-containing endosperm is that of castor bean (*Ricinus communis*). In the electron microscope, fat droplets present a sharply defined interphase with the surrounding cytoplasm, but lack a bounding membrane. Vacuolate inclusions are present in the fat globules of the endosperm of young seeds, although they disappear at maturity. A role for these vacuoles as sites for initiation of oil biogenesis has been postulated (Harwood *et al.*, 1971).

C. Perisperm, Nucellus and Other Tissues

The nutritive tissue of *Yucca* seed is the perisperm, and as such is appropriately considered here. This tissue exhibits a high degree of heterogeneity (Horner and Arnott, 1965, 1966). Histochemical and electron microscopical studies show that storage products of the cell occur almost exclusively as membrane-bound protein or oil bodies, while reserve carbohydrates are found in the cellulosic framework of the cell wall. Ultrastructurally, protein bodies appear to have a central core surrounded by an electron-dense matrix containing birefringent inclusions.

A few words should be added about the fine structure composition of cells of the nucellus which may serve as a potentially significant source of nutrition for the developing embryo. The only published studies are those of Jensen (1965a) and Norstog (1974) on the ultrastructure of the nucellus of cotton and barley, respectively. In cotton, storage materials are mainly conserved in the form of protein bodies in a group of cells known as the collar cells. Within the ER of the cell, protein bodies are built up from small granules to deposits of large size in a manner reminiscent of yolk formation in animals. Although nucellar cells of barley lack any characteristic storage materials, autolysis of these cells seen in the EM might signify the possible utilization of their breakdown products in nutrition.

The parenchymatous cells of the orchid embryo may be regarded as a nutritive tissue which has several interesting analogies to the endosperm. As such, the orchid embryo provides a favorable material to study the nature of reserve substances and to follow their mobilization during embryogenesis. In the embryo of *Vanda* (Ricardo and Alvarez, 1971) the parenchymatous cells accumulate large quantities of proteins, lipids and carbohydrates which are slowly depleted as the embryo commences to grow. The proteins initially appear as discrete bodies within a framework of small clusters of tubules. The fate of the tubules is aligned with that of the protein body as they eventually disperse throughout the cytoplasm and disappear following depletion of the proteins. The lipid body is attached as an appendage to a dense laminate inclusion which increases in size as the former is hydrolyzed. Starch accumulation centers around the plastids and eventually starch grains completely replace the plastid stroma. While these observations do not present definitive proof for the utilization of storage reserves of the parenchymatous cells by the meristematic region of the embryo, the underlying idea is an appealing one; just how they fit in with the evidence for what appears to be a remarkable way of digesting the storage products of the endosperm by the embryo will be briefly explored in the following discussion.

D. Mobilization

Interpretation of the nutritional relationship between the embryo and the storage products of the endosperm depends upon a proper understanding of the processes of digestion and absorption of food materials. Not much is known about the mechanism of digestion and even less is known about the ways the digested food is absorbed by the embryo. Certainly, digestion might take place by means of enzymes and transfer of soluble food materials might be accomplished by diffusion. According to Siegel (1952), extracellular secretion of phosphorylase in red kidney bean embryo has significance in the digestion of starch in the tissues surrounding the embryo during its early development. In this connection, the recent and provocative studies on the mobilization of starch during germination of barley grains are also of interest. This work, which is reviewed in greater detail in Chapter 17, has shown that α-amylase concerned with the digestion of starch in the endosperm of barley during germination is synthesized in the aleurone cells in response to a hormone GA coming from the embryo (Yomo, 1960; Paleg, 1960; Varner and Chandra, 1964). The embryo–endosperm relationship during progressive embryogenesis may be explicable on similar lines if we assume that the key to the digestion of starch in the endosperm is held by the embryo, probably in the form of a hormonal substance.

The presence of other storage products in the endosperm indicates the prospects of involvement of other specific enzymes in their digestion. Thus, one may perhaps view the protein body equipped with enzymes for its own breakdown in the crystalline lamellar structure (Ory and Henningsen, 1969). In castor beam endosperm, there is evidence that active lipase is associated with the membrane of the spherosomes (Ory et al., 1968). However, the specific conditions under which enzymes digest the storage products of the endosperm into soluble food for utilization by the developing embryo have not been elucidated.

II. GYMNOSPERMS AND PTERIDOPHYTES

A. Gymnosperms

The fundamental plan of distribution of ergastic substances in the cells of the megagametophytes of gymnosperms is the same as in the cells of the endosperms of angiosperms—a fact long apparent from descriptive embryology. Stopes and Fujii (1906) have followed the appearance and disappearance of starch and protein granules in the megagametophytes of several gymnosperms. These inclusions appear to be arranged in distinctive patterns within the gametophyte, their type and proportion often fluctuating with age. Generally, at a very early stage of growth of the gametophyte, carbohydrates and proteins are present in a soluble form in the vacuolar sap. In an older series of ovules, starch grains and protein granules appear as discrete solid bodies around the egg. At a still later stage of development, there is a decreasing gradient in their distribution from the peripheral cells of the gametophyte to the vicinity of the egg or the embryo. The megagametophytes of *Pinus sylvestris* and *Picea abies* are well differentiated into three concentric zones with regard to the distribution of starch grains. The peripheral layers of cells are densely packed with large starch grains. The middle zone has sparse inclusions and virtually none, or only simple grains, are present in the cells bordering the embryo (Håkansson, 1956). The density of distribution of lipids and lipoproteins in the gametophyte of *Ginkgo biloba* also decreases from the epidermal cell inwards (Favre-Duchartre, 1958; Dexheimer, 1973a, b). If the specific arrangement of reserve food materials in the megagametophyte has some functional significance, it is natural to relate these observations to the channeling of food to the growing embryo.

The amino acids and sugars of the female gametophytes of some species have been identified by paper chromatography. Bartels (1957b) has shown the presence of asparagine, glutamic acid, L-alanine, proline, threonine, serine, valine, leucine, phenylalanine, tyrosine and cystine in the gametophytes of germinating seeds of *Pinus nigra*, *P. sylvestris* and *Picea abies*; it is, however, not clear whether all of these compounds were present in the gametophytes before germination, or some of them were formed during germination. In *P. roxburghii* (Konar, 1958a) those same amino acids and amides present in the embryo (Chapter 3) are found in the gametophyte; moreover, metabolism of these compounds with age of the ovule appears to be similar in both the embryo and the gametophyte. As in the case of the embryo, the free amino acids gradually disappeared as the gametophyte developed, and finally in the mature gametophyte, only asparagine was present in the free state.

The decrease in the amount of free amino acids is expected to be linked with the enhanced protein accumulation that occurs at this stage. According to Tulecke (1967), the megagametophyte of *Ginkgo biloba* has large amounts of glutamic acid, aspartic acid, arginine, valine, leucine and alanine.

Sucrose, fructose and glucose are the main sugars present in the young gametophyte of *P. roxburghii*. As it matures, there is a decrease in fructose and glucose contents and a concomitant increase in the concentrations of starch and fat (Konar, 1958a, b). According to Ball (1959), ethanol extracts of the female gametophyte of *Ginkgo biloba* contain large quantities of sucrose, glucose, fructose, galactose, raffinose and probably stachyose, all of which may be utilized in varying degrees by the developing embryo. In the gametophyte of *Pinus banksiana*, stachyose appears to be the predominant form of soluble sugar (Durzan and Chalupa, 1968). The existence of correlations between the levels of sugars and nitrogenous compounds of the female gametophyte and embryo of this species makes a strong case for a role for food materials of the gametophyte in the growth of the embryo.

A discussion of the nutritional resources of the female gametophyte must inevitably consider any evidence for the existence of non-nutrient regulatory substances of the type of growth hormones. The only evidence is from *Ginkgo biloba* where Steward and Caplin (1952) have demonstrated the presence of an active substance capable of inducing growth in explanted carrot tissues. More recently, Banerjee and Radforth (1967) have suggested the possibility that this may be a gibberellin-like compound. It is to be hoped that as female gametophytes of more gymnosperms are examined, evidence for the existence of specific growth substances will be revealed.

B. Pteridophytes

In pteridophytes, remarkably little is known about the chemical nature of the nutritive substances in the prothalli on which embryos are nourished. The information available is drawn from circumstantial evidence which does not provide specific details. Carbohydrates are known to be abundantly present in the prothallus of *Lycopodium* (Treub, 1884) and in the megaspore of *Selaginella* where they accumulate more rapidly than proteins (Wardlaw, 1955). In eusporangiate ferns the visible reserves take the form of starch grains. In some species the prothalli are associated with endotrophic mycorrhiza which apparently provides them with soluble carbohydrates. From this evidence it seems safe to assume that the supply of nitrogen-containing substances to the sporophyte is inadequate. The notoriously slow growth of the primordial shoot system of the sporophytes of many lower vascular plants makes it likely that some specific growth hormones or nitrogenous compounds limit their embryonic development.

III. *IN VITRO* CULTURE OF NUTRITIVE TISSUES

With the advent of tissue culture techniques, successful attempts have been made to grow endosperms of angiosperms, and female gameto-phytes of gymnosperms in media of defined composition. These tissues, which develop within the confines of the sporophytic tissues, are not unique in their developmental potencies, but are so only in their nuclear constitution. One outcome of this interesting situation is that it is poss-ible to determine whether tissues definitely restricted to limited growth can be switched on to a pathway leading to unlimited growth, and how far such growth is a consequence of the ploidy level of the nucleus. From the practical point of view, regeneration of plants of a chromosome con-stitution different from that of the sporophyte will be of importance in plant breeding and horticulture. The long-range objective of *in vitro* culture of the endosperm of angiosperm and the gametophytic tissue of gymnosperm is, as indicated in the introduction to this chapter, to understand the fundamental biochemical needs of developing embryos and the complex and diversified metabolic interactions between the sporophyte and its nutritional milieu under controlled conditions. With a different objective in mind, somatic tissues of the ovule such as the nucellus which provide nutrients for the growth of embryos of some plants have also been grown in culture (Rangaswamy, 1958a). Con-sidering that the endosperm is the site of biosynthesis of specific storage products, investigations on their metabolism in cultured tissues seem to be in prospect, as shown by the results of a study of phytin biosyn-thesis in cultured castor bean endosperm (Sobolev *et al.*, 1971).

A. Endosperm Culture

The idea of growing excised endosperm in culture was perceived by LaRue, and by his pioneering efforts a successful culture of maize endo-sperm capable of unlimited growth as a callus-like material was estab-lished (LaRue, 1947, 1949). LaRue and his associates have explored in some detail the nutritional requirements and growth morphology of the cultured endosperm tissue (see Narayanaswami, 1956; Johri and Srivastava, 1973, for reviews).

Even in early trials, it was found that the margin between success and failure in raising cultures was confined to a relatively narrow range of selection with regard to variety of the grain and its age. Ability to proliferate in culture was restricted to the endosperm of sugary types of grains, and not to the starchy or waxy types (Fig. 6.1) (Sternheimer, 1954; Tamaoki and Ullstrup, 1958; Sun and Ullstrup, 1971). The endosperm excised from half-grown and large grains did not proliferate in culture, nor did very young endosperm (LaRue, 1947). Apparently, there is an optimum stage in the development of the endosperm when its potentialities for regeneration are at their peak.

Although the use of yeast extract, by itself or in combination with other substances, has served to bring cereal endosperm into culture (LaRue, 1947; Sternheimer, 1954; Straus and LaRue, 1954; Norstog, 1956b; Tamaoki and Ullstrup, 1958; Sehgal, 1969), Straus (1960) was

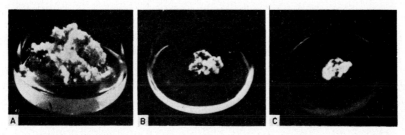

Fig. 6.1. Growth of excised maize endosperm *in vitro* in a medium containing yeast extract, 5 g/l. **A**, Thirty-five-day old culture of late yellow sugary hybrid. **B**, Thirty-day old culture of starchy hybrid ('Country Gentleman'). **C**, Twenty-four-day old culture of starchy and waxy hybrid. (After Tamaoki and Ullstrup, 1958)

able to develop a completely synthetic medium containing mineral salts, sugar, vitamins and asparagine which gave excellent growth of maize endosperm. In testing the effects of various sugars, sucrose was found to be markedly superior to any other sugar, closely followed by fructose and glucose. All of the other carbohydrates tested, namely, arabinose, galactose, glycerol, lactose and rhamnose, were inactive as carbon energy sources (Straus and LaRue, 1954). Thus, in its carbon energy requirements, maize endosperm does not appear to be unlike embryos of several species grown in culture.

The transformation of the endosperm into a rapidly proliferating tissue is due to the division of its peripheral or occasionally of some of its deep-seated cells (Straus, 1954). Straus (1958, 1959) has observed spontaneous changes in the cultured tissue due to the occurrence of strains which vary in their ability to synthesize anthocyanins. Remarkably enough, culture of the endosperm does not appear to diminish its growth-promoting effects on cultured embryos (Pieczur, 1952; Norstog, 1956b). This leads to the inference that the composition of essential metabolites of the cultured endosperm does not differ from that of the intact tissue from which it is derived, although the former is growing in the absence of restraint that applies normally.

Although a physiological basis for the effect of ploidy level on the endosperm is still mysterious, some data bearing on this question have come from the comparative behavior in culture of the embryo and endosperm of *Cucumis sativus* (Nakajima, 1962). For optimum growth, embryo and endosperm required a medium of high osmolarity, supplemented with yeast extract or IAA, diphenylurea and casein hydrolyzate. The essential difference between the growth of the embryo and the endosperm in culture was in the differentiation of tissues and organs in the former, while the latter continued to grow essentially as an undifferentiated tissue. Does a higher ploidy level in a tissue, therefore, inhibit its intrinsic capacity for differentiation? This is not clear at present. Perhaps it does so, by blocking the biosynthetic pathways concerned with the activation of certain essential metabolites.

Recognition of the fact that complex additives to the medium could promote proliferation of the endosperm in the absence of embryo growth followed from studies of Rangaswamy and Rao (1963) on the seeds of *Santalum album*. These investigators found that when seeds of *S. album* containing a massive endosperm and a small embryo were grown in a medium containing casein hydrolyzate and coconut milk, the embryo formed a normal seedling at the expense of the endosperm. Addition of yeast extract and low concentrations of kinetin and 2,4-D, on the other hand, led to proliferation of the endosperm and inhibition of embryo growth (Fig. 6.2). The inhibition of growth of the embryo is reversible since it can be separated from the callused endosperm, and grown in isolation in a medium of a different composition (Rao and Rangaswamy, 1971).

Following upon this demonstration of endosperm growth in the intact seed, it was possible to obtain continuous growth of the endosperm accompanied by inhibition of embryo growth in seeds of *Ricinus communis* (Satsangi and Mohan Ram, 1965; Mohan Ram and Satsangi, 1963; Srivastava, 1971b), *Exocarpus cupressiformis* (Johri and Bhojwani, 1965), *Oxalis dispar* (Sunderland and Wells, 1968) and *Tilia platyphyllos* (Zenkteler and Guzowska, 1970). These studies show that the extent of visible differentiation of the endosperm in culture depends not only upon an interplay of hormonal substances but also on the simultaneous inhibition of embryo growth. The inhibition of embryo growth may be a direct effect of substances added to the medium, or may be indirectly related to the altered metabolic status of the regenerated endosperm.

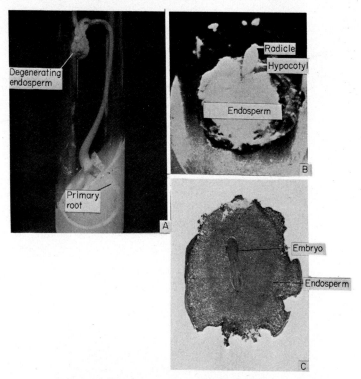

Fig. 6.2. **A**, When seeds of *Santalum album* are sown in a medium containing coconut milk and casein hydrolyzate, they germinate normally (the cotyledons are partially enclosed by the degenerating endosperm). **B**, Addition of yeast extract, kinetin and 2,4-D to the medium results in the proliferation of the endosperm and inhibition of embryo growth. **C**, The proliferation of the endosperm results in a population of parenchymatous cells. (After Rangaswamy and Rao, 1963; photographs courtesy of N. S. Rangaswamy)

In all of the above examples, the endosperm callus continued to grow without any signs of tissue or organ differentiation. The demonstration that such cultures could give rise to vascular tissues, roots and shoot buds followed from pioneer studies in Johri's laboratory. Endosperm cultures of *Exocarpus cupressiformis*, *Osyris wightiana*, *Putranjiva roxburghii* and *Jatropha panduraefolia* frequently gave rise, during growth, to groups of cells which were identified in squash preparation as tracheids (Johri and Bhojwani, 1965) (Fig. 6.3). The differentiation of the parenchymatous cells into thick-walled cells and embryoid-like structures has also been observed in endosperm cultures of *Ricinus communis* (Satsangi and Mohan Ram, 1965) and *Croton bonplandianum* (Bhojwani, 1966). A final step toward demonstration of complete totipotency is the differentiation

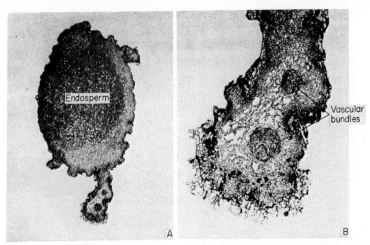

FIG. 6.3. **A**, Nodules of vascular tissues differentiating in endosperm culture of *Osyris wightiana*. **B**, Enlarged view of a portion from A showing vascular bundles. (After Johri and Bhojwani, 1965; photographs courtesy of B. M. Johri and S. S. Bhojwani).

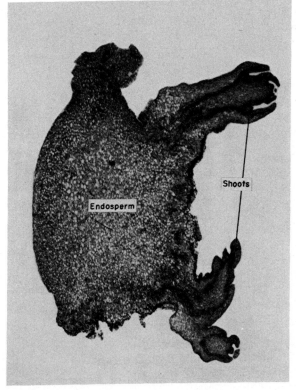

FIG. 6.4. Differentiation of shoot buds on endosperm callus of *Exocarpus cupressiformis*. (After Johri and Bhojwani, 1965; photograph courtesy of B. M. Johri and S. S. Bhojwani)

determine the organization of cell populations into shoot or root meristems and their subsequent development is also required. How subtle changes in medium composition can so profoundly influence the organization of cells remains one of the key problems to be studied in detail.

We see that some aspects of growth of the endosperm have been described in detail, but the physiological basis that determines the embryo–endosperm relationship is still puzzling. The availability of massive amounts of endosperm tissue growing under sterile conditions should provide an opportunity to integrate the metabolic and nutritional aspects of embryo–endosperm relationship into a comprehensive understanding of morphogenesis of the embryo with reference to specific chemical agents.

B. Culture of Megagametophytes of Gymnosperms

As indicated in the previous subsection on angiosperms, the culture of megagametophytes under defined conditions offers unique opportunity to study the nutritional aspects of embryogenesis *in vitro*, and the question of totipotency of cells of the gametophyte. Certainly, in the relationship of the gymnosperm embryo to its megagametophyte there are compelling similarities to the embryo–endosperm relationship observed in angiosperms.

LaRue (1948, 1954) gained considerable success in culturing the megagametophytes of several gymnosperms in relatively simple media. Tissues from *Zamia integrifolia* and *Cycas revoluta*, excised and cultured at the time of fertilization, possessed unusual powers of regeneration of apogamous roots and buds. In recent studies (Norstog, 1965b; Norstog and Rhamstine, 1967) it has been shown that by changing the composition of the medium, the tissue from *Z. integrifolia* can be perpetuated in an undifferentiated state or induced to form embryo-like outgrowths called "pseudobulbils" on apogamous regenerates. The notoriously slow growth of the tissues in culture has restricted refined studies on its growth physiology; it may be expected that as the growth rate of the cultures is enhanced by the development of more effective culture media, this gap will be filled. Perhaps the age at which the gametophyte is cultured is critical here also. Radforth and Bonga (1960) have stated that gametophytes of *Pinus nigra* var. *austriaca* cultured at two different times after fertilization of the egg show different responses to IAA. Other than a possible decrease in some essential constituent of the gametophytic tissue due to continued nurture of the embryo, little can be said to explain this.

Following the success in obtaining a haploid tissue from the micro-gametophyte of *Ginkgo biloba* (Tulecke, 1953; see p. 390), Tulecke (1964) obtained a continuously growing tissue from the megagametophyte of this species. When cultured in a medium containing 2,4-D and coconut milk, both apical and basal halves of the gametophyte proliferated into a friable callus which later became a hard and compact tissue. In the presence of kinetin in the medium, there was a manifest tendency in the tissue for root and shoot initiation (Tulecke, 1965). In contrast to the intact megagametophyte (see p. 150), the tissue obtained in culture had high concentrations of serine, glutamic acid, alanine, cysteic acid and γ-aminobutyric acid in the free amino acid pool (Tulecke, 1967). The success in obtaining a continuously growing tissue from the megagametophyte of *Ginkgo biloba* is interesting in view of the difficulty encountered by previous workers in formulating a suitable medium for rapid growth of the gymnosperm tissue.

In other studies, the possibility of obtaining callus growth *in vitro* from megagametophytes of *Ephedra foliata* (Sankhla *et al.*, 1967b), *Pinus lambertiana* (Borchert, 1968), *P. resinosa* (Bonga and Fowler, 1970), *P. nigra* var. *austriaca*, *P. mugo* var. *mughus* (Bonga, 1974a) and *Taxus baccata* (Zenkteler and Guzowska, 1970) has been demonstrated. However, success in growing the gametophyte is of strictly limited value and does not lead to any general conclusions on its regenerative powers.

IV. COMMENTS

From the survey presented in this chapter, it is seen that where the composition of the nutritive tissue from different sources can be compared, agreement is quite good to the extent that all sources investigated contain amino acids and growth substances. It seems likely that the complete explanation of the action of the endosperm on the embryo will ultimately reside in a relatively small number of discrete growth hormones, which may act alone, or in an interaction with one or more of the other components. In the latter case, a great deal more information must be obtained before a comprehensive picture of the interactions can be drawn. A question of enduring interest is whether there is a single substance common to endosperms of several plants which functions as a master hormone in regulating embryo growth. Experiments designed to answer this question will not be easy to perform with intact endosperms and it is here that their culture into continuously growing tissues attains significance. Although the composition of the nutritive tissues of only a few species is known, there are good reasons to believe that it may change during progression of ovules to maturity.

Data are thus needed particularly on the composition of the endosperm in ovules in the earlier stages of development, as it is here that the different constituents may exert subtle effects to initiate cell and tissue differentiation in the embryos. It would seem that these processes are open to experimental manipulation, one of the more promising lines of attack being the culture of excised embryos *in vitro*. The results of these studies are reviewed in the next chapter.

7. Embryo Culture

In the preceding two chapters, the main features of the environment of developing embryos in the different groups of vascular plants were examined in some detail. We have seen that tissues surrounding the embryos provide not only a physical framework that integrates the latter into organized structures, but also the nurture and nutriments for their early growth. Further complexity is added to the nutritional relationships of embryos by the presence of special structures which expose them in favorable contact with the source of food materials.

These conclusions, derived from studies of embryos growing *in vivo*, suffer from the drawback that the complexity of an amorphous nutritive tissue can obscure the nature of the real morphogenetic substances at work in the control of embryo growth. In fact, indications given by the chemical composition of the nutritive tissue may be totally misleading if the observed growth is due to one or two critical metabolites. In the absence of definitive biochemical evidence, it is also difficult to be certain that nutrients are absorbed and metabolized by the embryo in the form in which they are present in the surrounding tissue.

Some of these difficulties may be circumvented by separating embryos from the bulk of the maternal tissues and growing them in media of known chemical composition. This approach has from time to time tempted investigators to study the factors that are essential for the nutrition of embryos and has contributed information of a fundamental nature concerning the metabolic relationships of developing embryos. While such knowledge is far from complete, there is a considerable body of experimental information relating to it. It is the purpose of this chapter to bring together this information and some of the problems involved. In its preparation I have drawn upon some of the early literature on embryo culture to achieve a broad coverage and to present the older literature in its historical perspective for a proper appreciation of the evolution of ideas on the nutritional requirements of developing embryos. It is also the intention of this survey to show how some advances in embryo culture have awaited discoveries in other fields. It is worth noting that no information on the physiology of growth of embryos was sought in the early studies, but it was encountered incidentally as an unexpected bonus from investigations primarily directed toward other objectives. Several review articles on embryo culture and related problems have been published in the last few years (Rappaport, 1954; Narayanaswami and Norstog, 1964; Maheshwari and Rangaswamy, 1965; Wardlaw, 1965b; Degivry, 1966; Raghavan, 1966) and reference may be made to these sources for appropriate details. For the interested reader who wishes to try his own hand at the game, some of the general techniques employed in the culture of embryos are described in the articles by Sanders and Ziebur (1963), Raghavan (1967) and Torrey (1973). Reference is also made to papers by Newcomb and Cleland (1946), Gilmore (1950), Stokes (1952a), Keim (1953a), Bonga (1965), Rommel (1958), Tillett (1966), Miflin (1969a), Colonna et al., (1971), and Dahmen and Mock (1971) for notes on specialized culture techniques applicable to individual species.

A simple way to approach the control of growth of the embryo in culture is to ask whether, under natural conditions in the embryo sac, the stimulus to growth is located in the embryo itself or in the surrounding milieu. In answering this question, one cannot fail to notice that the increasingly complex external morphological development and internal biochemical specialization of developing embryos are broadly reflected in the changes in their nutritional requirements. Thus, a basic distinction between a heterotrophic phase and an autotrophic phase may be made with regard to the dependence of the embryo on the food substances stored in the cells of the endosperm. The fertilized egg and embryo in the division stages immediately following fertilization generally develop at the expense of the nutritional resources of the endosperm or of the accessory cells of the embryo sac which begin to degenerate at this time. They have low synthetic capacities and are thus hetero-trophic, being nourished by specialized nutritional substances includ-ing amino acids, carbohydrates, purines, pyrimidines, perhaps vita-mins, plant hormones and other essential metabolites present in the environment of the embryo sac. At the globular stage the embryo is still heterotrophic. Only in the late heart-shaped stage, with the begin-ning of cotyledonary development and the associated internal dif-ferentiation, does the embryo become sufficiently independent and autotrophic, possessing large and varied synthetic capacities, for it to be possible to remove it and culture it *in vitro* in a nutrient medium. It goes without saying that this critical age varies in the different species.

In reviewing the work on culture of embryos from a historical per-spective, I shall describe the results under two heads: (i) culture of dif-ferentiated and mature embryos, and (ii) culture of proembryos. The former corresponds to the post-germinal and the latter to the pre-germi-nal embryos of some authors. Unfortunately, there is no single morpho-logical character on which this hard distinction between embryos of two different levels of organization has been drawn. But this distinction has some merits because it calls attention to the basis of cellular activity as an indicator of morphological maturity.

I. CULTURE OF DIFFERENTIATED AND MATURE EMBRYOS

Current interest in experimental and cultural studies on embryos can be traced to two seemingly unrelated and obscure programs of research spanning a period of over 100 years. One was the experiment of Bonnet (1754), who attempted to germinate seeds of bean in the absence of the cotyledons. This investigator was intrigued by the role of the coty-ledons during germination of the seed, and his work, scarcely more than a beginning, provided suggestive evidence of their role in germination. This idea received the support of Sachs (1859, 1862), who found that embryos of *Phaseolus multiflorus* and maize, separated from the coty-ledons and endosperm, respectively, germinated poorly and formed short, stunted seedlings. The other landmark was a study by van Tieg-hem (1873) which had even deeper implications. He found that when the embryo of *Mirabilis jalapa* was separated from the endosperm and planted alone, after an initial period of rapid growth its growth declined; on the other hand, if the living endosperm was replaced by a paste of the crushed endosperm or by starch paste containing nitrate and phosphate, continued growth of the embryo ensued. The simplicity of this description gives rise to the important inference that the nutritive power of the endosperm does not depend upon its cellular organization, but can be substituted by synthetic substances. The better understanding that has come in recent years of the role of the endosperm in the nurture of the embryo has its beginning in this discovery.

Results in harmony with the above conclusions were presented between the intervening period and before the end of the century by other investigators. Since the endosperm was not appreciably digested during the early phase of germination of maize, Gris (1864) concluded that growth of the mature embryo was not dependent upon the storage products of the endosperm. Blociszewski (1876) succeeded in growing mature plants from embryos of oat, rye, clover, pea and lupin separated wholly or in part from their storage tissues and planted on moist earth. When storage tissues of rye and pea were replaced with macerated endosperm of cotyledons, starch, grape sugar, asparagine or mixtures of these substances, it was found that rye embryos absorbed all nutrients except asparagine, and pea embryos utilized all except its own macerated cotyledons. Largely as a result of these studies, the intervention of the storage tissues of the seed in the growth of the embryo became an important issue in subsequent studies.

Brown and Morris (1890) performed a number of ingenious experiments with barley embryos. They showed that if an excised embryo was grafted on to an embryo-less endosperm of a grain, the grafted embryo developed into a normal plant, as readily as if it were on its own endosperm. These workers also succeeded in obtaining limited growth of barley embryo on the endosperm of wheat. From these experiments, a relatively new idea emerged, that the embryo was not necessarily limited to the nutrients supplied by its own endosperm. Further work of these authors was dominated by their interest in the embryo–endosperm relationship during germination. They showed that a viable embryo grafted on to the endosperm that had been killed by chemical or heat treatment germinated normally. They also succeeded in rearing normal plants from embryos excised from mature grains and grown on glass wool in solutions containing gelatin and a source of carbohydrate. In another series of investigations, dissolution of starch and its eventual appearance in the embryo were successfully demonstrated when the latter was placed in contact with a moist endosperm. The importance of this discovery lies in the intervention of α-amylase secreted by the aleurone cells which metabolizes starch in the endosperm. In view of the current interest in the hormonal control of α-amylase synthesis in barley endosperm (Paleg, 1960; Yomo, 1960), these early results are of considerable significance.

The breakthrough that opened up the field of embryo culture was the discovery that embryos can be separated from the environment of the maternal tissues and cultivated aseptically in an artificial medium. At the turn of the century, Hannig (1904) published a paper that represents a cornerstone in the study of physiology of growth of embryos. If this paper is repeatedly quoted in the literature, it is not only because of its historical value, but also because experimental embryologists want to pay a tribute to Hannig, whose work opened up the field of embryo culture for exploitation and study. Using completely aseptic techniques, Hannig cultured relatively mature embryos of *Raphanus sativus*, *R. landra*, *R. caudatus* and *Cochlearia danica* of different ages in a mineral salt medium supplemented with sugar, and obtained transplantable seedlings. This investigation was also addressed to a number of basic problems, the outcome of which provided a fabric of background information on the controlling factors in the growth of cultured embryos. For example, can the liquid endosperm support growth of embryos *in vitro*? To reach an experimental decision on this point, embryos were cultured in the liquid extracted from the ovules with a fine capillary. Although the results were negative, the experiment brought into focus the possible role of the liquid endosperm of the ovule in the nurture of young embryos. In another experiment, by demonstrating the inability of embryos to grow in a mineral salt medium unless supplied with a concentration of sucrose far above that required as a carbon energy source, Hannig provided graphic proof of the requirement for a high osmotic concentration in the medium for successful growth of embryos. He also studied the effects of several nitrogen sources in a range of concentrations and emphasized their role in the growth of cultured embryos. All of Hannig's experiments were accompanied by quantitative data in contrast to the abstract and general arguments of some of the earlier workers. The final morphology attained by embryos in some of the media used is illustrated in Fig. 7.1.

Since Hannig's work, cultivation of embryos as isolated systems has never ceased to intrigue investigators and has expanded rapidly in both scope and depth, continuing to provide useful information on the physiology of their growth. Later attempts at culture of embryos utilized synthetic tissue culture media free from substances of unknown chemical nature. As a result, the picture was enlarged, and some of the apparent contradictions cleared, although at the same time new problems and new issues arose. Some of these early investigations will be

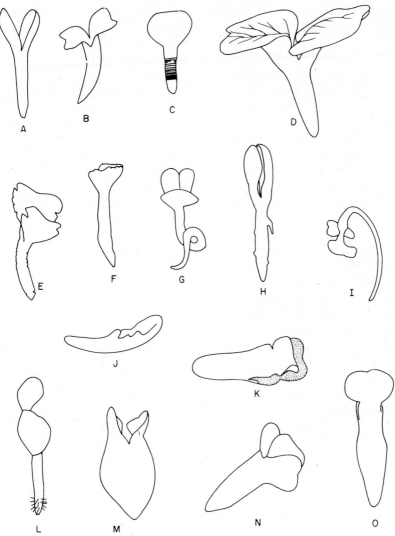

Fig. 7.1. Final appearance of excised embryos of *Cochlearia* and *Raphanus* after culture in different media. **A**, Embryo of *Cochlearia* grown in sucrose solution. **B**, **C**, Embryos of *Raphanus* grown in medium containing sucrose and KNO_3. Hatched zone is dead part of the stem. **D**, *Raphanus* embryo grown in glucose–asparagine medium shows vein formation on cotyledons. **E**, *Raphanus* embryo in 10% peptone, 1% sucrose medium. **F**, Same in 5% sucrose, 0·5% peptone medium. **G**, Same in 5% sucrose–KNO_3 medium showing germination. **H**, Side roots on *Cochlearia* embryo grown on 5% glucose–asparagine medium. **I**, Same as G. **J**, *Cochlearia* embryo in 5% glucose–asparagine medium. **K**, *Raphanus* embryo initially in 5·3% glucose, gradually transferred to 1·25% glucose. **L**, Side roots of cultured *Cochlearia* embryo. **M**, **N**, **O**, *Raphanus* embryos. (After Hannig, 1904)

referred to here in brief to complete the sequence, and will be described fully in appropriate chapters later. Brown (1906) studied the relative efficiency of various organic nitrogen compounds on the growth of excised barley embryos cultured in a medium containing mineral salts and sucrose. The dependence of the embryo on nutrients from its endosperm seemed less probable as a general rule from the findings that embryos of a number of species of Gramineae could grow well when grafted onto each other's endosperm (Stingl, 1907; Câmara, 1943). In other studies, the role of the storage tissues of the seed in the growth of embryos was explored by separating the latter from the endosperm, or by decotylating embryos, and planting them in nutrient solutions (Dubard and Urbain, 1913; Buckner and Kastle, 1917). A function for the scutellum of graminean embryos in the absorption of nutrients was implied in a study which showed that embryo devoid of this organ grew somewhat feebly (Andronescu, 1919).

Knudson (1922) succeeded in germinating orchid embryos into plantlets in the absence of the symbiotic fungus by growing them on nutrient agar medium containing sugar. In the absence of sugar, embryos failed to develop beyond the protocorm stage. The reality of a specific stimulation of growth in orchid embryos by a simple carbohydrate led to an appreciation of a suggestive role of the mycorrhizal fungi in the conversion of starch and other complex polysaccharides into simpler forms. The asymbiotic propagation of orchids, now practiced on an unparalleled commercial scale, had its inception in these experiments.

On the basis of the behavior of cultured embryos of plants belonging to the families Solanaceae, Linaceae, Cruciferae, Polygonaceae, Compositae, Cucurbitaceae and Gramineae, Dieterich (1924) pointed out two important generalizations, which have proved to be significant in understanding the physiology of growth of embryos. One was that the embryo grown *in vitro* usually skipped a rest period that was observed when it was part of the intact seed and germinated. In addition, it was found that a solid medium with Knop's mineral salts and 2·5–5·0% sucrose could support normal growth of embryos isolated from mature seeds, but in the same medium embryos from immature seeds tended to form malformed seedlings, omitting the regular stages in embryogenesis. Dieterich aptly introduced the term "künstliche Frühgeburt" to designate this type of growth, now well known as "precocious germination" in embryo culture studies. At about this time, Laibach (1925, 1929) envisaged the immense applications of embryo culture technique in rearing viable seedlings from otherwise unsuccessful crosses. This aspect of embryo culture is discussed in detail in Chapter 13.

The ability of excised embryos to grow in culture has been demonstrated in other genera and species. Deviations in detail concern the nature of the medium employed and the type of growth that ensued. These early attempts utilized full-grown embryos excised from mature seeds, and the evidence therefrom that mature embryos require an extremely simple medium is particularly impressive. Although mineral salts and a carbohydrate source in the form of sucrose were the only major components of the medium, compounds outside these classes were occasionally added, and were shown to have specific growth-promoting properties, but none of them was found to be an indispensable component of the medium. The suggestive feature of these studies is that the mature embryos are autotrophic, and that their subsequent development is to a large extent under the control of factors inherent in their cells.

II. CULTURE OF PROEMBRYOS

With refinements in tissue culture techniques, emphasis in embryo culture shifted from mature embryos to comparatively young, immature embryos, mostly in the early and late heart-shaped stages. The domestication of young isolated embryos to their new environment proved to be no easy task. For one thing, the medium used to grow mature embryos successfully fell short of being an adequate and balanced one for immature embryos. As mentioned earlier, the composition of the medium was much too simple; consequently, the small embryos failed to survive transfer to nutrient media or underwent only a few additional divisions before turning into undifferentiated callus. Another complication was that, not infrequently, young embryos grew precociously into tall spindly seedlings and considerable skepticism arose in accepting them as normal seedlings.

The key for the successful culture of small embryos turned out to be that they required additional substances in the medium to foster normal growth in culture, this being circumscribed by the concept that the optimum growing condition for *in vitro* culture of embryos is one which imitates closely the composition of the endosperm or the milieu of the embryo sac. Some of these requirements may be supplied from reserves in the embryo itself or by biosynthesis, but very often one or more of them may act as limiting factors. In succeeding attempts to culture *heart-shaped* and *torpedo-shaped* embryos, IAA, vitamins, amino acids, and natural products rich in amino acids, such as yeast extract, casein hydrolyzate and other complex products were tested with some degree of success. As early as in 1932, White isolated young heart-shaped embryos of *Portulaca oleracea* and grew them on a medium containing mineral salts, glucose and fibrin digest. In about 3 weeks, embryos grew from 0·12 to 1·84 mm in length and assumed the form of nearly mature embryos complete with visible root primordium and cotyledons (Fig. 7.2). Even globular embryos gave faint indications of

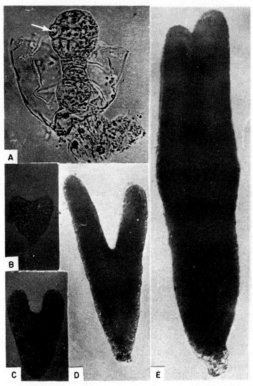

FIG. 7.2. Culture of proembryos of *Portulaca oleracea*. **A**, Globular embryo of about 60 cells with attached suspensor; except for a noticeable enlargement of one cell (arrow) during the culture period, the embryo did not grow. **B**, Heart-shaped embryo (120 μm long) at the time of culture. **C–E**, Final appearance of three embryos, initially heart-shaped, after culture for varying periods. (After White, 1932)

growth *in vitro*, but they did not seem to be viable for more than 4 days in the medium. In 1936, LaRue successfully grew small embryos of a number of dicotyledons, monocotyledons and gymnosperms. Inorganic salts with sugar, IAA and yeast extract supported growth of embryos as small as 0·5 mm long. Such embryos developed roots and green leaves on an agar medium, and eventually formed seedlings capable of living in the soil. Even when those additives were supplied either alone or in combination, it was not possible to culture heart-shaped embryos about 200–300 μm long, which at best expanded to a larger size, but did not grow. LaRue (1936) summarized the situation at that time with the prophetic remark that "at [embryo] length of about 0·5 mm we may have reached a new lower threshold which it will be difficult to pass".

A. Role of Natural Plant Extracts

Coconut Milk

A turning point in the culture of proembryos came from studies of van Overbeek *et al.* (1941a, 1942) on the physiology of growth of embryos of *Datura stramonium*. They demonstrated that whereas mature embryos were self-nourishing and grew into seedlings in a simple medium containing mineral salts and 1% dextrose, torpedo-shaped and heart-shaped embryos required a mixture of growth factors such as glycine, thiamine, ascorbic acid, nicotinic acid, pyridoxine, adenine, succinic acid and pantothenic acid for normal growth in culture. In this enriched medium still smaller embryos failed to grow, or grew only feebly before turning into undifferentiated callus. Culture of such embryos was possible only when the medium containing the physiological substances noted above was supplemented with nonautoclaved coconut milk (Fig. 7.3). It was observed, for example, that globular proembryos initially about 200–500 μm long increased to several times their original length in the course of a week in the coconut milk medium, without germinating precociously (Fig. 7.4). The unleashing of growth in the tiny embryos by coconut milk led to the logical conclusion that the liquid endosperm contained one or perhaps more critical metabolites, tentatively designated as "embryo factors", designed for active intervention in promoting growth of cultured embryos. In further work (van Overbeek, 1942; van Overbeek *et al.*, 1942), coconut milk was fractionated, and after eliminating the toxic principles, an active fraction was obtained in a relatively pure form. This preparation, presumably of the embryo factor, was still more dramatic in its effects, and induced growth in the cultured embryo when added in a dilution as low as 1:19 000 on a dry weight basis, compared with 1:110 for the crude milk.

The physiological effects of coconut milk vary in embryos of different species. For example, seedlings were reared from embryos of sugarcane initially 66 μm long by supplementing a standard embryo culture medium with coconut milk (Warmke *et al.*, 1946), but the extract had no beneficial effects on the growth of 10 day old (more than 300 μm

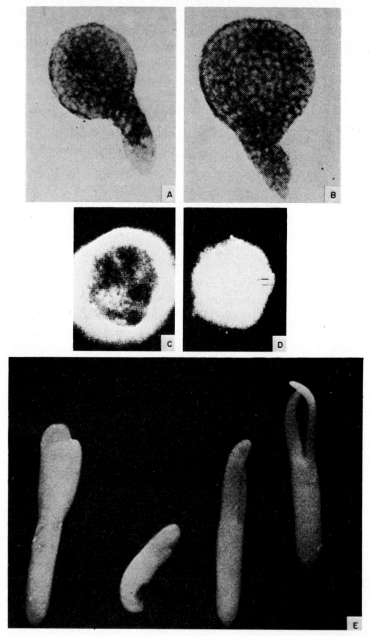

Fig. 7.3. Successful culture of proembryos of *Datura stramonium* is achieved in a medium containing embryo factor from coconut milk. **A, B**, Proembryos 100 μm and 130 μm in diameter which do not grow in the basal medium. **C, D**, Occasional formation of callus from proembryos grown in the basal medium. **E**, Normally differentiated embryos are obtained when proembryos are cultured in a medium containing embryo factor from coconut milk. The dimensions of the embryos at excision are, respectively, left to right, 100 μm, 130 μm, 170 μm, and 170 μm. (After van Overbeek, 1942; photographs courtesy of J. van Overbeek)

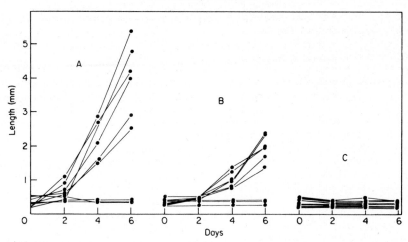

Fɪɢ. 7.4. The growth of embryos of *Datura stramonium*, isolated at the heart-shaped stage (200–500 μm long) and cultured in mineral salt–sucrose medium containing: **A**, Mixture of growth factors and nonautoclaved coconut milk. **B**, Similar to A with mixture of growth factors omitted. **C**, Similar to A with coconut milk omitted. Each line represents the growth curve for a single embryo. (After van Overbeek *et al.*, 1942)

long) maize embryos (Haagen-Smit *et al.*, 1945). Growth in length of the latter was, however, accelerated by the addition of asparagine and biotin (Haagen-Smit *et al.*, 1945) or of adenine, ascorbic acid, succinic acid, glycine, nicotinic acid, pantothenic acid and pyridoxine (Uttaman, 1949b) to a medium already containing 5% sucrose. In 2 week or 3 week old maize embryos, coconut milk had a depressing effect on growth of the shoot and root systems. When the extract was applied about 24 h after transfer of embryos, part of the growth inhibition was relieved, but the overall growth of embryos was still less than in the control (Uttaman, 1949c, e). No more should be read into these results than the simple inference that substances necessary for growth induction in isolated maize embryos are different from those present in coconut milk.

An embryo factor necessary for growth of 20 day old tomato embryos is reportedly present in nonautoclaved coconut milk and a temporary maintenance of embryos in a medium containing 50% coconut milk is advocated for rearing healthy plants (Choudhury, 1955b). According to Zenkteler *et al.* (1961), irrespective of the presence of accessory substances in the basal medium, coconut milk seemed to serve as a much improved substrate for the survival and growth of embryos of carrot excised at the very young cotyledon stage (450–600 μm long). The specificity of this requirement may be questioned since growth was not improved by addition of coconut milk to a basal medium of a different composition.

Although earlier investigators did not observe any favorable effects of coconut milk on the growth of immature barley embryos, later work (Norstog, 1956a, 1961) showed that undifferentiated barley embryos which generally failed to survive in White's medium could be successfully cultured by supplementing it with coconut milk (Fig. 7.5). The

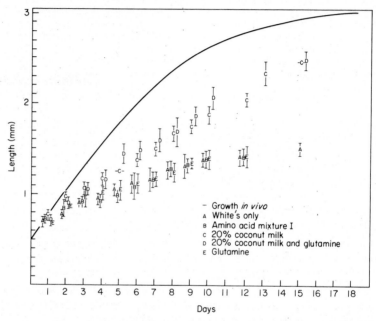

FIG. 7.5. The growth of 500 μm long embryos of barley in White's medium is promoted by coconut milk, or by coconut milk and glutamine. In this figure, growth of embryos *in vitro* in different media is compared with growth *in vivo*. Vertical bars indicate standard deviations. (After Norstog, 1961)

smallest embryo thus cultured had probably no more than 100 cells (about 60 μm long), and its survival and subsequent growth in culture were enhanced by the addition to the medium of glutamine or a mixture of amino acids, besides coconut milk. The interaction between coconut milk and amino acids or glutamine in this system is not entirely clear.

In the experiments described above, milk from young or mature coconuts was equally effective; paradoxically enough, the concept of coconut milk as a growth promoter is not borne out by its effects on cultured coconut embryos. Cutter and Wilson (1954) showed that milk from mature coconuts or an infusion of the solid endosperm markedly inhibited growth of mature coconut embryos, although milk from young and tender coconuts promoted growth to some extent (Abraham and Thomas, 1962). The question as to what factors in the endosperm of the mature drupe prevent growth of the cultured embryo is clearly not an easy one to answer and has received little attention so far, despite the widespread use of coconut milk in tissue culture media. The suggestion that the endosperm of the mature nut helps to maintain the embryo in a dormant state appears an attractive basis for investigating this question further.

The stimulatory effect of coconut milk does not seem to be restricted

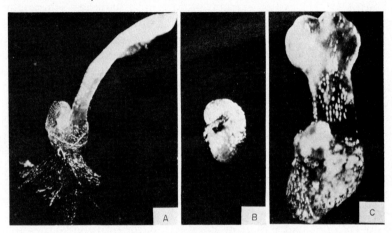

FIG. 7.6. Effect of coconut milk on the growth of embryos of *Todea barbara*. **A**, Normal sporophytic growth in an embryo excised 20 days after fertilization and cultured in a mineral salt medium for 1 month. **B**, Absence of adult organs in a 17 day old embryo grown in the same medium. **C**, Formation of normal sporophyte in a 17 day old embryo grown in a medium supplemented with 10% coconut milk. (After DeMaggio and Wetmore, 1961b; photographs courtesy of A. E. DeMaggio)

to angiosperm embryos, as seen from studies on embryos of the fern, *Todea barbara* (DeMaggio and Wetmore, 1961a, b). Given an adequate nutrition in the form of a simple mineral salt medium containing sucrose as the carbon energy source, embryos isolated 20 days after fertilization developed into normal sporophytes. When 17 day old embryos were cultured in the same medium, they increased in size, but never reached a point where distinctive adult organs were formed. Addition of 10% autoclaved coconut milk to the medium resulted in normal growth of the sporophyte, and in many cases leaf growth and maturation of cultured embryos exceeded that of their counterparts growing in a natural environment in the greenhouse (Fig. 7.6).

Other Endosperm and Plant Extracts

The role of natural plant extracts, most of which are of endospermic origin, in the culture of immature embryos has gradually emerged during the last few years, suggesting that embryo factors must be of very widespread occurrence. If small embryos of *Ginkgo biloba* were cultured in a medium containing an extract of *Ginkgo* endosperm, their growth was enhanced considerably (Li, 1934; Li and Shen, 1934). Much the same results were obtained in inducing growth of immature maize embryos by an extract of maize kernel (Voss, 1939). According to Kent and Brink (1947), water extracts of dates and bananas, wheat gluten hydrolyzate and tomato juice promoted growth of excised embryos of barley to the same extent as casein hydrolyzate. In further work (Ziebur and Brink, 1951) when immature embryos (0·3–1·1 mm long) were cultivated in a solidified medium containing 12·5% sucrose, mineral salts and casein hydrolyzate, improved growth was obtained by placing aseptically excised barley endosperm around the embryos. Some growth promotion was also achieved by incorporating endosperm extract in the culture medium, by autoclaving or by freezing the endosperm before application or by passing the activity from fresh endosperm through a cellophane bag or filter paper (Table 7.I). The observed activity of barley endosperm in stimulating growth of embryos of *Raphanus sativus* and *Capsella bursa-pastoris* is an important and as yet unexplained phenomenon. Although this may indicate a lack of specificity of the endosperm in growth induction, the limiting factors in barley endosperm tissue might conceivably be related to the chemical control mechanism involved in initiating cell division in the embryos of the two latter species. Similarly, embryos excised from young caryopses of *Hordeum* and *Triticum* have been cultured with success on a medium containing the milky endosperm of maize, or an active fraction precipitated therefrom. The active fraction was better than the crude milk or casein hydrolyzate in enchancing growth of embryos (Györffy *et al.*, 1955). A persistent growth-promoting effect of the maternal endosperm has been noted in the growth of small ($<200\ \mu$m) embryos of *Hordeum* × *Triticum* hybrid (Kruse, 1974).

In a study designed to discover natural plant extracts which could substitute for coconut milk in promoting growth of embryos, embryo factor activity on a level comparable to that exhibited by coconut milk was found in alcohol diffusates from young seeds of *Lupinus luteus* and *Datura stramonium*, and old seeds of *Sechium edule* (Matsubara, 1962, 1964c). The extract of *L. luteus* showed a wide spectrum of activity when tested against embryos of a number of species and was most effective

TABLE 7.I. The effect of barley endosperm on the growth of barley embryos *in vitro*. (From Ziebur and Brink, 1951)

Experiment	Treatment	No. of embryos	Length × width (% of average) and standard error
1	Control	18	13 ± 1
	Same-aged endosperms	16	75 ± 11
	Older endosperms	17	215 ± 19
2	Control	14	6 ± 1
	Older endosperms	14	210 ± 13
	Older endosperm extract	15	96 ± 6
3	Control	26	38 ± 6
	Older endosperms	28	203 ± 13
	Older endosperms autoclaved	28	62 ± 5
4	Control	20	31 ± 7
	Older endosperms	20	172 ± 29
	Older endosperms frozen	20	102 ± 15
5	Control	39	38 ± 5
	Older endosperms in cellophane box	40	128 ± 9
	Older endosperms in filter paper box	39	136 ± 8

for embryos of *Datura tatula, Pharbitis nil, Bidens biternata, Brassica campestris, Lupinus luteus, Capsella bursa-pastoris*, and of the hybrid *Brassica pekinensis × B. chinensis*, only slightly effective for embryos of *Stellaria media, Antirrhinum majus, Astragalus sinicus* and *Triticum aestivum*, more or less ineffective for embryos of *Vicia faba* and *V. sativa*, and inhibitory for embryos of *Iris pseudacorus*. Diffusates of young seeds and fruits of several other plants also showed embryo factor activity and promoted growth of young embryos of *D. tatula* (Table 7.II). The most effective extracts enhanced growth in length of embryos several-fold over their length at excision. It follows from the results that there is no species-specificity in the effects of diffusates in promoting growth of embryos or in the response of embryos to the diffusates. Since promotion of embryonic growth by diverse plant extracts appears widespread, it is

Table 7.II. Effect of ethanol diffusates from young seeds or fruits of various plants on the growth of young *Datura* embryos. Diffusates obtained from 5, 10 or 20 g materials were sterilized with Seitz-filter and added to 100 ml of the basal medium. *Datura* embryos were cultured at 30°C in darkness for 5 days. (From Matsubara, 1964c)

Diffusates from	g/100 ml	Number of embryos observed	Initial length (mm)	Final length (mm)	Growth value
Glycine soja	0	15	0·29	1·78	6·1
	5	7	0·27	4·44	16·4
	20	9	0·27	0·27	1·0
Phaseolus vulgaris	0	9	0·28	1·63	5·8
	5	11	0·24	3·06	12·7
	20	7	0·27	3·12	11·9
Phaseolus vulgaris 'Black Valentine'	0	11	0·25	1·01	4·0
	5	9	0·22	1·60	7·3
	20	11	0·25	1·72	6·9
Brassica campestris (fruit)	0	11	0·19	0·79	4·2
	5	10	0·21	1·24	5·9
	20	6	0·20	1·17	5·9
Pisum sativum	0	6	0·25	1·18	4·7
	5	7	0·26	2·97	11·4
	20	9	0·29	4·00	13·8
Arachis hypogaea	0	10	0·23	1·45	6·3
	5	9	0·24	3·61	15·1
	20	9	0·25	4·34	17·4
Canavalia ensiformis	0	10	0·23	1·45	6·3
	5	10	0·23	1·60	7·0
	20	9	0·24	3·14	13·1
Lupinus luteus	0	10	0·26	0·63	2·4
	5	10	0·21	5·10	24·3
	10	10	0·19	4·83	25·4

unfortunate that little information is available on the mechanism by which this growth is attained.

There is some evidence that undifferentiated tissues of embryonic origin may impart some growth-promoting factors for growth of isolated embryos. According to Thomas (1972, 1973a), when pine *(Pinus mugo, P. sylvestris* and *P. nigra)* embryos isolated at the time of cleavage were cultured in close contact with a nurse tissue of parenchymatous callus cells originating from the embryo, elongation of the suspensor, and division of the embryonal cell occurred during the first few days in culture (Fig. 7.7). This evidently suggests that potential growth factors for embryo growth are not necessarily confined to tissues of the megagametophyte. Far from representing the activation of an unchanged component within the original embryonic cells, the observed

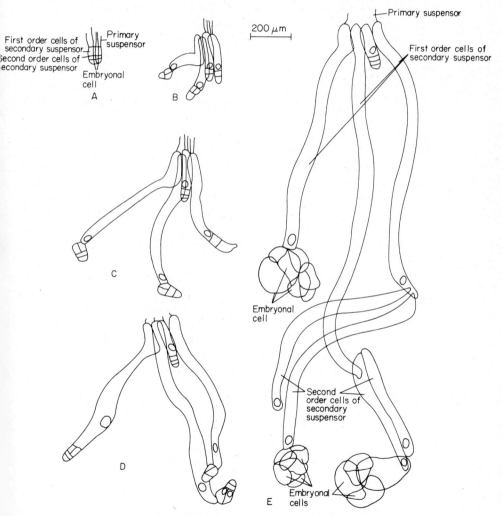

FIG. 7.7. Growth of embryos of *Pinus nigra* cultured in the presence of a nurse tissue. **A**, At the time of culture. **B**, **C**, **D**, **E**, After 1, 2, 3 and 9 days' culture. At 9 days cells of the suspensor are dead, and embryonal cells dissociate. (After Thomas, 1972)

growth effect must be ascribed to the synthesis of new metabolites under conditions which transform the embryo into an undifferentiated tissue.

Chemical Factors of Endosperm Extracts

Some authors have approached the question of growth of embryos induced by endosperm extracts, with a view to defining clearly the factors contained in them that control cellular growth and differentiation. Only a limited number of cases have been analyzed in detail, but these are all important in their various ways. The general concept is pressed that embryonic growth induced by complex fluids of natural origin is the result of an interaction of specific hormonal or chemical substances which control differential cell division and cell elongation. One of the most persuasive pieces of evidence for a link between chemical factors present in coconut milk and embryo growth stems from the work of DeMaggio and Wetmore (1961a, b). The success of these authors in inducing growth of 17 day old embryos of *Todea barbara* by the addition of coconut milk to the medium was described earlier. When a mixture of the hexitol, *myo*-inositol, and the sugar alcohol, sorbitol, was substituted for coconut milk, growth of embryos exceeded even that obtained by the addition of the most favorable concentration of the latter. It seems that *myo*-inositol and sorbitol, known to be the chief constituents of the neutral fraction of coconut milk (Pollard *et al.*, 1961), can adequately replace the need for whole coconut milk in the normal growth of excised embryos.

The ability of the endosperm extracts of pumpkin *(Cucurbita maxima and C. moschata)* and cucumber *(Cucumis sativus)* to promote growth of their respective embryos is probably the result of the action of a plant hormone, a cell division substance and several organic nitrogen compounds (Nakajima, 1962). When embryos were grown in a medium containing IAA, growth of embryonic leaves lagged behind, probably because of lack of cell divisions. Upon addition of 1,3-diphenylurea or kinetin, leaf growth was restored, although the overall growth of the embryo was poor. A medium containing IAA, 1,3-diphenylurea and casein hydrolyzate supported growth of embryos to the same extent as did a medium containing embryo factor. Coincidentally, the same set of factors that control growth of embryos was found to regulate the growth of endosperm *in vitro*. It is easy to visualize the interesting possibility that endosperm which controls the growth of the embryo by supplying it with specific nutrients is not capable of producing them within itself but receives them from the parent tissue. Attractive though this idea may be, there is minimal evidence to support it. More recently, Norstog and Smith (1963) have concluded that a phosphate-enriched

White's medium at an optimum pH of 4·9 fortified with glutamine (400 mg/l) and alanine (400 mg/l) as major nitrogen sources and lesser amounts of leucine (20·0 ml/l), tyrosine, phenylalanine, cysteine and tryptophan (all at 10·0 mg/l) can substitute for coconut milk in the culture of barley embryos as small as 60–90 μm in length. Embryos cultured in the modified medium for 10 days attained the same degree of morphological and anatomical differentiation as a normal full-term embryo. In a further modification of this medium, it has been shown that survival of embryos in culture can be greatly enhanced by a 5–10 × increase in the concentration of KNO_3 and KCl in the medium (Norstog, 1967). The conclusion seems inescapable that growth induction in cultured embryos by endosperm extracts is a measure of the effect of specific chemical components present in the endosperm.

It should perhaps be stressed at this point that evolution of synthetic media containing known chemical substances capable of supporting growth of proembryos has only been achieved in a few cases. Only when culture of proembryos is possible in such media will our understanding of their physiology and metabolism be furthered.

Effect of Protein Preparations

Particular interest may attach to the observations of some investigators that addition to the medium of malt extract, casein hydrolyzate and other commercial preparations containing amino acids is effective in promoting growth of immature embryos, and in some cases the additives may actually substitute for the requirement for coconut milk or other growth-promoting extracts. Work in this direction began with a report (Blakeslee and Satina, 1944) that an aseptically filtered solution of powdered malt induced growth of small hybrid embryos of *Datura* (less than 500 μm long) as effectively as coconut milk. Autoclaved malt extract was inhibitory, however, due to the presence of an inhibitor, possibly formed by the breakdown of some precursors during autoclaving and which masked the activity of the embryo factor (Solomon, 1950). In a recent work, Matsubara and Nakahira (1965a) found that among a number of compounds tested only casein hydrolyzate, tryptone and peptone approached the level of activity of *Lupinus luteus* extract in promoting growth of heart-shaped embryos of *D. tatula*. Addition of casein hydrolyzate to the basal medium has been shown to promote growth of immature embryos of cotton (Lofland, 1950; Mauney, 1961), barley (Ziebur *et al.*, 1950) and *Capsella bursa-pastoris* (Rijven, 1952). In the culture *in vitro* of proembryos of *Ginkgo biloba* (Radforth, 1936) and *Pinus nigra* var. *austriaca* (Radforth and Pegoraro, 1955), best growth was obtained in a medium containing sucrose or dextrose and yeast extract. According to Rangaswamy (1958a, 1961), supplementation of White's basal medium with 400 mg/l casein hydrolyzate enabled proembryos of *Citrus microcarpa*, even down to 28 μm in length, to grow and become fully organized. In contrast, most of the embryos cultivated in the basal medium alone made little or no growth during prolonged periods in culture. Among a number of substances tested, only casein hydrolyzate and yeast extract proved suitable for rearing young embryos of *Gnetum ula* (Vasil, 1963). It seems safe to assume that in all of the above cases it is the non-specific mixture of amino acids in casein hydrolyzate and other products that contributes in some measure to the growth induction in embryos. A prevalent notion that the growth-promoting effect of casein hydrolyzate is due to its high osmotic pressure is discussed below.

B. Role of High Osmotic Concentration

The experimental evidence discussed in the preceding subsection appeared to be a convincing lead to explain the mechanism that initiates growth in small embryos. Such evidence was, however, inconsistent with the hypothesis of a number of investigators on the possible role of high osmotic pressures in regulating embryo growth in culture. This view is clearly in harmony with the common observation that the amorphous liquid endosperm in which young embryos are constantly bathed has a high osmolarity (Ryczkowski, 1960, 1961, 1965, 1969; Kerr and Anderson, 1944; Mauney, 1961; Smith, J. G., 1973). Artificially increasing the osmotic concentration of the culture milieu by addition of sucrose or mannitol lent itself to the culture of embryos of a number of plants which did not grow previously even in the most complex media tried. A favorable osmotic concentration, besides preventing a possible osmotic shock to the embryo excised from an environment of high osmolarity, also inhibits cell elongation usually observed during precocious germination. By suppressing the germination potential of the embryo and switching cells from a state of elongation to one of division, the high osmotic value of the sap exerts its apparent growth-promoting effect. This is illustrated in some early studies where precocious increase in length and other deleterious effects of an inorganic salt solution on the growth of embryos of several species were overcome by the addition of high concentrations of sucrose to the medium (Hannig, 1904; Dieterich, 1924). Similarly, an osmotic effect appears to be the basis for a requirement for increasing concentrations of sucrose for growth of embryos isolated at progressively earlier stages from seeds of deciduous trees (Tukey, 1934b, 1938; Lammerts, 1942). While these reports are

indirect in their manifestation, and can thus hardly be considered conclusive, Kent and Brink (1947; see also Ziebur *et al.*, 1950; Ziebur,
1951) provided the first clear-cut evidence of a relationship between osmolarity of the medium and growth of small embryos. Their
work was the outcome of attempts to control precocious germination
of barley embryos excised and cultured 10–15 days after pollination.
The germination-inhibiting effects of 1% casein hydrolyzate and the
parallel effects of isotonic solutions of sucrose or mannitol on embryos

TABLE 7.III. The effects of various components of casein hydrolyzate tested
singly, in pairs and all together on growth of barley embryos. (From Ziebur
et al., 1950)

Treatment	Osmotic pressure of medium (atm)	Average shoot height ± standard error[a]
Control	1·3	276 ± 32
Amino acids	3·3	57 ± 8
KH_2PO_4 0·013 M	2·0	156 ± 24
NaCl 0·13 M	7·1	84 ± 13
Amino acids, KH_2PO_4 and NaCl	9·7	6 ± 3
Casein hydrolyzate 1%	9·7	4 ± 2
Control	1·3	175 ± 11
Amino acids + KH_2PO_4	3·9	268 ± 12
Amino acids + NaCl	9·1	5 ± 1
KH_2PO_4 + NaCl	7·7	34 ± 3
Amino acids, KH_2PO_4 and NaCl	9·7	12 ± 2

[a] Shoot height expressed as a percentage of the average value of that statistic for all
treatments in the whole replicate.

led them to attribute the effects of casein hydrolyzate to the high
osmotic pressure it produced in the medium. It was found, for example,
that immature embryos about 1·4–2·8 mm long, when planted on the
basal medium, did not generally continue embryonic growth but
germinated to form spindly seedlings. Normal embryonic growth was
sustained by the addition of 1% casein hydrolyzate to the medium.
After a growth period of one week, such embryos were slightly larger
than mature embryos in the seed and had a high percentage of dry
matter.

Amino acids, sodium chloride and inorganic phosphates are the usual components of commercial preparations of casein hydrolyzate. Additional evidence for the osmotic role of casein hydrolyzate comes from experiments in which growth of embryos was studied with the individual components of this preparation added singly, or in combinations of two and all three together, to the basal medium. As shown in Table 7.III, precocious germination was completely blocked in a medium containing both amino acid mixture and sodium chloride, with or without phosphate. Thus, the main function of casein hydrolyzate in this system is probably not nutritional in the sense of supplying combustible substances for metabolism, but to provide the correct osmotic conditions for the growth of embryos.

Datura is another example which illustrates the significance of osmotic value of the medium in the growth of embryos (Rietsema *et al.*, 1953a; Matsubara, 1962, 1964c; Matsubara and Nakahira, 1965a). Mature embryos (4·5 mm or longer) of *D. stramonium* grow even in the absence of sucrose in the medium, while successively younger embryos require progressively higher concentrations of sucrose (Table 7.IV). Is this due to the well known ability of sucrose to provide carbon energy source for growth processes, or is it an expression of the osmotic effect of the medium? In this system, the role of sucrose as an osmoticum was confirmed by substituting nutritionally inert substances like mannitol or glycerol in different concentrations in the medium. Considering the growth of the hypocotyl of the embryos of *D. stramonium* as a criterion for evaluation, Fig. 7.8 shows the results of an experiment on the effects of addition of different concentrations of mannitol to a basic medium containing 2% sucrose and 400 mg/l casein hydrolyzate on embryo growth. It appears that while growth of torpedo-shaped embryos (2 mm long) is inhibited even at an osmolarity of 2·0 atm (0% mannitol), optimum growth of late heart-shaped (0·5 mm long) and pre-heart-shaped embryos (0·25 mm long) is obtained at 3·2 atm (1% mannitol) and 6·9 atm (4% mannitol), respectively. When the osmotic values were kept constant and sucrose concentrations varied by changing the sucrose:mannitol ratio in the medium, irrespective of the age of the embryos, optimum growth was observed at 2% sucrose (Rietsema *et al.*, 1953a). Similarly, in media of constant osmotic value secured by the addition of different amounts of mannitol and NH_4NO_3, maximum growth of embryos of *D. tatula* was obtained in 0·0075 M NH_4NO_3, and further increase in concentration of NH_4NO_3 did not further enhance growth (Matsubara, 1964c). These results lead us to conclude that differences in carbon energy or nitrogen requirements of embryos of different ages can be due to differences in their osmotic requirement.

TABLE 7.IV. The number of embryos of *Datura stramonium* which grew on various sucrose concentrations in series of ten. Numbers in italics other than column headings are shown against the lowest sucrose concentrations at which more than five embryos grew. (From Rietsema *et al.*, 1953a)

Initial embryo size (mm) →

Sucrose concentration (%) ↓	Pre-heart		Heart			Early torpedo			Torpedo		Mature embryo	
	0·1	0·15	0·2	0·35	0·5	0·6	1·0	1·4	2·0	3·5	4·5	5·0
0	—	—	0	0	0	0	0	0	0	0	*10*	*8*
0·1	—	—	0	0	0	—	—	*9*	*10*	—	10	10
0·2	—	—	—	—	—	*7*	2	—	10	—	10	—
0·5	—	—	0	0	4	—	—	10	10	*10*	—	10
1·0	1	0	2	2	*9*	8	*10*	10	10	10	10	10
2·0	0	0	*6*	*10*	10	6	10	10	10	10	10	10
4·0	2	1	6	10	10	9	10	10	10	10	10	10
8·0	3	*6*	7	10	10	7	10	10	10	10	10	10
12·0	*7*	7	9	10	10	—	—	—	—	—	10	—
16·0	5	4	—	—	—	—	—	—	—	—	—	—

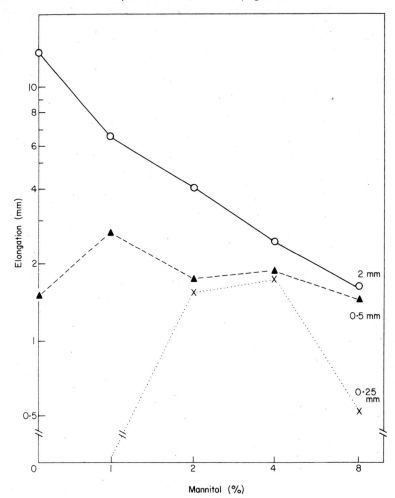

Fig. 7.8. Addition of 1–4% mannitol to a medium containing 2% sucrose promotes growth of the hypocotyl of heart-shaped (0·5 mm long) and pre-heart-shaped (0·25 mm long) embryos of *Datura stramonium*. The initial embryo size is indicated in the figure. (After Rietsema *et al.*, 1953a)

There are few other data as complete as those on barley and *Datura* on the relation between osmotic value of the medium and the growth of embryos of different ages. Several other investigators have also established the need to provide culture media of high osmotic values to induce growth of embryos which do not usually grow even in the most complex media tried, and it is to such studies that attention is now directed. Rijven (1952) and Veen (1961, 1962, 1963), for example, have routinely used 12–18% sucrose in a liquid medium in attempts to culture torpedo-shaped and heart-shaped embryos of *Capsella bursa-pastoris*. Later work (Raghavan and Torrey, 1963) has, however, shown that the need for high osmolarity can be dispensed with by growing embryos in a simple solidified medium containing the usual macro- and micronutrients, vitamins and 2% sucrose. When cultured in this medium even heart-shaped embryos went through the normal embryonic stages of growth without showing signs of germination and gave rise to small plantlets (Fig. 7.9). There is every reason to suppose that the solid versus liquid medium may account for these differences, but until experimental evidence for this is presented, the role of high osmotic pressure in growth induction in the embryos of *Capsella* must remain uncertain.

Rangaswamy (1961) has reported that nucellar proembryos of *Citrus microcarpa* reared in a medium containing 5–10% sucrose remained healthy for as long as 20 days, while no growth, beyond swelling, was obtained in a lower concentration of sucrose. Following Ziebur and Brink (1951), Norstog (1961) used 12% sucrose in his successful culture of small barley embryos in glutamine–coconut milk medium. High osmotic pressure of the medium reportedly has an inhibitory effect on root elongation in rice embryos (Amemiya, 1964). Finally, in long-term cultures of embryos, it is necessary to make provision for their changing osmotic requirements, and transfer of embryos periodically to media with progressively lower osmotic values is necessary for the production of transplantable seedlings (Mauney, 1961; Pecket and Selim, 1965; Rangan *et al.*, 1969). These observations are an interesting and important corollary to the work demonstrating a role for high osmoticum in the initiation of growth in small embryos.

Against the background of experimental work discussed above, it seems permissible to suggest that growth induction in proembryos of several species in artificial culture is achieved by the establishment of high osmotic value in the medium, which probably allows for an effective flow of metabolites. Unfortunately, even in species which have been extensively studied, no data on the uptake and utilization of specific nutrients as a result of increased osmotic pressure are available.

C. Effect of Growth Hormones

Although there is an impressive body of data on the effects of growth hormones on the morphogenesis of mature embryos (see Chapter 10), reports on their use in the induction of growth of proembryos are

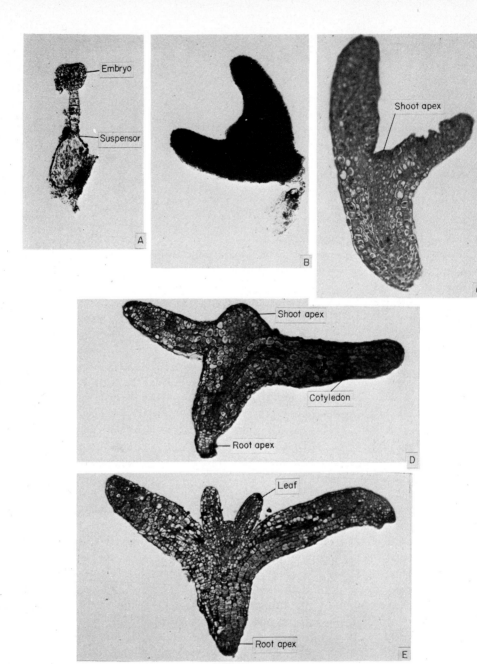

Fig. 7.9. Heart-shaped embryos of *Capsella bursa-pastoris* are successfully grown in a solidified mineral salt medium containing vitamins and 2% sucrose. **A**, Embryo is 81 μm long at the time of excision. **B**, Embryo after 2 weeks in culture in the dark. **C**, Section through an embryo after 3 weeks' culture, showing the shoot apex. **D**, Section through an embryo after 5 weeks' culture, showing root and shoot apices. **E**, Section through an embryo grown in alternating 12 h periods of light and dark for 5 weeks showing formation of the first pair of leaves. (After Raghavan and Torrey, 1963)

limited and many of them are negative (LaRue, 1936; Tukey, 1938; Sterling, 1949; Sanders, 1950). Some of these studies were undertaken in the early years following the discovery of auxin; no goal could have been more rewarding than to identify auxin as the single substance causing growth of very small embryos, and thereby add yet another significant effect of auxin to the growing list of its effects on plants. Among the successful attempts to use growth hormones in the culture of embryos may be mentioned the work of Loo and Wang (1943), who grew embryos of *Pinus* and *Keteeleria* one to several cells in size in a medium supplemented with auxin. In another case, fortification of the basal medium containing yeast extract with IAA and GA is reported to have led to successful culture of young embryos of *Corchorus olitorius* (Iyer *et al.*, 1959). Significant increases in length of torpedo-shaped embryos of *Capsella bursa-pastoris* have been attributed to the addition of GA and IAA to a medium containing 12–18% sucrose, but neither compound provoked growth in still smaller embryos (Rijven, 1952; Veen, 1962, 1963). On the other hand, addition of kinetin seemed to increase the chances of survival of embryos, probably by inducing a few rounds of cell divisions. As can be seen from Fig. 7.10, in a manner not clearly understood some of the embryos treated with kinetin grew abnormally and, unlike normal embryos, did not develop cotyledons. A partial resumption of growth in such embryos, manifest by an increased elongation of the hypocotyl, was possible by transferring them to a medium containing 12% sucrose and GA (Fig. 7.11) (Veen, 1963).

By the use of a balanced mixture of IAA, kinetin and adenine, some success has been achieved in inducing growth of proembryos of *Capsella bursa-pastoris* in an osmotically unadjusted medium in which slightly older embryos were grown (Raghavan and Torrey, 1963). The development of an early globular embryo, initially 54 μm long, in the basal medium supplemented with IAA (0·1 mg/l), kinetin (0·001 mg/l) and adenine sulfate (0·001 mg/l), is illustrated in Fig. 7.12. During the first 7 days in culture, the embryo increased in size as a sphere by irregular cell divisions and cell enlargement. After it had attained a few hundred cells, bilateral differentiation to form the cotyledons was observed. However, even this medium was inadequate to induce growth in embryos in the very early division stages. Securing a suitable range of growth hormones and other activating substances, the need for which seems highly probable in the light of general nutritional and metabolic studies, may be the crux of the problem in designing a medium for the growth of such embryos. The only other supplements to the medium which supported growth of embryos of comparable length were either a high concentration of sucrose or major salts.

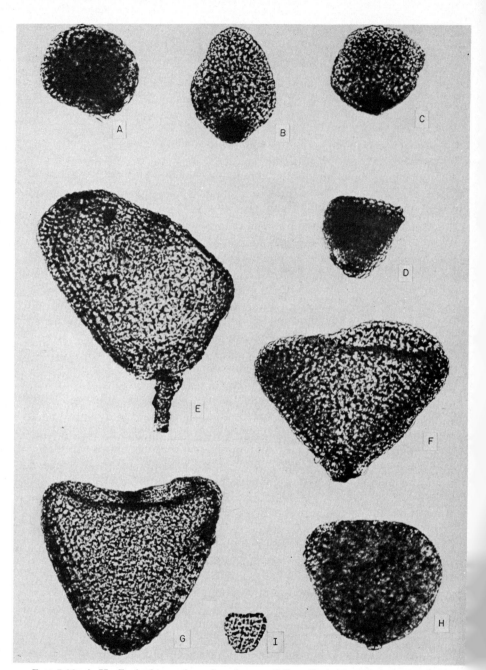

Fig. 7.10. **A–H**, Early heart-shaped embryos of *Capsella bursa-pastoris* grown in a medium containing 18% sucrose and different concentrations (10^{-9}–10^{-7} g/ml) of kinetin fail to form cotyledons. **I**, Embryo grown in the basal medium. (After Veen, 1963; photographs courtesy of H. Veen)

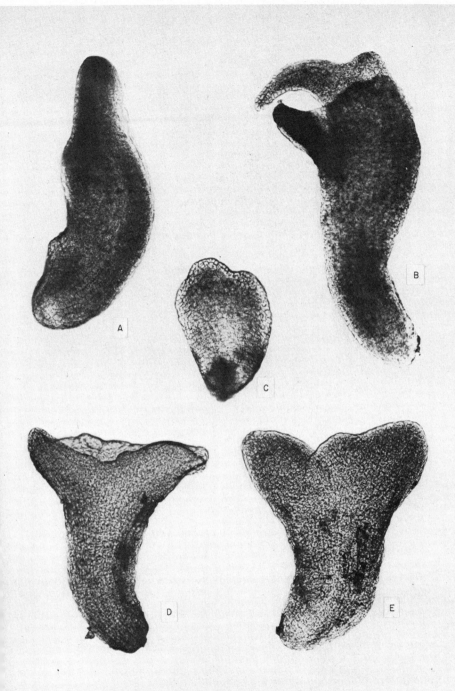

FIG. 7.11. Early heart-shaped embryos of *Capsella bursa-pastoris* showing abnormal growth in a medium containing 18% sucrose and 10^{-8} g/ml kinetin (see Fig. 7.10) partially resume normal growth when transferred after 8 days to a medium containing 12% sucrose and 10^{-5} g/l gibberellic acid. (After Veen, 1963; photographs courtesy of H. Veen)

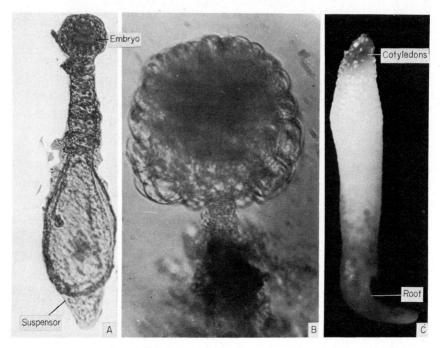

Fig. 7.12. Stages in the development of the proembryo of *Capsella bursa-pastoris* grown in a medium containing IAA (0·1 mg/l), kinetin (0·001 mg/l), and adenine sulfate (0·001 mg/l). **A**, The embryo is 54 μm long at the time of excision. **B**, Formation of a mass of cells from the embryo after 10 days in culture. **C**, Differentiation of the cotyledons and root meristem after 6 weeks in culture. (After Raghavan and Torrey, 1963)

The ease with which heart-shaped embryos of *Capsella* grow in a simple osmotically unadjusted mineral salt medium and the induction of growth in still smaller embryos by growth hormones or by increased concentrations of sucrose or major salts pose a problem of considerable magnitude on the precise role of osmotic milieu of the medium and growth hormones in the regulation of embryo growth. Obviously the physical control of a high osmotic pressure and the chemical control of hormones must be linked in some fashion. It is possible that the activity of one or more components of the balanced hormonal control system is in turn controlled by the high sucrose or high salt concentrations, perhaps through osmotic processes preventing cell elongation.

D. Nutrient Media for Culture of Proembryos

Derived from the above considerations, the optimal growing conditions for proembryos of several species are set out in Table 7.V.

TABLE 7.V. Nutrient media for growth of proembryos of different species

Species	Length of embryo at excision	Nutrient media			Reference
		Carbon energy	Mineral salts	Organic supplements	
Capsella bursa-pastoris	140–170 μm	Sucrose, 18%	Olsen's mineral salts in phosphate buffer	(mg/l): thiamine, 0·15; nicotinic acid, 1·0; pyridoxine, 0·2; calcium pantothenate, 0·2; inositol, 0·5; p-aminobenzoic acid, 0·5; riboflavin, 0·1; folic acid, 0·01; biotin, 0·0004; alanine, 43·1; valine, 57·0; leucine, 47·0; isoleucine, 75·0; proline, 42·5; phenylalanine, 54·5; cystine, 9·3; cysteine, 5·0; arginine, 167; histidine, 29·0; lysine, 2·4; aspartic acid, 120; glutamic acid, 207; amide-NH_3, 21·5; serine, 63·0; threonine, 38·5; tyrosine, 43·4; tryptophan, 14·8; and methionine, 24·0 (or glutamine 600, instead of the mixture of amino acids)	Rijven (1952)

Table 7.V (cont.)

Species	Length of embryo at excision	Carbon energy	Mineral salts	Organic supplements	Reference
			Nutrient media		
	40·5–67·5 μm	Sucrose, 2%	Modified Robbins and Schmidt's medium	(mg/l): thiamine hydrochloride, 0·1; pyridoxine hydrochloride, 0·1; niacin, 0·5; indoleacetic acid, 0·1; kinetin, 0·001; and adenine sulfate, 0·001	Raghavan and Torrey (1963)
	less than 50 μm	Sucrose, 12%	Modified Murashige–Skoog medium	glutamine, 400 mg/l; thiamine and pyridoxine, 10^{-7} M	Monnier (1973)
Citrus microcarpa	0·14–0·28 mm	Sucrose, 5 or 10%	Modified White's medium	(mg/l): glycine, 7·5; niacin, 1·25; thiamine hydrochloride 0·25; calcium pantothenate, 0·025; pyridoxine hydrochloride, 0·025; indoleacetic acid, 1·0; and casein hydrolyzate, 400	Rangaswamy (1961)
Corchorus olitorius	0·3 mm	Sucrose, 5%	White's medium with ferric	yeast extract, 0·1%; indoleacetic acid, 0·1 mg/l; and gibberellic	Iyer et al. (1959)

Species	Embryo size	Medium	Sugar	Additives	Reference
Datura stramonium	0·15 mm	Tukey's medium	Dextrose, 1%	...ryptophan, 2·5; and casein hydrolyzate (mg/l): glycine, 3·0; thiamine, 0·15; ascorbic acid, 20·0; nicotinic acid, 1·0; pyridoxine, 0·2; adenine, 0·2; succinic acid, 25·0; pantothenic acid, 0·5; and nonautoclaved coconut milk, about 21% by volume	...(1962) van Overbeek *et al.* (1942)
	0·10–0·16 mm	Modified Randolph and Cox's medium	Sucrose, 8–12%	400 mg/l casein hydrolyzate	Rietsema *et al.* (1953a)
Datura tatula	0·11–0·27 mm	Modified White's medium	Sucrose, 6%	alcohol diffusates from young seeds of *Datura stramonium*, *Lupinus luteus* and *Sechium edule*	
	0·13–0·28 mm	As above	Sucrose, 8%	50·0 mg/l casein hydrolyzate	Matsubara (1962)
	0·11–0·48 mm	As above	Sucrose, 8%	10% (by volume) nonautoclaved coconut milk	
	0·15 mm	As above	Sucrose, 4–12%	50–500 mg/l casein hydrolyzate or 50–500 mg/l peptone	Matsubara and Nakahira, (1965a)

TABLE 7.V (cont.)

| Species | Length of embryo at excision | Nutrient media | | | Reference |
		Carbon energy	Mineral salts	Organic supplements	
Gossypium hirsutum	0·1–0·2 mm	Sucrose, 2%	White's medium 5 times concentrated; 7·0 g/l sodium chloride	(mg/l): glycine, 10·0; pyridoxine hydrochloride, 2·5; nicotinic acid, 2·5; thiamine, 0·5; casein hydrolyzate, 250; adenine, 40·0; and nonautoclaved coconut milk, 150 mg/l	Mauney (1961)
Hordeum vulgare	0·3–1·1 mm	Sucrose, 12·5%	Randolph and Cox's medium	0·1% casein hydrolyzate, and surrounding the embryos with aseptically excised endosperm of Hordeum vulgare	Ziebur and Brink (1951)
	60–1 500 μm	Sucrose, 12%	White's medium	20% (by volume) nonautoclaved coconut milk; glutamine, 400 mg/l	Norstog (1961)
	90–100 μm	Sucrose, 9%	Phosphate-enriched White's medium	(mg/l): niacin, 1·25; thiamine, 0·25; pyridoxine, 0·25; calcium pantothenate, 0·25; malic acid, 100; glutamine, 400; alanine, 400; leucine, 20·0; tyrosine, 10·0; phenylalanine, 10·0; cysteine, 10·0; and	Norstog and Smith (1963)

Species	Stage	Carbohydrate	Basal medium	Supplements	Reference
		5·5%	with 3–10× increase in the concentration of KNO₃ and KCl, pH 4.9. Cultures kept in 50–100fc light	malic acid was added as ammonium malate; glutamine filter-sterilized	(1967)
	200 μm	Sucrose, 3·42%	(g/l): KH₂PO₄, 0·91; KCl, 0·75; MgSO₄·7H₂O, 0·74; CaCl₂·2H₂O, 0·74; (mg/l) MnSO₄·H₂O, 3·0; H₃BO₃, 0·5; ZnSO₄·7H₂O, 0·5; CoCl₂·6H₂O, 0·025; CuSO₄·5H₂O, 0·025; NaMoO₄, 0·025; ferric citrate, 10·0. All components filter-sterilized	(mg/l): meso-inositol, 50·0; thiamine, 0·25; calcium pantothenate, 0·25; pyridoxine hydrochloride, 0·25; L-glutamine, 400; L-alanine, 50·0; L-cysteine, 20·0; L-arginine, 10·0; L-leucine, 10·0; L-phenylalanine, 10·0; L-tyrosine, 10·0; malic acid, 1·0 g/l added as ammonium malate. All components filter-sterilized	Norstog (1973)
Keteleeria davidiana	One to few-celled	Sucrose, 2%	Modified Pfeffer's medium	(mg/l): indoleacetic acid; 10·0; and thiamine, 0·1	Loo and Wang (1943)
Pinus yunnanensis	One to few-celled	Sucrose, 2%	Modified Pfeffer's medium	(mg/l): indoleacetic acid, 10·0; or thiamine, 0·1	Loo and Wang (1943)
Pinus mugo	at the time of cleavage	Sucrose, 2%	Heller's, Murashige and Skoog's or Halperin's medium	(mg/l): arginine, 2·0; asparagine, 0·1; glutamine, 50·0; glycine, 2·0; folic acid, 0·5;	Thomas (1972)

TABLE 7.V (cont.)

Species	Length of embryo at excision	Nutrient media			Reference
		Carbon energy	Mineral salts	Organic supplements	
Pinus mugo (cont.)				nicotinic acid, 5·0; biotin, 0·05; pyridoxine hydrochloride, 0·5; thiamine hydrochloride, 0·5; myo-inositol, 100; embryos cultured in the medium surrounded by a nurse tissue of embryonic origin	
Todea barbara	160 μm	Sucrose, 1%	Knudson's ½ strength medium with 10 mg/l ferric citrate and Nitsch's trace element mixture modified by addition of 25 mg/l CoCl₂	5% or 10% nonautoclaved coconut milk or 50·0 mg/l each of sorbitol and myo-inositol	DeMaggio and Wetmore (1961b)
	35–55 μm	Sucrose, 3%	As above	50·0 mg/l each of sorbitol and inositol	
Zea mays	0·3–3·0 mm	Sucrose, 5%	Tukey's medium	(mg/l); glycine, 3·0; thiamine, 0·15; ascorbic acid, 20·0; nicotinic acid, 1·0; pyridoxine, 0·2; adenine, 0·2; succinic acid, 25·0; pantothenic acid, 0·5; biotin, 0·001;	Haagen-Smit et al. (1945)

While some insight has been gained into the growth requirements of proembryos, the reader should not fail to notice that there is a fundamental roadblock in the obvious dearth of information on the culture of fertilized eggs and embryos in early division phases. It is upon the availability of this knowledge that a complete understanding of the control of embryogenesis will ultimately depend. It should, however, be borne in mind that the fertilized egg and its early division stages are not easy targets for culture. In the majority of vascular plants, they are too minute to be accessible to isolation. The paucity of data on their cultural requirements is not so much a reflection of the complexity of the medium required to foster their growth, as it is a consequence of their inaccessibility to experimentation.

III. COMPARATIVE GROWTH OF EMBRYOS *IN VIVO* AND *IN VITRO*

When viewed in its entirety, the changing pattern of growth of embryos *in vivo* provides a frame of reference for assessing the extent of growth and developmental morphology they attain in culture. The ultimate aim in embryo culture studies is to raise seedlings closely comparable to those obtained from germinated seeds, but this goal is seldom achieved owing to the prevalence of precocious germination in culture (Yoshii, 1925; LaRue, 1936; Tukey, 1938; McLane and Murneek, 1952; Ziebur *et al.*, 1950; Guzowska and Zenkteler, 1969). Needless to say, seedlings from immature embryos, if transplanted to soil and allowed to grow, are smaller and weaker than those raised from mature seeds.

When embryos of different ages are cultured, the final morphology attained by seedlings run a wide gamut of morphogenetic aberrations. In several varieties of pear *(Pyrus communis)* there was no development in culture of embryos excised prior to 46 days after anthesis. Embryos excised 66 days after anthesis showed greening of the cotyledons, while in 69 day old embryos there was also elongation of the cotyledonary axis which terminated in a small rosette of leaves. Small, transplantable seedlings with slender, long stems, stipule-like leaves and elongate roots were obtained from culture of 75 day old embryos. Similar but larger plantlets with typical normal leaves and vigorously formed roots originated from culture of 81 day old and 97 day old embryos. Embryos of different ages of sour cherry, sweet cherry, apricot, plum, apple and peach behaved in a similar manner with respect to the final morphology of the seedlings (Tukey, 1938).

Growth in culture of comparatively immature embryos of *Zizania aquatica* was generally characterized by precocious development of the shoot and retarded growth of the primary root (LaRue and Avery, 1938). In mature embryos, elongation of the primary root was normal. Thus, one can visualize a progressive retardation of growth in culture of increasingly younger embryos, as shown in Fig. 7.13. Similar observations have been made in the embryos of *Hordeum sativum* (Merry, 1942), *Lychnis alba* (Devine, 1948) and cotton (Lofland, 1950).

Considerable variation exists in the development in culture of embryos of different ages of *Todea barbara* (DeMaggio and Wetmore, 1961b). The major difference is that while embryos isolated at advanced stages of development attained a normal proportion of growth in a suitable nutrient medium, those isolated after fertilization, before the formation of the first division wall, eventually developed into flat thalloid structures, more closely resembling gametophytes than sporophytes (see p. 283). An interesting correlation between the stage of embryonic development at excision and the extent of unorganized growth prevails in cultured embryos of *Cuscuta gronovii* (Truscott, 1966). Upon culture in an appropriate medium, relatively young embryos (less than 0·4 mm long) lose their original form and proliferate callus throughout their length. Slightly older, 0·4–1·2 mm long embryos, while retaining their original form in culture, differentiate callus at the radicular end, but not at the plumular pole. Embryos of 1·5 mm and longer form normal shoots in addition to a radicular callus. It thus seems that there is a stage in development at which an irreversible determination of the morphogenetic pattern of the embryo is reached. Accordingly, the formation of a callus would seem to be the simplest expression of growth of an isolated embryo capable of cell division. Only when it has acquired determination can the embryo form a normal shoot without slipping back to callus growth.

Several investigators have commented upon the slow growth made by embryos in culture relative to their growth *in ovulo*. A comparative study of the growth of barley embryos *in vitro* and *in ovulo* (Chang, 1963b) showed that embryos about 0·4 mm long required 10–15 days to reach a length of 3·0 mm under natural conditions, whereas they reached only 1·8 mm in length after 30 days in culture. At the morphological level, the scutellum was poorly developed, and the coleoptile appeared anomalous in cultured embryos. The poor growth in embryos in culture seems to be due to the relatively simple medium employed. This was confirmed by Norstog (1965a), who showed that embryos of the same size as those studied by Chang, if grown in a complex medium containing several amino acids and increased salt concentrations, exhibited a rate of growth nearly equal to that of embryos *in ovulo* and eventually surpassed the latter in length. Growth of embryos of different

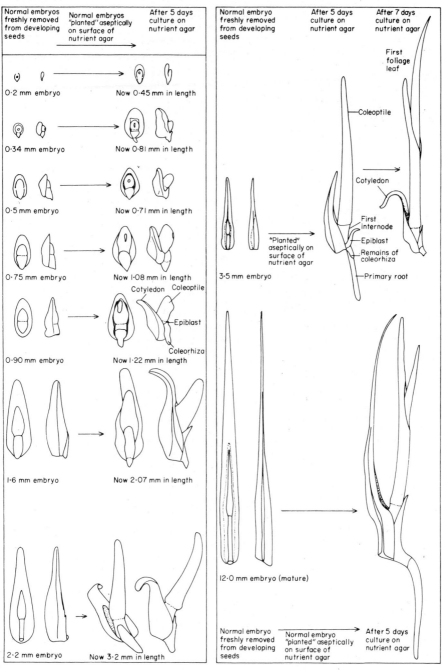

FIG. 7.13. Growth in culture of progressively older embryos of *Zizania aquatica*. In each case, diagrams of normal embryos at the time of excision and after 5 days in culture are shown. (After LaRue and Avery, 1938)

ages of *Capsella bursa-pastoris* was also slow in culture, although the development of the cotyledons was found to be characteristically out of proportion to that attained by other parts (Monnier, 1968a). Considering that the expansion of the cotyledons would be greatly restricted *in ovulo*, their aberrant behavior in culture is probably due to the release from restraint.

Although the above observations relate exclusively to the behavior of the embryo proper in culture, the possibility that certain morphogenetic syndromes might appear in the suspensor part of the embryo is by no means precluded. Since the suspensor completes its growth during the early phase of embryogenesis, morphogenetic studies on this structure can be made only through recourse to culture of fertilized egg and early division phase embryos; because of the manipulative difficulties involved in the operations, the suspensor has hardly been examined from this viewpoint. According to Poddubnaya-Arnoldi (1959), growth of the suspensor in the cultured zygote of the orchid, *Dendrobium nobile*, may be regulated experimentally to produce a vesicular structure rather than the branched haustorial form it attains *in vivo*. Exploitation of favorable materials that permit manipulation of very young embryos may be a fruitful method of studying the factors controlling the form of the suspensor in angiosperms.

The examples considered so far have sought to emphasize the effect of age of embryos at excision on their final morphology in culture. From a developmental point of view, one would like to know whether, irrespective of culture conditions, the preferred developmental pattern of embryos of a particular age group can be considered implicit in their age. Is it conceivable that, as the embryo matures, the total milieu of the embryo sac exercises less and less influence on the final embryonic pattern attained in culture? Does the biochemical development of the embryo determine its morphological development in culture? An examination of these questions will serve to present in a new light the issues opened by the bewildering variations observed in the morphology of cultured embryos of different ages.

IV. COMMENTS

The main import of the discussion presented in this chapter bears upon the problems involved in the regulation of growth in culture of embryos of different ages. The results support the view that growth of embryos is continually open to external physical or chemical control mechanisms which vary with their age.

The successful culture of the isolated embryo was the starting point of a forward movement in the understanding of the physiology of their growth, as revealed by the documentation of nutritional requirements of embryos of a considerable number of species. Even in the present state of our knowledge, the requirements for the culture of fertilized eggs, and embryos in early division phases, that are so perfected in the embryo sac of each species, remain deeply mysterious. For handling this problem, the investigator will probably have to evolve very different methods, involving a great deal of manipulative skill and a great many delicate combinations of hormonal and chemical substances. If and when he is armed with the tools and tricks, the prospects of establishing a firm basis for the understanding of the complexities of embryonic differentiation will be clearly within view. It should remain a challenging field in the years ahead.

8. Embryogenesis in Cultured Ovaries, Ovules and Seeds

Tissue culture methods have made a profound contribution to our understanding of the physiology of reproduction in plants and in providing a critical approach to the determination of mutual relationships of individual reproductive parts. This has been achieved by culturing entire inflorescences, flowers or isolated floral parts in media of known chemical composition. Much has recently been done in extending these studies to the point where it is even possible to follow fertilization and embryogenesis in a test tube. What now seems clear from these investigations is that organization of individual floral parts is sufficiently loosely integrated to permit them to be separated from the flower and maintained on their own for varying periods of time. Insofar as it relates to the physiology of growth of the embryo, cultivation of the flower, ovary and ovule under controlled conditions has made it possible to study outside the confines of the parent plant the behavior and growth of the fertilized egg and early division phase embryos which have hitherto defied attempts at excision and culture. Since the seed encloses the embryo which initiates growth under suitable conditions, it is not difficult to envisage that in certain cases meristem function and organ differentiation in the embryo can be studied by seed culture.

Our main concern in this chapter will be to consider experimental data relating to growth and morphogenesis of embryos in cultured flowers, ovaries, ovules and seeds. Ideally, we should like to relate changes in growth and morphogenesis of the contained embryo to changes in the nature and composition of nutritive materials of the medium, but this is hardly achieved. The embryo housed in the ovule is an integrated cellular system, in physical union with the surrounding milieu. From this point of view the enclosed embryo must be regarded as a highly coordinated organ, in which the different component cells may respond to external stimuli differently from those of an isolated embryo. Changes observed in the growth of the embryo may depend not only upon critical metabolites provided exogenously, but also on those present within the embryo sac. Nevertheless, it should be recognized that as the nutritional requirements of reproductive parts are fully met and clearly defined, new dimensions will be added to our understanding of growth interactions during embryogenesis and of the controlling factors in the growth of very small embryos.

Recent work on the controlled growth of reproductive organs of angiosperms has been extensively reviewed (Maheshwari, P., 1959; Maheshwari, P., and Rangaswamy, 1965; Maheshwari, P., and Kapil, 1966; Johri, 1962a; Johri and Guha, 1963a; Johri and Sehgal, 1967) and reference is made to these sources for adequate details.

I. CULTURE OF FLOWERS AND OVARIES

The technique of culture of flowers and ovaries introduced by LaRue (1942) has opened up an avenue of promise to study growth interactions during development of the embryo. This investigator found that pollinated flowers of tomato, *Kalanchoe*, *Forsythia* and *Caltha* cultured with a piece of the pedicel in a mineral salt medium remained alive and grew appreciably in size during the culture period, resulting in normal fruits. Nitsch (1949, 1951) extended these studies to the culture of flowers of gherkin, bean, strawberry, tobacco and tomato excised from plants before and after pollination. It was found that ovaries of pollinated flowers of gherkin and tomato cultured in a relatively simple medium containing sucrose and mineral salts formed fruits which ripened and produced even seeds. Later workers confirmed the results by reporting success with flowers of other plants (Jansen and Bonner, 1949; Leopold and Scott, 1952; deCapite, 1955). In these experiments, effects of cultural conditions on the growth of embryos enclosed in the ovules were not studied.

A. Role of Accessory Floral Parts

Recent studies have led to a realization of the role of accessory floral parts in the development of the embryo in cultured ovaries. In the ovaries of *Triticum aestivum* and *T. spelta* excised from the plant soon after pollination and cultured, growth of proembryos was impaired if the floral envelopes were removed before culture, while in intact florets enveloped by lemma and palea (hull) normal growth of embryos occurred (Rédei and Rédei, 1955b, c). More or less similar results were obtained in barley, where dehulling was found to affect growth of younger embryos more markedly than that of older ones (LaCroix *et al.*, 1962; LaCroix and Canvin, 1963). In cultured spikes, retention of a single leaf was sufficient to induce normal growth of proembryos even when all the florets were dehulled. At the morphological level, growth of the meristematic shoot and root was greatly retarded in the embryo growing in the dehulled floret due to lack of cell divisions. DNA content of cells of the embryo of the dehulled floret determined micro-spectrophotometrically appeared to be abnormally high, conceivably because of blockage of mitosis following DNA duplication. The remarkable effect on embryonic growth of lemma and palea, tentatively called "hull factor", warrants further study, particularly with regard to the nature of the substances that are transmitted to the embryo which preserve its ability for continued growth.

The calyx lobes of flowers of dicotyledonous plants have been shown to exert beneficial effects on the growth of cultured ovaries (Maheshwari, N. and Lal, 1958, 1961a; Chopra, 1958, 1962; Sachar and Baldev, 1958; Sachar and Kanta, 1958; Bajaj, 1966a; Guha and Johri, 1966), but their effects on the growth of the contained embryo have been recorded only in a few cases. In *Althaea rosea* (Chopra, 1958, 1962), normal embryo development occurred in flowers excised at the stage of the globular proembryo only when they were planted with the calyx lobes intact. The beneficial effects of the calyx were not replaced by growing the calyx-free flowers on media containing growth substances such as IAA, IBA, GA or a combination of IAA and kinetin, or anti-mitotics like colchicine (Fig. 8.1). In *Aerva tomentosa*, where there is a complete elimination of pollination and fertilization resulting in parthenogenesis, embryos formed apomictically in cultured flowers devoid of the perianth are underdeveloped, as compared to the nearly normal development of embryos in cultured complete flowers (Puri, 1963).

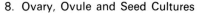

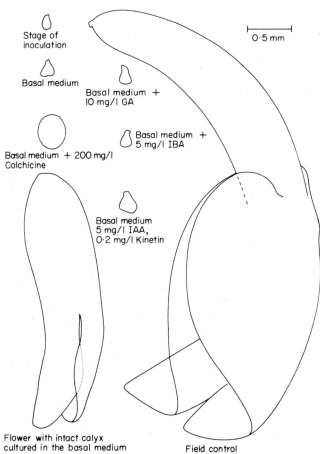

FIG. 8.1. Comparative growth *in vivo* of embryos of *Althaea rosea*, in flowers cultured in the basal medium with the calyx intact and in calyx-free flowers cultured in media supplemented with different additives. (After Chopra, 1962)

These examples indicate that calyx or perianth lobes contain substances necessary for growth induction in the embryo, which should be something quite different from the usual array of growth hormones. It has been suggested (Nitsch, 1963) that accessory floral lobes render inorganic nitrates into a utilizable form, probably by reducing them to the level of ammonia.

B. Role of Growth Hormones

Several studies have led to attributing the morphological type of growth of the embryo encountered in cultured ovaries and flowers to auxin, gibberellin, kinetin and other growth adjuvants provided in the medium. However, direct evidence for a stimulation of growth of

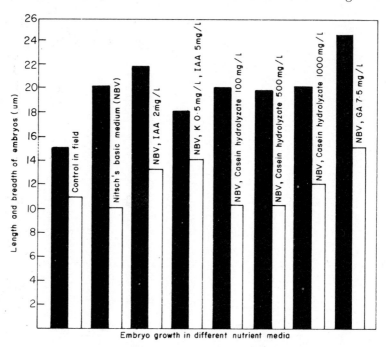

FIG. 8.2. Best growth of embryos in cultured ovaries of *Ranunculus sceleratus* is obtained in media supplemented with IAA or GA. Solid bands, length; hollow bands, breadth. K, kinetin. (After Sachar and Guha, 1962)

embryonic organs by the applied substances is lacking, and not infrequently, growth adjuvants employed are found to be totally ineffective or inhibitory (Sachar and Kanta, 1958; Sachar and Baldev, 1958). The latter proved to be the case when auxins were applied to pollinated flowers either as a spray or in lanolin paste (Heslop-Harrison, 1957; Monnier, 1968b). In the ovaries of *Ranunculus sceleratus* cultured in a mineral salt medium, embryos were longer and broader than those *in vivo*, and their growth was further enhanced by supplementing the medium with IAA or GA (Sachar and Guha, 1962). Combination of IAA and kinetin was inhibitory for embryo growth (Fig. 8.2).

It is interesting to note that in a medium containing casein hydroly-zate, growth of the embryo was not significantly different from that in the basal medium, while the percentage of achenes with viable embryos was very high. Although there is suggestive evidence for a role for IAA and kinetin in producing viable seeds in ovaries of *Reseda odorata* excised and cultured 10–15 days after pollination (Sankhla and Sankhla, 1967), stages of embryo development at the time of culture or at the termination of the experiment are not given for a proper evaluation of the results. In Chapter 10, the reader will find an account of some deeper implications of the effects of growth hormones on the organogenesis of cultured embryos.

When flowers of *Allium cepa* were cultured 2 days after pollination, viability and growth of embryos were appreciably enhanced by trypto-phan (Guha, 1962; Guha and Johri, 1966). The development of the embryo through the globular stage to the mature stage was completed in about 16 days after inoculation and a good percentage of seeds reached maturity in about 35 days. Addition of IAA to the medium induced relatively slow growth of embryos, and the percentage of viable seeds remained low. This does not appear to make it likely that trypto-phan is serving as a precursor for auxin biosynthesis in the flower.

In *Iberis amara*, satisfactory growth of embryos was obtained in flowers cultured 1 day after pollination in a mineral salt medium sup-plemented with B vitamins (Maheshwari, N. and Lal, 1961a). Addition of kinetin and IAA or supplementation of the medium containing IAA and kinetin with maleic hydrazide (1 mg/l) did not affect the rate of growth of the embryo. At a high concentration of maleic hydrazide (5 mg/l), which nearly doubled the rate of ovary growth, development of the embryo was retarded beyond the octant stage. In the ovaries of *Anethum graveolens*, containing the fertilized egg and free endosperm nuclei at the time of excision and culture, mature embryos were formed in about 31 days in the basal medium. Embryos formed in ovaries cultured in a medium containing IAA were nearly twice as long as those formed in nature, although auxin was ineffective in enhancing the maturity of embryos. A remarkable feature of the responses of the embryo of *Anethum* to various concentrations of casein hydrolyzate and yeast extract was the induction of polyembryony (Johri and Sehgal, 1963b). The responses of embryos in cultured ovaries of *Foeniculum vul-gare* and *Trachyspermum ammi* to growth adjuvants in the medium were similar to those exhibited by *Anethum* (Johri and Sehgal, 1966; Sehgal, 1964).

Various abnormalities in the growth of embryos have been docu-mented when ovaries of certain members of the Gramineae were grown in media containing IAA, kinetin and adenine, individually or in combination (Narayanaswami, 1963). For example, adenine or a mixture of IAA and kinetin induced callus proliferation from the scutellum, and formation of multiple shoots from the mesocotyl of *Avena sterilis*

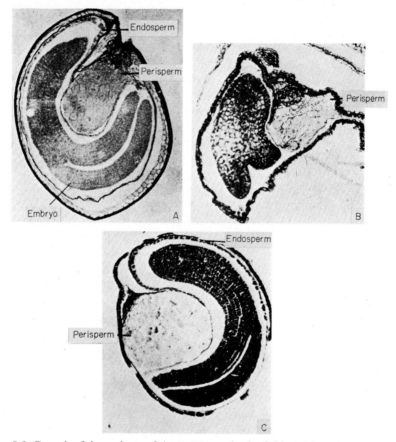

Fig. 8.3. Growth of the embryo of *Aerva tomentosa* in the field and in culture. **A**, Field control, 2 weeks after anthesis. **B**, Twenty-five day old culture of the ovary in a medium supplemented with yeast extract. **C**, Embryo of a seed originating from a piece of inflorescence cultured in a basal medium with 4% sucrose for 20 days. (After Puri, 1963)

embryos. In *Echinochloa frumentacea*, multiple root primordia were characteristically observed on the embryo when ovaries were cultured in a medium containing IAA, while supplementation of the medium with IAA and adenine additionally induced twinning of the shoot apex.

     Some understanding of the factors that control the activation of the egg in the apomict, *Aerva tomentosa*, has been obtained by culturing ovules, ovaries and portions of the spike (Murgai, 1959; Puri, 1963). Remarkable development of the embryo occurred when portions of the unpollinated spike were cultured in a basal medium alone or supplemented with yeast extract, casein hydrolyzate, kinetin or coconut milk (Fig. 8.3). When individual flowers were cultured in these same media, only a low percentage of them formed mature embryos, while in excised ovaries, growth of embryos was poor even in the best medium employed. The mechanism of the "spike effect" implied here is almost completely unknown. We have the inviting prospect, therefore, that exploration of the spike to detect specific compounds, and studies on the effects of other untested substances on the growth of excised ovaries, may yield important information for a more direct analysis of the basis for parthenogenetic activation of the egg, although the massive size of the explants employed, and the presence of considerable amount of chlorophyllous tissues in them, will make interpretations difficult.

II. CULTURE OF OVULES

A. Effect of Growth Hormones

From culture of flowers and ovaries it is one further step to excise ovules and grow them *in vitro*. There are undoubtedly various interactions of culture conditions of ovule on the growth of the contained embryo. Relatively small, 16-celled embryos of potato *(Solanum tuberosum)* conditioned for a few days by culturing ovules lent themselves more readily to subsequent culture than freshly excised embryos (Haynes, 1954). Muzik (1956) showed that when mature ovules of *Hevea brasiliensis* were planted in White's medium supplemented with coconut milk and IAA, embryos promptly began to grow. This appeared significant since embryos excised from ovules of the same age failed to grow in isolation. N. Maheshwari (1958) was among the first to culture successfully excised ovules at precisely timed stages following pollination to the point of producing viable seeds. The significance of this work has been heightened by the demonstration that ovules of *Papaver somniferum* containing even fertilized eggs or two-celled proembryos can be grown to maturity in culture. Addition of kinetin, casein hydrolyzate or yeast extract accelerated the initial rate of growth of proembryos in varying degrees, although the final length of embryos was no greater than in ovules growing *in vivo* (Maheshwari, N. and Lal, 1961b). IAA and GA tended to inhibit the growth of the embryo.

Ovules of *Zephyranthes* excised at the zygote stage, containing the undivided primary endosperm nucleus, were found to grow to the stage of formation of globular embryo upon culture in distilled water containing 5% sucrose and solidified with agar (Sachar and Kanta, 1958). Cultivation of the ovules in a mineral salt medium supplemented with B vitamins, calcium pantothenate, glycine and sucrose, or additionally with auxin, kinetin or GA also, did not improve their growth nor that of the enclosed embryo (Fig. 8.4). Since division of the primary endosperm nucleus did not take place in any of these media, it is

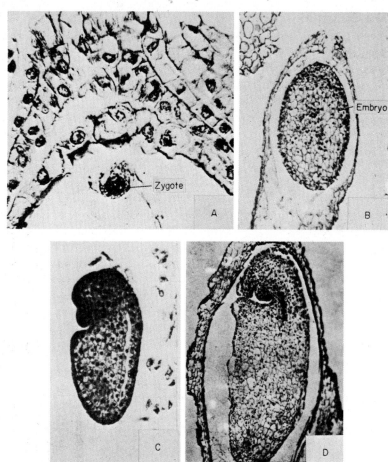

Fig. 8.4. Supplementing the medium with "Casamino acids" induces normal growth in embryos of cultured ovules of *Zephyranthes*. **A**, Zygote 2 days after pollination, at the time of inoculation. **B**, Four week old embryo grown in a medium containing kinetin and indolebutyric acid. **C**, Eleven day old embryo from a field control. (After Sachar and Kapoor, 1959) **D**, A normal embryo from an ovule cultured in a medium containing Casamino acids. (After Kapoor, 1959)

safe to assume that the initial round of divisions of the zygote has taken place at the expense of its own stored reserves, and reserves of the embryo sac. The ease with which ovaries of *Cooperia pedunculata* and *Zephyranthes* can be grown in culture compared with excised ovules is indeed striking. Ovules containing the fertilized egg cell and endosperm nucleus enclosed in the ovary form fully differentiated embryos in a medium supplemented with vitamins, glycine and sucrose. Excised ovules grow little on this medium, nor on a medium fortified with growth substances, unless they are removed from the ovary at the globular embryo and free nuclear endosperm stage (Sachar and Kapoor, 1958, 1959). It is clear that simultaneous endosperm development would render more or less complete development of the embryo in the ovule, and that the ovary wall contributes something which causes the endosperm nucleus to initiate division.

Later work (Kapoor, 1959) showed that supplementing the basal medium with coconut milk or "Casamino acids" supported development of viable seeds from cultured ovules of *Zephyranthes* excised at the stage of the zygote (Fig. 8.4). The favorable effects of the complement of amino acids present in Casamino acids were completely replaced by histidine, arginine and leucine, and to some extent by valine.

In other investigations dealing with this problem, it has become clear that interaction of growth hormones should be regarded as a frequent, although not an invariable, physiological requirement for growth of embryos in cultured ovules. In an attempt to grow 6 day old ovules of cotton at the 12-celled proembryo stage by supplying various growth adjuvants to White's medium, it was found that in the presence of kinetin, embryos with folded cotyledons were formed in about 81 days after culture (Joshi, 1962; Joshi and Johri, 1972). Growth of the embryo was even slower in media containing other growth additives such as IAA, GA, yeast extract, casein hydrolyzate, and ovule extract than in kinetin. The response of embryos in cultured ovules of cotton to growth adjuvants seems to depend upon the availability of high amounts of inorganic nitrogen in the medium, as later workers (Eid *et al.*, 1973) obtained pronounced embryo growth in 5 day old cotton ovules cultured at the 2–10-celled proembryo stage in the nitrogen-enriched Murashige–Skoog medium. Generally, when a growth medium fails to stimulate growth of an explanted organ, we may postulate that this is due to its failure to attain self-sufficiency in certain critical metabolites; this reasoning may explain the relatively slow growth of embryos in the ovules of *Nicotiana tabacum* cultured in a simple medium (Siddiqui, 1964). An extreme case is in *Capsella bursa-pastoris* where there is virtually complete inhibition of growth of embryos in cultured ovules compared to growth of embryos isolated from ovules of the same age (Veen, 1963). Reasons for this are not clear. Germination of embryos *in situ* was observed when ovules of *Abelmoschus esculentus* containing proembryos or globular embryos were cultured in White's medium. Supplementation of the medium with IAA, casein hydrolyzate or coconut milk was beneficial for the overall growth of the ovule, but there was no apparent effect on the embryo (Bajaj, 1964).

B. Effect of Placental Tissues

While a need for growth-promoting materials present in the endosperm for embryo growth is now clearly established, other tissues are active only rarely. There is some evidence for an essential role of the placenta in the growth of embryos in cultured ovules. When ovules of *Gynandropsis gynandra* with adherent placenta were cultured at the stage of globular embryo and free nuclear endosperm, irrespective of the growth adjuvants present, the rate of embryo development and its final size were enhanced. Consequently, as in nature, normal seeds were produced in the ovule 13–14 days after culture (Chopra and Sabharwal, 1963). Poppy is another example in which a role for placental tissue in embryo growth appears probable. In contrast to the successful culture of excised ovules of poppy described earlier (Maheshwari, N., 1958; Maheshwari, N., and Lal, 1961b), later work in another laboratory (Pontovich and Sveshnikova, 1966) failed to elicit embryo differentiation in ovules cultured at the zygote or proembryo stage. Several additives to the medium such as kinetin, adenine, casein hydrolyzate, coconut milk, etc., either singly or in combination, were ineffective in promoting growth of embryos in isolated ovules which at best produced an amorphous mass of cells accompanied by proliferation of the integument or nucellus. Normal differentiation of the zygote or proembryo into an embryo was obtained in the ovule cultured with attached placenta. The mechanism by which the placental tissue influences growth of the embryo is not elucidated; possibly, it is related to the path of entry of metabolites into the embryo or has morphogenetic substances of its own. No stimulation of embryonic growth occurred when ovules of *Opuntia dillenii* were grown along with adhering placenta in a medium supplemented with auxin and cytokinin, although the placenta itself proliferated into a callus (Sachar and Iyer, 1959).

C. Other Studies

In most of the work described above, when unpollinated ovules were cultured under the same conditions as pollinated ones, they failed to grow and aborted; growth hormones have, however, been shown to induce a limited enlargement of unpollinated and unfertilized cotton ovules in culture (Beasley, 1973). In a recent innovation, ovule culture has been adopted as an experimental system in which unpollinated ovules are grown along with pollen grains, and fertilization of the egg accomplished in the test tube by encounter with the germinated pollen grains. In another significant effort, intraovarian pollination of emasculated flowers was accomplished by injecting a pollen suspension into ovules. The development of the embryo after intraovarian pollination and *in vitro* fertilization has been described in many cases (Kanta, 1960; Kanta and Maheshwari, 1963a, b; Kanta *et al.*, 1962; Dulieu, 1963, 1966; Shivanna, 1965; Usha, 1965; Rao, 1965a; Rangaswamy and Shivanna, 1967; Zenkteler, 1965, 1967; Wagner and Hess, 1972; Zdruikovskaia-Rikhter and Babasiuk, 1974) and there is a general measure of agreement in the descriptions that suggests no deviations from the normal pattern. Because of incompatibility, embryogenesis was, however, arrested beyond the two-celled proembryo stage when pollen grains of other genera were used for *in vitro* pollination and fertilization (Zenkteler, 1970). Techniques now being developed, such as placental pollination involving culture of ovules attached to the placenta and depositing pollen grains directly on ovules (Rangaswamy and Shivanna, 1971), and the use of precultured pollen (Balatková and Tupý, 1968, 1972a, 1973), are providing the foundations for what should be effective approaches to obtain normal embryogenesis in incompatible crosses and to overcome other genetic problems.

In other studies employing culture techniques, it has become evident that culture conditions can modify the growth not only of zygotic embryos but also of embryos originating from accessory tissues of the ovule. For example, nucellar proembryos contained in cultured ovules of *Citrus microcarpa* differentiated in White's medium containing 5% sucrose or tomato juice. Growth was relatively rapid as seen from the fact that embryos with well developed cotyledons were formed after about 2 weeks in culture (Rangaswamy, 1959). Extensions of these studies using explanted nucelli are discussed in Chapter 14.

III. CULTURE OF SEEDS

As a basis for discussion of the morphogenesis of embryos in cultured seeds, it is suitable to look at some of the overall morphological features of the embryo in a seed. Although the degree of development of the embryo in seeds varies, it generally appears as an elongate, bipolar

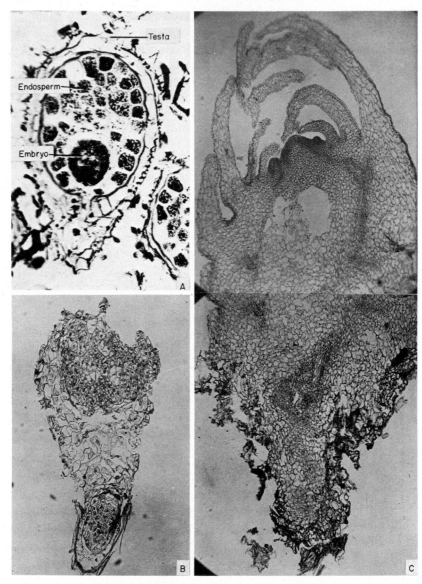

FIG. 8.5. Formation of a seedling from the radicular end of the embryo of *Orobanche aegyptiaca* is possible when seeds are cultured in a medium containing casein hydrolyzate. **A**, Section of a seed at the time of culture showing the globular embryo. (After Rangaswamy, 1967) **B**, Section of a seed cultured for 3 weeks in a medium containing casein hydrolyzate showing proliferation of the radicular end. **C**, Section through a shoot apex differentiated on a callus. (After Rangaswamy, 1963; photographs courtesy of N. S. Rangaswamy)

structure, consisting of a meristem at each pole, and one or two lateral appendages, the cotyledons. The point of attachment of the cotyledons separates the embryonic axis into a hypocotyl–root region and an epicotyl–plumule region. The former has at its lower end the primordial root, the radicle, while the primordial shoot, the plumule, is attached to its stem part called the epicotyl. As a result of resumption of metabolic activity during germination, cells of the embryo undergo division, elongation and differentiation, leading to the progressive development of the root and shoot axes.

Studies on the germination of seeds with fully differentiated embryos have provided us with much information about fundamental mechanisms in the transformation of the embryo into a seedling, but it is quite evident that they can tell us little about the ways by which the various morphogenetic processes during embryogenesis are integrated. The use of seeds which contain undifferentiated embryos for such problems is quite advantageous as factors influencing their differentiation into complete embryos can be followed more easily than in cultured ovules. In seeds of stem and root parasites, saprophytes and even several autotrophs, the embryo may not be structurally differentiated and such seeds provide useful experimental objects for morphogenetic studies.

In recent years, a beginning has been made at studying morphogenesis of embryos in the seeds of angiosperm root parasites. Here, except for the fact that the embryo lacks tissue and organ differentiation, we have a system which consists of all the parts of a normal seed in close contact with each other, and which undergo rapid changes during germination. Moreover, contrary to reports in the literature, it seems certain that seeds of root parasites germinate in significant numbers in the absence of the host plant if they are sown in a medium containing a suitable organic growth supplement.

When seeds of *Orobanche aegyptiaca* were cultured in a medium containing yeast extract, coconut milk or casein hydrolyzate, a callus which developed from the radicular end of the globular embryo gave rise to a shoot bud from its superficial cells (Rangaswamy, 1963, 1967); here, the seedling was formed without the intervention of cells of the plumular pole (Fig. 8.5). On the other hand, in the presence of kinetin or GA in the medium, the radicular end of the embryo produced roots instead of a callus and the plumular pole formed scale leaves (Fig. 8.6). Thus, to induce initiation of root and shoot axes from the respective poles in the embryo of the cultured seed, the medium must contain a specific growth hormone.

A similar pattern of morphogenesis was observed in the embryo of the cultured seed of *Cistanche tubulosa* (Rangan, 1965; Rangan and

Rangaswamy, 1968) where the radicular end formed a massive callus in the presence of casein hydrolyzate and coconut milk. Differentiation of a shoot bud on the callus occurred in the dark on a complex medium containing IAA, GA and coconut milk, and in no instance was meristematic activity induced at the plumular end (Fig. 8.7). In the seed of *Striga euphrasioides* (Rangaswamy and Rangan, 1966), the radicle elongated normally in a medium containing kinetin, but the cotyledons

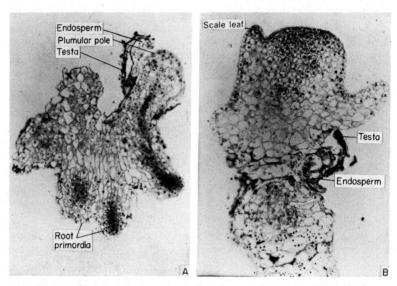

Fig. 8.6. Both radicular and plumular poles of the embryo are activated when seeds of *Orobanche aegyptiaca* are cultured in a medium containing gibberellic acid. **A**, Section of a 15 day old seed cultured in this medium showing root primordia, and plumular pole. **B**, Twenty-five day old culture showing the shoot apex and the scale leaves at the plumular end. (After Rangaswamy, 1967; photographs courtesy of N. S. Rangaswamy)

formed unorganized callus which did not differentiate further. Taken together, these observations focus attention on the organic supplement added to the medium as the determinative factor in the activation and differentiation of cells of the functional meristems of the globular embryo of the seed. The significance of the activity of the radicular meristem of the embryo is to be accounted for wholly or partly on the basis of the fact that seedlings normally establish haustorial contact through their roots.

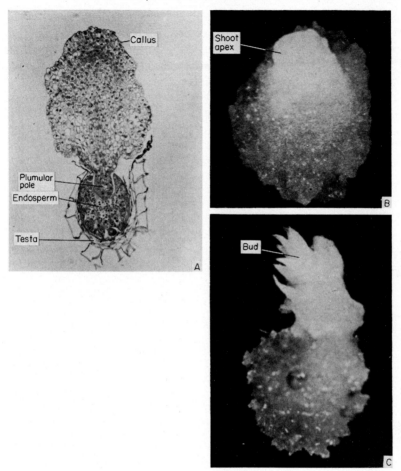

Fig. 8.7. Radicular end of the embryo proliferates when seeds of *Cistanche tubulosa* are cultured in a medium containing coconut milk and casein hydrolyzate. Subculture of the proliferated embryo tissue in the dark in a medium containing IAA, GA and coconut milk leads to shoot bud initiation. **A**, Tissue formation from the radicular end of the embryo. **B** and **C**, Initiation and differentiation of shoot bud from the callus. (After Rangan and Rangaswamy, 1968; photographs courtesy of N. S. Rangaswamy)

IV. COMMENTS

It is clear that facts about embryogenesis in cultured floral organs are many, but generalizations are few. Beyond the well established fact that ovaries and ovules of a variety of plants can be grown to maturity in culture, no further conclusions of equal firmness can be made. Although new generalizations are not at hand, culture of reproductive parts has entered upon a phase of great promise. Thanks to the application of tissue culture techniques, ovary and ovule culture has provided a new testing ground for the culture of fertilized egg and early division phase embryos outside the parent plant. Recent studies leading to test-tube fertilization have given added importance to the culture of reproductive parts to suggest that reproduction in angiosperms is continually open at all levels to external controls which can be exploited to a remarkable extent. It is in these categories that some of the data reviewed in this chapter attain their major importance.

9. Nutrition and Metabolism of Cultured Embryos

A natural consequence of the success achieved in growing plant organs in isolation was an elaboration of their nutritional and metabolic relationships. The food requirements which in nature are provided by the parent plant are supplied in culture, and by manipulating the constituents of the culture medium, an analysis of the nutritional relationships of the individual organs is possible which it will be difficult to perform in the entire plant. Since synthetic potentialities of the cultured organs are reflected in their nutritional requirements, broad generalizations on their metabolism have been justifiably attempted from these studies. Although there is no dearth of data on the nutrition and metabolism of cultured organs and tissues, it is not yet clear to what degree these results can be extrapolated to systems *in vivo*.

Unlike other plant organs, cultured embryos have not been used to any great extent for nutritional and metabolic studies, but their potential value should not be overlooked. Indeed, the advantages of studying growth and metabolism of the primordial shoot and root of a plant in the same system within the confines of a culture tube should stimulate greater use of embryos for biochemical investigations than before.

In the present chapter, studies on the carbohydrate, nitrogen and vitamin nutrition and metabolism of embryos are reviewed. No attempt is made to relate the results to the more extensive data obtained with other organs and tissues. Since appropriate nutrition of cultured organs eventually determines their form, this discussion will emphasize, whenever possible, morphological changes in the embryos brought about by the added nutrients.

I. CARBOHYDRATE NUTRITION

Almost everyone who has grown isolated plant organs or tissues in aseptic culture has assumed that they have a requirement for a carbon energy source to sustain continued growth. Because carbon lends itself to the formation of indefinitely repeating chains and links by virtue of its ability to bind other atoms and to link up with other atoms, it is at the very kernel of all life processes in the cell. For cultured plant tissues the carbon energy source is usually supplied as a carbohydrate, and over the years it has become almost axiomatic in plant tissue culture work to provide the medium with a suitable carbohydrate at an appropriate concentration.

By now there has developed a mass of literature dealing with the effects of a variety of carbohydrates on an equally bewildering variety of cultured plant cells, tissues and organs. Much of this work is concerned with finding the most suitable carbohydrate for the successful growth of the culture. The generalization emerging from these studies is that plant tissue cultures grow best when supplied with sucrose, and rarely has any other sugar been conspicuously as successful. Two recent reviews (Street, 1966, 1969) have explored at some length the question of carbohydrate uptake, metabolism and translocation in plant tissue cultures and we are reminded that at the present time the problem is full of imponderables.

A. Growth Effects

The osmotic effects of carbohydrates in inducing growth of embryos of some plants were discussed in Chapter 7. As pointed out there, the recognition of the role of carbohydrates as osmotica has required their use in culture media in concentrations higher than normally necessary as a nutrient. Here we are concerned with the nutritional effects of carbohydrates on cultured embryos with a view to assessing the relative efficiencies of individual compounds. The role of carbohydrates in the organogenesis of embryos is considered later.

For embryos of different species, the nature of the carbohydrate required might be expected to vary from simple monosaccharides to complex polysaccharides. The disaccharide sucrose seems to be the best carbon source for embryos of a number of plants. The study of heterotrophic carbohydrate nutrition of isolated embryos was initiated by

TABLE 9.I. Effects of different carbohydrates on the growth of excised barley embryos. (From Brown and Morris, 1890)

Results are given in terms of dry weight of 50 embryos after growth for 7 days in a medium containing 3·5% of the carbohydrate.

Medium	Weight (mg)
Before germination	89·0
Solution culture (no carbohydrate)	52·0
Natural endosperm	436·2
Cane sugar	195·5
Dextrose	164·5
Levulose	162·5
Maltose	155·0
Invert sugar	132·5
Milk sugar	99·0
Raffinose	91·0
Mannitol	89·5

Brown and Morris (1890) when they observed that cane sugar had the highest nutritive value for growth of excised barley embryos (Table 9.I). Embryos supplied with cane sugar had a relatively high dry weight, and showed the presence of abundant starch in their cells. By the same token, substances like milk sugar, invert sugar, raffinose and mannitol were ruled out as unsatisfactory carbohydrate sources.

Sucrose has served as a satisfactory carbohydrate for embryos of several, but not all, species. For embryos of some species of *Zea* freed from the endosperm, sucrose was better than lactose, or a combination of sucrose, lactose and maltose (Andronescu, 1919). Superiority of sucrose over glucose has been demonstrated in the culture of embryos of *Datura stramonium* (van Overbeek *et al.*, 1944), *Solanum nigrum* (Hall, 1948) and *Pinus nigra* var. *austriaca* (Radforth and Pegoraro, 1955). In experiments in which sucrose, glucose, levulose, mannose and glycerol were tested as carbon energy source for embryos of 10 species of *Datura*, significant differences in favor of sucrose over other substances were observed (Doerpinghaus, 1947). Sucrose was also favored over maltose, glucose, fructose, raffinose, lactose, mannitol and glycerol by embryos of *Capsella bursa-pastoris*, and little if any success was experienced with the latter seven compounds (Rijven, 1952).

The most detailed studies on the carbohydrate nutrition of cultured embryos are those of Amemiya *et al.* (1956a, b), who emphasized that growth of immature embryos of *Oryza sativa* was dependent on the presence of an adequate sugar supply in the medium, in the absence of which they degenerated, without showing any appreciable cellular and morphological development. In terms of relative effectiveness, fructose, glucose, sucrose and maltose proved superior to xylose, lactose, galactose and mannose. Optimum concentrations of the different sugars varied, and were dependent upon the growth parameter measured. Thus, as shown in Fig. 9.1, in terms of fresh weight of embryos, both sucrose and fructose shared a higher optimum than maltose and glucose. Probably related to this is the curious synergism between some of the sugars: at concentrations of either sucrose or fructose lower than the optimum, equimolar concentrations of maltose and glucose gave better growth than the individual sugars. When glucose was used, there was a sharp optimum for embryo growth, and concentrations lower or higher than the optimum affected longevity of the leaf (Bouharmont, 1961).

The growth of immature sedge *(Carex lurida)* embryos on a medium containing glucose was as good as, or better than, that on a medium containing sucrose, and cultured embryos produced best shoot and root growth at a range of concentrations of sugar from 2% to 6% (Lee, 1952). Glucose has been successfully used for growth of excised embryos of several species of the family Rosaceae and of hybrids of *Lilium* (Tukey, 1938; Skirm, 1942; Asen, 1948). For growth of embryos excised from after-ripened seeds of *Heracleum sphondylium*, the nature of the sugar incorporated into the medium appeared relatively unimportant since glucose, fructose, galactose, mannose and sucrose were almost identically effective (Stokes, 1953b). In connection with the carbohydrate nutrition of embryos, particular interest is attached to the ability of embryos of some species to utilize substances such as starch which are not ordinarily metabolized by cultured tissues. This property is characteristic of graminean embryos (Brown and Morris, 1890;

Stingl, 1907; Dieterich, 1924; Esenbeck and Suessenguth, 1925) where digestion of starch is thought to be regulated by the scutellum.

From work with excised embryos of the stem parasite *Cassytha fili-formis*, it has been claimed that incorporation of a range of concentrations (0–8%) of sucrose into a medium containing 1·0 mg/l IAA does not introduce any beneficial effects on their growth (Rangan and

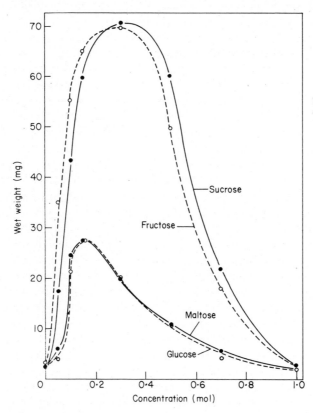

FIG. 9.1. For the growth of cultured embryos of *Oryza sativa*, sucrose and fructose are required at higher optima than maltose and glucose. (After Amemiya *et al.*, 1956a)

Rangaswamy, 1969). At the highest concentration of sucrose used, embryo growth was delayed and the plumule failed to open even after 10 weeks in culture. Occasionally embryos could grow better on a non-sucrose medium than on a sucrose medium (Ohta and Furusato, 1957). In most of the above experiments, sugar was generally added to the basal medium and autoclaved. Comparison of the relative effectiveness of autoclaved and cold-sterilized sucrose on the growth of embryos in cultured flowers of *Allium cepa* has shown that the former is better utilized than the latter (Johri and Guha, 1963b). Whether this is due to the partial breakdown of sucrose that occurs upon autoclaving is not determined.

Carbohydrate Nutrition of Orchid Embryos

A number of studies have been carried out on the carbohydrate nutrition of immature or mature embryos of plants belonging to the family Orchidaceae. The limited scope of this book does not permit a full consideration of all the relevant papers, which are too numerous to list, ranging from reports of amateur orchid growers to those of careful scientific studies; thus, the following information is intended to be representative rather than inclusive. For those seeking more than a superficial treatment of this topic, reference is made to Withner (1959) and Arditti (1967a) and the original papers cited therein.

It is worthwhile to state at the outset that some of the results on carbon nutrition of orchid embryos are difficult to interpret because far too little attention has been given to dose–response relationships and far too many observations appear to be puzzling rather than meaningful. Nevertheless, a great weight of evidence indicates that orchid embryos are able to draw upon a wider range of carbohydrates than embryos of other angiosperms. Among the sugars tested, sucrose seems undoubtedly to be best for germination or growth of embryos of several epiphytic species (Breddy, 1953). Germination of embryos of a few terrestrial orchids is reported to be greatly enhanced by sucrose, or a mixture of glucose and fructose (Stoutamire, 1963, 1964). For growth of embryos of *Brassocattleya*, *Cattleya* and *Vanda* sucrose was best utilized in the presence of niacin in the medium (Bahme, 1949). Maltose and mannose proved superior to fructose and sucrose for germination and growth of embryos of *Cattleya trianaei*, but D-galactose, L-arabinose, L-rhamnose, and L-xylose failed as carbon energy source (LaGarde, 1929; Wynd, 1933). Observations on embryos of other genera have also confirmed the general conclusion that maltose is a significantly superior sugar (Noggle and Wynd, 1943; Breddy, 1953; Arditti, 1967a). For embryos of *Odontoglossum* and *Miltonia*, sucrose or maltose was best, while for *Cattleya*, *Vanda* and *Phalaenopsis*, glucose was as good as sucrose (Breddy, 1953).

According to Knudson (1922), embryos of *Laeliocattleya* hybrid favored fructose over glucose, with the latter giving rise to chlorosis. For the germination of embryos of various epiphytic genera, and for two species of *Orchis*, sugars ranked in the following descending order: glucose, fructose, sucrose, maltose, mannose, galactose and lactose. Citric acid, malic acid, tartaric acid, oxalic acid and succinic acid were not suitable as carbon energy sources (Quednow, 1930). Glucose also appeared to be best for germination of embryos of *Cymbidium* and *Vanda* (Burgeff, 1936). According to F. Smith (1932), fructose, glucose and sucrose used separately showed no differences in the growth of seedlings obtained from cultured embryos of some orchids, but a combination of glucose and fructose was no better than either substance alone in stimulating growth. Ito (1951) has made the interesting observation that embryos of *Dendrobium, Cattleya, Vanda, Phalaenopsis* and *Epidendrum* can utilize corn starch provided tomato juice is present in the medium.

Ernst (1967b) has explored the effects of a selection of mono-, di-, and trisaccharides and sugar alcohols as energy sources for growth of freshly germinated embryos of *Phalaenopsis* ('Elinor Shaffer' × 'Doris') and *Dendrobium phalaenopsis* with special reference to the comparative effectiveness of D and L enantiomers of the different compounds. Maximum growth increments in embryos of *Phalaenopsis* were observed in media containing fructose, D-xylose, and D-glucose, while ribitol, D-ribose and D-arabinitol gave barely noticeable growth increments. Other substances tested were intermediate in their growth effects. For embryos of both species, generally L forms of sugars, and all deoxy sugars, were inhibitory for growth, while all D forms except D-galactose were readily assimilated. Galactose inhibition of growth of embryos is of great interest, in view of its extreme toxicity to cultured roots (Street, 1969). One possible mode of action of galactose on the embryo is indicated by the recent observation that in the treated embryo, the tonoplast was ruptured and the nuclear membrane had evaginations (Ernst *et al.*, 1971). This suggests that galactose feeding might affect some factors responsible for membrane permeability in such a way as to inhibit osmotic processes concerned with growth.

B. Effects of Carbohydrate Supply on Growth of the Root and Shoot

Angiosperms and Gymnosperms

Attempts have been made to relate the growth of the root and shoot systems of embryos to the nature of the carbohydrate incorporated into the medium. Buckner and Kastle (1917) noted that growth of the root of lima bean *(Phaseolus lunatus)* embryo was negligible on mannitol or starch, as compared to that on sucrose or glucose. In the embryos of *Datura stramonium* (Rietsema *et al.*, 1953a), *Phaseolus vulgaris* × *P. acutifolius* hybrid (Honma, 1955) and *Citrus* (Ozsan and Cameron, 1963), the most effective root growth occurred in media containing sucrose. Incorporation of glucose into White's medium is reported to promote

growth of root and shoot primordia of isolated wheat embryos
(Augusten, 1956). In the cultured embryos of oil palm *(Elaeis
guineensis)*, addition of sucrose above or below the optimum level led
to inhibition of growth of the root and the cotyledonary leaf; this is
clearly seen in Fig. 9.2 (Buffard-Morel, 1968). In the absence of sucrose,
embryos of early-ripening selections of peach appeared flaccid and
transparent, and seldom formed roots and leafy shoots in culture

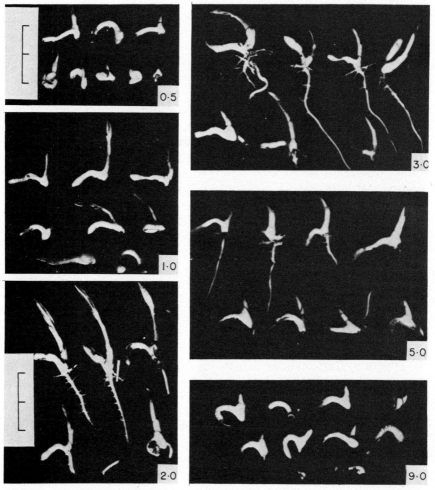

Fig. 9.2. Culture of embryos of *Elaeis guineensis* in media containing sucrose above or
below the optimum (2–3%) leads to inhibition of growth of the root and cotyledonary
leaf. Concentration of sucrose in each case is indicated at bottom right-hand corner.
(After Buffard-Morel, 1968)

(Brooks and Hough, 1958). The presence of sucrose in the medium thus appears to tip the balance toward or away from the production of healthy seedlings that survive transplantation in the soil by the embryo culture method. At another morphological level, growth of the epicotyl and hypocotyl of cultured bamboo *(Bambusa)* embryo was found to be markedly retarded in the complete absence or in the presence of suboptimal concentrations of sucrose in the medium (Alexander and Rao, 1968).

As part of a series of studies on the vernalization of excised cereal embryos, Purvis (1944) found that root and coleoptile primordia of the embryo of winter rye *(Secale cereale* var. 'Petkus') were extremely sensitive to the presence of carbon energy source in the medium. Compared to growth in a medium containing sucrose, elongation of the root and coleoptile was retarded in the absence of a carbohydrate, as well as in the presence of arabinose, fructosan, mannose, mannitol and pyruvic acid. Raffinose, maltose, glucose and fructose were intermediate in their effectiveness. Ribose, xylose and glycerol were inactive in promoting root growth, but promoted coleoptile growth to some extent. In considering these results, it should be kept in mind that embryos were cultured at 1°C, which is ideal for vernalization, but far from optimal for normal growth processes, and it is not determined whether the low temperature treatment affects the utilization of some of the compounds tested.

The root:shoot ratio in embryos of several orchids is markedly affected by the sugar concentration in the medium (Yates and Curtis, 1949). As an example may be cited *Epidendrum nocturnum*, which is tolerant to a wide range of sucrose concentrations. In the embryo of this species, high concentrations of sucrose induce formation of relatively long roots while shoot growth requires a lower sucrose optimum. A requirement for high sucrose concentration for the growth of roots in the embryos of *Dendrobium* and *Brassolaeliocattleya* has also been established (Kano, 1965).

The relationship between the site of application of sugar and the growth of the root and shoot systems of cultured embryos of gymnosperms has received some consideration. In *Ginkgo biloba* the plumule completely failed to elongate, or elongated only feebly, unless the embryo was planted with the cotyledons in contact with a medium containing sucrose or glucose (Bulard, 1952; Ball, 1956a). It appears that sugar is effectively metabolized and translocated when it is supplied through the cotyledons. Best root growth was obtained in a medium supplemented with glucose, followed in order of decreasing effectiveness by sucrose, levulose, raffinose and galactose (Fig. 9.3). Some of these sugars have been detected in the mega-gametophyte of *Ginkgo* by

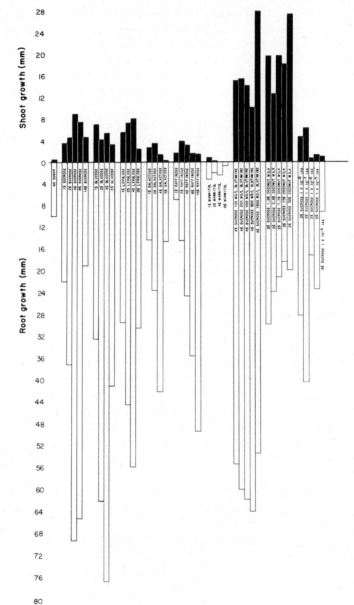

FIG. 9.3. When embryos of *Ginkgo biloba* are grown with the cotyledons in the agar medium, addition of sugars supports better root growth than shoot growth. Effects of addition of glutamine, coconut milk and indoleacetic acid to a medium containing sucrose on growth of embryos are also shown in this figure. (After Ball, 1959; print courtesy of E. Ball)

paper chromatography. Although there is some likelihood of chemical modification of carbohydrates in long-term cultures, utilization by the cultured embryo of those same sugars normally present in the gametophyte is a matter of considerable interest.

In *Pinus lambertiana*, the root failed to grow when the embryo was planted with its radicle end submerged in the nutrient medium unless some nutritive tissue of the megagametophyte was retained around the cotyledons (Sacher, 1956; Baron, 1962). However, if the embryo, completely devoid of gametophytic tissue, was planted with its cotyledons in the medium, and the tube was inverted so that the root grew with gravity into air or into a tube of plain agar, growth of the root and hypocotyl was more rapid than when planted in the normal way

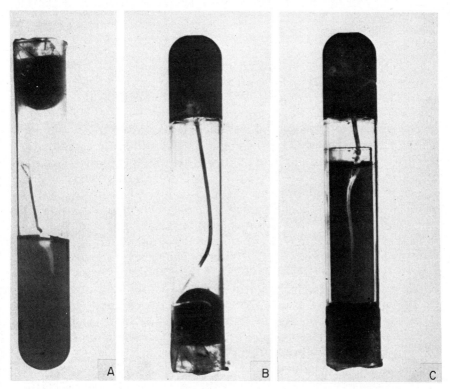

Fig. 9.4. Orientation of the rest of the embryo of *Pinus lambertiana* with respect to cotyledons in the medium affects the subsequent growth of the root and hypocotyl. **A**, Embryo oriented normally. **B**, Cotyledons embedded in sucrose medium. **C**, Cotyledons embedded in the sucrose medium and root growing with gravity in plain agar. (After Brown and Gifford, 1958; photograph from a lantern slide, courtesy of C. L. Brown)

(Brown and Gifford, 1958) (Fig. 9.4). In cultured embryos of *P. nigra*, the cotyledons are believed to absorb and translocate nutrients from the medium for growth of the hypocotyl and the terminal bud, but not for growth of the root (Bartels, 1957a). Other studies (Asakawa, 1961; Berlyn, 1962; Engvild, 1964) have also shown that for optimum growth of pine embryos, carbohydrates and other nutrients must be supplied through the cotyledons. Nonetheless, it is doubtful whether cotyledons supply any unique morphogenetic substances for the growth of the embryo, and it is not at all certain that they can be considered analogous to a haustorium. Since the pine embryo is found to grow normally in culture when it is decotylated and metabolites are allowed to diffuse through the cut surface of the cotyledons, it seems that the importance of the cotyledons lies in their large surface area (Berlyn and Miksche, 1965).

Pteridophytes

Effects of sucrose on the morphogenesis of "contained" and "freed" embryos of *Todea barbara* have been considered by DeMaggio and Wetmore (1961b) and DeMaggio (1963). In the former type, the optimum concentration of sucrose for growth was 1% or 2%, and the response of the embryo to concentrations of sucrose above and below the optimum varied with age. For example, in a medium lacking sucrose, embryos in the division phase of growth (7–15 days after fertilization) attained nearly the same degree of morphological complexity as those growing in an optimum sucrose concentration, while in 4% or 10% sucrose, growth was accompanied by a delay in organ initiation. On the other hand, the maturation pattern of slightly older embryos (15–30 days after fertilization) was better in 4% sucrose than in a medium lacking a carbohydrate supply. In isolated embryos of all ages, concentrations of sucrose higher or lower than the optimum affected their survival or delayed organ initiation in them. Thus, in a subtle way, changes in the sugar concentration determine the developmental sequence in fern embryogeny. These results somewhat extend the observations from other studies which have demonstrated that such phases in the life history of ferns, like sporangia formation in isolated leaves (Sussex and Steeves, 1958), and induction of apogamy (Whittier and Steeves, 1960), can be modified by alterations in the carbohydrate content of the medium.

Effect of Carbohydrate Supply on the Growth of Embryos of Irradiated Seeds

Brief mention should be made of the effects of some carbohydrates on the morphogenesis of the embryos of normal and of heavily irradiated wheat ("gamma plantlets") (Long and Haber, 1965a, b). Embryos excised from both types of grains grew well on a medium containing optimum concentrations of D-glucose. Addition of D-galactose to the medium containing D-glucose inhibited growth of the first leaf of normal embryos, but failed to limit the growth of the leaf of the gamma plantlet. Since there is a preferential promotion of cell elongation in the embryos of the gamma plantlet (Haber and Luippold, 1960), this is regarded as indicating that conditions favoring the elongation-oriented state are not compatible with D-galactose action. The fact that addition of galacturonic acid to a medium containing D-glucose resulted in an increase in the number of seminal roots formed and in a decrease in the elongation of the roots, leaf and coleoptile of normal embryos adds to the interest of the results and to the complexity of the system.

II. NITROGEN NUTRITION

The well established role of nitrogen as the raw material for protein synthesis need not be emphasized here. The need to provide nitrogen in the form of inorganic nitrates for successful growth of excised organs was recognized by the pioneer workers in the field of plant tissue culture (Robbins, 1922; Gautheret, 1935; White, 1943). Over the years, a great deal has been learned concerning the uptake and metabolism of nitrogenous compounds by plant tissue cultures. An evaluation of this information is necessarily complex, since the accumulated knowledge can be discussed in relation to several different categories of biological organization. For example, we may be concerned with isolated organs of plants, such as roots which have proved in particular to be favorite materials for such studies. At another level we may be concerned with the metabolism in culture of undifferentiated callus tissues which could be induced by wounding, hormonal treatments or bacterial agents. At a third level of organization, we may discuss the nutrition of suspensions of free cells or cell aggregates. Since generalizations concerning nitrogen nutrition have been evolved mostly with vegetative organs and tissues of plants, certain reservations should be maintained in applying the results to embryos.

Experiments with cultured embryos have revealed the role of some inorganic and organic nitrogen compounds in their growth and metabolism. Despite the small number of species so far surveyed, it would appear that most embryos receive and utilize nitrogen in an organic form or make their own protein from inorganic nitrates and sucrose supplied in the medium. Enough is not known at present about the biochemical events associated with nitrogen utilization by embryos to permit a meaningful picture interpretable in metabolic terms.

A. Inorganic Nitrogen Compounds

Utilization of Nitrates, Nitrites and Ammonium Ions

Only in embryos with potentially active enzyme systems to reduce nitrates to nitrites and thence to ammonium nitrogen will their capacity to utilize nitrates ever become evident. Although several nitrates are widespread constituents of culture media, embryos of only relatively few species are known to meet their amino acid requirements by synthesis from nitrate nitrogen supplied in the medium. An attempt to grow embryos of *Datura stramonium* by addition of nitrate to the medium containing casein hydrolyzate was not successful (Paris *et al.*, 1953). This led to the view that nitrate was not a requirement for the growth of embryos, but the possibility that the enzymes necessary to metabolize nitrates were inactive was not eliminated. It is also possible that the complete nitrogen requirement of embryos was met by the amino acid complex of casein hydrolyzate, rendering added nitrate superfluous. In this connection, it is noted that in a medium devoid of external sources of nitrogen, growth of embryos of *D. tatula* was favored by KNO_3, $NaNO_3$, $(NH_4)_2HPO_4$, and NH_4NO_3 (Matsubara, 1964b). This study also revealed that NH_4NO_3 was significantly superior to other compounds as a source of nitrogen. When embryos of different ages were grown on NH_4NO_3, younger embryos responded to the additive more vigorously than the older ones. Evidently, mature embryos have reached a metabolic autonomy as far as their nitrogen requirement is concerned or they have, perhaps, progressively adapted to an entirely different mode of nitrogen utilization with age.

For growth of embryos of some plants, there is an absolute requirement of nitrate nitrogen in the medium. Embryos of jute *(Corchorus capsularis)* appeared to grow better when KNO_3 was the sole nitrogen source instead of $(NH_4)_2SO_4$; in the latter medium, malformations of the cotyledons occurred (Mitra and Datta, 1951a, b). Similarly, growth of young embryos of rice was appreciably promoted by several nitrates which functioned as better nitrogen sources than ammonium ion (Amemiya *et al.*, 1956a). Harris (1953, 1956) studied the nitrogen nutrition of excised oat *(Avena sativa* var. 'Victory') embryos, and noted substantial increases in length of the first order root and shoot in the presence of sodium nitrate in the medium. Growth increments of the root and shoot in the nitrate medium were of the same magnitude as those exhibited by embryos growing in ammonium sulfate, casein hydrolyzate or a mixture of amino acids. Particularly interesting was the observation that in embryos growing in a medium containing sodium nitrate, addition of casein hydrolyzate or individual amino acids did not produce additive growth effects. From this we must infer that embryos meet their amino acid requirements by synthesis from nitrate nitrogen. To reconcile the various observations, it would be desirable to have for embryos of this species some information on the relevance of utilization of nitrates to the distribution of enzymes concerned in their breakdown.

A more direct approach to this problem was made by Rijven (1958) by feeding isolated embryos of *Capsella bursa-pastoris*, *Arabidopsis thaliana*, *Sisymbrium orientale* (Cruciferae) and *Anagallis arvensis* (Primulaceae) with intermediates of nitrate metabolism. The puzzling aspect of the work with embryos of Cruciferae was that nitrite was well utilized, while ammonium ion, presumably in the pathway of utilization of nitrite, was not favored. For embryos of *Capsella* sodium nitrate was also not favored as a source of nitrogen; it was, however, possible to induce the synthesis of nitrate reductase by incubation of embryos in the substrate, but there was no parallel growth promotion. Here one is confronted with a situation where the enzyme is not functional, although it is produced in quantity under the conditions of culture.

Inorganic Nitrogen Nutrition of Orchid Embryos

Published accounts on the inorganic nitrogen nutrition of orchid embryos are dominated by the fact that they grow best in media containing ammonium salts. The affinity for ammonium salts is, no doubt, related to the peculiar nutritional habitat of the orchids. In nature, it is known that germination of embryos and early growth of seedlings are invariably dependent upon their successful association with a symbiotic fungus. While the exact nature of the nutrients transmitted by the fungus to the host is not clear, there is reason to believe that sugars, minerals and nitrogenous materials are included.

Curtis and Spoerl (1948) found that in a medium of constant pH, ammonium salt was best for embryos of *Cymbidium eburneum* × *C. peach pauwelsii*, and slightly better than nitrate for *Cattleya mollie* and *C. trianaei*, while embryos of *Vanda tricolor* grew best in a medium containing nitrate. In a comparative study of the effects of several ammonium, nitrate, and nitrite salts on the growth of embryos of *Cattleya labiata* 'Wonder' × *C. labiata* 'Treasure' (Raghavan and Torrey, 1964a), ammonium salts were found to promote growth in terms of increasing morphological complexity, while nitrates and nitrites appeared hardly effective in advancing growth beyond the protocorm stage (Fig. 9.5).

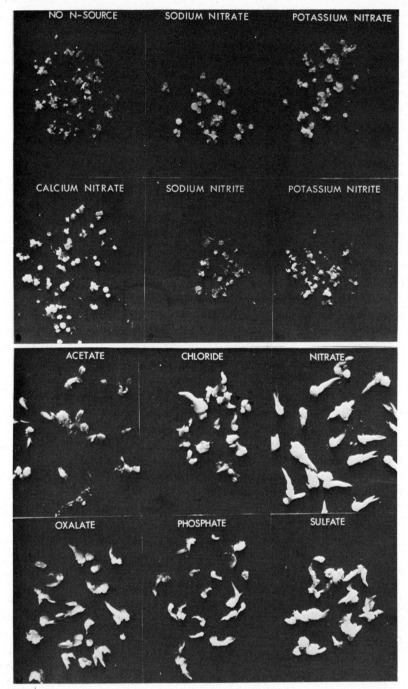

Fig. 9.5. Embryos of a hybrid *Cattleya* grow well on a medium containing ammonium salts (acetate, chloride, nitrate, oxalate, phosphate and sulfate), while nitrate and nitrite salts are ineffective. (After Raghavan and Torrey, 1964a)

A striking observation made in the course of this work was that although young (20 and 40 day old) embryos of *Cattleya* failed to utilize nitrate nitrogen, older embryos in which leaves and roots had emerged (60 and 80 day old) resumed normal growth if transferred to a nitrate-containing medium (Fig. 9.6). The ability of germinated embryos to

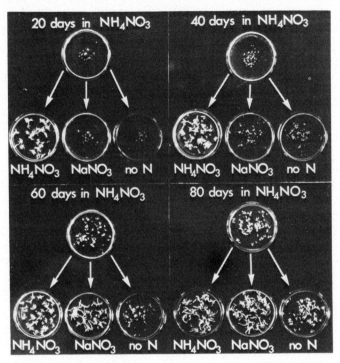

FIG. 9.6. While 20 and 40 day old embryos of *Cattleya* fail to grow in a medium containing sodium nitrate, 60 and 80 day old embryos resume normal growth in this medium. Embryos were grown in a medium containing ammonium nitrate for different periods (shown at the top in each group of figures) before transfer to fresh supply of the same medium, medium containing sodium nitrate or medium lacking nitrogen supply. Cultures photographed 14 days after transfer to the latter media. (After Raghavan and Torrey, 1964a)

grow in a medium containing nitrate was also paralleled by their ability to form nitrate reductase, while younger embryos which did not utilize nitrate did not form the enzyme even after prolonged periods of contact with the substrate. Clearly, a crucial biochemical differentiation was established in the older embryos and it may well be that selective activation or suppression of enzyme systems operate to control progressive embryogenesis in this species.

B. Organic Nitrogen Compounds

Investigations on the organic nitrogen nutrition of embryos occupy a prominent place in the literature on embryo culture. A commonly accepted generalization is that by comparison of the effects of closely related compounds on a system, one can get a meaningful idea of the pathway of utilization of different compounds interpretable in metabolic terms. Before we attempt to examine the available information, it is necessary to add that knowledge of the organic nitrogen nutrition of embryos cannot be complete until more is known about the nature of the enzymes and intermediary substances that function in the metabolism of the compounds tested.

Asparagine, Glutamine annd Related Compounds

Because of the significance and universality of asparagine and glutamine in the nitrogen metabolism of plants (Steward and Street, 1947), a number of investigations have dealt with their utilization by embryos. For embryos of *Cochlearia*, asparagine was superior to leucine or tyrosine as nitrogen source (Hannig, 1904). Among a number of organic substances tested, including choline, allantoin, betaine, aspartic acid, glutamic acid and asparagine (Brown, 1906), the last named was most effective in enhancing dry weight and nitrogen content of excised embryos of barley (Table 9.II). Paris *et al.* (1953) made a detailed study

TABLE 9.II. Effects of different sources of nitrogen on the growth of barley embryos. (From Brown, 1906)

Composition of nutrient solution	Dry weight of 35 embryos (mg)	Nitrogen in 35 embryos (mg)
Mineral salts + cane sugar only	135·0	3·00
Tyrosine	53·1	1·72
Phenylalanine	59·1	2·62
Leucine	77·6	3·44
Malt-peptone bodies of malt (crude "unclassified" bodies)	93·0	
Choline	152·0	4·39
Allantoin	143·0	4·48
Betaine	135·0	4·51
Ammonium sulfate	143·0	4·93
Aspartic acid	153·0	5·53
Glutamic acid	156·0	5·86
Potassium nitrate	154·0	6·54
Asparagine	155·0	8·10

of the effects of glutamic acid, aspartic acid and their respective amides on the growth of embryos of *Datura stramonium*, and showed that glutamine stimulated growth while glutamic acid, aspartic acid and asparagine were inhibitory. Matsubara (1964b) tested, in addition to glutamine and asparagine, 18 amino acids as nitrogen sources for growth of embryos of *D. tatula*. These compounds were incorporated aseptically into autoclaved White's medium in which $CaCl_2$ was substituted for $Ca(NO_3)_2$ and KNO_3. Glutamine proved superior for growth of embryos, which also exhibited a high level of growth in a medium containing L-cystine. The rest of the compounds were either mildly inhibitory or marginally promotory in their effects. A comparative study (Rijven, 1955, 1956) of the utilization of glutamine and asparagine by torpedo-shaped embryos of 12 species of plants distributed within nine families revealed without exception a greater growth-promoting effect of glutamine over asparagine (Fig. 9.7). In the embryos of *Capsella*

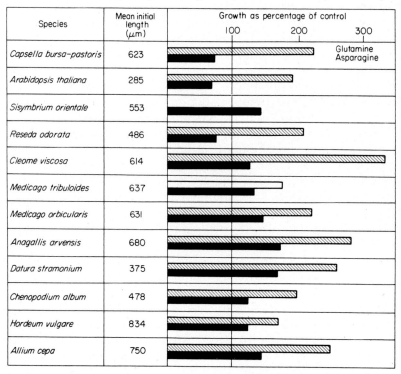

Fɪɢ. 9.7. Glutamine (400 mg/l) is superior to asparagine (400 mg/l) for the growth of embryos of 12 species of angiosperms. Growth is expressed as percentage of control after 48 h in culture, except for *Chenopodium album* (24 h) and *Hordeum vulgare* (72 h). (After Rijven, 1956)

bursa-pastoris, Arabidopsis thaliana and *Reseda odorata*, asparagine drastically inhibited growth; in the rest of the species investigated, it slightly
stimulated growth. Interpretation of the contrasting effects of asparagine and glutamine on *Capsella* embryos was difficult since glutamic
acid was not as effective as glutamine in promoting growth of embryos,
while aspartic acid, unlike asparagine, was only slightly inhibitory.
Moreover, decarboxylation products of glutamic acid and aspartic
acid, γ-aminobutyric acid and β-alanine, respectively, could not substitute for their respective amides. From a superficial analysis, differences
between the effectiveness of asparagine and glutamine unmistakably
point to the amide groups which appear to be crucial in determining
the behavior of these compounds. In studies originating with animal
systems it has been demonstrated that certain tissues can absorb and
translocate glutamine more rapidly than glutamic acid (Meister, 1965)
and that while asparagine is hydrolyzed during absorption, glutamine
is absorbed unchanged (Fridhandler and Quastel, 1955). Judging from
the results with embryos of different species, one suspects that their
ability to respond differently to asparagine and glutamine is more likely
to be influenced by a combination of various mechanisms than by a
single factor.

Results somewhat different from *Capsella* embryos were obtained in
wheat embryos where both asparagine and glutamine were utilized
(Rijven, 1960). γ-Aminobutyric acid was also taken up by the embryos,
and incorporated into proteins, but its utilization was sluggish in comparison to nitrate and glutamine or amino acids like proline, alanine
and glutamic acid (Table 9.III). Obviously, the precise course of utilization of any amino acid will be determined by the appropriate
transaminases present. Thus, the presence of a highly active alanine–
glutamic acid transaminase may explain the rapid utilization of alanine. The idea that a similar transaminase reaction is not a significant step
in the utilization of γ-aminobutyric acid is borne out by the absence
of γ-aminobutyric acid–glutamic acid transaminase in the embryo.

Another aspect of amino acid metabolism in wheat embryo also calls
for comment. Although glutamyl transferase, which catalyzes the transfer of glutamic acid residue to form glutamine, was absent in the
embryos of germinating wheat grains, it was readily induced in isolated
embryos by soaking (Rijven, 1961; Rijven and Banbury, 1960). The
activity of the enzyme was inhibited by treating embryos with glutamine, but the simultaneous addition of glutamic acid reversed this
inhibition. Although the results do not prove that glutamine is synthesized from glutamic acid, proportionate changes in the ratio of glutamine to glutamate concentration in embryos during ripening to a level

TABLE 9.III. Effects of γ-aminobutyric acid and some other amino acids on insoluble nitrogen content and growth of wheat embryos cultured in darkness. (From Rijven, 1960)

Mean values per seedling after 120h of incubation at 25°C in the dark. The grains from which the embryos were excised weighed 60·0–64·9 mg. The mean value of the insoluble nitrogen content of the embryos at the time of excision was 109·5 μg per embryo and the standard error 2·6 μg.

Substance added	Concn (mM)	Insoluble nitrogen content per seedling (μg)		Coleoptile length (mm)	First leaf length (mm)	Root growth	
		Mean	SE			Number	Total length (mm)
Nil		105·4	4·4	61·1	50·1	4·7	168·4
DL-α-Aminobutyric acid	3·0	123·1[a]	1·8	45·7[a]	33·7[a]	7·9[a]	60·1[a]
L-α,γ-Diaminobutyric acid	1·5	103·2	3·1	4·9[a]	4·8[a]	3·6	16·2[a]
γ-Aminobutyric acid	3·0	124·9[a]	2·8	61·9	57·2[a]	5·2	156·7
Monosodium L-glutamate	3·0	174·9[a]	0·7	61·8	73·0[a]	5·0	229·3[a]
L-Proline	3·0	158·0[a]	4·9	68·2	81·5[a]	5·2	146·4
L-Alanine	3·0	181·8[a]	9·4	64·4	80·1[a]	5·0	192·7[a]
L-Asparagine	1·5	207·8[a]	7·8	69·8	89·6[a]	4·9	244·0[a]
L-Glutamine	1·5	235·6[a]	10·4	68·9	94·7[a]	4·8	250·6[a]

[a] This value differs significantly, at least at the 1% level, from the value recorded for the treatment in which no substance was added to the basal medium.

commensurate with the concentration of glutamyl transferase appear likely from these results. It is of obvious interest to examine whether the example cited above is representative of other control mechanisms operating during embryogenesis, or whether it is an exceptional case. The central role of glutamine in protein metabolism of wheat embryo is also indicated by the fact that inhibition of embryo growth induced by methionine sulfoximine was negated by glutamine, while methionine, the natural base of the inhibitor, produced additive effects with the latter (Hinton and Moran, 1957). Presumably, methionine sulfoximine inhibits protein synthesis in which glutamine synthesis is an essential step.

At the level of organ formation, addition of glutamine to the medium remarkably enhanced shoot growth in the embryos of *Ginkgo biloba* (Fig. 9.3). This effect was evident not only in growth in length but also in growth in diameter of the shoot (Ball, 1959). Elongation of the root, on the other hand, was slightly inhibited by the amide. Insofar as roots grow best in the presence of sugar additives in the medium (see p. 229), a basic difference in the metabolism of root and shoot systems of embryos is obvious in this study.

Metabolic Pathways in Amino Acid Utilization

Excised and cultured embryos are potentially of great value for analysis of the pathways of biosynthesis of amino acids. Barnes and Naylor (1962) used cultured embryos of longleaf pine *(Pinus palustris)* and slash pine *(P. elliottii)* to map out the pathway of β-alanine biosynthesis. These studies, which involved stepwise feeding of successive intermediates between β-alanine and orotic acid, appear to indicate that the former is formed through a reductive pathway from orotic acid → uridine → uracil → dihydrouracil → N-carbamyl-β-alanine → β-alanine.

According to Oaks and Beevers (1964), maize embryos contain a large pool of amino acids in a dynamic equilibrium with the surrounding endosperm. During the growth of excised embryos in a medium containing inorganic nitrates and glucose, in the absence of exogenous source of amino acids, the soluble pools of basic and neutral amino acids decreased, but the concentration of glutamic acid and aspartic acid remained close to the control level. The magnitude of changes in the amounts of glutamic acid and aspartic acid is such as to suggest the existence in the embryo of enzymatic mechanisms for the conversion of carbohydrate carbon to protein carbon of glutamic acid and aspartic acid. These results recall earlier studies of Kandler (1953) and Kandler and Fink (1955) who suggested that glucose taken up by excised embryos of different species is incorporated into their proteins.

In a later work, Oaks (1965) has shown that α-ketoisovaleric acid and acetate are the precursors of leucine biosynthesis in maize embryos and that α-isopropylmalic acid, β-isopropylmalic acid and α-ketoisocaproic acid are intermediates in this conversion. An interesting feature of leucine metabolism of the embryos is that specific activities of two leucine biosynthetic enzymes, β-isopropylmalate dehydrogenase and alanine transaminase, increase with age, probably because of release from repression. But unlike the phenomenon of enzyme repression and end product inhibition so familiar in the control of many biosynthetic processes in microorganisms, this derepression was not due to the end product leucine. The significance of this finding is somewhat obscure, but it certainly points towards a greater diversity in biosynthetic pathways in embryos than hitherto suspected.

Antagonism and Synergism between Amino Acids

A major outcome of investigations on the amino acid nutrition of embryos is the discovery of interesting mutual antagonism and synergism between different amino acids. Sanders and Burkholder (1948) found that addition of amino acids to the medium influenced growth and differentiation of pre-heart- and early heart-shaped embryos of *Datura stramonium* and *D. innoxia* (Fig. 9.8). In the basal medium, growth of the cotyledons was mostly affected in both species; by supplementing the medium with casein hydrolyzate, L-cysteine and L-tryptophan, marked promotion in embryo growth was registered. A mixture containing 20 amino acids in the proportion present in casein hydrolyzate acted instead in promoting growth of embryos, but this favorable effect was not reproduced by incomplete mixtures or by single amino acids. Somewhat similar results were obtained for growth increments of the root and shoot systems of oat embryos grown in media containing single amino acids or their mixtures (Harris, 1953, 1956). These results are interesting when it is noted that the complete mixture of amino acids employed had some readily utilized, some partially utilized and others apparently toxic. Considering these facts, the net growth-promoting effects of a mixture of amino acids appear to be a balance between promotory and inhibitory substances. In cultured embryos of *Capsella bursa-pastoris* where a mixture of amino acids prepared to conform to the composition the plant globulin edestin enhanced growth to the same extent as casein hydrolyzate, it was shown that the activity of the complete amino acid mixture could be reproduced or surpassed by glutamine (Rijven, 1952).

Addition of single amino acids such as valine and leucine to the medium inhibited growth of barley embryos (Joy and Folkes, 1965;

Miflin, 1969b). Growth inhibition by valine was relieved by isoleucine and that by leucine by the addition of both isoleucine and valine. Strong antagonism also existed between tyrosine and phenylalanine and between lysine and ornithine or arginine in the growth of barley embryos (Miflin, 1969b). Although antagonism between amino acids of closely similar structures involves competition at specific sites, these

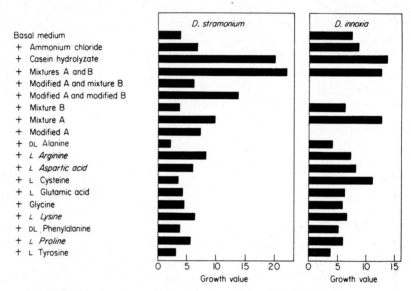

Fig. 9.8. Effect of ammonium chloride and various organic nitrogen sources on the growth of embryos of two species of *Datura*. Amino acid mixture A (mg/l): DL-alanine (6·0), *L-arginine* (12·0), *L-aspartic acid* (18·0), L-cysteine (20·0), L-glutamic acid (60·0), glycine (1·5), *L-lysine* (18·0), DL-phenylalanine (12·0), *L-proline* (24·0) and L-tyrosine (18·0). Amino acid mixture B (mg/l): *L-cystine* (1·5), *L-histidine* (9·0), L-hydroxyproline (1·5), DL-isoleucine (15·0), L-leucine (15·0), *DL-methionine* (9·0), *DL-serine* (18·0), DL-threonine (12·0), L-tryptophan (20·0) and *DL-valine* (24·0). Amino acids printed in italics are those which are included in modified mixtures A and B. (After Sanders and Burkholder, 1948)

studies have not beeen carried far enough to relate them to the meta-bolic pathways involved. The picture provided is further complicated by the fact that the different compounds produce different effects on growth of the root and shoot and on dry weight increases. Nevertheless, the data seem to underscore the fact that the design of a completely satisfactory amino acid mixture for growth of embryos is a problem of relative complexity, and one which may be nearer solution only after the metabolic pathways of compounds in the system in question have been worked out.

TABLE 9.IV. Growth of embryos of an inbred strain of maize, excised 21 days after pollination, in media containing arginine and canavanine. Values represent mean dry weights of seedlings in milligrams (±standard error) after 10 days in culture. (From Wright and Srb, 1952)

Canavanine (mg/l)	Arginine (mg/l)					
	160	80	40	20	10	0
80	5·0±0·7	4·6±0·4	2·4±0·4	1·6±0·4	1·1±0·2	0·1
40	6·0±0·5	8·1±1·5	4·3±1·2	2·6±0·5	2·1±0·3	0·8±0·2
20	7·6±1·2	10·5±1·2	7·6±0·8	6·5±0·4	5·1±0·2	1·2±0·2
10	8·1±1·8	13·6±2·1	13·4±3·1	9·5±2·1	4·2±0·6	1·3±0·5
0	14·5±3·6	15·2±3·4	14·5±1·7	16·4±2·3	15·0±2·2	14·0±2·4

Wright and Srb (1950) have demonstrated a marked antagonism between arginine and canavanine in the growth of mature maize embryos. Supplementing the growth medium with canavanine inhibited the growth of embryos, the degree of inhibition increasing with concentrations of canavanine. Within limits the inhibition by canavanine was mitigated by application of arginine (Table 9.IV). Besides

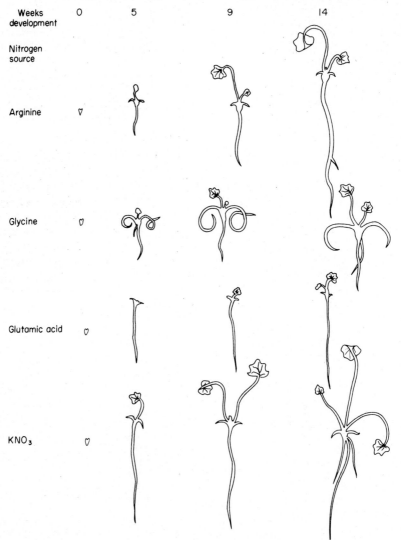

FIG. 9.9. Morphology of growth of cultured embryos of *Heracleum sphondylium* is modified by the nature of nitrogen source incorporated into the medium. (After Stokes, 1953b)

arginine, other amino acids such as citrulline, ornithine, glutamic acid and lysine were effective to some extent in nullifying the inhibition by canavanine. In *Heracleum sphondylium*, morphological changes resulting in the formation of normal seedlings followed the transfer of embryos from after-ripened seeds to a medium containing arginine, but growth was poor when serine and alanine were supplied along with arginine (Stokes, 1953b). A relationship of the substances used in the above experiments to arginine biosynthesis is implied by the results, but except for citrulline, ornithine and possibly glutamic acid, which are well established precursors of arginine, biochemical evidence on others is lacking.

Embryos of *Heracleum* exhibited characteristic morphology when supplied with different forms of organic nitrogen in the medium (Stokes, 1953b). In the presence of arginine, the embryo formed a thickened root and a bulbous base, closely resembling the one grown in KNO_3. In glycine medium, the embryo was somewhat stocky and distorted, but with normal cotyledons. The bulk of growth in glutamic acid medium consisted of root elongation, with little shoot or cotyledon development (Fig. 9.9). The influence of amino acids on the morphology of specific parts of the embryo seems to suggest that the restricted growth of certain embryonic regions observed in media containing combinations of amino acids cannot always be attributed to toxicity due to injuriously high levels, but is probably due to lack of a specific amino acid in the medium.

Organic Nitrogen Compounds in the Growth of Orchid Embryos

The role of organic nitrogen compounds in the nutrition of orchid embryos has been investigated to some extent, the principal rationale for this being the fact that embryos have a specific requirement for nitrogen. A comparative evaluation of the data is hardly possible, however, because of the range of species and horticultural varieties used. Even in the same pod, embryos show different degrees of response to added substances, probably due to their different degrees of maturity and consequent changes in their enzyme complement.

It is known from the work of Burgeff (1936) that when orchid embryos are grown in media containing only amino acids as source of nitrogen, they fail to grow or grow only feebly. A detailed study on the amino acid nutrition of orchid embryos was made by Spoerl (1948) and Spoerl and Curtis (1948). These investigators cultured embryos of *Cattleya amethystoglossa, C. trianaei, C. mollie, Epidendrum cochleatum* and *Zygopetalum mackayi* aseptically in media supplemented with ammonium nitrate or different amino acids. To determine the effect of amino acids upon growth, each compound was supplied at different concentrations under conditions of light or darkness. Arginine and aspartic acid were the only two amino acids which supplied nitrogen as well as ammonium nitrate to the embryos. Aspartic acid was good for embryos from ripe pods and inhibitory for embryos from unripe pods of *C. trianaei*, while arginine was good for growth of embryos from unripe pods and inhibitory for embryos from ripe pods of *C. amethysto-glossa*. The finding that arginine and aspartic acid are suitable for growth of embryos suggests that they should be more usefully tested as a source of nitrogen in orchid embryo culture media.

In the embryos of *Cattleya labiata* 'Wonder' × *C. labiata* 'Treasure' there is evidence for the operation of ornithine cycle since arginine and ornithine induced rapid growth in the absence of ammonium salts (Raghavan, 1964a). γ-Aminobutyric acid, presumably related to ornithine cycle through ornithine and glutamic acid, was also utilized to some extent. The favorable growth effects exhibited by compounds of the ornithine cycle indicate a high degree of specificity in the nitrogen nutrition of *Cattleya* embryos, and serve to emphasize the relatively narrow range of compounds which serve as raw materials for protein synthesis in the embryos as effectively as ammonium salts.

Urea has served as an effective source of nitrogen for embryos of some species, but is inhibitory in others. Growth of embryos of *Laeliocattleya* (Magrou *et al.*, 1949), *Cattleya* (Curtis, 1947b; Raghavan, 1964a) and *Vanda* (Burgeff, 1936) was enhanced by urea. On the other hand, presence of urea in the medium was positively inhibitory for growth of embryos of *Dendrobium phalaenopsis* and *Phalaenopsis* sp. (Burgeff, 1936). The reported effects of this compound on embryos of *Cymbidium* are contradictory, and both promotory and inhibitory effects have been claimed (Cappelletti, 1933; Burgeff, 1936). Seeds of *Vanilla planifolia*, which germinate only in very weak solutions of inorganic nitrates, are tolerant to high concentrations of organic nitrate when this is supplied as urea (Lugo-Lugo, 1955a, b).

III. COMPLEX ORGANIC NUTRIENTS

There are several reports of favorable effects of extracts of roots, tubers, fruits, seeds, endosperms and other parts of the plant on the growth of embryos. In this section, attention is directed to a brief analysis of what is currently known about the effects of these substances as they relate to the various nutritional, metabolic and morphogenetic activities of embryos. Perhaps the greatest disadvantage in the use of these substances is that they constitute combinations of such a great variety of unique nutritional factors that it is possible to ascribe only vaguely their observed effects to any single substance or to a group of related substances.

Coconut milk has enjoyed widespread application as a growth promoter in the culture of embryos. The dramatic effect of this and other substances of endospermic origin in inducing growth of proembryos of several plants was discussed in detail in Chapter 7 and effects of these substances on the morphogenesis of mature and differentiated embryos will be described briefly here. Some factors necessary for inducing growth of embryos isolated from ripe or nearly ripe seeds of *Cyclamen persicum* are reportedly present in coconut milk, peat extract and yeast extract (Gorter, 1955). Norstog (1956a) has shown that addition of 90% by volume of coconut milk induced normal development in barley embryos, initially 0·3–0·8 mm long, whereas in lower concentrations callus growth was initiated. In certain gymnosperm embryos, pronounced stimulation of growth of the shoot, root or leaf is caused by coconut milk (Ball, 1959; Berlyn, 1962; Gjønnes, 1963). Puri (1963) has shown that addition of coconut milk to the medium stimulated parthenogenetic activation of embryos of the apomictic species, *Aerva tomentosa*. Best results were obtained when flowers or spikes were cultured in media containing 20% coconut milk. Further effects of coconut milk are noted in the inhibition of differentiation of the haustorium in embryos of the stem parasite, *Scurrula pulverulenta* (Bhojwani, 1969; Johri and Bhojwani, 1970), and promotion of adventitious root formation in embryos of *Coffea canephora* var. *robusta* (Colonna et al., 1971). A different situation exists in embryos of *Dendrophthoe falcata*, in which development of the plumular leaves is inhibited by the addition of an extract of its own endosperm to the medium (Bajaj, 1968). It has been reported (Augusten, 1956) that endosperm extracts from wheat, maize, barley, oat and rye incorporated into an agar medium stimulate development of isolated wheat embryos, although root growth is inhibited. Interesting as these observations are, information concerning the factors in coconut milk and other endosperm extracts which regulate morphogenesis of embryos is almost totally lacking in these studies.

Suitability of coconut milk for successful germination and growth of orchid embryos has been reported (Hegarty, 1955; Niimoto and Sagawa, 1961). In *Dendrobium*, coconut milk did not affect germination of embryos, but inhibited their growth in the early germination phase (Kotomori and Murashige, 1965). In another instance, embryos of a hybrid *Phalaenopsis* showed callus-like proliferation and retarded organ differentiation when coconut milk was present in the medium (Ernst, 1967a).

A number of other substances have found their way into tissue culture media commonly employed to grow orchid embryos. Tomato juice is reported to be a good nutrient for the germination and growth of embryos of some orchids (Meyer, 1945c; Vacin and Went, 1949; Griffith and Link, 1957). Since ashed tomato juice, in contrast to whole juice or filtered juice, was ineffective as a growth promoter in some experiments, organic substances of the juice have been invoked as causal agents of growth. Some support for this view comes from the finding that commercial protein hydrolyzates can satisfactorily substitute for tomato juice in the culture media (Vacin and Went, 1949). The reported growth-promoting effects of tomato juice are to be accepted with some reservation since a recent re-examination of the effects of this additive on embryos of *Cattleya* Sudan × *C. percivaliana* showed that when growth was measured in terms of organ differentiation, multiple meristem formation and general proliferation, whole tomato juice or ether, acetone, ethanol or water extracts of the juice were unsatisfactory (Arditti, 1966a).

Professional orchid growers have repeatedly used other materials, including beef extract, extracts of wheat, barley, carrot, mushroom, potato and yeast, peptone and juices of various fruits (Knudson, 1922; Lami, 1927; Curtis, 1947b; Mariat, 1952; Withner, 1953). It is unfortunate that an understanding of the unique activity of some of these substances as nutrients for embryos has not yet been achieved, but there is little doubt that highly specific chemical stimuli capable of inducing normal growth in excised embryos of a variety of plants are present in them. Since orchid embryos invariably have a specific requirement for reduced nitrogen, conceivably the nitrogenous constituents of the extracts may be operating in promoting growth.

IV. VITAMIN NUTRITION

Vitamins early attracted the attention of investigators interested in plant tissue culture, and a great deal of interest was aroused by the demonstration of a requirement for these accessory growth factors by cultured roots. Although addition of vitamins to embryo culture medium is now a routine practice, no real evidence has been found to justify the view that some or all of them are "essential". Uncertainties also prevail concerning the most useful vitamins for culture of embryos. Reliable information is available to show that vitamins function in living cells in a strategic role as prosthetic groups of certain enzymes, and this basic concept of their role is no doubt applicable in the growth of embryos also.

The relative importance of individual vitamins for growth of embryos may perhaps be assessed from the extent to which they have been used in culture media. Defined in this way, biotin, thiamine, pantothenic acid, nicotinic acid, ascorbic acid and pyridoxine appear to be more commonly used than others.

In cultured embryos of a few species, addition of vitamins seems to affect the morphological type of growth. For example, in apparent agreement with the results from root culture investigations, several workers proceeding independently have confirmed the importance of thiamine in enhancing root growth in embryos (Kögl and Haagen-Smit, 1936; Bonner and Axtman, 1937; Lammerts, 1942; Mitra and Datta, 1951a; Helmkamp and Bonner, 1953; Bartels, 1957b; Sircar and Lahiri, 1956; Ohta and Furusato, 1957; Chatterji and Sankhla, 1965). In isolated pea embryos, pantothenic acid (Bonner and Axtman, 1937) and biotin (Kögl and Haagen-Smit, 1936; Helmkamp and Bonner, 1953) were beneficial for shoot growth and the influence of these substances on the root was less marked. Pyridoxine was generally inhibitory for growth of pea embryos, especially the shoot (Helmkamp and Bonner, 1953), but it enhanced coleoptile and root growth in rice embryos (Sircar and Lahiri, 1956). Addition of a mixture of thiamine, nicotinic acid and pyridoxine was necessary for root elongation in cultured embryos of *Cicer arietinum*, *Cajanus cajan* and *Dolichos biflorus* (Sen and Mukhopadhyay, 1961). In pea embryos, *meso*-inositol stimulated growth, while D- and L-forms of this compound were ineffective (Fleury *et al.*, 1951). Although there is no exact recognition of the role of vitamins in the growth of embryos of gymnosperms, in the absence of vitamins, the rate of elongation of the embryonic organs is reduced (Bulard, 1968; Thomas, 1970a, b).

The potential role of endogenous vitamins of the embryo in growth processes has been emphasized (Bonner and Bonner, 1938). Embryos of pea variety 'Wrinkled Winner', which synthesize ascorbic acid in the shoot, did not respond to the added vitamin, while those of 'Perfection', which has low ascorbic acid content, showed pronounced growth stimulation in the presence of exogenous supplies of the vitamin (Fig. 9.10). It is unlikely that the spontaneous growth effects induced by ascorbic acid in the embryos are specific, since other accessory growth factors also produce growth stimulation of equal magnitude. Thiamine, which is as effective as ascorbic acid in enhancing growth of embryos, presumably acts by increasing their ascorbic acid content.

A general potency of nicotinic acid in stimulating shoot growth in excised pea embryos has been reported (Bonner, 1938). There are also consistent reports of enhancement of germination and growth of orchid embryos by nicotinic acid (Schaffstein, 1938; Withner, 1943, 1955; Burgeff, 1959; Noggle and Wynd, 1943; Mariat, 1949, 1952; Bahme, 1949; Hegarty, 1955; Arditti, 1966b). Although nicotinic acid is

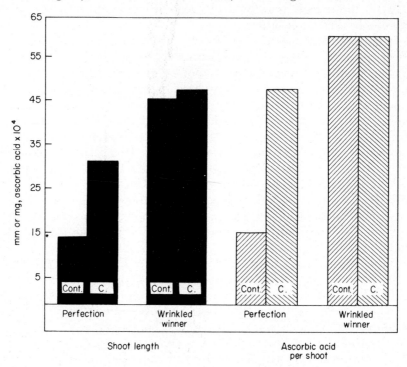

FIG. 9.10. The response of excised embryos of two varieties of pea to exogenous ascorbic acid in the medium is dependent upon their endogenous ascorbic acid content. The rate of growth and ascorbic acid content of embryos are measured after 4 weeks in culture. Cultures receiving ascorbic acid are designed as C. (After Bonner and Bonner, 1938)

apparently involved in the formation of nicotinamide of the coenzymes NAD and NADP, orchid embryos do not show enhanced growth when supplied with NAD and NADP (Arditti, 1966b). By feeding intermediates of nicotinic acid to embryos of *Laeliocattleya*, a probable pathway in the biosynthesis of this compound by way of tryptophan, kynurenine, 3-hydroxyanthranilic acid and quinolic acid has been established (Arditti, 1967b). Despite the number of workers who have studied the effects of vitamins on germination and growth of orchid embryos, it appears that a generalized trend in vitamin requirements cannot be demonstrated in them (Arditti, 1967a).

At the biochemical level, there is some evidence to show that biotin may stimulate lipid synthesis in cultured embryos. When flax embryos were grown in the presence of ^{14}C-acetate, addition of biotin to the medium stimulated the incorporation of radioactive substrate into lipids. Generally older embryos appeared to be more tolerant to higher doses of biotin than younger ones (Miramon, 1957). In the embryos of both flax and cotton, biotin promoted lipid synthesis to a greater extent in light than in dark (Kurtz and Miramon, 1957; Brown and Kurtz, 1959); this is probably due to differences in the assimilation rate between light-grown and dark-grown embryos. The precise stage in lipid biosynthesis at which biotin acts is not borne out by these results. Furthermore, the greatly increased requirement for biotin by older embryos is difficult to understand in terms of normal metabolic processes.

V. MISCELLANEOUS SUBSTANCES

A. Sterols and Organic Acids

Several miscellaneous substances, some of which do not have a direct bearing on the metabolism of plant cells, have been tested on embryos, but the results have been equivocal. Helmkamp and Bonner (1953) have shown that growth stimulation in pea embryos by steroids and steroidal materials is chemically specific. Estrone was highly effective in enhancing growth while related compounds such as testosterone, dihydroxyestrin, methyl- androstenediol and cortisone were growth inhibitory. Naturally occurring plant sterols such as ergosterol and stigmasterol were without effect. However, in both pea and wheat embryos, steroidal saponin markedly promoted growth (Balansard and Pellessier, 1943). Since questions concerning biosynthesis, metabolism and significance of steroidal materials in plants are not clearly understood (Heftmann, 1963), it is difficult to evaluate these results, and further work on their regulatory effect is much needed.

There is suggestive evidence for a role for organic acid salts in the growth of young cotton embryos (Mauney et al., 1967; see also Norstog, 1967). Based on the chromatographic identification of malic acid in the endosperm of cotton, it was found that addition of ammonium or calcium malate to the medium enhanced growth and viability of cultured embryos. Apart from its possible involvement as an energy source or as an effective buffering agent the role of malate in this system remains obscure.

B. Effect of Basal Nutrient Medium

In most of the published studies, the medium employed for embryo culture has been selected empirically from one of the several media standardized for general plant tissue culture work. It is a truism that a medium suitable for one organ or tissue might prove unsuitable for another organ or tissue of the same plant; the nutrient requirements of the same organ or tissue of different species are also normally different. These considerations prompted Randolph and Khan (1960) to study the merits of five basic nutrient media on the growth of excised embryos of *Iris* and wheat. These media—Knudson's orchid agar, Randolph and Cox's, Nitsch's, Street's and Rappaport's—have in common several major salts, a source of iron and 2% sucrose. Although no dramatic differences in the growth of embryos were observed in the various media, Nitsch's medium appeared best for shoot growth of *Iris* embryos. Wheat embryos grew uniformly well in all media, except in Knudson's agar medium where root elongation was poor. Since the media employed differed from one another only in minor details such as presence or absence of trace elements, and the nature of the source of iron, the differences observed in the growth of embryos must, therefore, depend upon some factors other than medium composition. Concentration of agar in the medium also influences growth of embryos to some extent, perhaps through the regulation of available moisture (Randolph, 1959; Khan and Randolph, 1960). In a preliminary study of the growth of embryos of certain legumes in White's and Tukey's media, the former was shown to be superior (D'Cruz and Kale, 1957).

In considering the effects of different salt media, one has also to consider the possibility of a changing salt requirement for embryos of different ages. For heart-shaped embryos of *Capsella bursa-pastoris*, Murashige–Skoog medium appeared superior to Knop's, Heller's, White's and Nitsch's; yet it was significantly inferior to the others for the survival of globular embryos in culture (Monnier, 1970). Evidently there is a change in the tolerance of embryos to ionic balance in the medium with age. It appears that before the organic nutrient requirements of proembryos can be studied, it might well be necessary to work out the composition of the inorganic salt medium that will assure their survival.

VI. COMMENTS

In conclusion, the evidence shows that addition of a suitable carbon energy source, a nitrogen compound, and a mixture of vitamins to a basal medium of major salts and trace elements is important for fostering continued growth of cultured embryos. Despite several reports of the relationship between the presence of certain nutrient factors in the medium, and growth of a specific organ of the embryo, it is doubtful whether these substances have a morphogenetic role other than affecting growth of primordial organs already blocked out. However, at the fundamental level, the more important question is the mode of action of the nutrients and it is in this area that the usual gap in our knowledge is wide. A proper understanding of the metabolic aspects of the added nutrients relies heavily on biochemical methods, and the hope is that it will not be too long before such methods can be applied to embryos of a larger number of species than hitherto possible. If the approach is successful, progress in our understanding of the biochemistry of nutrition and metabolism of embryos should be rapid because of the accumulation of pertinent morphological and physiological data.

10. Hormonal and Intercellular Control of Organogenesis in Embryos

Morphogenesis and differentiation usually result from the impact of a host of impinging physical and chemical influences on the cellular environment of the organism. These influences, by regulating the internal physiological processes, may be instrumental in modifying the potency of cells in regard to their polarity, gradients, biochemical circuitry, and inductive effects which cause the directed transformation of undifferentiated cells. The common trend of thought is that it is possible to superimpose upon the genetic potentialities of the organism special environments which result in a greater or lesser concentration of a particular chemical stimulus and thus modify its growth. This forms a basis for the concept of correlation and morphogenesis of Sachs

(1887), who formulated that chemical or metabolic substances are present in specific loci in plants where organs are eventually initiated.

Students of plant embryogenesis have attempted to modify the growth of embryos by specific chemical treatments or by altering their physical environment. These studies have been largely undertaken with excised and cultured embryos where medium composition and environmental factors are easily and accurately controlled. Although this approach is much too simple, results of these studies have provided particularly instructive information on the external factors controlling growth and organogenesis in embryos and their response mechanism in chemical terms.

For the experimental embryologist an important morphogenetic consideration is that concerned with the initiation, growth and modification of the root and shoot systems in embryos because of the divergence in function of these organs and their bearing on the question of organizational relationships. An embryo excised from a mature seed generally begins to grow within 24–48 h after transfer to a medium. The first readily observed change is the elongation of the primary root from what appears to be a primordial organ. After root growth is under way, elongation of the shoot and formation of leaves occur. Although a simple medium will support overall growth and differentiation of the embryo, it does not define clearly the requirements for growth and intercellular correlation of component parts. We therefore need to know what additional supplements are to be added to the medium to promote or inhibit growth of the different organs, and it is from this point of view that the theme of this chapter is developed and some of the results examined. In the following sections, we will discuss morphogenesis of embryos when they are grown in media supplemented with different plant growth hormones. Several types of growth hormones appear to exercise control in embryo growth, including auxins, gibberellins and cytokinins. Although evidence is sparse, there is reason to believe that abscisic acid, a recently discovered plant hormone which is particularly active in growth-inhibiting processes in plants, is also involved in the control of organogenesis in embryos. Increased or decreased response to an applied substance is not necessarily an endorsement of the view that the substance in question controls growth of the particular organ in the adult plant.

I. EFFECTS OF AUXINS

A. Modifications of Growth of Embryonic Organs

In appraising the effects of auxins on embryo growth, it is important to recognize their effects on cell division and cell elongation separately. However, this is not easily done because, except for specialized organs like the epiblast and coleorhiza, the rest of the embryo grows by both cell division and cell elongation.

Root and Shoot Primordia

Although in some early work excised embryos are reported to show inconsistent responses to auxins (Tukey, 1938), there are many definitive reports of their growth-modifying effects. One such effect is the inhibition by auxins of the ability of embryos to initiate roots (Solacolu and Constantinesco, 1936; Gautheret, 1937; Voss, 1939; Gardiner, 1940; Nuchowicz, 1965). A suggestive auxin-mediated morphogenetic effect was described by van Overbeek *et al.* (1942), who noted marked root growth inhibition and stem elongation in embryos of *Datura stramonium* growth in a medium containing nonautoclaved coconut milk. After extraction of coconut milk with alcohol and ether, its root growth-inhibiting properties disappeared (van Overbeek, 1942). It was thought that an auxin suspected to be present in coconut milk was responsible for the observed effects, although from present-day tissue culture literature it appears that free auxin is released from coconut milk upon autoclaving. Later work in which embryos of various ages were grown in media containing different concentrations of IAA confirmed the inhibitory effects of auxin (Rietsema *et al.*, 1953b). This work also revealed that growth of the primary root was stimulated by extremely low concentrations of IAA. From reasoning somewhat similar to that of van Overbeek (1942), Nakajima (1962) suggested a role for auxin to account for the inhibition of root growth in embryos of *Cucurbita maxima* and *C. moschata* by an endosperm extract of the same species.

Some workers (Sircar *et al.*, 1955; Sircar and Lahiri, 1956) have explored the relationship between the auxin content of the endosperm and growth of embryos of rice. By eliminating fractions of the endosperm and substituting IAA during growth of the embryo, indirect evidence was obtained to show that growth of the coleoptile and root was regulated by the amount of auxin available from the endosperm. In general agreement with the postulated role for auxin of endospermic origin in the growth of embryos, it was found that rearing excised embryos aseptically in media containing IAA or tryptophan facilitated their growth. In cultured embryos of Japanese varieties of rice, IAA, NAA, and 2,4,5-trichlorophenoxyacetic acid promoted growth of root and shoot systems with corresponding increases in dry weights (Amemiya *et al.*, 1956a).

The involvement of auxins in formative processes is further illustrated by experiments which have yielded evidence suggestive of their role in the growth of root and shoot systems of embryos of several angiosperms, gymnosperms and pteridophytes. Kruyt (1952, 1954) made careful measurements of the root and shoot systems of pea embryos treated with potassium salts of IAA, NAA and 2,4-D which were administered directly in the nutrient medium or to seeds during soaking preparatory to germination. This work showed that irrespective of the mode of administration, low concentrations of auxins stimulated growth of the primordial root and shoot. Although growth in length of the shoot and root rapidly declined at higher auxin concentrations, their fresh and dry weights tended to increase. The diagram reproduced in Fig. 10.1 shows the response of embryos to IAA, which is also typical of others. An excellent illustration of the auxin sensitivities of the different organs of the embryo of *Phaseolus vulgaris* was provided by Furuya and Soma (1957), who found that the embryonic shoot had a higher auxin optimum than the root and that intermediate regions between the shoot and root primordia had optima intermediate with respect to promotion or inhibition. An apparently similar situation is encountered in the embryo of *Avena sativa* in which extremely low concentrations of IAA stimulated root growth, while at a much higher range of auxin which promoted coleoptile and mesocotyl growth, root elongation was arrested (Guttenberg and Wiedow, 1952). The inception of coleorhiza in the embryo of this species is also stimulated by IAA (Norstog, 1955).

Although the primary root of the embryo of *Pinus lambertiana* cultured with the radicle end in the medium lost its capacity for growth, meristematic activity was potentiated by the addition of NAA, which induced formation of small protuberances resembling short roots (Sacher, 1956). However, when continued growth of the root was secured by supplying nutrients through the cotyledons addition of IAA appeared to inhibit this growth (Brown and Gifford, 1958). In the embryo of *Ginkgo biloba* there was no stimulation of growth of the root and shoot systems by IAA treatment except at the lowest concentration tried (Ball, 1959; Bulard, 1967b).

The question of auxin promotion of root growth in the embryo of the fern *Thelypteris palustris* has been approached from a different point of view, starting with the concept that auxin diffusing from the apical

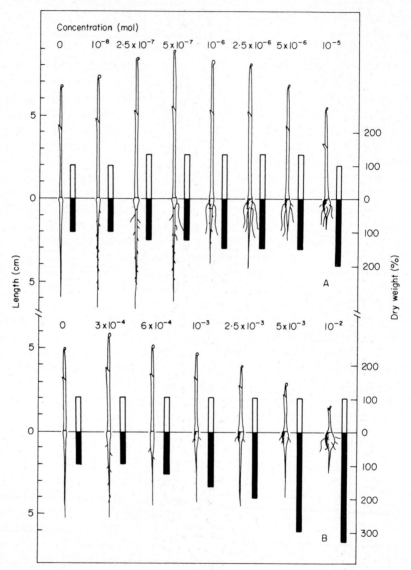

F<small>IG</small>. 10.1. Effect of different concentrations of the potassium salt of IAA on growth in length and dry weight changes of the root and shoot systems of pea embryos. Vertical bars represent dry weights. **A**, Auxin applied through the nutrient medium. **B**, Auxin applied to seeds during soaking. (After Kruyt, 1954)

meristem accelerates differentiation in the embryo and in particular the formation of the first root (Jayasekera and Bell, 1959). In the normal embryogeny of this plant, the leaf was the first organ to break open the calyptra and emerge, and the first root appeared at the same time or soon after. When the embryo embedded in a small cube of dead prothallial tissue from the apical region was cultured, leaf initiation preceded root growth by several days. If the apical notch of the gametophyte was replaced by a source of IAA, priority of organ initiation was strikingly changed, and invariably the root was the first organ to emerge. The relation between growth of the root and the auxin concentration might serve as a basis for further work, especially on species with filamentous gametophytes, where root growth is normally retarded, and those with subterranean gametophytes, where the root is well developed.

Age-dependent differences in the responses of embryos of *Capsella bursa-pastoris* to auxin, and their modifications by illumination or its lack thereof, have been demonstrated (Raghavan and Torrey, 1964b). Dark-grown heart-shaped and older embryos ($>80\ \mu$m long) formed long primary roots and shoot systems composed of linear leaves. When grown in a 12 h light–12 h dark cycle, root growth was inhibited in all but embryos $>1000\ \mu$m long and supplementation of the medium with low concentrations of IAA promptly offset this inhibition. This may perhaps imply that light causes the inactivation of an auxin-like substance necessary 'for root initiation in embryos. Age-dependent responses of embryos of *Allium cepa* to IAA (1 mg/l) were manifest by the initiation of lateral roots on the radicle of mature embryos and callus formation in proembryos (Guha and Johri, 1966).

Embryos of orchids vary considerably in their responses to auxins. Burgeff (1934, 1936) did not detect any stimulation of growth of embryos when orchid pollinia presumed to contain auxin were added to the medium or by the addition of IBA. Curtis and Nichol (1948) found a definite inhibition by IAA of growth of orchid embryos during the early stages of development after germination. Several later authors have observed promotion of growth (Withner, 1951, 1955; Mariat, 1952; Kano, 1965) or promotion of both germination and growth (Meyer and Pelloux, 1948) of embryos of different hybrids. In *Epidendrum*, auxin (α-naphthalene acetamide) did not affect germination of embryos, but its presence was stimulatory for growth of the root and shoot (Yates and Curtis, 1949). In a few published experiments auxins are reported to be active when they are added along with growth-promoting extracts such as coconut milk (Hegarty, 1955), tomato juice (Meyer, 1945a, b) and pineapple juice (Hamilton,

1965), and so it is difficult to ascribe the observed effects to auxins alone.

Scutellum

Since several nutritional and correlational factors contribute significantly to the functional organization of the scutellum of graminean embryos, references to the effects of auxins on the growth of this organ are appropriate here. In the embryo of *Pennisetum*, application of a mixture of 2,4-D and IAA was necessary to induce root formation on the scutellum, but surprisingly, either auxin alone was ineffective. Scutellum of maize embryo represents a different level of control where a low concentration of 2,4-D promptly induces rooting. An extreme case is represented by the embryo of *Echinochloa* in which endogenous roots are formed abundantly on the scutellum even in the absence of specific hormonal stimulation (Narayanaswami, 1962; Narayanaswami and Rangaswamy, 1959). To formulate principles that are more generally applicable, it will be useful to determine the response category in which scutella of some of the commonly studied graminean embryos are included.

Non-specific Effects of Auxins

A few examples will illustrate the fact that applied auxins also cause non-specific growth effects on embryos. Rappaport *et al.* (1950b) noted an irregular twisting and bending, reminiscent of an epinastic curvature, when embryos of *Datura stramonium* were grown in a medium containing 0·5–1·0 mg/l NAA. Soaking freshly collected seeds of *Eranthis hiemalis* containing undifferentiated embryos in 0·1% 2,4-D, NAA or 2,4,5-trichlorophenoxyacetic acid for 12–14 h resulted in grades of fusion and separation of cotyledons in virtually all embryos (Haccius, 1955b; Haccius and Trompeter, 1960). In addition, there was some twinning in embryos upon treatment with auxins, probably due to regeneration from old embryos. Low concentrations of IAA and NAA inhibited water uptake by dry embryos of *Lupinus luteus*; in this respect the action of auxins was analogous to inhibitors of respiratory metabolism such as azide and fluoride (Czosnowski, 1962). In long-term cultures of oil palm embryos, addition of auxin enhanced the growth of the haustorium and prolonged its survival (Rabéchault, 1962). Embryos of several plants are also known to respond to the presence of auxin in the medium by the production of callus (Carew and Schwarting, 1958; Ball, 1959; Mitra and Kaul, 1964; Raghavan and Torrey, 1964b; Chatterji and Sankhla, 1965;

Sankhla *et al.*, 1967a; Bulard, 1967a; Button *et al.*, 1971; Mehra and Mehra, 1974).

B. Modifications of Embryo Growth by Antiauxins

The role of auxins in the root growth of embryos described earlier has been indirectly substantiated by studies on the effects of auxin antagonists, such as maleic hydrazide and TIBA. Mericle *et al.* (1955) found

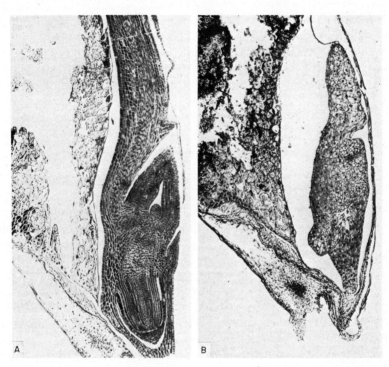

FIG. 10.2. Embryos of oat plants sprayed with maleic hydrazide exhibit degeneration of the root primordium due to mitotic inhibition. **A**, Control embryo at 15 days post-treatment. **B**, Embryo 15 days after treatment with 0·3% maleic hydrazide. (After Mericle *et al.*, 1955; photographs courtesy of L. W. Mericle)

that embryos collected from oat plants, sprayed once 3–4 days after fertilization with maleic hydrazide, exhibited varying degrees of degeneration of the root primordium as a result of mitotic inhibition. This is discernible in Fig. 10.2, which shows a control embryo and one after treatment. The extent of inhibition was dependent upon the concentration of the chemical and the age of the embryo.

Higher concentrations were more effective than lower ones and younger embryos were more sensitive than older ones. Recovery from inhibition during a subsequent growth period was also rapid in plants sprayed with lower concentrations of the chemical. Maleic hydrazide may well be interfering with auxin metabolism of the developing caryopsis, but in the absence of experiments on the reversal of inhibition by auxin, a firm conclusion is not possible.

Growth of cultured embryos of *Pinus lambertiana* was retarded by TIBA, resulting in the dimunition of root and hypocotyl elongation (Brown and Gifford, 1958). This observation assumes greater credibility as a result of the demonstration that the treatment led to a loss of ability of the hypocotyl to transport auxin basipetally, which apparently affected the normal growth of the embryo. Narayanaswami (1959b, 1962, 1963) observed obliteration and collapse of the root initials upon treatment of graminean embryos with maleic hydrazide. The deleterious effects of this compound on the embryo of *Pennisetum* were overcome by the simultaneous addition of IAA to the medium, thus showing competitive inhibition.

II. EFFECTS OF GIBBERELLINS

Much interest has centered around the effects of gibberellins on the growth of plants, and the dramatic growth-promoting effects of this class of hormones, especially GA, have bestowed on this substance the great biological interest it currently enjoys. However, references to work on the effects of GA on the growth of embryos are unfortunately limited. Dure and Jensen (1957) found that cotton embryos of two different physiological ages responded differently to GA supplied in the medium. The only noticeable effect of this compound on relatively immature embryos (37 mg fresh weight) was in enhancing cotyledon maturation and cell elongation. In older embryos (64 mg fresh weight), GA accelerated cell division, cell elongation and axis growth. Although IAA appeared inhibitory to growth of embryos, addition of GA along with IAA tended to offset the inhibition to some extent.

As Fig. 10.3 shows, GA greatly stimulated the growth of torpedo-shaped embryos of *Capsella bursa-pastoris* and, in particular, promoted the activity of the root meristem when grown in hanging drop cultures containing 12% sucrose (Veen, 1961, 1963). When cultured on a solid medium with 2% sucrose and optimum concentrations of GA, the destiny of the root meristem of embryos of different ages was found to be variable (Raghavan and Torrey, 1964b). From relatively older embryos emerged roots with abundant laterals, while younger embryos

Fig. 10.3. Gibberellic acid-induced growth stimulation in torpedo-shaped embryos of *Capsella bursa-pastoris*. **A**, Embryo at the time of excision. **B**, After 7 days' growth in the basal medium. **C**, After 7 days' growth in 10^{-5} mg/l GA. **D**, After 3 weeks' growth in the basal medium. **E**, After 3 weeks' growth in 10^{-5} mg/l GA. The medium contained 12% sucrose in all cases. (After Veen, 1963; photographs courtesy of H. Veen)

formed roots with fewer laterals. Other workers (Schooler, 1959, 1960c; Skene, 1969; van Staden *et al.*, 1972a; Kochba *et al.*, 1974) have also reported GA stimulation of growth of the root or of both shoot and root primordia in cultured zygotic embryos and adventiously formed embryoids. Of interest is the observation that coleorhiza of barley embryo which does not elongate in culture is stimulated to do so by addition of GA to the medium (Norstog, 1969b). In mature oil palm embryos, depending upon culture conditions, root growth was un-affected or inhibited by GA (Bouvinet and Rabéchault, 1965a, b). The dose-response relationship of excised embryos of *Avena fatua* to GA is consistently linear suggesting their possible use for bioassay of the hor-mone (Naylor and Simpson, 1961b).

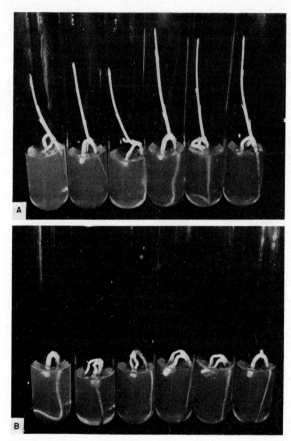

FIG. 10.4. Culture of embryos of *Ginkgo biloba* with cotyledons in contact with the medium containing GA results in inhibition of epicotyl growth. **A**, Control. **B**, With 1·0 mg/l GA. Both after 30 days in the dark. (After Bulard, 1967b; photographs courtesy of C. Bulard)

A dramatic growth inhibition of the epicotyl occurred when em-bryos excised from mature seeds of *Ginkgo biloba* were cultured with their cotyledons in contact with media containing GA_3, GA_4 and GA_7 (Fig. 10.4). The effect of gibberellins is attributable to the physiological age of the embryo and the site of absorption of the hormone, as absence of growth inhibition when the hormone was applied to the epicotyl of partially germinating embryo, or when the hormone was absorbed through the root, leaves little doubt in that matter (Bulard, 1967a, b; Bulard and Le Page-Degivry, 1968). Application of GA through the cotyledons also interferes with the full expression of the effect of the hormone in the growth of the embryo of cold-requiring seeds of *Euonymus europaeus* (Monin, 1959; Bulard and Monin, 1960a, b) (see Fig. 16.5). Interestingly, growth correlations observed in the intact plant may be obscured when GA is applied to excised embryos. Although elongation of the first leaf of the maize seedling is a characteristic GA effect, the same organ on an excised embryo failed to respond to the hormone (Bulard, 1960). Any apparent ineffectiveness of GA on the excised embryo cannot be ascribed to its failure to enter the transport stream, since growth of the first internode of the embryo was greatly enhanced by the hormone.

Some information bearing on the effects of GA on the growth of orchid embryos has accumulated, but it does not yield a satisfactory understanding of the role of the hormone in embryonic growth. Though experimental details differ from one investigator to another, it can be stated in a general way that gibberellins enhance the overall growth of embryos of various hybrid orchids (Moir, 1957; Smith, D. E., 1958; Blowers, 1958; Humphreys, 1958; Hirsh, 1959), other symptoms of the treatment being leaf elongation and chlorosis (Smith, D. E., 1958; Harbeck, 1963). In some cases, root growth of embryos was affected by application of gibberellins (Kano, 1965). An extreme response resulting in the death of the embryo upon treatment with gibberellins has also been recorded (Hyatt, 1965).

The classical demonstration of the role of GA of embryonic origin in the mobilization of stored food reserves of the endosperm during germination of cereal grains gives rise to added interest in the above results obtained with excised embryos. The bulk of the available evidence is consistent with the interpretation that the effect of the hormone is a direct one on the embryo itself and is not indirectly connected with the digestion of the starchy endosperm (Chen, 1970; Chen and Chang, 1972; Chen and Park, 1973). A comparative study of the effect of GA on γ-irradiated wheat embryo attached to or isolated from the endosperm showed that any compounding effect of GA on embryo growth due to its action on the endosperm was confined to a barely significant promotion of root growth (Kefford and Rijven, 1966). Since the embryo of γ-irradiated wheat grows by cell elongation in the absence of cell division (Haber and Luippold, 1960), one is led to attribute to GA an exclusive role in the growth of the embryo by stimulating cell elongation.

III. CYTOKININS AND RELATED COMPOUNDS

A. Cytokinin Effects

The evidence available at present is interpreted as showing that auxins and gibberellins play an important role in the control of organogenesis in isolated embryos, but their influence is subject to the presence of other growth hormones. Some observations on the effects of cytokinins on the growth of embryos tend to support this view. In the heart-shaped and older embryos of *Capsella*, suppression of root growth seemed to be a primary effect of kinetin, the degree of inhibition depending upon their age and conditions of culture (Raghavan and Torrey, 1964b). Precocious leaf expansion at low concentrations of kinetin and undifferentiated callus growth at high concentrations were also characteristically observed (Fig. 10.5). In the embryo of *Ginkgo*, both kinetin and 6-benzylaminopurine led to callus proliferation from the cortical cells of the epicotyl and inhibition of leaf and epicotyl growth. The permanence of the latter effect is doubtful since transfer of kinetin-treated embryos to a normal medium led to some reversal of growth inhibition (Bulard, 1967a). The general growth-inhibiting effects of cytokinins are not surprising in view of their relationship to nucleic acids which themselves are inhibitory for growth of cultured embryos (Rappaport *et al.*, 1950b). Two rather unusual effects of cytokinins are the enhancement of nitrate reductase activity in excised embryos of *Agrostemma githago* by *de novo* synthesis (Borriss, 1967; Kende *et al.*, 1971; Hirschberg *et al.*, 1972) and the induction of formation of typical leaf hairs and chlorophyll in the coleoptile and scutellum of barley embryos (Norstog, 1969a). The latter effect provides some proof of the foliar nature of these controversial structures of the grass embryo.

In a few cases studied, cytokinins seem to be inhibitory for growth of orchid embryos. In the embryo of a hybrid *Dendrobium*, kinetin was inhibitory for growth when added singly or along with IAA (Kano, 1965). The same hormonal treatments are also reported to be inhibitory for growth of roots in cultured *Laeliocattleya* embryos. In both these species, embryos were relatively slow-growing and noticeable effects were observed after nearly 4 months in culture. The specificity of these effects is therefore doubtful.

It is of considerable physiological significance that interactions exist between kinetin and other hormones in organ initiation in embryos. In *Capsella* embryos, combined IAA-kinetin treatments showed dominant promotive effects of IAA on growth of the shoot and hypocotyl and dominant inhibitory effects of kinetin on growth of the root (Raghavan, 1964b). However, kinetin had no effect on GA-induced root growth of the embryo. Thus, IAA-induced root initiation seems to be quite different from that induced by GA, at least insofar as it relates to inhibition by kinetin. Further studies along these lines to resolve the intriguing possibility that IAA and GA act at separate sites in the embryo to promote root growth might be of great interest.

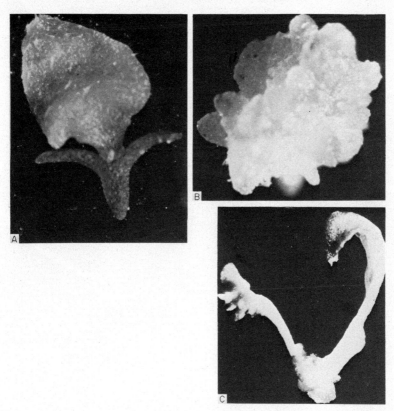

FIG. 10.5. Kinetin-induced modifications of ontogenetic pattern of embryos of *Capsella bursa-pastoris*. **A**, Precocious leaf expansion in 0·0001 mg/l kinetin. **B**, Callus growth in 1·0 mg/l kinetin. **C**, Same as in B, photographed 10 weeks later, showing leaf bud formation from callus. (After Raghavan and Torrey, 1964b)

The growth of isolated embryo of sycamore *(Acer pseudoplatanus)* is of interest because it involves developmental systems that respond differently to kinetin and GA (Pinfield and Stobart, 1972). Thus, elongation of the radicle was stimulated by kinetin, but not by GA, while unrolling of the cotyledons was accelerated by GA, but not by kinetin (Fig. 10.6). This would be an extremely favorable object for analyzing the interaction of exogenous growth hormones, since correlative

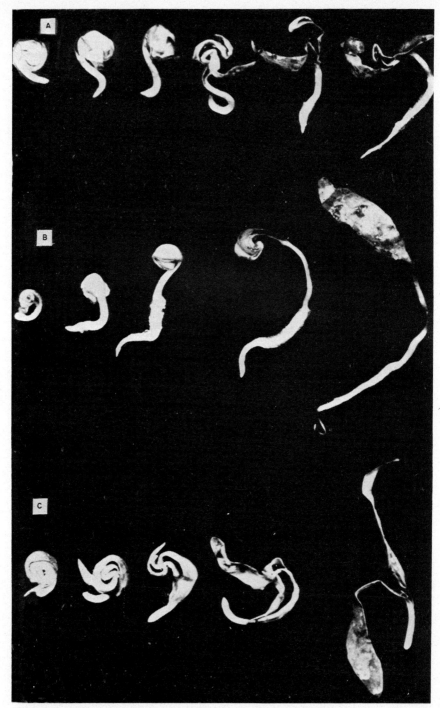

FIG. 10.6. The developmental stages of isolated embryos of sycamore *(Acer pseudopla-tanus)* incubated in (**A**) water, (**B**) kinetin and (**C**) GA overnight. (After Pinfield and Stobart, 1972)

data on the changes in the endogenous hormone contents of the embryo are available.

B. Effects of Adenine

Some interesting growth modifications in the embryo of *Pennisetum typhoideum* by the application of adenine have been recorded (Narayanaswami, 1959a, 1963). Addition of low concentrations of the purine to the growth medium induced spectacular overgrowth of the scutellum while root–shoot axis remained dormant (Fig. 10.7).

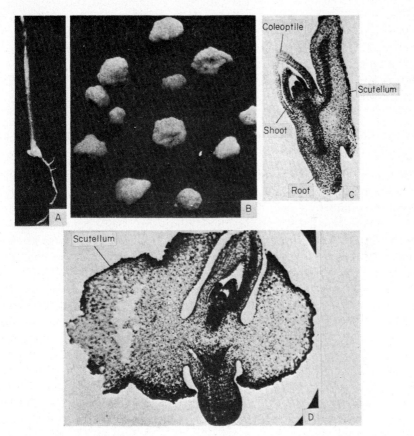

Fig. 10.7. Addition of adenine induces spectacular overgrowth of the scutellum of *Pennisetum typhoideum* embryos. **A**, Embryo growth in the basal medium. **B**, Overgrowth of the scutellum in a medium containing 40 mg/l adenine. **C**, Section through an embryo cultured in the basal medium. **D**, Section through an embryo cultured on a medium containing adenine (20 mg/l). (After Narayanaswami, 1959a)

Furthermore, adenine effects were dramatically enhanced by the addition of coconut milk or IAA to the medium. A case for interaction of a substance originating from the embryonic shoot in the morphogenesis of the scutellum has been suggested from the observation that excision of the coleoptile and shoot prevented the response of the scutellum to adenine. Like adenine, the related purine guanine was also equally effective in inducing proliferation of the scutellum.

In the embryos of *Capsella*, adenine effects are similar to those reported for kinetin, namely, inhibition of root elongation and precocious leaf expansion (Raghavan and Torrey, 1964b). In orchid embryos, effects of adenine are imperfectly understood and they range from incipient growth promotion to complete growth inhibition (Withner, 1942, 1943, 1951; Arditti, 1966b). Regarding the mode of action of adenine, attention has been drawn to the fact that in pine *(Pinus elliottii* and *P. palustris)* embryos, its breakdown and utilization follow the usual pathway involving hypoxanthine, xanthine, uric acid, allantoin and allantoic acid (Barnes, 1961).

IV. HORMONAL EFFECTS IN THE GROWTH OF EMBRYOS OF STEM PARASITES

Morphogenesis in the embryos of stem parasites belonging to the Loranthaceae and the role of hormones therein have been made special objectives of studies by Johri and associates (Johri and Bajaj, 1962, 1963, 1964, 1965; Bajaj, 1967, 1968, 1970; Bhojwani, 1969; Johri and Bhojwani, 1970). Aside from their massive size, other interesting features of embryos of stem parasites are the absence of a conventional root system and the presence of a haustorium at the radicular end which forms a graft with the stem of the host plant (Fig. 10.8 A, C, D). Upon culture, the first visible sign of growth in the embryo is elongation of the hypocotyl followed by the formation of a haustorium by the activity of the marginal meristem around the hypocotyl (Fig. 10.8 B). Normal morphogenesis of the embryo is complete with the appearance of a pair of plumular leaves. Mature embryos of *Dendrophthoe falcata* (Johri and Bajaj, 1962, 1963), *Amylotheca dictyophleba, Amyema pendula* and *A. miquelii* (Bajaj, 1968, 1970; Johri and Bajaj, 1964) go through this sequence of events when they are planted in a medium containing appropriate concentrations of casein hydrolyzate and IAA. Embryos of *Scurrula pulverulenta* (Bhojwani, 1969; Johri and Bhojwani, 1970) grown in a medium containing casein hydrolyzate produced normal seedlings, while addition of IAA as well induced callus formation in about 45% of the embryos (Fig. 10.8 E, F). A medium containing casein

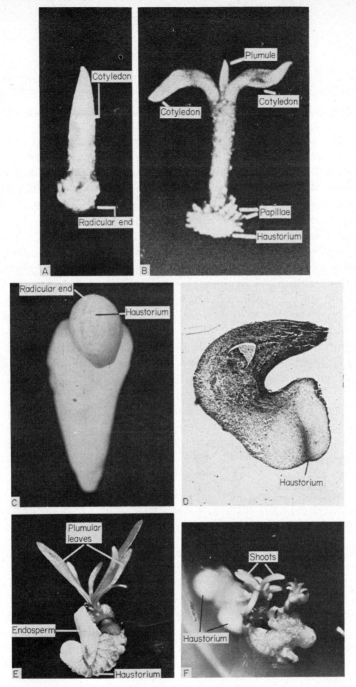

FIG. 10.8. Growth in culture of embryos of two members of Loranthaceae. **A**, **B**, *Amylotheca dictyophleba*: **A**, mature embryo without endosperm; **B**, 5 week old seedling from an embryo cultured in White's medium containing casein hydrolyzate and IAA. (After Bajaj, 1970) **C–F**, *Scurrula pulverulenta*: **C**, growth of the embryo planted with the endosperm showing the hypocotyl–radicular end and a lateral outgrowth of the haustorium; **D**, transection of the embryo at the stage shown in C; **E**, 9 week old seedling grown in a medium containing casein hydrolyzate; **F**, callus formation in an embryo grown for 20 weeks in a medium containing casein hydrolyzate and IAA. (After Johri and Bhojwani, 1970; photographs **C**, **D**, **E**, **F** courtesy of B. M. Johri and S. S. Bhojwani)

hydrolyzate, IAA and kinetin was best for growth of mature embryos of *Phoradendron tomentosum* (Bajaj, 1970). Embryos of *Arceuthobium pusillum* grew well in a medium containing coconut milk and casein hydrolyzate; addition of IAA consistently reduced their growth in length and the number of haustoria formed (Bonga, 1968; Bonga and Chakraborty, 1968). Variations in the morphological appearance of the radicle apex and haustorium were also observed when seeds were cultured in media containing different auxins (Bonga, 1969, 1971, 1974b).

In contrast to the aggressive behavior of the radicular pole of embryos of several members of the Loranthaceae, in the embryo of the stem parasite, *Cassytha filiformis* (Lauraceae), there is a rapid necrosis of the primary root meristem and production of adventitious roots from the base of the hypocotyl. There was no direct evidence for the involvement of a common identifiable agent such as auxin in the formation of adventitious roots on the embryo, although addition of high concentrations of IAA led to fasciation of the root and rooting of cotyledons in a small percentage of the embryos (Rangan and Rangaswamy, 1969).

An important observation which indicates the versatility of cultured embryos of stem parasites was made by Maheshwari and Baldev (1961, 1962) when they found that in a medium containing casein hydrolyzate and IAA, embryos of *Cuscuta reflexa* differentiated a callus upon which numerous embryos were adventitiously formed. In the embryos of all stem parasites studied, formation of callus and induction of adventive embryos therefrom have been achieved by subtle changes in the composition of the medium; the results of these studies will be taken up in a later chapter.

V. HORMONES AND PROTEIN AND NUCLEIC ACID METABOLISM OF EMBRYOS

Considering the effects of growth hormones on embryos, we have seen that they have much in common and much that is different. Recognition of the diverse morphological effects of hormones raises the question of their mode of action. For all the interest that is elicited by the manifold effects of hormones on embryos, mechanism of their action is mysterious here as in other systems where hormone effects have been extensively studied. As part of a search for a common denominator to explain the action of hormones, many experiments have been conducted on the relation between hormone-induced growth in plants and nucleic acid metabolism in some limiting aspect of protein synthesis. These experiments have forced a conviction that hormones promote growth directly or indirectly by increasing the nucleic acid and protein contents of target organs. Although embryos were not the objects of choice for this research, some information on the effects of hormones on nucleic acid metabolism have been derived incidentally from studies using embryos but directed primarily to other objectives.

Because of its striking growth-promoting effects on embryos, GA has figured in a few studies designed to relate hormonal effects to protein and nucleic acid metabolism. However, reports on the effects of this compound on protein synthesis are conflicting. Since GA-treated embryos of cotton had the lowest carbohydrate and nitrogen contents, it was concluded that a dilution of the cell wall material and protoplasm had occurred during the accelerated rate of cell elongation and cell division following hormone treatment (Dure and Jensen, 1957). In contrast to cotton, in the embryos of *Capsella bursa-pastoris*, GA treatment was associated with a high rate of protein accumulation (Veen, 1963). It should be remembered that the magnitude of differences in the protein content of control and GA-treated embryos depends greatly on how the rates are expressed. Some of the large differences in the protein content following GA application might disappear if values are expressed on a per cell rather than on a per embryo basis. Since cyclic AMP is known to be involved in the modulation of hormonal effects, a recent report (Rao and Khan, 1975) of the enhancement of polyribosome formation in barley embryos by both cyclic AMP and GA is of interest as a basis for investigating the mechanism of action of this hormone. In an operational sense, there is little difference between GA-induced growth of corresponding organs in a seedling and in an embryo, and so it seems reasonable to expect that mechanisms involved in hormone-induced elongation of embryonic organs may include those which are apparent in seedling organs.

A series of investigations has been undertaken on the effects of hormones on the nucleic acid metabolism of embryos of seeds which normally require a cold treatment for germination. The situation that is open for experimentation is almost ideal, namely, certain hormones can substitute for the cold requirement and promote germination of seeds. For example, it has been shown that the requirement for a cold treatment for germination of hazel seeds *(Corylus avellana)* can be overcome by GA. Application of GA to excised embryos is associated with an enhanced synthesis of DNA and RNA (Bradbeer and Pinfield, 1967; Jarvis *et al.*, 1968a; Pinfield and Stobart, 1969). The increased RNA synthesis is largely accounted for by an increased RNA polymerase activity and increased availability of DNA template for RNA transcription (Jarvis *et al.*, 1968b). Further studies with embryo slices from hazel seed indicate that GA may promote protein synthesis by increasing the synthesis of new tRNA and by changes in the aminoacyl tRNA-synthe-tases (Jarvis and Hunter, 1971). In somewhat oversimplified terms, the position is that GA may be considered to induce embryo growth by derepressing genes which are active in cold-stimulated seeds but in-active in dormant ones; however, we cannot yet define how GA exerts this control and how changes in RNA metabolism lead to growth pro-motion. In contrast, in non-dormant wheat embryos in which the first proteins of germination are coded on preformed mRNA, GA acts at the translation level in enhancing growth and protein synthesis during the early hours of imbibition (Chen and Osborne, 1970).

In the breaking and reimposition of dormancy in cold-requiring pear embryos, the problem is compounded by the participation of prob-ably more than one hormone acting on the nucleic acid syndrome. Abscisic acid inhibits the growth of the non-dormant embryo as well as the synthesis of tRNA, light rRNA and a fraction represented as DNA-RNA in the ^{32}P-incorporation profile of nucleic acids (Khan and Heit, 1969). With minor differences, this is also true of the effects of abscisic acid on embryos of *Fraxinus excelsior* (Villiers, 1968b) and *Phaseolus vulgaris* (Walton et al., 1970). Although the effect of abscisic acid on pear embryo is reversed by kinetin, a combination of GA and kinetin is more effective than kinetin alone in reinstating embryo growth and negating the inhibition of synthesis of the DNA-RNA fraction. Thus, when normal embryonic growth is induced, the synthesis of the DNA-RNA fraction is also reinstated. This, coupled with a later demonstration that abscisic acid alters the nucleotide composition of RNA synthesized in the embryo, suggests the ob-vious possibility that hormonal effects may be related to changes in gene activation (Khan and Heit, 1969; Khan and Anojulu, 1970). The situation is certainly more complex than would have been anticipated, and the very substantial literature on hormone effects on more versatile systems such as oat coleoptile or pea internode sections does not make the issues involved any less complex. What now seems necessary for a fuller understanding of the basic problem concerned with the action of growth hormones upon the nucleic acid system of embryos is some information on the sequential changes in RNA populations during embryogenesis and their modifications by hormones.

VI. EFFECTS OF ANTIMITOTIC SUBSTANCES

The normal ontogenetic pattern of organogenesis in the embryo can be modified by antimitotic substances such as colchicine, nitrogen mus-tard, nucleic acid analogs and antibiotics. Among the major manifestations of these substances is a capacity to suppress rather than sustain cell division, eventually leading to complete cessation of cellular activities. Just as the selective stimulation of cell division by hormones is a prominent feature of differentiation, so too are the mechanisms enabling cells to suppress mitotic activity relevant to developmental processes.

Inhibitory effects on cell division in rice embryo were seen following treatment of the spike with colchicine or nitol [methyl-bis-(p-chloroethylamine) hydrochloride], a derivative of nitrogen mustard (Mizushima *et al.*, 1956). If spikes were injected with 0·2% colchicine 24 h after anthesis, cleavage was delayed in the most abnormal types of embryos which remained largely undifferentiated. Nitol (0·1%) treatment led to even more drastic effects and complete growth inhibition of the embryo. Treatment of flowers on the fifth day of anthesis with the chemical induced rapid and irregular outgrowth of the scutellum but differentiation of seedling organs was inhibited. Abnormalities in growth were accompanied by corresponding anomalies in the distribution of nucleic acids, polysaccharides and enzymes. Since both colchicine and nitol are well known antimitotics, such abnormalities are scarcely surprising. Gross symptoms of growth inhibition of red kidney bean embryos by cadmium derivatives resemble those induced by nitol (Imai and Siegel, 1973).

Rau (1956) studied the course of early embryogenesis in excised ovaries of *Phlox drummondii* grown in a medium containing colchicine. Although embryos grown in the presence of the drug maintained a developmental pattern very similar to those grown *in vivo*, various abnormalities like chromosome scattering at metaphase and formation of bizarre giant cells with supernumerary nuclei were observed. Prolonged culture of the ovary in colchicine led to abortion of the embryo. In experiments in which colchicine was injected into ovules of *Argemone mexicana*, early maturation of the embryo at lower concentrations and its arrested development at higher concentrations were observed (Chopra and Rai, 1958). Colchicine effects on the growth of excised embryos of *Chlorophytum laxum* included promotion at lower concentrations and inhibition at higher concentrations. The latter was associated with abnormal morphogenetic symptoms such as callusing of the hypocotyl and initiation of numerous adventitious roots (Thomas, 1963). Occasionally, application of colchicine resulted in the loss of polarity of the embryonic root primordium (Roy, 1972).

Culture of pea embryos in low concentrations of pyrimidine analogs (5-fluorouracil, 5-fluorodeoxyuridine, 5-fluorouridine, 5-fluoroorotic acid and 5-fluorodeoxycytidine) resulted in growth inhibition as seen in the arrest of elongation of the root primordium (Paranjothy and Rag-havan, 1970). Although both cell division and cell elongation are in-volved in embryo growth, only the former is affected by the analogs. When pyrimidine bases, their ribosides and ribotides were applied along with analogs, inhibition of embryo growth was maximally re-versed by thymidine and thymidylic acid; thymine was additionally effective in negating the growth inhibition due to 5-fluorouracil. Put-ting these results together, one might conclude that fluoropyrimidines affect growth of the embryo by interfering with DNA synthesis and cell division. Apparently some effect on RNA synthesis is also the cause of growth inhibition induced by 5-fluorouracil. In striking contrast, there was an increase in seed set in poppy when ovaries were injected with 5-bromouracil or uracil (Balatková and Tupý, 1972b). When fertilized eggs of *Pteridium aquilinum* were treated with thiouracil, retardation of embryogenesis at lower concentrations and a reversion of the sporophyte to gametophytic growth at higher concentrations were observed (Jayasekera snd Bell, 1972).

The processes of multiplication of some cells and death of others in the presence of antimitotic substances may lead to radical changes in the morphology of embryos, as seen in phenylboric acid-induced mono-cotyledony in normally dicotyledonous embryos of *Eranthis hiemalis* (Haccius, 1959a, 1960) and dicotyledony in normally monocotyle-donous embryos of *Cyclamen persicum* (Haccius and Lakshmanan, 1967). Treatment of seeds of *E. hiemalis* with lithium carbonate (Haccius, 1956) and morphactins (Haccius, 1969) results in a partial fusion of cotyledons into a single unit.

The incidental application of other drugs in embryo culture studies also deserves mention. Addition of barbiturates [(sodium ethyl-1-methyl-butyl) barbiturate, sodium cyclopentenyl-allyl barbiturate and phenyl-ethyl barbituric acid] to the culture medium resulted in the col-lapse of the primary meristem of *Vanda tricolor* embryos and led to pro-liferation of the embryonic axis into an undifferentiated tissue (Curtis, 1947a). Thalidomide (α-N-phthalimido glutarimide) is a synthetic drug which was used as an excellent sleeping tablet and tranquilizer in Europe until its disastrous malformative effects in human embryos were discovered (Taussig, 1962). When flowers of *Arabidopsis thaliana* were treated with the drug following fertilization, division of the zygote was inhibited in a good percentage of the flowers (Arnold and Cruse, 1967). Thalidomide treatment of flowers containing first division phase of embryos was less effective and flowers with slightly older globular embryos were virtually unaffected by the drug. The available informa-tion does not enable us to speculate on the mode of action of the drug in this system. Reports on the effects of other drugs are much more circumstantial and both promotory and inhibitory effects on embryo growth have been ascribed to antibiotics such as terramycin (Iyer *et al.*, 1959), phleomycin (Tepper *et al.*, 1967) and mitomycin (Roy, 1972), and alkaloids such as caffeine (Rabéchault and Cas, 1973).

Fusaric acid is a nonspecific fungal toxin produced in the host plant by species of *Fusarium* and *Gibberella*. Incorporation of fusaric acid into the culture medium markedly affected the development of excised embryos of *Phaseolus vulgaris*, resulting in callus growth at low concentrations and arrested shoot growth at higher concentrations. In line with the known specialization of the toxin in interfering with water uptake by the host plant, characteristic wilting of the embryonic leaves was also observed (Padmanabhan, 1967). On the other hand, ascochitine, an antibiotic with phytotoxic properties produced by *Ascochyta pisi*, specifically inhibited growth of root primordia of cultured embryos, while shoot meristems exhibited nearly normal growth (Lakshmanan and Padmanabhan, 1968). These observations are of interest because sensitivity of the embryo to fungal metabolites would permit the study of their role within the plant under strictly controlled conditions of nutrition.

VII. COMMENTS

Although analysis of the morphogenetic effects of growth hormones and antimitotic substances in terms of control of growth and organ initiation in embryos has not advanced very much, it has revealed certain modes of behavior that apply generally. The responses of embryos to growth hormones sufficiently resemble their well known effects in other systems to encourage the belief that the inherent modes of their action may be the same in both. This also bolsters the conviction that embryo culture, with the opportunity that it offers for control of the cellular environment, will be increasingly useful in biochemically oriented developmental studies.

11. Extracellular Control of Embryogenesis

From the material presented in Chapters 7 and 10, we can conclude that growth hormones exert a major influence as controlling factors in the growth and organogenesis of cultured embryos. In order to understand the orderly and harmonious development of the embryo during progressive embryogenesis, we must extend our considerations to other forces which influence organogenesis. Some of these forces act outside the environment of the embryo and, in contrast to the intercellular nature of hormone action, they exert extracellular control on embryogenesis. In this category we can visualize at least three types of forces operating on embryos of vascular plants. First, there are physical factors such as restraint and pressure presumably acting on embryos *in vivo*. Secondly, there is the collective influence of the usual parameters of the environment such as radiation, temperature, pH, supply of gases, etc., which might affect growth of embryos. The third type of force of importance in the extracellular control of embryo growth is induced by microsurgery and wounding of the embryo and isolation of its constituent parts. These operations may lead to changes in the mutual interaction of the embryo with its normal environment, and the morphogenetic restoration of the ontogenetically formed organs can be analyzed under such supervening extrinsic conditions. Although physical and environmental influences might conceivably act at the extracellular level of control, there is good reason for believing that their effect is achieved at the intercellular level through the action of some subtle chemical messenger substances.

The purpose of this chapter is to review our knowledge about the control of embryogenesis by extracellular factors in the light of the generalizations presented above, and make whatever inferences seem to be justified about their role in the normal growth of the embryo.

I. RESTRAINT AND PRESSURE

A. Pteridophytes

In vascular plants, the egg is housed in a privileged geographic location in the female gametophyte. In both pteridophytes and gymnosperms, as we have already seen, the structure which bears the egg is the archegonium, which develops in the gametophytic tissue. It is a flask-shaped unit containing a basal cell and a variable number of cells constituting the neck of the flask. The egg is located at the base of the archegonium with only the neck protruding above the general level of the surrounding tissue. The cells surrounding the archegonium divide actively, acquire dense cytoplasmic contents and form a protective jacket around its base. In angiosperms, the egg, and the embryo which is subsequently formed from it, are lodged in the embryo sac and are bathed in a liquid of high osmotic concentration. While the biochemical constituents of the milieu surrounding the egg may determine its ability to grow, the immediate physical environment of the archegonium or of the embryo sac may determine the direction of this growth.

In studying the problem of physical factors in embryogenesis, fern embryos are found to be specially suitable because fertilization of the egg can be artificially controlled in culture and embryos of desired stages obtained at will. This advantage exists despite the fact that the fern egg is no larger than eggs of other groups of vascular plants. In homosporous ferns, the first recognizable event after fertilization is not the division of the zygote, but the division of the jacket cells of the archegonium to form a complex embryonic envelope of relatively turgid cells constituting the calyptra, which covers the young embryo completely for about 2–3 weeks. There is little doubt that the multilayered calyptra imposes a physical restraint upon the developing embryo. The question that has been debated by pteridologists for a long time (Lang, 1909; Bower, 1928) is whether this has any effect on the developmental pathway of the fertilized egg into a normal sporophyte.

We owe to Ward and Wetmore (1954) a valuable study on *Phlebodium*

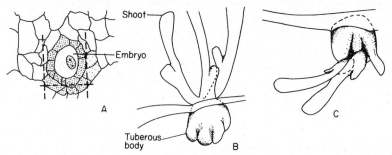

Fig. 11.1. Changes in the mode of growth of the embryo of *Phlebodium aureum* upon removal of the archegonial tissue. **A**, Cross-section of a prothallus showing the single-celled embryo and the remnants of the archegonium after removal of the neck. Dotted lines indicate places where cuts were made. **B**, A tuberous body bearing three young shoots on the upper part above the surface of the prothallus. **C**, Young shoots on the under surface of the tuberous body, becoming oriented toward the apical notch to the left. (After Ward and Wetmore, 1954)

aureum which was the first of its kind inspired by the prospect of resolving the question concerning the effect of the physical environment of the archegonial apparatus on embryo development. By various patterns of vertical and tangential cuts, these workers surgically removed portions of the calyptra and neck of the archegonium, thus relieving the contained embryo from restraining influence. When isolated pieces of the prothallus containing surgically treated archegonia were grown in a suitable medium, remarkable changes were observed in the mode of growth of the embryo, which broke through the cover tissue at the apex of the calyptra as a slow-growing irregular mass, initially lacking clearly delimited primordial appendages and a definite shape. Eventually, normal leaves each with an associated apical dome developed on the embryonic aggregate, giving it the conformation and constructional features of a normal sporophyte. Root formation did not occur or was postponed to a later stage of development. A curious effect of an extreme type of surgical treatment which effectively removed the arche-gonial neck was the formation of a tuberous mass of parenchymatous cells, lacking vascular elements, which later regenerated plants with normal leaves and stem (Fig. 11.1). Here, partial release of the develop-ing embryo from mechanical stress interfered with the timing and orderliness of events of early embryogeny without affecting the ultimate form of the sporophyte.

For the most part, growth modifications induced by surgical opera-tions in embryos of *Pteris longifolia* (Rivières, 1959) and *Thelypteris palustris* (Jayasekera and Bell, 1959) resemble those described in *Phle-bodium*. The greatest variation in embryo development in the latter species was observed when the archegonial neck was excised with a con-comitant release of pressure on the unicellular embryo. Irrespective of the orientation of the surgically treated gametophyte with the dorsal or ventral surface in contact with the nutrient medium, the embryo always emerged from the surface away from the medium and grew upwards to form multiple shoots (Fig. 11.2). The mechanical pressure of the archegonial neck thus appeared to modify the response of the embryo to gravity by affecting the organization of cells. The surgical operations are subject to the criticism that they inevitably introduce the possibility of wound responses and perhaps nutritional changes, the importance of which is not easily determined.

Since the cut tissues were still attached to the embryo during its growth, it must not be inferred from the above account that the mech-anical stress of the calyptra on the embryo was eliminated completely in these experiments. To understand the activity of the archegonial apparatus on the growth of the embryo, it is the ultimate goal to isolate the fertilized egg free from any archegonial tissues and study its mor-phogenetic potency in culture. To date, this has been successfully accomplished only in *Todea barbara* (DeMaggio and Wetmore, 1961a, b), but the morphogenetic principles that have emerged are of wide general interest. When unicellular embryos of *Todea* prior to the appearance of the first division wall were cultured in a suitable medium, they went through the usual stages of early segmentation with precise regularity like embryos growing *in vivo*. During further growth, embryos did not develop in the manner characteristic of those contained in the archegonium, but produced irregularly branching gametophyte-like entities, markedly different from the normal sporophyte (Fig. 11.3). The failure of the isolated unicellular embryo to organize as it grows contrasts with the normal development of relatively older embryos in a medium of the same composition (see Chapter 7). This tends to indicate that in normal embryogeny, but for restrictions imposed by the calyptra, uncontrolled growth of embryos would have resulted, leading to the formation of structures resembling gametophytes rather than

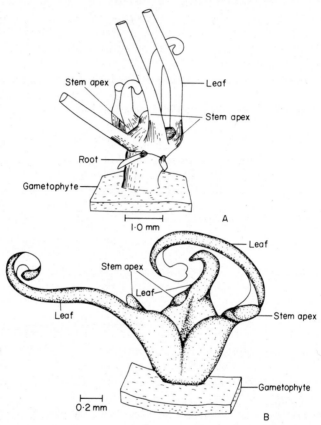

FIG. 11.2. Removal of the archegonial neck of *Thelypteris palustris* affects the orientation of the embryo in relation to the prothallial surface. **A**, Ventral surface of the prothallus in contact with the medium; embryo arising from the dorsal surface. **B**, Dorsal surface of the prothallus in contact with the medium; embryo arising from the ventral surface. (After Jayasekera and Bell, 1959)

sporophytes. It would appear that the acquisition of biochemical specialization which allows relatively older embryos to attain normal sporophytic form upon culture is in some way related to the development of three-dimensional morphology within the confines of the archegonial chamber. Such a view does not negate the established fact that the embryo is dependent upon a balanced set of nutrients provided by

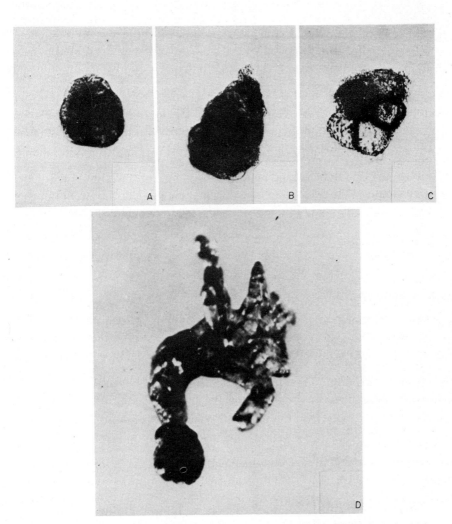

Fig. 11.3. Stages in the growth of unicellular embryos of *Todea barbara* in culture, in a liquid medium containing 3% sucrose and 50 mg/l sorbitol and inositol. **A**, One week after culture, after several divisions. **B**, **C**, Two weeks after culture, showing evidence of random divisions and bulging of cells. **D**, Four month old culture, showing the formation of a structure resembling a gametophyte. (After DeMaggio and Wetmore, 1961b; photographs courtesy of A. E. DeMaggio)

the gametophyte, but it does suggest that at a certain stage in the ontogeny which, of course, may vary in different species, the restraining influence of the surrounding tissue imposes its effects on the embryo.

This leads to the question whether an asymmetric flow of food substances does influence the shape of the embryo. This has been found to be the case. Thomson (1934) found that when *Marsilea* embryo was grown attached to the megaspore, the haustorial part of the embryo developed as a concave structure snugly fitting over the rounded apex of the megaspore with its rich endowment of food materials. After excision of the megaspore, culture of the embryo enclosed in the prothallus led to the formation of a convex haustorial structure very much different from that formed in the untreated control. In terms of the final shape of the embryo, it seems that the direction of flow of metabolites is an important factor to be considered, but a lot more experimental support should be marshaled before the concept can be used as a secure generalization.

B. Gymnosperms and Angiosperms

The mechanical stress provided by the tissues surrounding the egg seems to be relatively unimportant in the early embryogeny of gymnosperms and angiosperms, although it must be admitted that, as far as the author is aware, this has not been assessed experimentally in any published account. The physical environment which places the control of embryogenesis in angiosperms in a new perspective is the high osmotic pressure of the liquid endosperm, which, with its rich endowment of nutrient materials, surrounds the embryo. Very pertinent to this subject is the account presented in section II of Chapter 7. As pointed out there, in the absence of high osmotic pressure in the milieu, owing to a condition predominantly favoring cell elongation, the cultured embryo skips the normal stages of embryogenesis and germinates precociously. On the other hand, high osmotic pressure in the medium promotes a phase of growth in which cell division is predominant. There is a good correlation between the osmotic concentration of the ovular sap of progressively younger ovules and the requirement for high concentrations of sucrose or mannitol in the medium for successful culture of progressively smaller embryos. Given an appropriate physical environment in the form of a high osmoticum, what is the impetus for the cells of the embryo to undergo division and differentiation? There is as yet no evidence to identify one specific reaction or substance as the limiting factor in the observed effect and clearly much more work is required before we can evaluate the significance of the osmotic environment on the growth of embryos in metabolic terms.

II. RADIATION

Under this more inclusive title we will consider the action upon embryos of all types of radiations, both nonionizing (ultraviolet, visible and infrared) and ionizing (γ-rays, X-rays, α-particles, β-particles, protons and neutrons). In biological terminology, the action of the latter is designated as radiobiology.

A. Effects of Nonionizing Radiations

Unlike seeds with their contained embryos (see Chapter 16), excised embryos have not been exploited for studies on the effects of nonionizing radiations and therefore little evidence has accumulated in the literature regarding their growth responses to light. Rijven (1952), working with embryos of *Capsella bursa-pastoris* grown in hanging drop cultures, has reported that visible light from a high pressure mercury lamp has an inhibitory effect on their growth. However, as described in Chapter 10, when embryos of this species are grown in agar culture, morphogenetic effects of light seem to depend upon their length at excision and on the presence or absence of growth hormones in the medium (Raghavan and Torrey, 1964b). This may suggest that the effects of light on embryos are somehow related to their hormone metabolism, perhaps by controlling cell division and cell elongation. Support for this line of thought is provided by recent studies (Norstog, 1972b; Norstog and Klein, 1972) which have shown that high light intensity prevents precocious germination of barley embryos by switching cells to a division-oriented phase and that GA reverses this effect by promoting an elongation-oriented phase of the cells.

At the metabolic level the most obvious effect of light on the embryo is induction of chlorophyll synthesis (Bogorad, 1950; Lofland, 1950; Mauney, 1961). This has been quantitatively demonstrated in embryos of *Pinus jeffreyi* (Bogorad, 1950), an acetone extract of which showed absorption maxima characteristic of chlorophyll. In contrast, embryos grown in the dark for the same length of time showed absorption maxima typical of carotenoids. Since some workers have obtained substantial chlorophyll synthesis in dark-grown embryos by suitable modifications of the medium (Schou, 1951; Engvild, 1964), there is some question whether a true photochemical reaction is involved in the greening process.

B. Effects of Ionizing Radiations

Ionizing radiations have, as the name implies, the general property of ionizing atoms of the media in which they traverse. Although different forms of radiation, namely α-particles, β-particles, protons, neutrons, γ-rays and X-rays, have qualitatively the same effect on cells, they differ in the nature of the damage they inflict per unit of energy absorbed. By subjecting developing cells, tissues and organs to appropriate doses of radiation one can induce stable nuclear changes that can be perpetuated during subsequent cell divisions, or nongenetic physiological disturbances, which revert to normal growth during recovery periods.

Seeds of various crop plants containing mature embryos have long been used for the production of radiation-induced mutations, and there is now ample evidence that such mutations can give rise to improved yields. The volume of literature on the radiobiology of seeds is so great that it will be consciously omitted from this account. In what follows, we shall be concerned only with radiobiological studies on embryos designed to analyze the effects on organogenesis.

Age-dependent Effects of Irradiation on Embryos

The first impressive contribution of radiobiological study to embryogenesis was that of Mericle and Mericle (1957, 1961), who exposed *in situ* embryos of an inbred line of barley at specific stages to acute X- or γ-ray irradiation and followed the effects of radiation histologically and physiologically. Since embryos were not observed at the time of irradiation, for easy reference the developmental stages were designated by their chronological age, as shown in Fig. 11.4. Although the level of radiation used did not lead to lethality, the histological heterogeneity of embryos made them a prime target for radiation, which, depending on their age at the time of irradiation, induced several cell and tissue abnormalities, and disturbances leading to abnormal growth of the scutellum, root and shoot (Eunus, 1955) (Fig. 11.5). These disturbances can be traced to differential cell behavior in specific organs, where certain cells are killed, others have their mitotic rates reduced and still others go on dividing at the same rate as before. A particular abnormality—"cleft coleoptile"—was extremely stage specific and appeared only following irradiation of the spike at the mid-proembryo stage; irradiation either before or after the critical period failed to produce the cleft. In the early proembryo stage, the scutellum and shoot were most sensitive, but in the differentiating embryo, they lost their sensitivity and the ontogenetically younger root moved up on the list. Thus, so far as these organs are concerned, radio-sensitivity varies inversely with their degree of differentiation.

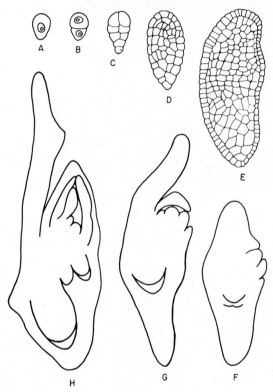

FIG. 11.4. Developmental stages in barley embryogeny designated by their chronologi-cal age. **A**, Zygote. **B**, **C**, Early proembryos. **D**, Mid-proembryo. **E**, Transition from late proembryo to early differentiating embryo. **F**, Early differentiation. **G**, Mid-differentiation. **H**, Fully differentiated mature embryo. (After Mericle and Mericle, 1957)

If embryos were irradiated at the early proembryo stages, when the damaged cells were capable of rapid division and differentiation, con-siderably less abnormal histological syndromes were observed than did treatment at later stages. However, the reverse was true with regard to physiological abnormalities such as embryo abortion, lethality at germination, and lethality at seedling stage and at maturity. Irradia-tion at proembryo stages also yielded a far greater number of isomutants than irradiation at later stages (Mericle and Mericle, 1962); this was

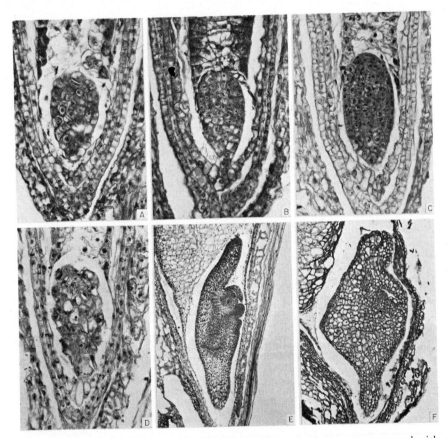

Fig. 11.5. Histological effects of X-irradiation on barley embryos are compared with unirradiated controls. **A**, Embryo 2 days after X-irradiation (700 rad), at stage C of Fig. 11.4; **B** is the corresponding control. **D**. Embryo 4 days after X-irradiation (700 rad) at stage C; **C** is the corresponding control. **F**, Embryo 3 days after X-irradiation (500 rad) at stage E; **E**, is the corresponding control. (After Mericle and Mericle, 1961; photographs courtesy of L. W. Mericle)

also true of *Nicotiana tabacum* where irradiation of gametes or proembryos yielded mutants for leaf character and flower color (Saccardo and Devreux, 1967). One further point is of interest. Does exposure of the plant to a dose of radiation make its progeny more sensitive, less sensitive, or differentially sensitive to subsequent radiation? Campbell (1966) found no change in the pattern of sensitivity to irradiation in developing embryos even when spikes from successive generations were irradiated recurrently.

Results similar to those obtained in barley have been reported for embryo abortion and developmental abnormalities in other plants after irradiation. By timing the stage at which plants were X-irradiated, Reinholz (1959) found it possible to control organ formation in embryos of *Arabidopsis thaliana* at any developmental period between fertilization of the egg and tissue differentiation. Embryo abortion occurred when flowers containing the egg, zygote or two-celled proembryo were irradiated. The transition of the young embryo from radial to bilateral symmetry was the most radiosensitive phase, and irradiation at this period resulted in abnormalities such as syncotyly, tricotyly, anisocotyly, tetracotyly and multiple embryos. The full-grown embryo was less vulnerable to a dose of radiation which was effective in earlier stages. This impression has been fortified by a more recent observation of γ-irradiation effects on the embryo–gametophyte complex of *Pinus resinosa* (Banerjee, 1968). Embryos irradiated at the zygote stage aborted in large numbers before organogenesis could occur, whereas the percentage of abortion was less when the same dose of radiation was applied at later stages of development. In a tetraploid strain of *Oenothera berteriana* the egg cell appeared to be more sensitive to γ-irradiation than the zygote (Seufert, 1965). Although not involving embryos in early stages of development, this is a convenient point to mention that seeds of oat, wheat, barley and rye in younger ontogenetic stages are more radiosensitive than older ones for nongermination, growth inhibition and morphological differences in the seedling (Sarić, 1957, 1958, 1961). It appears that radiosensitivity and radioresistance during embryogenesis are determined not only by the histogenic differentiation in the embryo, but also by the accompanying biochemical differentiation.

Some workers have made valid comparisons of the effect of different doses of radiation on the development of the embryo when given at a specific stage. According to Devreux (1963), embryos of *Capsella bursa-pastoris* chronically irradiated with γ-rays at the zygote or the proembryo stage bear various cellular anomalies such as vacuolization, disorganization, cell separation, lacuna formation and plasmolysis which increase with increasing exposure rate. Similar growth anomalies were observed in the proembryo when plants of *Nicotiana rustica* were exposed to acute γ-irradiation (Devreux and ScarasciaMugnozza, 1962). By careful analysis of the DNA content of the zygote of *Nicotiana tabacum* at various times after pollination, it was found (Devreux and Scarascia Mugnozza, 1965) that the resting zygote was in the G_1 phase of the mitotic cycle at the beginning of the third day after pollination and in the G_2 phase on the fifth day. If the zygote was irradiated with various doses of γ-rays at different times during the mitotic cycle, irradiation at the G_2 phase resulted in more abnormalities in embryonic development than when given at the G_1 phase (Devreux and ScarasciaMugnozza, 1965; Saccardo and Devreux, 1971). This suggests, but does not rigorously prove, that radiation is most effective in the cell when its DNA content is at the 4C level. However, the possibility does seem open that the variability in irradiation effects generally encountered reflects the failure of cells to achieve synchrony in the timing of DNA synthesis.

Not uncommonly, experimental results present the anomalous situation in which irradiation produces a transient growth stimulation in the embryo. An increased growth of the embryo followed by its arrested development occurred when acute γ-irradiation was applied to gametes, resting stage zygotes, or spherical proembryos of *Nicotiana tabacum* (Devreux and ScarasciaMugnozza, 1964). Although survival of the proembryos of barley following acute X-irradiation was generally erratic, a growth increment was registered as early as the first day following radiation, before mitotic inhibition and degenerative processes set in (Mericle and Mericle, 1969). Some stimulation of growth of the proembryo was also observed when the female gamete of *Triticum durum* was irradiated and pollinated with normal pollen (Donini and Hussain, 1968). The basis of radiation-induced growth promotion remains to be established, but it seems possible that it occurs through a route not dependent on cell division, but on accelerated cell enlargement.

Of particular interest in studies in which proembryo stages were irradiated was the behavior of the suspensor. There is strong evidence to indicate that, in general, the suspensor cells show a tendency to divide under conditions which damage the cells of the embryo proper. In chronically irradiated proembryos of *Capsella bursa-pastoris* (Devreux, 1963) and *Nicotiana rustica* (Devreux and ScarasciaMugnozza, 1962) degeneration of the embryo was often followed by division of the suspensor to form additional cells. According to Haccius and Reichert (1964), when rudimentary embryos of freshly harvested seeds of *Eranthis hiemalis* are damaged by X-rays, small groups of cells or single cells of the damaged embryo or cells of the suspensor burst into growth and establish new centers of mitosis which in a large measure become adven-

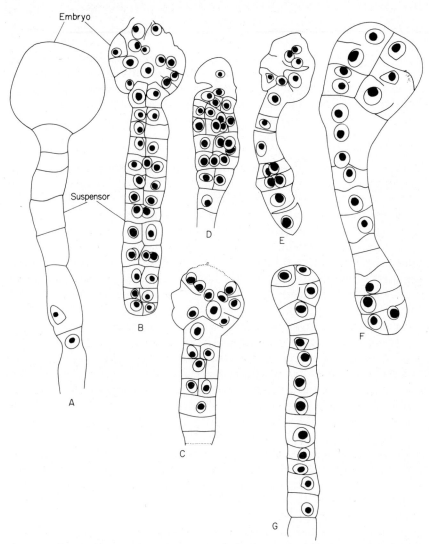

Fig. 11.6. Division of the suspensor in chronically X-irradiated embryos of *Arabidopsis thaliana*. **A**, Unirradiated globular embryo 4 days after pollination. **B–E**, Irradiated (12 krad) embryos 8 days after pollination: **B**, massive suspensor; **C**, degeneration of the embryo; **D**, embryo-like group of cells in the apical region of the suspensor; **E**, degeneration in the embryo seen as hollow areas. **F**, Irradiated (4 krad) embryo 8 days after pollination, showing the division of cells of the micropylar region of the suspensor. **G,** Irradiated (2 krad) embryo 6 days after pollination with elongate suspensor. (After Gerlach-Cruse, 1969)

tive embryos. In chronically X-irradiated plants of *Arabidopsis thaliana* (Gerlach-Cruse, 1969) almost all of the cells of the suspensor begin to divide long before the arrest of embryo development occurs. Although divisions of the suspensor cells were aberrant, in one type, the new cells formed at the micropylar end strongly resembled quadrant stage of the embryo (Fig. 11.6). One sees in the renewed activity of the suspensor an expression of totipotency when it is released from the restraint imposed by an actively developing embryo. As we have seen already, sensitivity of the cells to radiation depends to a great extent on the phase of the mitotic cycle in which they exist at the time of irradiation; lack of division in the suspensor cells as frequently observed during embryogenesis may render them radioresistant in comparison to the rapidly dividing cells of the embryo.

How can we explain the striking differences in radiosensitivity between embryos of different ages? Mericle and Mericle (1961) believe that neither changes in the moisture content of barley embryos nor variations in their nuclear volume can adequately account for embryo sensitivity during development. As might have been expected, there are some suggestions that nucleic acids of irradiated embryos exhibit stage-specific variability (Chang, 1968; Chang and Mericle, 1964). It was mentioned previously that early proembryos are more prone to physiological damage by radiation than late proembryos; yet they rarely show abnormal organogenesis. Following administration of an acute dose of X-irradiation, the early proembryos also display marked decreases in DNA and RNA contents at embryonic maturity as well as some alterations in RNA base ratios, but little change in DNA base ratios. Irradiation during late proembryo development, in contrast, leads to normal DNA values, but DNA has an altered base ratio due to a decrease in thymine. Miah and Brunori (1970) attribute differences in the radiosensitivity of different organs of the mature embryo to differences in the $2C:4C$ ratio of their cell populations. This thesis is supported by the occurrence of a high proportion of $4C$ cells in the embryonic root of *Vicia faba*, which is the most radiosensitive region. Although the objective of a full explanation of stage specificity in radiation damage in the embryos is far from being achieved, these findings are of interest because they raise the possibility of a causal relationship between nucleic acid metabolism and radiation damage.

Gamma Plantlets

The damaging effects of radiation with which we have been concerned above are those manifested by low doses which produce cell lethality in developing embryos. Low doses of irradiation also induce somatic mutations in seeds, leading to stable chromosomal rearrangements in cells that divide and perpetuate the mutation. In a few instances, however, large doses of radiation given to seeds can prevent subsequent mitosis and cytokinesis in the cells of embryos without killing them. The possibilities presented by embryos of seeds given massive doses of γ-irradiation and which grow in the absence of cell division ("gamma plantlets") have been exploited by Haber (1968) in a number of developmental studies concerned with the interrelationships between cell division and cell elongation on the one hand, and growth and differentiation on the other.

Embryo–endosperm Transplants

In all of the examples considered so far, embryos were shielded during irradiation by the ovular tissues and the endosperm, and so it is difficult to decide from the results how much of the destructive effect of radiation is due to direct action on the embryo itself and how much is indirect. Although this aspect of radiobiology of embryos has not been subject to critical examination, some recent reports in the literature indicate that radiation effects on the endosperm may indirectly affect growth of the embryo. The evidence bearing upon this is found in the reduced shoot and root growth and survival of embryos excised from unirradiated cereal grains, which are transplanted into their irradiated endosperms (Meletti and D'Amato, 1961; Fonshteyn, 1961). At the cytological level, γ-irradiation certainly confers on the endosperm mutagenic ability which is expressed in some plants as chromosome aberrations or as chlorophyll mutations (Yanushkevich, 1963; Avanzi et al., 1967; Meletti et al., 1968; Floris et al., 1970). Limited as these data are, they nevertheless show that the effect of radiation on extraembryonal tissues is to be kept in the background in considering much of the evidence on the radiobiology of the contained embryo. Indeed, if the mutagenic ability of the irradiated tissues persists for a time, this will have major importance in practical food irradiation procedures. Using a different approach, this point has been highlighted by some workers (Natarajan and Swaminathan, 1958; Swaminathan et al., 1962; Chopra and Swaminathan, 1963) who found several cytological abnormalities in cereal embryos planted on X-irradiated nutrient medium and potato mash or on potato mash prepared from tubers which were irradiated with γ-rays and stored for prolonged periods. Along the same lines, Ammirato and Steward (1969) have reported that embryoids originating from free cells of carrot and water parsnip (*Sium suave*) are particularly vulnerable to the addition of irradiated sucrose to the medium. These single cells organize into embryoids simulating normal stages of zygotic embryogeny in a medium containing unirradiated sucrose, but if supplemented with γ-irradiated sucrose, embryoids assume abnormal configurations, including fusion and formation of multiple roots and shoots. The biologically-active agents formed by radiolysis of sucrose, which induce morphogenetic effects on embryoids as they develop from free cells, have not been identified.

Pollen Irradiation

Embryo growth is also impaired following fertilization of a normal egg with sperm originating from an irradiated pollen grain. Despite the fact that mitosis in the generative nucleus of the irradiated pollen is often abnormal, the latter functions as the male gametophyte and the pollen tube continues to grow into the embryo sac. Embryos developing from the union of one set of normal chromosomes and one set of irradiated chromosomes show various cytological abnormalities which have been reviewed (Brewbaker and Emery, 1962).

Cave and Brown (1954) followed the development of the embryo and endosperm of lily *(Lilium regale)* fertilized by sperms from pollen irradiated with 400 rad X-rays. Differentiation of the embryo was disturbed to the extent that no viable seeds were obtained at this dosage which did not inhibit pollen tube growth and division of the generative cell. Frequent abnormalities in the growth of the embryo included failure of the egg and the sperm nuclei to fuse, degeneration of the zygote and developmental arrest beyond the proembryo stage. At the physiological level, the capacity of the embryo to hydrolyze stored reserves of the endosperm seems to be affected by the introduction of irradiated sperm. Comparative studies of the effects of irradiation of the male and female gametophytes on the development of the embryo in *Hordeum vulgare* (Donini, 1963) and *Triticum durum* (Donini and Hussain, 1968) have shown that irradiated sperm affects embryo development more than an irradiated egg cell. Necrosis of cells and eventual degeneration of the proembryo were frequently observed after fertilization of the normal egg of *T. durum* with irradiated sperm.

There is obviously a relationship between embryo development following fertilization of the egg with irradiated pollen and dosage of radiation applied to the pollen (Vassileva-Dryanovska, 1966a, b). In *Tradescantia paludosa* seed formation occurred at exposures of up to 1 krad X- or γ-rays. At a dose of 10–50 krad, the embryo degenerated either at the zygote stage or after producing a structureless mass of a few cells. At a still higher dose (150–500 krad) which reduced the sperm nucleus to a dense pycnotic mass of chromatin, complete degeneration of the zygote was observed. At this dose, two-celled embryos appeared occasionally, due to a stimulation of division of the female cell by the male chromatin. Clearly, the physiological state of the egg must have changed radically by the presence of the pycnotic male chromatin to favor its division in the absence of fertilization. This situation might well give a clue to the formation of haploid embryos observed in several plants as a consequence of pollen irradiation (Vassilev-Dryanovska, 1966c). More typically, in *Lycopersicon pimpinellifolium* the frequency of abnormally developing seeds increased with increase in the dosage of X-ray applied to pollen grains. At extreme doses of radiation mortality of seeds stemmed from a variety of causes, but mainly due to the malfunctioning of the endosperm, hyperplastic growth of the endothelium and degeneration of the embryo (Nishiyama and Uematsu, 1967).

Another side effect from the use of irradiated pollen is the induction of multiple embryos. In *Zea mays* the percentage of twin and triplet embryos increased from $0 \cdot 1\%$ in the control to about 18% at 2720 rad pollen dosage. In *Lilium regale* the effect was much less significant and the incidence of twinning was only $0 \cdot 3\%$ at 500 rad pollen dosage (Morgan and Rappleye, 1951). A different situation prevails in the cultivated lines of pepper *(Capsicum frutescens)* where polyembryony is frequent, but perhaps not excessively so. In this plant, fertilization with irradiated pollen did not lead to an increase in the incidence of multiple embryos (Campos and Morgan, 1960). The basic assumption is that use of irradiated pollen will result in polyembryony only if the paternal genotype controls the process; in pepper, polyembryony appears to be under the control of the maternal genotype.

To sum up this section, at present there is no firm hypothesis for the molecular basis of action of ionizing radiation on embryos. Nuclear magnetic resonance (n.m.r.) spectra similar to those given by free radicals are known to be produced by irradiated embryos (Conger and Randolph, 1959; Ehrenberg *et al.*, 1969; Randolph and Haber, 1961). The general assertion is that free radicals constitute a link in the chain of events leading from the absorption of radiation to the biologically observed effect. This is the popular theory to explain the action of radiation in biological tissues, but it is probably not the whole truth.

III. TEMPERATURE

Despite the significant effect of temperature on some aspects of growth and development of plants, it is remarkable that the basic pattern of differentiation in the embryo is little affected by changes in temperature. In most instances, temperature indirectly affects the overall growth of the embryo, in a relative way without inducing any morphogenetic effects (Iwahori, 1966). Temperature is vitally important in determining the rate of respiration and in controlling the overall metabolic rates of chemical reactions, but the temperature limits for growth of the whole plant or of its isolated parts are probably somewhat narrower than the limits set for individual metabolic processes.

A. Temperature Effects on Cultured Embryos

In many published studies, cultured embryos have been maintained at an arbitrarily selected temperature and only a few attempts have been made to explore the optimum temperature characteristics for their growth. From these studies, it can be concluded that best growth of the embryo occurs between 27° and 32°C (van Overbeek *et al.*, 1944; Rijven, 1952; Choudhury, 1955a; Norstog and Klein, 1972). As Fig. 11.7 shows, no matter for how long embryos of *Datura stramonium* were

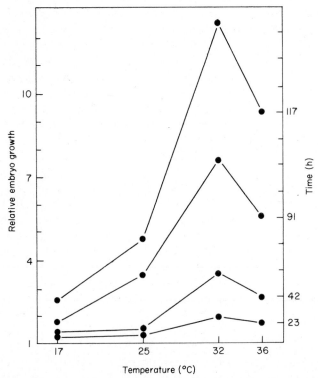

Fig. 11.7. For heart-shaped embryos (between 0·2 and 0·5 mm in length) of *Datura stramonium*, the optimum temperature for growth is 32°C. Embryos were grown in a mineral salt medium containing 2% dextrose and embryo factor from coconut milk, and were measured at different periods after culture. (After van Overbeek *et al.*, 1944)

maintained at the specified temperature, the highest optimum temperature of 32°C was found to be most suitable for growth. Embryos of other plants require a different optimum temperature for their growth; embryos of *Cyclamen persicum*, for example, grew well at 20° and 23°C and temperatures lower than 20°C were inhibitory for growth (Gorter, 1955). According to Meletti (1960), embryos of *Triticum* and *Hordeum* cultured on the endosperm of the whole grain grew best at a temperature as low as 5°C. Growth of the embryo of potato is relatively more rapid at 20°C than at 15° or 25°C (Haynes, 1954). Although cotton embryos do not respond to changes in temperature between 20° and 30°C (Lofland, 1950), at temperatures above 30°C they become spindly (Mauney, 1961). In the growth of barley embryos high temperature (25–30°C) acts like high intensity light in preventing precocious germination and thus increasing the potential for cell division (Norstog and Klein, 1972).

It is not easy to come by examples of embryos in which their tolerance to very high temperatures has been studied. In one instance (Siegel, 1953), it has been reported that exposures of red kidney bean embryos to a temperature of 100°C for about 35 min inhibited their subsequent growth. Embryos subjected to shorter periods of heat injury were reactivated by light and exhibited varying degrees of recovery in their respiration and enzyme activities. In some respects there are strong parallels between the deterioration of the heat-injured embryos and the deterioration of seeds with age at low temperature.

B. Vernalization

A phenomenon of considerable interest involving temperature effects on embryos is vernalization, which denotes a specific qualitative or quantitative promotion of flower initiation by a previous low temperature exposure of seeds, excised embryos, flower heads with fertilized ovules, or entire plants. Most of what is known about the role of the embryo in vernalization has been derived from the work of Gregory and Purvis (Purvis, 1961) on the winter strain of rye, the grains of which require chilling at 1°C for several weeks in order to flower subsequently when they were germinated and grown under appropriate conditions. Plants originating from unchilled seeds lapse into prolonged vegetative growth before they can flower.

One of the most important facts that has been established with regard to vernalization is that excised embryos grown in a nutrient agar medium containing a carbon energy source respond to vernalization treatment in a similar manner and to a comparable degree to whole grains (Gregory and Purvis, 1936, 1938). Later (Purvis, 1940), it was shown that even the shoot apex of the embryo isolated and grown *in vitro* was susceptible to vernalization. The key role of the embryo in the perception of vernalization gives a basis for interpreting the effects of cold treatment on ears of different ages. Especially remarkable was the fact that an ear vernalized on the fifth day after anthesis, probably containing an embryo no bigger than the globular stage in the floret, was responsive to chilling treatment. Obviously, the ability of a few-celled embryo to "remember" the cold spell received early in its ontogeny and to flower weeks or months later is of salient interest, and so are the even more elusive self-perpetuating changes induced in the embryo by the low temperature.

Although excised embryos of some species may undergo vernalization even when cultured in distilled water (Sen and Chakravarti, 1947), the responses of rye embryos to cold treatment vary with the concentration and type of carbon energy source provided in the medium (Gregory and de Ropp, 1938; Purvis, 1944). Unless the embryos were depleted of their endogenous sugar by starvation, a few embryos responded to cold treatment even in the complete absence of added sugar, while the majority remained unvernalized (Purvis, 1947). Apart from its role as a carbon energy source, sugar presumably mediates in the synthesis and accumulation of a substance in the embryo necessary for the progress of vernalization (Purvis, 1948). Although the identity of this substance is a matter of conjecture, preliminary experiments have shown that a chloroform extract of vernalized embryos can cause a slight acceleration in the flowering of excised unvernalized embryos (Purvis and Gregory, 1953). If this is true, it is a question of prime interest to developmental physiologists to determine how cold treatment can result in the synthesis of a transmissible stimulus. Curiously enough, although this discovery was made in the mid-1950s, little has been reported since then. Vernalization of excised wheat embryo has been shown to be associated with the appearance of new low molecular weight proteins (Teraoka, 1967).

IV. EFFECTS OF pH AND ATMOSPHERIC GASES

A. pH Effects

Owing to the selective uptake of ions by cultured tissues, and the poor buffering capacity of some of the well known culture media, the maintenance of pH of the medium is of some importance in plant tissue cultures. In devising media for the culture of embryos, it is therefore essential to prevent ionic imbalance by stabilizing the pH at the level of the optimal value throughout the experiment.

In embryo culture studies, where a relatively constant pH is desired, the standard technique is to adjust the pH of the medium before autoclaving to a value near neutrality. From investigations of the effects of a range of pH on the growth of embryos of a few species, it appears that there is no sharply defined optimum pH. According to van Overbeek *et al.* (1944), the responses to pH of *Datura stramonium* embryos initially 0·3–0·5 mm long vary according to their age in culture. Growth of embryos during the first 2–4 days is best at pH near neutrality, and at later stages a more acidic medium favors growth. Interesting as these results are, they do not tell us whether there is a changing pH requirement for embryos of different ontogenetic stages. With embryos of *Gossypium hirsutum*, pH could be varied over a relatively wide range (5·0–8·0) without impairing their survival and growth (Lofland, 1950). Although pH of the ovular sap of *Capsella bursa-pastoris* was about 6·0, excised and cultured embryos were capable of tolerating pH from 5·4 to 7·5 (Rijven, 1952). Tomato embryos grow better in a slightly acidic or neutral medium (Choudhury, 1955a). Rice embryos grown in a medium adjusted to a range of pH between 3 and 12 are reported to show optima at both pH 5 and pH 9, with a somewhat better growth occurring at pH 5 (Sapre, 1963).

Norstog and Smith (1963) encountered some difficulty in inducing differentiation in barley embryos 0·4 mm long at excision when they were cultured in a medium above pH 5·2, although growth in length by cell elongation occurred. Cell division and differentiation occurred in embryos grown in a medium adjusted to a pH range between 4·0 and 5·2, with the optimum at pH 4·9. These observations appear to provide an interesting approach for studying pH effect on the basic cellular processes of embryos, and it is unfortunate that supporting histological evidence has not been supplied to complement the growth data. A low pH (pH 3·4–4·5) in the medium affected cell division and differentiation in *Eranthis hiemalis* ovules by selectively killing undifferentiated embryos and inducing adventive embryony in the undamaged suspensor (Haccius, 1963). This phenomenon is further considered in Chapter 14 (see Fig. 14.4.).

Arditti (1967a) has summarized the effects of pH on the growth of several orchid embryos. For the majority of species studied, a pH around 5 appears suitable. Hydrogen ion concentration of the medium is apparently crucial at the time of germination of the embryo, since lowering the pH after germination is not deleterious for growth

(Knudson, 1951). The wide variety of nutrient media employed for culture of orchid embryos precludes any meaningful discussion on the effects of pH on their growth.

B. Effects of Atmospheric Gases

Little work has been done on the culture of embryos involving the use of any particular gas phase, since a great many of them survive in ordinary air, or are not affected by changes in the gaseous environment. To this latter category belong immature embryos of tomato which grow as well in nitrogen as in air (Choudhury, 1955b). However, the presence of nitrogen or low concentrations of oxygen in the ambient atmosphere appears to prevent precocious germination that normally occurs in cultured embryos (LaRue and Merry, 1939; Andrews and Simpson, 1969; Norstog and Klein, 1972); in this sense these conditions cause the same effects as those caused by high intensity light or high temperature. In cultured embryos of oat, low concentrations of oxygen in the atmosphere (down to 2%) produced elongated coleorhiza while higher concentrations (20%) promoted formation of epidermal hairs on the latter (Norstog, 1955).

V. MICROSURGICAL EXPERIMENTS ON EMBRYOS

Although embryogenesis is essentially a phenomenon which encompasses the entire embryo, there is obviously a great deal of interaction between its different parts. One can gain some useful information on this nexus of interaction from experiments designed to study the potentialities of different parts of the embryo for growth and development under particular conditions. Experimental modification of organogenesis of the embryo may also serve as a means to study the factors leading to normal embryonic development after such profound disturbances. Apart from the culture of excised embryos in media containing different additives already described, other direct experiments on embryos are of two kinds: (1) microsurgical experiments in which the embryo is split or individual parts removed and the partially mutilated embryo cultured; (2) culture of parts of the embryo. There can be little doubt that surgical disturbances of the embryo can lead to intercellular changes although artifacts to be expected in such treatments are to be kept in the background in interpreting the results.

A. Mutilation of Embryos

The relationship of the embryo axis to the scutellum, a structure characteristic of the graminean embryo, has figured in some microsurgical
experiments. When mature embryos of maize and other cereals were
cultured without scutellum in a mineral salt medium, shoot and root
primordia did not attain normal growth in length (Andronescu, 1919;
Narayanaswami, 1963). In a subsequent study on the culture of maize
embryo axis, shoot growth was enhanced by using high concentrations
of nitrate nitrogen in the medium (Smirnov and Pavlov, 1964). These
observations suggest a role for the scutellum in the growth of the
embryo, probably by channeling soluble food from the medium. In the
embryos of certain Gramineae there is also a close dependence of the
scutellum on the shoot and root meristems for its normal development
in culture. LaRue (1952) found that the scutellum of maize embryo
devoid of the shoot and root meristems when grown in culture did not
attain the size it attained in the fully mature germinated embryo. In
Pennisetum typhoideum, proliferation of the scutellum which occurs when
the embryo is grown in a medium containing adenine is inhibited if
the shoot meristem is removed (Narayanaswami, 1959a). Clearly, the
meristematic regions of the embryo hold sway over the potential morphogenetic capacity of the scutellum.

   Although there are occasional reports to the contrary (Nickell, 1951),
normal growth of the shoot and root systems in embryos seems to be
dependent upon the cotyledons, since in decotylated embryos these
organs appear abnormal. Completely decotylated embryos of 'Lovel'
peach, however, failed to grow in culture, whereas embryos which had
some part of the cotyledons attached formed essentially normal seedlings (Kester, 1953). A study of the role of auxins in the growth of
cultured pea embryos showed that the hormones apparently interacted
with the reserve substances of the cotyledons in enhancing elongation
of the root and shoot primordia, as the growth of these organs was inhibited when cotyledons were not attached to the embryo axis (Kruyt,
1952, 1954). Alternatively, when embryos excised from seeds presoaked
in auxins were cultured with the cotyledons attached to them for different periods of time, the longer the cotyledons were left on the embryo,
the greater was the growth stimulation. When excised embryos of *Phaseolus vulgaris* with attached cotyledons were exposed to vapors of
methyl-2,4-dichlorophenoxyacetate, the first foliage leaf became malformed and the subsequent leaves became 1–7-foliate (Furuya and
Osaki, 1955). In contrast, decotylated embryos treated with the test
substance (Furuya, 1956) showed a syndrome of abnormalities mainly

Pattern of   Feature of seedling
development   (one week old) → (five weeks old)

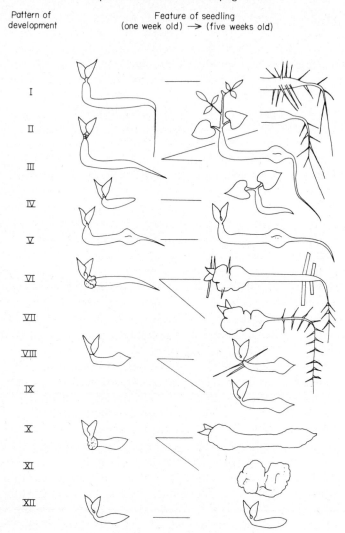

Fig. 11.8. Abnormal growth patterns exhibited by decotylated embryos of *Phaseolus vulgaris* exposed to vapors of methyl-2,4-dichlorophenoxyacetate for 0–32 days, and then cultured for 1 week in the dark and 5 weeks in light. Abnormal patterns 1–9 were found mostly in embryos exposed for 0–4 days while patterns 10–12 were characteristic of embryos exposed for 10–32 days. (After Furuya, 1956)

confined to inhibition of growth of the plumule and radicle, formation of stunted foliage leaves, root initiation on leaves and callus growth (Fig. 11.8). In the embryos of *Vigna sesquipedalis* (Hotta, 1957), decotylation resulted in the inhibition of development of the shoot and formation of lateral roots; varying degrees of inhibition of growth of embryonic organs were also noted when embryos of cherry and peanut were cultured without the cotyledons (Nuchowicz, 1965; Abou-Zeid, 1972). In *Hevea brasiliensis*, it was necessary to culture the embryo axis with at least two-thirds of the cotyledons for normal growth, as growth was slow with lesser amounts (Muzik, 1956). Less uniquely under cotyledon control is the formation of small outgrowths on the hypocotyl of the embryo of *Pharbitis nil*. Despite the close resemblance between the

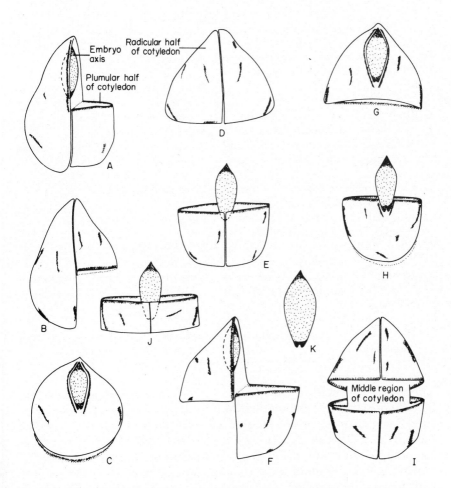

Fig. 11.9. Decotylating patterns used in microsurgical experiments on embryos of *Cassytha filiformis*. In **C**, **G**, and **H**, the cotyledon is lying on its concave surface. For description see Table 11.I. (After Rangaswamy and Rangan, 1971)

cotyledons and the hypocotyl outgrowths, the latter originated most rapidly on the embryo cultured without the cotyledons or after they became senescent (Matsubara, 1964a; Matsubara and Nakahira, 1966).

Morphogenesis of the decotylated embryos of the parasitic angiosperm *Cassytha filiformis* has been the subject of recent studies (Rangaswamy and Rangan, 1963, 1971) and so requires separate consideration.

TABLE 11.I. Growth of decotylated embryos of *Cassytha filiformis*. (From Rangaswamy and Rangan, 1971)

Portions of cotyledons severed	Responses
A. Radicular half of one cotyledon	Seedling well developed
B. Plumular half of one cotyledon	Seedling well developed
C. One entire cotyledon	Both hypocotyl and plumule inhibited
D. Plumular half of each cotyledon	Seedling well developed
E. Radicular half of each cotyledon	Plumule quiescent
F. Radicular half of one cotyledon and plumular half of the other cotyledon	Hypocotyl growth limited, plumule inhibited
G. One entire cotyledon and plumular half of the other cotyledon	Plumule quiescent
H. One entire cotyledon and radicular half of the other cotyledon	Plumule quiescent
I. Small portion from the equatorial region of each cotyledon	Seedlings formed occasionally
J. Almost the whole of both cotyledons, retaining only a belt around the plumular pole of embryo axis	Growth inhibited, plumule quiescent
K. Both cotyledons, leaving behind the embryo axis	Growth suppressed, plumule quiescent, rooting rare

The embryo is enclosed within the two massive cotyledons which can be fully or partially removed without injury to the embryo. The principal decotylating patterns studied are depicted in Fig. 11.9, in which the narrow region of the cotyledon near the radicle is referred to as the radicular half, and the opposite somewhat broad region as the plumule half. These studies show that, in general, the larger the portion of the cotyledon removed, the more inhibited was shoot morphogenesis in the embryo axis during its subsequent growth in White's medium (Table 11.I). The minimum amount of intact cotyledonary tissue

necessary for shoot morphogenesis was the radicular halves of both cotyledons or the radicular half of one cotyledon together with more than
one half of the other cotyledon. In the completely decotylated embryo
grown in a medium containing coconut milk, some growth of the plumule occurred.

The sum of the results of these and other studies reviewed earlier
shows that a complex of nutritional and hormonal effects is brought
to bear upon the organization of adult organs in the embryo. A recent
work has suggested that decotylation might affect the embryo axis in
a very specific way, by interfering with the synthesis of isoenzymes in
the latter (Khan *et al.*, 1972). Pertinent to this discussion also is the
effect of decotylation in substituting for cold treatment of the embryo
and, in a reverse way, the effect of excision of the embryo axis on the
longevity of the cotyledons; these are discussed in Chapters 16 and 17,
respectively. The cotyledons are also important in the growth of the
gymnosperm embryo, because of their large surface area for translocation of food substances (see Chapter 9).

Another class of mutilation experiments is exemplified by splitting
longitudinally the whole embryo or parts of it and growing them in
culture to determine whether there is a requirement for a minimal
number of initial cells for the formation of any particular organ or of
the whole plant. For example, in the embryonic root and shoot apices
there are groups of meristematic cells which give rise to mature tissues
and organs. Is this entire complement of meristematic cells indispensable for initiation of the particular organ? Is the destiny of the initial
cells already determined to produce particular organs? Hanawa and
Ishizaki (1953) found that when the shoot apex and subjacent regions
of the embryo of *Sesamum indicum* were divided longitudinally, each portion regenerated shoots with normal phyllotaxy. Similarly, Ball (1956b)
divided into two parts the root apex of the excised embryo of *Ginkgo
biloba* and followed the whole process of regeneration in exquisite detail
in culture, tracing the formation of the different tissues. The fundamental observation was that, during growth of the embryo in culture,
those portions of the root containing half or less than half of the original
initial cell group regenerated new root apices. If the root apex is divided
unequally, the final result will be a fast-growing long root and slow-
growing short root (Fig. 11.10). In somewhat the same way, if the
radicular end of the embryo of *Dendrophthoe falcata* which forms a holdfast is split longitudinally without affecting the plumule, new meristems
differentiate in both segments and twin holdfasts are formed, but plumule growth is arrested (Johri and Bajaj, 1963; Bajaj, 1966b). If the
incision is deep enough to divide the plumule as well, twin shoots and

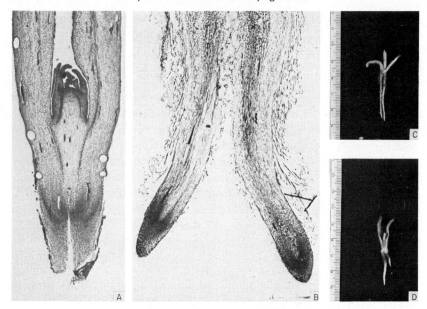

FIG. 11.10. When the root apex of the embryo of *Ginkgo biloba* is split longitudinally, those parts of the root containing half or less than half of the initials regenerate new roots. **A**, Longitudinal section of an embryo fixed immediately after operation. **B**, Longitudinal section of the twin roots 9 days after operation. **C**, An embryo in culture showing normal roots from the halves of the initial root. **D**, An embryo in culture showing two unequal roots as a result of unequal division of the original root. (After Ball, 1956b; photographs courtesy of E. Ball)

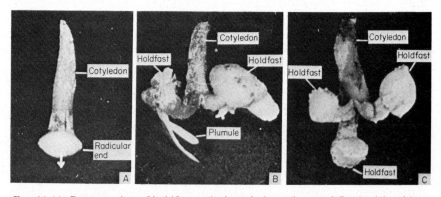

FIG. 11.11. Regeneration of holdfast and plumule in embryos of *Dendrophthoe falcata* split longitudinally. **A**, Mature embryo longitudinally split (arrow). **B**, Six week old culture showing regeneration of twin plumules and holdfasts. **C**, Formation of three holdfasts by two perpendicular longitudinal incisions; 2 week old culture. (After Bajaj, 1966b)

twin holdfasts regenerate (Fig. 11.11). If the radicle is split into eight segments, up to five of these may form holdfasts. Thus, the regimentation of cells of the root meristem to produce a viable holdfast seems to depend upon a relatively small minimal number of initials whose derivatives differentiate into normal tissues and regenerate the organ. These results provide impressive evidence to show that embryonic initials of the root and shoot are not predetermined to produce specific tissues, but function according to the position that they occupy. The results of the surgical operations of Clowes (1953) and Wardlaw (1949) on the root apices of grasses and shoot apices of ferns, respectively, lend support to this view.

B. Growth of Embryo Segments

Many surgical operations combined with cultural studies have been carried out on embryos with a view to investigate the capacity of embryonal segments to undergo growth and development in the absence of the subjacent tissues. It should be kept in mind that the segments at the time of excision have reached the stage at which growth can begin immediately as part of the whole embryo. The simplest experiment consisted of the division of the embryo into root, shoot, cotyledon and hypocotyl regions and determination of their potentialities for growth. Results gained from the culture of embryo segments of several plants (Molliard, 1921; Lee, 1955; Furuya and Soma, 1957; Sen and Verma, 1959, 1963; Johri and Bajaj, 1963; Nuchowicz, 1965; Bajaj, 1966b; Modrzejewski et al., 1970) have uniformly shown that in the regeneration following surgery, even in the most favorable medium employed, only the shoot segments developed into complete seedlings. Developmental hazards were, however, great when the tiny primordial meristems alone, in the absence of subjacent tissues, were grown (Bulard, 1954). The cotyledons and the region below the shoot apex usually formed callus, while the root and hypocotyl produced numerous lateral roots, but no shoots or buds, and thus retained their typical organization. Figure 11.12 illustrates the growth of the cotyledons, plumule and radicular segments of the embryo of *Dendrophthoe falcata*. Summarily, these results confirm the observations of Ball (1946) on pieces of the shoot apical region of mature lupin *(Lupinus albus)* plants and indicate the high degree of autonomy of this region. As far as the other parts are concerned, their pattern of development appears to be already laid down in the embryo, and is faithfully reproduced even in isolated segments.

Embryo segments of *Cajanus cajan* containing unorganized axillary bud initials possess considerable regenerative powers (Kanta and Pad-manabhan, 1964). Bud initials which did not grow out in cultured intact embryos were stimulated to grow rapidly when segments of embryo without plumule or without plumule and radicle were cultured. It was possible to obtain up to three seedlings from a single embryo by culturing separately the plumule, and the two axillary bud primordia of the cotyledonary node. The failure of lateral buds to grow out in the normal embryo is an indication that apical dominance is prevalent even in the

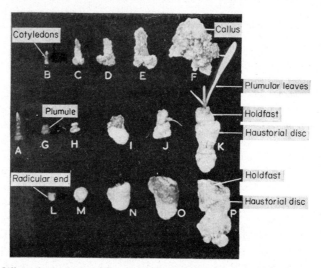

Fig. 11.12. When the embryo of *Dendrophthoe falcata* is segmented into cotyledons, plumule and radicle, and segments grown in culture, only the plumular segments regenerate a new plant. **A**, Mature embryo. **B–F**, Progressive growth of cotyledon resulting in callus formation. **G–K**, Formation of a normal seedling from the plumular segment. **L–P**, Holdfast formation from the radicular segment. (After Bajaj, 1966b)

embryonic stage. A greater morphogenetic plasticity is exhibited by the decapitated plumule of *Azadirachta indica* embryo which differentiates up to 20 shoot buds even when cultured in distilled water (Rangaswamy and Promila, 1972).

The formation of lateral roots on the hypocotyl segments of embryos described earlier may suggest participation of endogenous auxin in the process. This definitely seems to be the case in the hypocotyl segments of *Pinus lambertiana* embryos which regenerate roots when cultured in a mineral salt medium containing a relatively high concentration (4–8%) of sucrose (Greenwood and Berlyn, 1965, 1973). Addition of IAA to the medium enhances the rooting ability of the hypocotyl segments, and TIBA, an inhibitor of polar transport of auxin, inhibits it (Greenwood and Goldsmith, 1970). Thus auxin is believed to be involved in the regeneration of roots by hypocotyl segments of pine embryos and further experiments in others may well substantiate this belief.

Although experiments described above do not indicate whether regeneration of a root in the explanted plumule segment is necessary for its subsequent growth, this seems to be the case for growth of the shoot apices of wheat embryos (Smith, C. W., 1965, 1967, 1968). Generally, development of the excised shoot primordium was not influenced by addition of amino acids, coconut milk, or yeast extract to the medium. Addition of GA and adenine to the medium promoted growth to some extent, although it was less than that obtained normally. Best growth was obtained in a medium containing ammonium salts which also induced rooting in the explants, or when a root extract was added. It thus seems that the subjacent root system has an important function in controlling the full expression of the plumule development.

VI. COMMENTS

It is not intended in this chapter to provide a comprehensive treatment of the varied extracellular forces that operate in embryonic development, but rather the objective has been to outline the problem and illustrate its various facets with suitable examples. While there is evidence that some aspects of early embryogenesis are referable to physical pressure of the archegonial jacket or to the osmotic pressure of the embryo sac contents, to what extent temperature, radiation, pH, etc., lumped together as environmental forces are related to development of the embryo *in vivo* is quite unknown. It may mean no more than that the forces merely modify the detailed course of development of the embryo without producing any intrinsic changes in its basic pattern. The structural modifications produced in the embryos in response to wounding may have adaptive significance and survival value in their growth, but here again, the basic pattern of altered development may be determined by forces of a secondary nature. The general conclusion is that the control which extracellular factors exert upon embryogenesis is probably mediated through some hormonal relationships or requirement for nutrients; it is remarkable, however, that so little is known in physiological terms about the more fundamental nature of the intermediate substances which link the perception of a stimulus and the observed morphogenetic effect.

12. Biochemical Ontogeny of Embryogenesis

The previous chapters have dealt with the post-fertilization development of the egg through time and space into the embryo and the factors controlling embryonic growth and differentiation. These chapters have brought us to a point where we may usefully discuss in a general framework a common biochemical mechanism to explain progressive embryogenesis in plants. Such a discussion, however speculative, may prove to be fundamental to our understanding of the ways by which an embryo encased in the ovule develops structurally and functionally from stage to stage with amazing precision. In the biochemical ontogeny of embryogenesis, the theoretical interest lies not so much in the requirements for progressive form change in the embryos, as in the mechanism by which such form changes are accomplished.

At present it seems clear that growth and differentiation of the fertilized egg leading to the formation of the embryo with its independent, genetically controlled form is strongly influenced by the metabolites present in the embryo itself and in its surrounding environment. Experiments with isolated embryos have forced a recognition that during the heterotrophic phase of growth, the embryo requires some special and unique kinds of nutrients, provided not only by its own cells but also by the embryo sac and the endosperm; these substances may include amino acids, growth hormones, and perhaps purines and pyrimidines and their precursors. Attempts to culture heterotrophic embryos have centered upon developing media to match the nutrients present in the embryo sac and the endosperm and only in a few cases have these attempts been successful. Many arbitrary mixtures of growth hormones, amino acids and natural endosperm extracts have been used with varying degrees of success to grow embryos in the heterotrophic phase of growth. The younger the embryo, the more difficult it is to satisfy its needs synthetically in a nutrient medium.

The autotrophic phase of growth of the embryo begins only after it has developed within the ovule to the stage of cotyledonary initiation and the consequent attainment of bilateral symmetry. At this stage, the embryo can be excised and cultured in media containing major salts, vitamins and trace elements. Progressively older embryos require less and less complex media and thus become increasingly self-sufficient.

These facts constitute the basis for a biochemical ontogeny associated with progressive form change and functional differentiation in the embryo. As we have seen in Chapter 3, the differentiation of specific organs in the embryo is a frequent sequel to a high degree of enzyme activity in the presumptive cells. Since enzymes are proteins, it seems not unreasonable to suggest that embryogenesis involves not only a structural rearrangement of the cells but also a physiological reconstitution involving the synthesis of specific proteins.

Confronted with such a high degree of synchrony between structural differentiation and physiological activity, a fundamental problem of interest to the embryologist is the mechanism by which the embryo becomes progressively self-sufficient. It might be assumed that the nature of the changes in a developing embryo which represents an accomplishment of its transition from heterotrophy to autotrophy is concerned with the development of biochemical mechanisms to sustain the physiological activities and the associated structural changes. This idea has been elaborated in great detail for animal embryos by studies which have provided substantial evidence to show that evolution of structure and function and formation of enzymes are correlated events. If this idea is valid for plant embryos, it would appear that enzyme systems involved in the synthesis of critical metabolites are formed prior to, or synchronously with, the development of autotrophy. We still have much to learn about the enzyme systems that are unblocked during embryogenesis and the task of identifying them remains formidable indeed. There is some suggestive evidence, but the available data are limited and the discussion speculative. In order to understand the problems underlying the biochemical ontogeny of developing embryos it is necessary to turn again to a consideration of some of the work discussed in the previous chapters.

I. FROM HETEROTROPHY TO AUTOTROPHY

An early insight into the biochemical basis of embryogenesis was provided by the pioneering work of van Overbeek and associates (van Overbeek *et al.*, 1941a, 1942), who set the theme for one of the most rigorous analyses of the nutritional needs of developing embryos. These workers found that mature embryos of *Datura stramonium* were completely self-sufficient and could grow into normal seedlings if planted on a mineral salt medium containing dextrose. Proceeding from this background of information, the potential of several additives to the medium to induce growth of still smaller embryos was examined. Younger torpedo-shaped and heart-shaped embryos did not grow in the medium in which older embryos were successfully cultured. However, when this medium was fortified by the addition of a broad array of intermediary metabolites of possible nutritional significance such as glycine, thiamine, ascorbic acid, nicotinic acid, pyridoxine, adenine, succinic acid and pantothenic acid, it effectively insured growth of embryos as little as 500 μm long. In seeking conditions that might favor successful culture of still smaller proembryos, it was found that supplementation of the medium containing the accessory growth factors mentioned above with the natural endosperm extract of coconut milk provided the mitotic stimulus for initiating cell divisions in the limited population of cells of the very small embryos. Since a requirement for specific compounds by embryos in culture is an indication of their inability to synthesize them, one can reason that in the proembryos of *Datura* biosynthetic systems concerned with the formation of vitamins and other physiologically active substances and the expanded array of growth factors contained in coconut milk are blocked. Accompanying the increase in size and morphological development of the embryo, there is a sequential unblocking of metabolic pathways for autotrophic existence: for example, first the unmasking of the pathways in the synthesis of coconut milk factors before the embryo enters the heart-shaped stage, and later the capacity to synthesize vitamins and other intermediary metabolites when the embryo passes the torpedo-shaped stage. Thus the activity of genes for enzyme synthesis is controlled in a sensitive way during embryogenesis, and it is an interesting commentary that the synthetic capacities represented in the genome of the least specialized cells of the very young proembryos are not expressed fully.

The basis for progressive embryogenesis can be explored in other systems too, and for this purpose the analysis of the growth requirements of the embryos of *Todea barbara* is useful (DeMaggio and Wetmore, 1961a, b). Twenty day old embryos which have already undergone internal differentiation can be grown to maturity in a medium containing mineral salts and sucrose. Slightly younger embryos, excised 17 days after fertilization, could not be grown in this medium; evidently, the biochemical requirements essential for embryos to initiate cell division were not completely met by this medium. These requirements were,

TABLE 12.I. Progressive embryogenesis in *Capsella bursa-pastoris*. In column 3, additional nutrients added and/or the biosynthetic pathways probably becoming active at different stages of embryogenesis are italicized. (From Raghavan, 1965)

Developmental stage	Length, μm	Nutritional requirements
Early globular	20–60	Unknown for embryos <40 μm
Late globular	61–80	Basal medium + *indoleacetic acid* (0·1 mg/l), *kinetin* (0·001 mg/l), and *adenine sulfate* (0·001 mg/l)
Early heart-shaped	81–150	
Late heart-shaped	151–250	
Intermediate stage	251–450	Macronutrient salts[a] + 2% sucrose + vitamins[b] + *trace elements[c]* (basal medium)
Torpedo-shaped	451–700	Macronutrient salts + 2% sucrose + *vitamins*
Walking-stick-shaped	701–1000	
Inverted-U-shaped	1001–1700	
Mature embryos	>1700	Macronutrient salts + 2% sucrose

[a] Macronutrient salts (mg/l): 480 $Ca(NO_3)_2.4H_2O$; 63 $MgSO_4.7H_2O$; 63 KNO_3; 42 KCl; 60 KH_2PO_4.
[b] Vitamins (mg/l): 0·1 thiamine hydrochloride; 0·1 pyridoxine hydrochloride; 0·5 niacin.
[c] Trace elements (mg/l): 0·56 H_3BO_3; 0·36 $MnCl_2.4H_2O$; 0·42 $ZnCl_2$; 0·27 $CuCl_2.2H_2O$; 1·55 $(NH_4)6Mo_7O_{24}.4H_2O$; 3·08 ferric tartrate.

however, satisfied by coconut milk or by equal concentrations of sorbitol and inositol. Here again, the available information favors the view that form change in the embryo is associated with an unblocking of its synthetic capacities.

Nutrient media devised for growth of embryos of different ages of *Capsella bursa-pastoris* follow a pattern that is consistent with the assumption of an ordered unblocking of biosynthetic systems (Raghavan, 1965). For example, the minimal medium for growth of torpedo-shaped and older embryos is a mixture of macronutrient salts and 2% sucrose solidified with agar. In this medium, embryos developed into small plantlets as they did in a more complex medium. Addition of vitamins such as niacin, pyridoxine and thiamine to the macronutrient salt solution proved capable of supporting growth of smaller intermediate stage embryos into small plants. For growth of early and late heart-shaped embryos, a combination of trace elements and vitamins with the macronutrient salt solution was essential. Still smaller embryos required a balanced mixture of IAA, kinetin and adenine sulfate for continued cell division and growth. Table 12.I summarizes the results of these experiments which show that acquisition of autotrophy is paralleled by an increasing independence on exogenous nutrients.

A number of investigations from the older literature on embryo culture also support to a varying extent the fact that when the embryo is large, the growth medium can be simple, and if small, the growth medium must be complex. It would be fruitless to discuss these studies here since they differ only in minor details from the works already considered. From these various studies it can be concluded that at some point during its ontogeny, the cells of the embryo become metabolically self-sufficient to carry out the synthetic processes necessary for its growth and differentiation into a small plantlet. Young embryos lack one or more of the substances considered essential to sustain their autotrophic existence. With time, the capacity to synthesize these substances is expressed in a precise sequential pattern which is reflected in the increasingly complex morphology of the embryos.

II. BIOCHEMICAL SPECIALIZATION OF EMBRYOS

One of the facts that stands out in the foregoing discussion is the high degree of specialization of mature embryos and the relative lack of specialization of immature embryos. Since no extensive study of the biochemical basis for cellular activities in developing embryos has been undertaken, we have to follow an indirect approach to understand the basis for changes which take place during embryogenesis. The effects of different additives to the medium on the growth of embryos of different ages are of obvious interest, since it is to be expected that the development of a response mechanism is linked with the biochemical age of the embryos. Such an approach is somewhat crude and is likely to yield conclusions of only limited value. For instance, it would be a mistake to believe that a lack of response of embryos of a certain age to a specific cell division factor is due to its endogenous synthesis by the embryo. Other limitations such as the availability of nucleotide precursors or enzymes which catalyze the synthesis of the precursors and their incorporation into the macromolecules are also important considerations.

With these reservations in mind, it is worth while to spend a little time recapitulating the growth responses of embryos to nutrient additives in the medium. We have already pointed out in Chapter 7 that when embryos of different ages are cultured, younger embryos are retarded in growth, while the more mature ones grow into perfectly normal seedlings. Development of the younger embryos was for the most part restricted to processes of unorganized growth, but in some cases, abnormal leaves and roots also grew out irregularly from the embryo. Rietsema *et al.* (1953b) found that growth of embryos of *Datura stramonium* was inhibited by the addition of IAA to the medium and young embryos were found to be more sensitive to the auxin than advanced stage embryos. In this case, the influence of auxin on the young embryos was seen in the inhibition of meristematic growth and promotion of abnormal cell enlargement, suggestive of a loss of control of cellular integrity. Along the same line, according to their age at excision, embryos of *Datura* are also reported to show contrasting patterns of behavior in media containing casein hydrolyzate (Paris *et al.*, 1953) and ammonium nitrate (Matsubara, 1964b). Generally the stimulatory effect of these substances decreased when the stage of excision of the embryo advanced, and torpedo-shaped and older embryos were practically insensitive to their presence in the medium.

Other experiments dealing with the effect of growth inhibitors and extracellular factors on embryo growth have demonstrated an acquisition of independence from growth inhibition with progressive growth of the embryo within the ovule. For example, a lack of specialization in the young embryos of oat is indicated by their greater susceptibility when spikes of different ages are sprayed with maleic hydrazide (Mericle *et al.*, 1955). When embryos of different ages are subjected to critical exposures of ionizing radiations, it is frequently observed that the mature embryo is less sensitive to a dose of radiation that arrested cell division and growth in earlier stage embryos (Mericle and Mericle, 1961; Banerjee, 1968). Since functional and morphological differentiation may occur at the cellular level, the idea that these well defined differences in responsiveness of embryos to hormonal or physical factors are based on specific cellular adjustments required to implement normal growth is an attractive one. These adjustments can take various forms, such as a highly efficient IAA-synthesizing system in embryos that makes them insensitive to exogenous IAA, a capacity to synthesize amino acids that will make casein hydrolyzate a superfluous constituent of the medium or the absence of DNA synthesis in a large proportion of cells which imposes a resistance to radiation damage. These metabolic differences between embryos of different ages have yet to be proved experimentally, but they are of considerable interest as possible developmental models.

III. COMMENTS

With our emphasis on the nutritional needs of embryos of different ages, the point of view that emerges from this brief examination of embryogenesis is that advanced stage embryos have unusual physiological properties that might have a bearing on autotrophy. A logical step in the development of this point of view is to correlate the metabolic state of embryos of different ages to their nutritional requirements. Results from these studies may make an important contribution to our understanding of the biochemical ontogeny of embryogenesis.

13. Applied Aspects of Embryo Culture*

The promise inherent in technical advance in any branch of plant science generally raises our expectations for the utilization of knowledge for the enhancement of the quality of our agriculture and in overcoming the natural barriers in the breeding cycles of plants. This is especially true when, traditionally and tactically, these problems are initially attacked in plants that are of little economic importance. Although the biological knowledge gained from the successful culture of embryos has not become the principal ameliorating factor in any aspect of our agriculture, it appears to have sufficient potential to generate a feeling of foreseeable application in practical plant breeding programs and in the elimination of conditions that disturb normal reproductive processes.

In a general way, when reviewed in retrospect, the principal applications of techniques directly related to isolation and culture of embryos appear to be in the areas of overcoming embryo abortion in inviable hybrids and dormancy in seeds. In this chapter these will be illustrated by a series of examples, but it should be realized that these examples are intended to be illustrative and make no attempt to be comprehensive.

* In preparing this chapter, the author has drawn upon an article he wrote (Raghavan, 1975c) for "Applied and Fundamental Aspects of Plant Cell, Tissue and Organ Culture", edited by J. Reinert and Y. P. S. Bajaj. Springer-Verlag, Berlin (in press).

I. EMBRYO ABORTION IN HYBRIDS

In horticultural and breeding practices, embryo abortion is normally encountered in unsuccessful crosses. Although fertilization occurs normally in such crosses and the resulting embryos begin to develop in a relatively healthy way, a number of irregularities accompany the subsequent development of the embryos which bog down, so to speak, resulting in their eventual death and collapse of seeds. In a number of cases, these physiological disturbances have been overcome through the establishment of embryo cultures. For several reasons, this approach to the problem of overcoming inviability is impractical in large-scale field applications, but it presents certain advantages and field applications may be developed in the future. Before undertaking a survey of some of the contributions of culture techniques in overcoming barriers to viability of embryos, let us gain some perspective by a brief examination of the problems associated with seed failure and embryo abortion in plants. Such an examination might reveal those causes that actually warrant further investigation and focus on what has been accomplished to date. For a comprehensive review of hybrid inviability and sterility in plants, reference is made to Stebbins (1958).

According to Brink and Cooper (1947), most hazards to embryo growth in fertilized ovules are either due to heredity-oriented problems resulting in a disturbance of the equilibrium between growth of the maternal tissue, embryo and endosperm, or due to physiologically oriented developmental problems leading to tissue breakdown. Fertilization of the egg by the sperm is a specialized process and results in the perpetuation of the progeny of the partaking individuals. Specialization imposes the requirement that every step in the process is intertwined with that of others, and hence any alteration in the genetic make-up of one of the partaking individuals is bound to result in genetic imbalance in the progeny. On the basis of a detailed analysis of hybrid mortality in plants, Brink and Cooper (1947) have established six different types of hereditary alterations leading to embryo abortion, as follows: (i) enforced self-fertilization in a normally cross-fertilizing species; (ii) crosses involving parents of the same ploidy level but belonging to different species and genera; (iii) crosses between parents belonging to the same species but differing essentially in the degree of polyploidy; (iv) crosses involving parents belonging to different species and different ploidy levels; (v) unbalanced chromosome condition, especially aneuploidy in the endosperm; and (vi) maternal genotype causing arrested growth of the embryo irrespective of the source of the male gamete. In short, limitation of fertilization to union of the egg and sperm from members of the same species with identical chromosome constitutions appears mandatory if subsequent development is to be normal.

The act of fertilization resulting in the formation of a zygote implies something more than the transfer of genetic information, since it also sets in motion the train of physiological and developmental processes that lead to progressive form change in the zygote and its eventual transformation into the embryo. It is clear that any disturbance in the normal processes of tissue interaction during embryogenesis is bound to lead to abnormalities in the development of the embryo, resulting in a particular type of infertility. Part of the problem in this section is to discern the kinds of disturbances that occur and to determine how they modify the normal developmental processes in the complex setting of the ovule, resulting in embryo abortion.

A. Development of the Embryo and Endosperm in Inviable Hybrids

Embryo Development

Before we enter into a serious discussion of the problem posed above, let us look at the developmental and structural peculiarities of hybrid embryos of some crosses. Although accounts of the ontogeny of hybrid embryos differ in details, all seem to agree in identifying an early embryonic stage when developmental anomalies begin to appear. In reciprocal crosses between *Oenothera biennis* * × *O. muricata* and *O. biennis* × *O. lamarckiana*, the initial rate of growth of embryos was not appreciably affected, but it slowed down later, and eventually did not advance beyond a few cells (Renner, 1914). In crosses between species of *Epilobium*, proembryos appeared as abnormal multinucleate structures due to failure of wall formation after cell divisions (Michaelis, 1925). The embryological history of crosses between different species of *Datura* illustrates strikingly the range of developmental abnormalities in hybrid embryos. In the cross between *D. stramonium* × *D. metel*, there was no truly significant development of the embryo which formed no more than eight cells before it disintegrated (Satina and Blakeslee, 1935). Seeds of the hybrid between *D. metel* and *D. discolor* were generally devoid of embryos, or if present, they were too minute to be seen by dissection. In *D. discolor* × *D. stramonium*, *D. leichhardtii* × *D. ceratocaula* and *D. leichhardtii* × *D. discolor* crosses, division of the zygote was considerably delayed, resulting, not infrequently, in completely undifferentiated embryos. In still others (for example, *D. innoxia* × *D. stramonium*, *D. leichhardtii* × *D. metel*, *D. innoxia* × *D. metel* and many

---

* In the designation of crosses, the first-named is the female parent.

combinations involving *D. ceratocaula* as the parent); embryos were de-
cidedly abnormal and varied from completely undifferentiated masses
to differentiated misshapen structures (ghost embryos) (Fig. 13.1A).
Another variation is observed in seeds of the cross between *D. stramonium*
and *D. meteloides* which had transparent, heart-shaped or torpedo-
shaped embryos that lacked cytoplasm in their cells. Abnormalities in

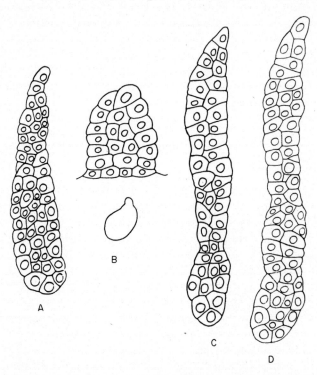

Fig. 13.1. Hybrid embryos from inviable crosses in *Datura*. **A**, Undifferentiated
embryo from *D. leichhardtii* × *D. metel*, 11 days old. **B**, Embryo with abnormally broad
and short suspensor from *D. pruinosa* × *D. stramonium*, 13 days old; detailed structure
of the suspensor is also shown. **C, D**, Abnormally long suspensor from *D. meteloides* ×
D. innoxia, 13 days old (**C**), and *D. metel* × *D. meteloides*, 16 days old (**D**). (After Rietsema
and Satina, 1959)

the form of the suspensor may also be associated with underdeveloped
embryos in *D. meteloides* × *D. innoxia*, *D. metel* × *D. meteloides* and *D.
pruinosa* × *D. stramonium* crosses (Fig. 13.1 B–D). The abnormalities,
which are probably derived from altered rates of cleavage or orientation
of the mitotic spindle, make the suspensor unusually long, short or
broad (Blakeslee and Satina, 1944; McLean, 1946; Sachet, 1948;

Sanders, 1948; Rietsema and Satina, 1959). The behavior of embryos of inviable crosses in the genus *Dianthus* appears to be comparable in many ways to that of *Datura* crosses, although in detail the timing of developmental events may be markedly different. In embryos of *Dianthus chinensis* × *D. plumaris*, the most consistent deviation from normal development occurred in the basal cells which frequently enlarged, followed by thinning, vacuolation and disintegration of the cytoplasm (Buell, 1953). Multiple shoot development, branching of cotyledons and other aberrations appeared in embryos of *Gossypium arboreum* × *G. hirsutum* hybrid (Pundir, 1972). In *Hibiscus costatus* × *H. aculeatus* and *H. costatus* × *H. furcellatus* crosses, paucity of organelles and vacuolization of the cytoplasm of embryos may be traced to fertilized eggs which failed to shrink and undergo polarization of cytoplasmic organelles to the same extent as in self-fertilized *H. costatus* zygotes (Ashley, 1972). In *Pinus*, embryo inviability is a barrier of major importance to crossability, and extensive observations on the subgenus *Strobus* show that although fertilization takes place normally in interspecific crosses, the embryo degenerates as the suspensor tiers elongate and thrust the apical unit into the female gametophyte (Hagman and Mikkola, 1963; Kriebel, 1972). Thus, embryos of inviable hybrids appear to possess the potential for initiating development but are somehow prevented from reaching the adult size and character in their full multicellularity and structure. Similar failures of embryo development are known to occur in seeds of other hybrids which are too numerous to list here. In certain crosses, disorganization of the embryo following successful pollination and fertilization is due to self- and cross-sterility (Bradbury, 1929; Brink and Cooper, 1939, 1940).

Endosperm Development

Successful development of the embryo, as distinct from formative processes of growth, depends upon a continuous supply of nutrients. Because the endosperm is the primary mechanism for nourishing the embryo, a relationship of embryo abortion to endosperm activity is apparent in a general sense. Our lack of understanding of the causes of embryo abortion in any particular cross might reflect an inadequate understanding of endosperm function and behavior. Renner (1914) threw this question into an illuminating framework when he first called attention to the fact that collapse of the embryo in unsuccessful crosses between species of *Oenothera* was preceded by the disintegration of the endosperm beginning soon after fertilization, thus depriving the embryo of the immediate source of food supply. Histological examination of aborted seeds resulting from unsuccessful interspecific crosses in *Epilobium* (Michaelis, 1925), *Iris* (Sawyer, 1925; Werckmeister, 1934), *Prunus* (Bradbury, 1929), *Galeopsis* (Müntzing, 1930), *Avena* (Kihara and Nishiyama, 1932), *Pyrus* (Bryant, 1935), *Datura* (Satina and Blakeslee, 1935; Sansome *et al.*, 1942; Sachet, 1948), *Triticum* (Boyes and Thompson, 1937), *Gossypium* (Beasley, 1940; Pundir, 1972), *Medicago* (Brink and Cooper, 1940), *Nicotiana* (Brink and Cooper, 1940, 1941), *Lycopersicon* (Cooper and Brink, 1945), *Secale* (Håkansson and Ellerström, 1950), *Dianthus* (Buell, 1953), *Melilotus* (Greenshields, 1954) and *Arachis* (Johansen and Smith, 1956), and intergeneric crosses

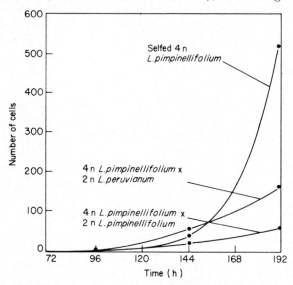

Fig. 13.2. Number of cells present in the endosperm of crosses involving tetraploid *Lycopersicon pimpinellifolium* as the female parent is plotted against time after fertilization. (After Cooper and Brink, 1945)

between *Hordeum jubatum* and *Secale cereale* (Cooper and Brink, 1944) and *Saccharum officinarum* and *Zea mays* (Hrishi *et al.*, 1969), have verified this in sufficient detail. The varied forms which the endosperm takes throughout these crosses make it impossible to describe the histological aspects of their development. Generally, the initial rate of development of the endosperm in inviable crosses does not differ appreciably from that observed in successful crosses, but afterwards the endosperm of the successful cross dramatically outstrips the other in its growth. This is shown by a time course study of increase in cell number in the endosperm following matings involving *Lycopersicon pimpinellifolium* as the female parent (Fig. 13.2). Cooper and Brink (1945) observed that in the cross between a diploid strain of *L. pimpinellifolium* and *L. peruvianum*, at about 144 h after pollination, cells of the endosperm at the chalazal end became vacuolate and less dense than those at the micropylar end. Although the endosperm survived for some more time, it underwent no further divisions. Instead, as a result of dissolution of the cell wall and fusion of the protoplast and nuclei with those of contiguous cells,

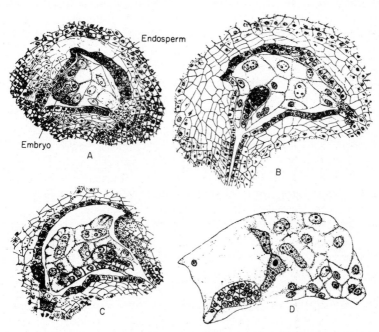

FIG. 13.3. Development of the embryo and endosperm in a 2N *Lycopersicon pimpinellifolium* × *L. peruvianum* cross. **A**, At 144 h, endosperm and surrounding cells at the chalazal end look impoverished of cell contents. The endothelium on the dorsal surface of the endosperm is conspicuous. **B, C**, Same at 192 h, with cells of the endosperm acquiring giant nuclei. **D**, Breakdown of endosperm and fusion of nuclei at 288 h. (After Cooper and Brink, 1945)

the endosperm became reduced to a few giant cells surrounding the embryo (Fig. 13.3). The first visible sign of deterioration of the endosperm in inviable crosses in *Datura* is the enlargement of cells and the formation of islands of large cells scattered throughout the tissue. During further deterioration of the endosperm these cells acquire pycnotic nuclei and granular cytoplasm. The cell walls eventually disappear and the endosperm collapses before the appearance of aleurone grains (Rietsema and Satina, 1959). Not infrequently, cells of an unhealthy endosperm might possess mitotic abnormalities such as lagging chromosomes, fused nuclei and nuclei of high or low chromosome number, unusual sizes, abnormal structure and density, and highly irregular shapes (Renner, 1914; Landes, 1939; Boyes and Thompson, 1937; Cooper and Brink, 1944, 1945; Thompson and Johnston, 1945; Beaudry, 1951; Brock, 1954, 1955; Lenz, 1956; Weaver, 1958; Rietsema and Satina, 1959; Pundir, 1972). Embryo degeneration may commence when the endosperm is still healthy or along with or after the disappearance of the endosperm. The extent to which disturbances in the endosperm in an incompatible cross make it incapable of supporting the growth of the embryo remains elusive indeed, although a frequent correlation of embryo abortion with the onset of endosperm deterioration observed in many plants indicates that it is due, in part, to the activity of the latter.

It has been claimed that endosperm failure in certain crosses might result from abnormal behavior of the antipodals. Brink and Cooper (1944) and Beaudry (1951) found abnormalities in the antipodals and a possible impairment of their function, respectively, in *Hordeum-jubatum* × *Secale cereale* and *Elymus virginicus* × *Agropyron repens* crosses. They postulated a mechanism of endosperm failure involving antipodals as intermediaries in the transport of food materials for the growth and division of the endosperm nucleus. Since some development of the endosperm occurs even in the presence of abnormal antipodals, it seems likely that malfunctioning of antipodals is a secondary rather than a primary effect. A point of view has evolved that the presence of a growing embryo weakens the endosperm and induces abnormalities in its development. In 2x × 4x crosses in *Citrus*, it is frequently observed that many embryoless seeds have normal endosperm, while in those containing an embryo, the endosperm invariably degenerates (Esen and Soost, 1973). Chromosomal imbalance is believed to result in physiological changes in the embryo, leading to the breakdown of the endosperm. Similar observations have been made in seeds derived from *Gossypium hirsutum* × *G. arboreum* crosses (Weaver, 1957). It is not known, however, how the developing embryo provokes the inhibition of growth of the endosperm and how the nutritional demands of the embryo are met in the absence of the endosperm.

The abnormal behavior of the somatic tissues of the ovule following pollination and fertilization has also engaged the attention of embryologists to explain the causes for embryo abortion in inviable crosses. For example, a weak endosperm development in inviable crosses in *Oenothera* is associated with the proliferation of a tissue from the nucellus (Renner, 1914). Intrusive growth of the nucellar tissue has also been described in interspecific crosses in *Epilobium* (Michaelis, 1925), *Nicotiana* (Kostoff, 1930) and *Medicago* (Ledingham, 1940).

In a comparative histological examination of ovules in crosses between *Nicotiana* species, Brink and Cooper (1941) emphasize the differences in the extent of development of the endosperm, embryo, and somatic tissues as being crucial for the success of any particular cross. The conceptual framework of this comparison is that the continued development of the ovule following fertilization demands a balance in the rate of growth of the endosperm and the adjacent somatic tissues. In this work it was found that following matings between *N. rustica* and *N. tabacum* and between *N. rustica* and *N. glutinosa,* most of the hybrid seeds collapsed at various stages of maturity, whereas, upon being self-pollinated, *N. rustica* yielded a full complement of seeds. Details of the events leading to seed collapse in the species crosses vary somewhat, but are basically similar in both types of crosses. In summary, there was a retardation of growth of the endosperm, which attained only a fraction of the growth attained in the selfed series, and a pronounced hyperplastic growth of the nucellus. The term "somatoplastic sterility" is used to describe this type of hyperplasia of the maternal tissue. At the same time, the integumentary cells lying between the apex of the vascular bundle and the chalazal pocket failed to differentiate into conducting elements, foreshadowing a suppression of nutrient transport to the endosperm. By contrast, normal development of the seed in selfed *N. rustica* appeared to be closely related to the presence of a well developed endosperm, absence of nucellar hyperplasia and presence of conducting elements in the integument.

Reciprocal crosses between 4N *Lycopersicon pimpinellifolium* and 2N *L. peruvianum,* 2N *L. pimpinellifolium* and 4N *L. peruvianum* (Cooper and Brink, 1945), and *Arachis hypogaea* and *A. diogoi* (Johansen and Smith, 1956) show a type of tumor formation from the inner epidermis of the integument, the endothelium (Fig. 13.4). In *Lycopersicon* crosses, after fertilization, the endothelium becomes a densely cytoplasmic layer which surrounds the endosperm except for a small gap at the chalazal end. The tumor begins to proliferate near the chalaza in the dorsal region of the endothelium about 144 h after poilination, extending into the embryo sac as a voluminous tissue mass. The cells of the endosperm surrounding the tumor deteriorate progressively and eventually the embryo is surrounded by the overgrown endothelial tissue. The deterioration of the endosperm and growth of the endothelial tissue are associated, but distinct, events. Increase in tumor size and decrease in endosperm size do, however, serve the same end, namely, abortion of the already undernourished embryo.

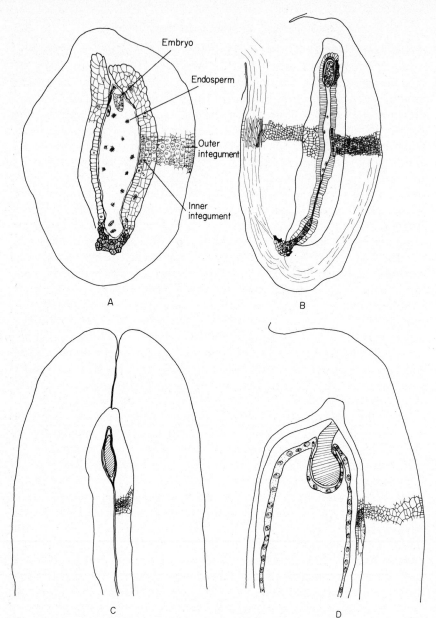

F IG. 13.4. Tumor formation from the inner integument in *Arachis hypogaea* × *A. diogoi* cross. **A**, Cell divisions in the inner integument in a young seed containing 12-celled embryo and 32-nucleate endosperm. **B**, A seed with 43-celled embryo and a chalazal cellular endosperm, 44 days after pollination. Hyperplastic growth of the inner integument is seen. **C**, Integumentary growth in a cross-pollinated seed of the same length as in D; embryo is undifferentiated and endosperm collapsed. **D**, *Arachis hypogaea* seed, 20 days after pollination. The embryo has a well differentiated suspensor, the endosperm is multinucleate, and the inner integument has two to three cell layers. (After Johansen and Smith, 1956)

The most spectacular type of tumor with regard to its final size and rapidity of growth is seen in both interspecific and diploid × polyploid crosses in *Datura*. Here the initial disintegration of the embryo or the endosperm is followed by enlargement of the endothelium and its division in various planes to form an overgrowth of several layers of cells in the embryo sac. As Fig. 13.5 shows, the proliferating tissue invades the embryo sac cavity leading to the final digestion of the endosperm (Sansome *et al.*, 1942; Sanders, 1948; Sachet, 1948; Satina *et al.*, 1950; Rappaport *et al.*, 1950a, b). The general course of growth of the endothelium in unsuccessful crosses between species of *Solanum*

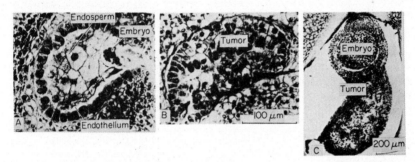

FIG. 13.5. Ovular tumor growth in incompatible crosses in *Datura*. **A, B,** *D. stramonium* × *D. metel*. Endothelial cells are enlarged in **A**, which also shows the embryo and endosperm (7 days after pollination); 11 days after pollination, the embryo sac is filled with tumor tissue (**B**). **C,** *D. metel* × *D. ceratocaula* cross, 29 days after pollination. The endosperm is entirely replaced by the tumor, which encircles an undifferentiated embryo at the micropylar end. (After Satina *et al.*, 1950).

(Beamish, 1955; Lee and Cooper, 1958) is very similar to that described for *Datura* crosses.

The mechanism by which the endosperm deteriorates in inviable crosses is much less debated than the role of the tumor in the process. The basic question is that of cause and effect. Is seed collapse, in a broader sense, the result of tumor formation in the embryo sac cavity? To begin with, we have to admit that the proposal that the endosperm disintegrates passively by the activity of a burgeoning tumor tissue has not been adequately tested. In crosses between species of *Datura* (*D. pruinosa* × *D. metel* and *D. metel* × *D. discolor*) where tumors are absent, endosperm deterioration accompanied by embryo abortion takes place shortly after fertilization. In other crosses (for example, *D. innoxia* × *D. discolor*), endosperm deterioration is complete before endothelial activity is seen. In still others (for example, reciprocal *D. stramonium* diploid and tetraploid crosses, *D. stramonium* × *D. metel*, and *D.*

stramonium × *D. innoxia* crosses), deterioration of the endosperm is accompanied by degeneration of the endothelial layer, preventing the latter from forming a tumor (Sansome *et al.*, 1942; Sachet, 1948; Satina *et al.*, 1950). Thus any general hypothesis concerning the role of the tumor must consider the fact that tumors fail to appear, or appear after the collapse of the endosperm has been initiated.

In certain instances it is difficult to characterize whether embryo abortion in inviable matings is due to the disintegration of the endosperm, or to the action of substances from the tumor itself on the embryo. Using an embryo culture assay, tumor tissue of the inviable *Datura innoxia* × *D. discolor* cross has been shown to possess a substance identical to IAA, which inhibits growth *in vitro* of embryos of selfed *D. stramonium* (Rietsema *et al.*, 1954). Despite the presence of the tumor, occasional embryos attained considerable proportions, casting doubt on whether flooding the embryo sac cavity with inhibitors from the tumor could completely account for the arrested growth of the hybrid embryo.

B. Culture of Hybrid Embryos

The analysis presented above leads us to conclude that without the continuous support of the endosperm, the embryo of an inviable cross cannot proceed with its normal development. In some species, the hostile environment of the ovule in the form of a tumor is probably responsible for retardation of embryo growth. The special nutritional relationship between the embryo and the endosperm tends to suggest that the latent capacity of a hybrid embryo for full development may be expressed in an artificial milieu supplied with exogenous nutrients conforming to the composition of the endosperm. The triumph of the embryo culture method is nowhere better seen than in the dramatic demonstration of the ability of embryos removed from seeds of inviable crosses to grow in culture. The present subsection is devoted to an account of the growth of embryos of inviable crosses in which investigators mainly employed the technique of embryo excision and culture, and is illustrated by examples drawn from interspecific and intergeneric crosses. Like all other methods of practical application, this method has its weakness as well as its strength, and it is recognition of this limitation that permits its most effective use.

Interspecific Hybrids

The pioneering work of Laibach (1925, 1929), who demonstrated that embryos of seeds obtained from crosses between *Linum perenne* and *L.*

austriacum which were traditionally condemned as being incapable of germination could be reared to maturity in culture, laid the foundation for a series of attempts to surmount barriers to crossability in plants' by excision and culture of embryos. This was followed by another report claiming limited success in culturing embryos from seeds of a cross between *Galeopsis (Ladanum) pyrenaica* and *G. (L.) ochroleuca* (Laibach, 1930). These results, when considered in the light of embryo–endosperm relationship, must serve as strong evidence for the view emphasized earlier that the immediate cause of failure of hybrid seeds is the lack of proper nutrition for their embryos. A number of successful attempts at culture of hybrid embryos during subsequent years reflect this view even more succinctly.

In a study mainly devoted to cytomorphology of heteroploid plants in *Solanum*, Jørgensen (1928) found that seeds of a hybrid between *S. nigrum* and *S. luteum* have a low percentage of survival. A large number of fruits dropped before maturation, and fruits which matured were empty or contained a few dried, shriveled seeds enclosing tiny embryos. Transplantable seedlings were obtained from seeds by culturing embryos in Knop's solution for several days.

The prospect of a widespread application of embryo culture technique in overcoming barriers to crossability is provided by attempts to culture embryos of horticultural varieties of deciduous fruits which ripen early but have a low percentage of viable seeds (Tukey, 1934b; Blake, 1939; Skirm, 1942; Danielsson, 1950). In breeding programs of fruit trees where heterozygous varieties are crossed, embryo culture is the most reliable method to preserve progenies of all potentially viable seeds which may have characteristics of horticultural value when they mature. Lammerts (1942) showed that seeds of several early hybrids of apricot and peach which contain abortive embryos do not respond to stratification; however, embryo culture was the method of choice to obtain mature plants from such seeds. In other hybrids, plants raised by embryo culture proved superior to those raised by stratification with regard to earliness in flowering and number of flowers per tree. In crosses between distant genera and species of fruit trees, for example *Cerasus vulgaris* × *C. tomentosa* and *Ribes nigrum* × *Grossularia reclinata*, it was possible to raise a second generation of plants from seeds which do not normally germinate, by growing the aborted embryos under aseptic conditions (Kravtsov and Kas'yanova, 1968). Thus, embryo culture might help in the wide hybridization of plants and in large-scale commercial fruit-breeding programs by shortening the breeding cycle of deciduous fruits and giving a higher percentage of seed germination.

Attempts to produce a hybrid between American tetraploid and

Asiatic diploid cotton go back to the 1920s, and to date, cotton geneticists have not succeeded in hybridizing *Gossypium arboreum* $(2n=26)$ with pollen from *G. hirsutum* $(2n=52)$. This is somewhat surprising since there are numerous varieties of both species, and because of the economic importance of the crop there is an obvious need for production of hybrids that involve combinations of desirable features. As in other interspecific crosses, the main stumbling block appears to be the endosperm, which collapses about 15 days after pollination, resulting in retarded growth of the embryo. Further, only marginal success has been attained in culturing hybrid embryos. In one case, culture of embryos in White's medium yielded seedlings with rudimentary cotyledons (Beasley, 1940). In another case, when embryos from capsules of varying ages from 20 days after pollination to maturity were cultured, they showed variable growth, and only one normal seedling was obtained (Weaver, 1958). Differentiation of relatively small late globular or heart-shaped embryos into seedlings is reported to occur in a relatively complex medium containing casein hydrolyzate, GA and kinetin (Joshi and Pundir, 1966). In contrast, embryos of the reciprocal hybrid grow reasonably well on nutrient media and afford promise of forming vigorous seedlings if they are cultured at a stage large enough to be handled (Beasley, 1940; Weaver, 1957). Successful hybrids of the cross between *G. davidsonii* and *G. sturtii* were obtained by rearing embryos in culture (Skovsted, 1935).

Production of hybrid tomato with respect to disease resistance is another example of what can be gained by embryo culture. Many strains of wild tomato *(Lycopersicon peruvianum)* are resistant to mosaic virus, leaf mold, *Fusarium* wilt, *Septoria* blight, *Alternaria* blight, root knot nematode and spotted wilt. It is possible, but most difficult, to hybridize cultivated tomato *(L. esculentum)* with *L. peruvianum*. In the *L. esculentum* × *L. peruvianum* cross, although fruit development is normal, seeds often harbor underdeveloped embryos which do not readily germinate. By excision and culture of embryos, it is possible to raise healthy hybrid seedlings, some of which have been nurtured to the stage of flowering (Smith, P. G., 1944; Choudhury, 1955c; Alexander, 1956). The margin of success in these cultures is very narrow, which perhaps emphasizes a need for improvements in the composition of the medium. In view of the general demand for disease-resistant tomato plants, growing large populations of the hybrid by this method will be a most significant contribution in any tomato breeding program.

Since barley crops suffer enormous losses from attack by mildew *(Erysiphe graminis* f. *hordei)* and spot blotch *(Helminthosporium sativum)*, the experimental approach to the problem of hybridization in barley

has been concerned with the transfer of genes conditioning disease resistance from wild barley to cultivated barley. Most species of wild barley do not cross readily with the cultivated species; nonetheless, the
former have desirable disease-resistance qualities. For this reason,
embryo culture technique has been resorted to in rearing hybrid progenies of barley. Many winter strains of cultivated barley, *Hordeum vulgare* and *H. sativum*, are deficient in winter hardiness and are susceptible
to attack by the mildew, whereas wild barley *(H. bulbosum)* is reasonably winter hardy and resistant to mildew attack. Embryos of the cross
H. sativum × *H. bulbosum*, which failed to grow in the ovule beyond 15
days after pollination, resumed normal growth when transferred to a
culture medium and formed healthy seedlings in soil (Konzak *et al.*,
1951). Interspecific hybrids have also been obtained in crosses between
H. vulgare, *H. bulbosum*, *H. brachyantherum*, *H. depressum*, *H. jubatum*, *H.
spontaneum* and *H. californicum* by culture of embryos (Morrison *et al.*,
1959; Davies, 1960). Embryo culture is also a promising method for
raising haploids which occur in high frequency in crosses between
diploid strains of barley (Kasha and Kao, 1970). In all of these crosses,
only a small percentage of the cultured embryos proceed to form normal
seedlings; apparently there is some unusual physiological condition
which limits growth of the developing hybrid embryos in culture. With
proper attention to nutritional requirements, the culture of hybrid
embryos could be improved enormously to assure a greater survival
rate.

Wild barley is also resistant to the spot blotch disease. To circumvent
difficulties in crossing wild barley with the cultivated species, interspecific hybrids have been reared between wild species to cross with
the cultivated species. Successful hybrids between *H. compressum* × *H.
pusillum*, and *H. marinum* × *H. compressum*, were obtained by culture of
hybrid embryos (Schooler, 1960a, b). Recently, culture techniques
have been further modified to raise seedlings from *H. vulgare* × *(H. compressum* × *H. pusillum)* and *H. vulgare* × *H. hexapodium* crosses (Schooler,
1962).

An immense effort in many laboratories has led to the study of species
relationship in the genus *Oryza* with the ultimate aim of evolving
hybrids tailored to withstand unfavorable environmental conditions,
offer resistance to diseases and pests, and give high yields. Indeed, the
current food situation in the world makes the demand for such hybrids
very significant. Although fertilization occurs normally in most of the
species crosses in the genus, failure of caryopses to produce viable
embryos has continued to be the principal obstacle in the way of large-
scale successful adoption of old crosses and attempting new crosses.

Niles (1951) found that when sterile kernels from crosses between cultivated varieties of rice were planted in solidified mineral salt medium,
embryos grew into transplantable seedlings. Certain difficulties recognized in rearing plants from crosses involving *O. sativa* as one of the
parents were also overcome by culturing embryos (Butany, 1958; Bouharmont, 1961; Nakajima and Morishima, 1958). In several other
interspecific crosses, supplementation of the medium with coconut milk
was necessary for growing hybrid embryos (Li *et al.*, 1961; Iyer and
Govila, 1964). The efficiency of embryo culture varies markedly with
the hybrid combination, and invariably the percentage of survival of
seedlings in some of the crosses was very low due to failure of embryos
to withstand transplantation in the soil. Transplantation of agar-
cultured embryos to a liquid medium before transfer to soil has led to
a marked improvement in the growth potential and survival of seedlings
and there are promising indications that seedlings so obtained may form
the basis for production of favorable lines of hybrid rice (Sapre, 1963;
Iyer and Govila, 1964). In crosses between tetraploid and triploid races
of maize, culture of embryos excised from young kernels is far more
efficient in securing F_1 plants than the ordinary method of pot germination of ripe seeds (Uttaman, 1949d).

An example of hybridization work on the jute plant illustrates the
success of embryo culture where all other methods had failed. Demand
for jute fibers centers primarily on two species, *Corchorus olitorius* and
C. capsularis, each of which has its own good and bad features. *C. capsularis* yields the white fiber of commerce, and being drought- and flood-
resistant, it can be cultivated in low lands. In contrast, *C. olitorius*, which
produces the stronger red jute, is relatively disease-resistant and
adapted to highland cultivation. The objective of jute breeders is to
produce disease-, flood- and drought-resistant hybrids which will yield
strong, white fiber, but because of the early abscission of pollinated
flowers or low fruit-set and premature abortion of the young embryo,
development of a hybrid with these characters by reciprocal crossing
between *C. capsularis* and *C. olitorius* has not been as successful as some
breeders had hoped. Cytological study of seed formation in *C. olitorius* ×
C. capsularis cross led to the conclusion that embryo abortion in the
young seed was due to impairment of the capacity of the endosperm
to grow into a cellular tissue (Ganesan *et al.*, 1957). The technique of
using reciprocally grafted plants as parents has provided a way to increase the percentage of fruit-set in *C. olitorius* × *C. capsularis* hybrid
(Sulbha and Swaminathan, 1959). Application of IAA to pedicels of
the pollinated flowers from *C. olitorius* × *C. capsularis* cross not only
increased fruit-set, but it also counteracted the degenerative processes

in the embryo and endosperm (Islam and Rashid, 1961). As a result a few fully developed seeds with normal embryo and endosperm which germinated to form F_1 plants were obtained. A slight increase in fruit-set was obtained in the reciprocal cross, but seeds germinated with diffi-culty. However, transplantable seedlings were obtained if hormone application of the pedicel was combined with culture of embryos in a medium containing yeast extract, IAA and kinetin (Islam, 1964).

There are other promising developments in the hybridization of crop plants that may link the eventual success of the program with culture of embryos or other reproductive parts. Wall (1954) found that embryo culture was superior to direct planting of seeds in soil for the production of vigorous seedlings of reciprocal crosses between *Cucurbita pepo* and *C. moschata*. By the same technique, viable seedlings between Chinese cabbage and cabbage *(Brassica pekinensis* × *B. oleracea)* have been ob-tained for use in breeding work to produce a Chinese cabbage hybrid with the disease-resistant qualities of cabbage (Nishi *et al.*, 1959). In crosses between species of *Abelmoschus* where embryo abortion occurred at the heart-shaped or torpedo-shaped stage, both embryo culture and ovule culture were used to raise hybrid seedlings. Limited numbers of viable seedlings were obtained from cultured embryos of *A. tubercu-latus* × *A. moschatus*, *A. esculentus* × *A. moschatus*, *A. esculentus* × *A. manihot* crosses and from cultured ovules of *A. esculentus* × *A. ficulneus* cross (Patil, 1966; Gadwal *et al.*, 1968).

Difficulties usually encountered in the breeding of legumes have given rise to numerous applications of embryo culture. As in the other examples considered above, the trend in all unsuccessful crosses in legumes has been toward embryo abortion, resulting in nonviable and physiologically sterile seeds. While several species of *Trifolium* behave as perennials under field conditions, the common forage species of the genus lacks this quality. Since hybridization between *Trifolium* species has been hampered by embryo abortion, embryo culture was attempted to obtain hybrids combining perennial habit with forage quality. By this means, sterility obstacle was overcome in the 2N *T. ambiguum* × 2N *T. hybridum* cross and to a lesser extent in 2N *T. nigrescens* × 2N *T. repens* and 4N *T. repens* × 4N *T. nigrescens* crosses (Keim, 1953b). Later, using plants of different ploidy levels, Evans (1962) extended culture methods to salvage embryos from reciprocal crosses between *T. repens* × *T. uniflorum*. Within the species combinations tried, seedlings from 14 successful crosses were perpetuated in culture, of which nine were grown in soil. Although a simple mineral salt medium such as Randolph and Cox's proved suitable for growth of embryos of successful crosses, embryos of other crosses failed to grow or grew abnormally, possibly because of specific deficiencies in the medium.

A particularly challenging aspect of the breeding work with legumes has been to produce a fine-stemmed, leafy sweet clover hybrid relatively low in coumarin, a compound which is harmful to cattle. The common cultivated sweet clover, *Melilotus officinalis*, has a high coumarin content and it has not been possible to hybridize this species with others. Some success was achieved in rearing hybrids between *M. officinalis* and a low-coumarin line of *M. alba* by culturing embryos before they aborted (Webster, 1955; Schlosser-Szigat, 1962). Since there are a number of wild species of *Melilotus* which possess properties lacking in the culti-vated species, embryo culture offers promise for improvement of this crop to yield not only low-coumarin hybrids, but also hybrids with in-creased nitrogen content for use as green manure, and resistance to sweet clover weevil attack. In recent years, species of *Lotus*, especially *L. corniculatus*, have been increasingly used as a forage crop, as a result of which interest has centered on the production of hybrids with charac-teristics such as indehiscent and soft seeds and disease resistance. In view of the difficulty experienced by earlier workers in experimentally hybridizing species of *Lotus*, due to sterility, hybrids from crosses between *L. corniculatus* and related diploid species and between *L. tenuis* and *L. uliginosus* were raised exclusively by embryo culture (Grant *et al.*, 1962; Davies, 1963).

Embryos of the cross *Phaseolus vulgaris* × *P. acutifolius* respond to culture in solidified White's medium with 4% sucrose by producing transplantable seedlings in about 60–70 days. An unexpected difficulty in nurturing seedlings in the greenhouse was overcome by initially cul-turing embryos in liquid medium and transferring seedlings of trans-plantable age to media containing stepwise lower concentrations of sucrose (Honma, 1955). In other similar studies, viable plants have been raised by embryo culture from interspecific crosses in *Medicago* species (Fridriksson and Bolton, 1963) and between *Lathyrus clymenum* and *L. articulatus* (Pecket and Selim, 1965).

Reports on breeding of gymnosperm hybrids by embryo culture are rare. Stone and Duffield (1950) have described a case of breeding hybrid seedlings from *Pinus lambertiana* × *P. armandi* and *P. lambertiana* × *P. koraiensis* crosses by planting embryos encased in the gametophytic tissue on agar slants. Since both *P. armandi* and *P. koraiensis* are blister rust resistant, these attempts constitute a significant step toward breed-ing a disease-resistant line. Successful rearing of excised embryos of *P. lambertiana* completely free of the megagametophyte (Haddock, 1954) makes it possible to grow hybrid embryos similarly.

A limited number of attempts have been made to propagate, by embryo culture, progenies of crosses between species of garden plants. Even before embryo culture technique was introduced to overcome delayed germination of *Iris* seeds (see p. 339), it was shown that malformed embryos of inviable crosses within the genus which would otherwise perish in the ovule could be grown normally in a solidified mineral salt medium containing sucrose (Werckmeister, 1934, 1936). In recent years, embryo culture has become the chosen method for propagating seeds of rare species and important hybrids of this genus (Lenz, 1954, 1956). Accomplishments in producing hybrid seedlings from *Lilium henryi* × *L. regale*, *L. speciosum* 'Album' × *L. auratum*, *L. speciosum* 'Rubrum' × *L. auratum* crosses are promising (Skirm, 1942; Emsweller and Uhring, 1962; Emsweller *et al.*, 1962). Techniques for rearing interspecific hybrids in *Chrysanthemum* have shown that successful hybrids originating from cultured embryos are possible between *C. boreale* (4N) and *C. pacificum* (10N) (Kaneko, 1957).

In investigations on the nature and evolutionary relationships of various genera of plants and speciation within them, chromosome analyses of interracial and interspecific hybrids occupy a prominent place, but securing hybrids has been a stumbling block for cytological studies. Nowhere is this more clearly illustrated than in *Datura*, a herbaceous genus of approximately 10 species. Especially interesting is *D. ceratocaula*, a semiaquatic species, endemic to Mexico, characterized by whorled branches, smooth capsules, hollow stems and reduced vascular tissues. Attempts to hybridize *D. ceratocaula* with any of the other species within the genus have failed because of the formation of abortive embryos in every case when *D. ceratocaula* was used as a parent. By excision and culture of embryos in a medium containing cold-sterilized powdered malt extract, McLean (1946) was able to grow to maturity seedlings from eight out of nine possible hybrids involving *D. ceratocaula* as the pollen parent and one involving *D. ceratocaula* as the female parent. Embryos from other inviable crosses involving *D. discolor*, *D. innoxia*, *D. stramonium*, and between *D. innoxia* and tree *Datura* (Brugmansia?) have also lent themselves to culture (Blakeslee and Satina, 1944; Sanders, 1950). The success obtained in hybridizing *D. innoxia* with the tree *Datura* is significant in view of the controversial taxonomic position of the latter.

Intergeneric Hybrids

Some interesting potentialities of embryo culture have been brought to light in hybridization programs involving species belonging to different genera. Cooper and Brink (1944) have given an account of the developmental behavior of a hybrid between *Hordeum jubatum* and *Secale cereale*. In this cross, although the embryo was normally formed and potentially functional, seeds ceased to grow from 6 to 13 days after fertilization, and thus did not attain a germinable condition. Practicability of rearing hybrids from this cross was demonstrated by dissecting embryos from 9–12 day old seeds and growing them in White's medium (Brink *et al.*, 1944). By the excised embryo method, hybrids have also been obtained from *H. californicum* × *S. cereale, H. vulgare* × *S. cereale, H. depressum* × *S. cereale* and *H. jubatum* × *Hordelymus europaeus* (Morrison *et al.*, 1959).

Wide hybridizations between cereal grains and their wild relatives have been attempted for a variety of reasons, the chief of which is the possibility of introducing genetic material which will be of practical value in improving the crop. Arising from the success or failure of the cross, and the degree of ease in accomplishing it, is the theoretical consideration of relating results to the origin of the crop plant and its taxonomic position. Ivanovskaya (1946) attempted a cross between *Triticum* and *Elymus* with a view to transmitting from the latter to the former such features as a strong root system and a higher grain yield. However, the caryopses formed were so small and malformed, and had such low viability, that it was not possible to raise hybrids except by embryo culture. Transplantable hybrid seedlings have thus been obtained from crosses between *T. durum* × *E. arenarius, T. durum* × *E. giganteus, T. compactum* × *E. giganteus, T. durum coerulescens* × *E. arenarius* and *T. vulgare* × *E. arenarius* (Ivanovskaya, 1946, 1962). Embryo culture method has also been applied to raise hybrid seedlings from crosses between *T. durum* var. *abyssinicum* × *Secale cereale* (Rédei, 1955). A problem with these hybrid seedlings was their sterility, which was overcome by colchicine treatment. In a cross between *Tripsacum dactyloides* and *Zea mays*, higher percentages of plants were brought to maturity by culture of embryos than by normal germination (Farquharson, 1957).

In another line of investigation, the embryo transplantation technique has been used to ensure success of intergeneric hybrids. This involves the use of female parents raised from embryos implanted in the endosperm of the species that contributes the pollen. By this means, a statistically significant increase in the crossability between wheat and rye was obtained if wheat plants used for the cross were derived from embryos nurtured on rye endosperm (Pissarev and Vinogradova, 1944; Hall, 1954). Another study showed that the endosperm of a variety of wheat which offered no crossability barrier with rye had the same effect as rye endosperm on the embryo of a variety of wheat that was difficult to cross with rye (Hall, 1956). In crosses between wheat and

Elymus, best results were obtained when both parents were raised from embryos grafted onto the opposite endosperm types (Pissarev and Vinogradova, 1944). The method of embryo transplantation in general appears to be of limited value since transplantation of certain diploid, tetraploid and hexaploid varieties of wheat on wheat or rye endosperm did not improve their crossability with plants of the donor endosperm type (Rommel, 1960). A modified technique in which hybrid embryos are implanted on cultured maternal endosperm has yielded high percentages of survival of embryos of barley and wheat crosses (Kruse, 1973, 1974).

From this survey of the application of excised embryo method in overcoming the crossability barrier in hybrids, it is obvious that an effective way of raising hybrids in numbers required for field trials has not been achieved. In a considerable measure, this reflects the inherent practical difficulties of applying sterile culture techniques for large-scale field applications, and the lack of a broad systematic program to achieve this. Considering that great economic gains will be realized from the evolution of successful hybrids for the breeder, a more rapid tempo of investigation is called for in this direction.

II. SEED DORMANCY AND LOW VIABILITY OF SEEDS

The phenomena of dormancy and low viability of seeds pose the kinds of problems for which the embryo culture method is suitable. In this area, application of the method may be concerned with shortening the breeding cycle of plants and with overcoming dormancy itself. Seeds in which dormancy cannot be broken by any of the known methods and which germinate only after a rest period make ideal objects for embryo culture. The use of embryo culture in overcoming seed dormancy is often equated with the work of Randolph (1945) on tall bearded *Iris* seeds, which require a period of dormancy varying from a few months to many years. Recognition of the need for an effective method of breaking dormancy of seeds and obtaining flowering of plants in a shorter period of time than normal led to attempts to study seedling production by embryo culture. Within 2–3 months after culture, young seedlings with well developed roots and leaves appeared in culture tubes, ready to be transplanted into soil. The application of embryo culture has done much to reduce the cycle from seed to flowering in *Iris* to less than a year in contrast to the 2–3 years normally required for flowering, and its full potential has by no means been reached.

It is significant that although excised embryos of *Iris* grow readily in culture, growth is inhibited when they are cultured with even a small portion of the endosperm intact (Randolph and Cox, 1943; Werckmeister, 1962). This has led to the view that a stable inhibitor of embryo growth is present in the endosperm. Embryos excised from species of *Iris* belonging to the subsection *Hexapogon* or from interspecific hybrids where the female parent is a member of this subsection fail to grow in nutrient agar under conditions suitable for growth of embryos of tall bearded *Iris* (Werckmeister, 1952; Lenz, 1955). Here the inhibitory substance probably resides in the embryo itself since supplementation of the medium with an extract of the *Hexapogon* type of embryo inhibits growth of a normal embryo. In the seed of *I. douglasiana*, the inhibitor is present in the seed coat; addition of a leachate of the seed to the culture medium was highly inhibitory to growth of embryos of other *Iris* species which grow normally in an unsupplemented medium (Lenz, 1955). The importance of the role of inhibitors in the embryo, or in the seed coat, in regulating dormancy of seeds of other plants has also been emphasized by embryo culture studies (Cox *et al.*, 1945; Bulard and Degivry, 1965; Villiers and Wareing, 1965a; Dore Swamy and Mohan Ram, 1967; Bradbeer, 1967; Le Page, 1968; Le Page-Degivry, 1970a; Le Page-Degivry and Garello, 1973; Bewley and Fountain, 1972; van Staden *et al.*, 1972a). The relative ease of extraction and culture of embryos from oil palm seeds has made it possible to examine in detail a range of factors involved in dormancy, in particular the age of seeds (Rabéchault and Ahée, 1966), water content of seeds (Rabéchault, 1967; Rabéchault *et al.*, 1969) and the duration of soaking (Rabéchault *et al.*, 1968).

Seeds of some of our common cereals remain dormant immediately after harvest, even though embryos of dormant seeds are fully mature and are capable of protein synthesis as well as those of nondormant ones (Chen and Varner, 1969, 1970). If seeds are stored at dry temperatures, they gradually overcome dormancy and begin to germinate. In certain types of seeds such as those of wild oat *(Avena fatua)* which do not attain full germinability until after several years of storage, this type of dormancy may pose a major problem to the planter. Culture of embryos isolated from dormant seeds of wild oat at different times after harvest has shown that a major defect of completely isolated embryos is overcome by supplementation of GA in the medium (Naylor and Simpson, 1961a; Simpson, 1965). Considering that the metabolism of reserve carbohydrates of the endosperm into simpler compounds is of paramount importance for germination of the seed, it seems likely that release from physiological dormancy in wild oat is due to the synthesis of GA which promotes the formation of necessary enzymes for the hydrolysis of the endosperm reserves. From a practical point of view, culture of embryos of dormant grains in appropriate hormone-supplemented media might be advantageously used to raise a new crop of seedlings from grains immediately after harvest.

In breeding practice, when dormancy of seeds and slow growth of seedlings necessitate long breeding seasons, the embryo culture method is of value to the breeder in reducing the breeding cycle of new varieties. Cultivated varieties of rose generally take about a year to flower, and 2–3 months for the formation of fruits. Although excision of the embryo from the seed is somewhat tedious, seedlings originating from cultured embryos flower in 2–3 months. These flowers can serve as the male parent for further crosses, thus enabling the breeder to produce two generations in one year or shortening the breeding cycle to 3–4 months (Lammerts, 1946; Asen, 1948). The value of embryo culture in circumventing the slow germination of the seed and slow growth of the seedling is also illustrated in weeping crabapple (Nickell, 1951). In this plant, by embryo culture, seedlings about 4 ft tall are obtained in about 9 months, which is, incidentally, the time it takes for seeds to germinate in the soil.

There is a hard core of information on the germination of seeds of some early-ripening fruit trees which have very low viability and which fail to germinate even after appropriate after-ripening treatments. Tukey (1934b) has aptly stated that "were it not for the horticultural practice of budding and grafting, many of our finest sorts [of early ripening deciduous fruit trees] would be lost to cultivation". In general it appears that embryos excised from such seeds grow normally *in vitro* into healthy plants. It is thus possible to preserve these varieties for development of still earlier ripening fruit characteristics which would be of practical application to fruit breeders.

Certain varieties of sweet cherry which ripen in less than 50 days after full bloom either abort or fail to complete development of their embryos, and are thus nonviable. Embryos from seeds of such fruits germinate when cultured *in vitro* and produce normal plants (Tukey, 1933a). Embryos of other early-ripening deciduous fruits behave in a similar manner in culture, the initial size of the embryo being the only limitation for growth in culture (Tukey, 1934b, 1938; Davidson, 1933, 1934). Often excised embryos of progressively earlier-ripening varieties show decreased growth potential in culture. Storage of fruits at a low temperature for a certain period can bring about growth of embryos excised from such fruits (Lesley and Bonner, 1952; Hesse and Kester, 1955; Kester and Hesse, 1955). As we shall see in Chapter 16, dormancy of seeds which require light or cold treatment for germination has been overcome by the embryo culture method. Such studies have helped to localize endogenous inhibitors or promoters in the embryo which are responsible for maintaining seeds in the dormant or nondormant state.

The possibility of raising seedlings of crop plants which are traditionally propagated vegetatively has been explored by embryo culture method. The classical example of this category of plants is the banana, of which there are many seeded varieties. *Musa balbisiana* is a wild relative of the commercial banana, the seeds of which do not germinate in nature; however, if embryos are excised and grown in culture in a simple mineral salt medium, seedlings are readily obtained (Cox *et al.*, 1960). With appropriate modifications, this method could be applied to obtain seedlings from other varieties of banana. Similarly, the tuber crops *Colocasia esculentum* and *C. antiquorum* are propagated only by vegetative means, and seeds are never known to germinate in nature. The natural sterility barrier in the seed could be overcome by resorting to culture of embryos (Abraham and Ramachandran, 1960). Although these results are preliminary, the success of the embryo culture method with these plants opens up new avenues for their improvement by interspecific and intervarietal hybridization.

In seed-testing practice, the embryo culture method has figured as a rapid means of determining the viability of particular lots of seeds. Tukey (1944) found a good correspondence between growth of excised embryos of non-after-ripened peach seeds and germination of after-ripened seeds, and suggested culturing embryos from randomly selected mature seeds as an accurate method of testing the germination of after-ripened seeds. By this means the planting value of germination of any seed lot can be obtained several weeks earlier, while it might take months for completion of a normal germination test including the after-ripening treatment. From a practical point of view, nurserymen could use this as a quick test for predicting the viability of the current season's supply of seeds for specific planting dates and thereby eliminate planting failures resulting from the use of seeds of low viability and expedite commercial movement of seeds of known germinability. In testing samples of maize of local origin in India, the growth of seedlings originating from shriveled grains was found to be similar under normal germination practice as well as by the embryo culture method; this perhaps justifies the application of the culture method to test the viability of samples of food grains for seed purposes (Mukherji, 1951). By the embryo culture method it has also been demonstrated that cold storage at freezing point will preserve the viability of relatively young immature embryos of maize (Uttaman, 1949a). A wider application of the embryo excision method in seed-testing practice has been predicted as a result of the success by which the degree of viability of seeds of several conifers, broad-leaved trees, vegetable and flower plants, shrubs, vines and fruit trees was accurately determined by this method (Heit, 1955). The possible role of endosperm factors in the viability of wheat grains aged for several years has been studied by embryo culture (Aspinall and Paleg, 1971).

III. OTHER APPLICATIONS OF EMBRYO AND OVULE CULTURE

The above two sections have passed in review examples of possible applications of embryo and ovule culture methods in horticulture and plant-breeding investigations. Additionally, the special attributes of tissue culture methods promise to be useful in studying some very fundamental problems of embryo and ovule growth. The growth of individual parts of an ovule reflects an integrated response to a nexus of interacting factors of environmental and hormonal origin, which by influencing internal physiological processes contribute variously to the formation of the adult organs. As growth of any particular part of the ovule is controlled by these interdependent intercellular and extracellular interactions, it is extremely difficult to appraise the specific growth requirements of individual parts, and of the contributions of individual factors. It is from this point of view that studies on culture of embryos, ovules, ovaries and whole flowers, severed from all organic contact with the parent plant, attain their greatest theoretical interest. The progress of sophistication in culture methods, and the increasing familiarity with their use, have made it feasible to study the growth requirements of progressively younger embryos, the effect of environmental factors and hormonal substances in the organogenesis of embryos, and the nutrition and metabolism of embryos. Results of studies on the culture of embryos of stem and root parasites, insectivorous plants, and of seeds with underdeveloped embryos have given valuable clues to their morphogenetic potential. The demonstrated ability of orchid embryos to grow in culture in the absence of a symbiotic fungus has led to an unprecedented utilization of this method to produce a steadily increasing number of hybrids. All of these studies have been reviewed in sufficient detail in earlier chapters and so will not be repeated here. Suffice it to state at this point that these studies arouse the feeling that it is possible to relate embryonic growth more closely to intercellular and extracellular stimuli by applying techniques commonly used in tissue culture. In other developments the use of cultured embryos offers promise of future in studying host–pathogen interactions, for example ergot formation on rye by *Claviceps purpurea* (Tonolo, 1961) and *Fusarium* wilt of bean (Padmanabhan, 1967).

In biochemical investigations, the use of embryos having adult organs present as primordial structures has certain advantages. Apart from the fact that nutrition of isolated embryos is more readily controlled than when they are in the seed, metabolic and enzymic changes take place earlier than in whole seeds. Consequently, isolated embryos have been the chosen material for a number of biochemical studies related to metabolism of nucleic acids, proteins, carbohydrates and respiratory enzymes during the initiation of developmental potencies such as occurs at germination. Although freshly excised embryos of cereal grains and pulses have been used frequently, the most favored embryo material for these studies has been commercial wheat germ. Since commercial wheat germ has serious drawbacks for physiological research due to inviability, methods have been developed for mass isolation of intact wheat embryos giving high germination (Johnston and Stern, 1957).

Using cultured ovules, a particularly promising line of research has recently been initiated to study the physiology and biochemistry of formation of cotton fibers. Cotton fiber is the single-celled extension of the epidermal cells of the ovule. Generally, not all cells of the epidermal layer produce fibers, and not all fibers elongate to produce long lint fibers. Thus here is an immense area of study of epidermal cell differentiation and factors involved in fiber elongation under strictly controlled conditions. Preliminary experiments have indicated a favorable effect of GA and IAA in promoting fiber development in fertilized (Beasley, 1971, 1973; Beasley *et al.*, 1971; Beasley and Ting, 1973) or unfertilized ovules (Beasley and Ting, 1974). Further studies on the culture of unfertilized ovules have shown a requirement for boron in fiber elongation in response to IAA, thus suggesting a basic relationship between boron deficiency and fiber quality (Birnbaum *et al.*, 1974). By establishing precise cultural conditions to study the physiology of fiber elongation and primary cell wall formation, it will be possible to gather the information necessary to increase the yield and quality of cotton fibers.

IV. COMMENTS

The survey presented in this chapter does not afford us with more than an entry into some of the possible ways by which embryo culture techniques may be applied. Attempts at adapting the culture method to solve the problems of inviability of hybrid seeds in large-scale field trials have not been notably successful owing to the technical difficulties involved. While the method thus presents its own difficulties, there is also the inherent challenge here because of the far-reaching economic gains to be accrued from successful field applications. In short, seen in this way, embryo culture is an area ripe for exploitation, investigation and application.

Section II

Adventive Embryogenesis

14. Adventive Embryogenesis: Induction of Diploid Embryoids

Although the embryo is usually derived from the product of fertilization of the egg by a sperm, it may also arise from the cells of the sporophyte or gametophyte by means not involving sexual union. Embryos which arise from the vegetative or reproductive cells of the plant by a developmental pathway that is an alternative to sexual union are known as adventive embryos and the process itself is referred to as adventive embryogenesis or adventive embryony. In the more recent botanical literature, however, terms like embryoids and pseudoembryoids are used to refer to such embryo-like structures. Embryoid is the preferred term used here to describe an asexually produced bipolar structure which lacks a vascular connection with the mother tissue and which resembles a zygotic embryo or functions like a zygotic embryo in giving rise to a new plant (Vasil and Hildebrandt, 1966a; Haccius, 1971; Haccius and Lakshmanan, 1969). Origin of embryoids from single cells has been suggested as an additional criterion (Street and Withers, 1974); since this has not been unequivocally demonstrated in many cases where perfectly normal bipolar embryoids are formed, this criterion is excluded from the usage of the term in this chapter. Implicit in the definition of an embryoid adopted here is the fact that meristems and growing points found in organized callus cultures which give rise to shoot buds are not embryoids. Foliar buds will also not be considered as embryoids as they have well established vascular connection with the mother tissue.

The investigations in adventive embryony that might be logically considered in this chapter are numerous and diverse in nature. They encompass the spontaneous development of embryoids from vegetative and reproductive cells of the plant, regeneration of embryoids by physical and chemical treatments, and embryoid formation in cultured organ and tissue explants and suspension cultures. Owing to the diversity of the subject matter, I plan to be selective in this chapter, emphasizing areas in which potentialities for further work exist. The intent is to dramatize the recent innovations in techniques that have altered the strategy of our approach to the study of adventive embryogenesis and to convey a sense of the accomplishments of the past and the outstanding problems of the future. Formation of haploid embryoids from microspores by anther culture is considered in Chapter 15.

Since the ability to form adventive embryos has radiated successfully into a number of angiosperms, members of this group exhibit much more variability in the manner of origin of embryoids than any other group of vascular plants.

I. SPONTANEOUS ORIGIN OF EMBRYOIDS

A. Origin of Embryoids from Vegetative Cells

Asexual reproduction, which is the basis for vegetative propagation, enables plants to regenerate new offsprings and is due to the latent meristematic activity of cells of certain parts of the plant. The methods of vegetative propagation are many, and clonal propagation of plants by runners, offshoots, stem tubers, tuberous shoots, corms, and leaf and stem cuttings are all based on localized meristematic activity of the dif-

ferentiated cells of the plant. However, embryo-like stages are not involved in the formation of new plantlets by these methods and so they
will not fall within our preview.

There are nevertheless some well documented cases in which spontaneous meristematic activity of mature differentiated cells results in
the development of specialized structures resembling zygotic embryos.
The occurrence of apomictic reproductive structures on the leaf of the
bog orchid, *Malaxis paludosa* (Taylor, 1967), is of interest in that in their

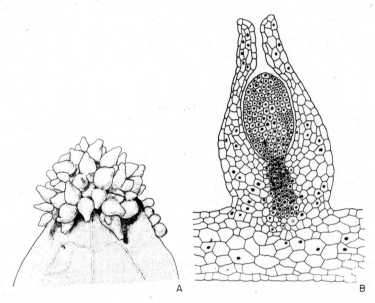

Fig. 14.1. Apomictic embryos formed on the leaf tip of *Malaxis paludosa* look very much
like zygotic embryos. **A**, Leaf showing a cluster of embryoids. **B**, A section through
the embryoid showing the jacket layer of cells. (After Taylor, 1967)

final form they look very much like zygotic embryos (Fig. 14.1). In this
plant, the mature leaf becomes meristematic and produces numerous
flask-shaped embryoids partially enclosed within a jacket layer of cells
which themselves may become active and produce new embryoids. Unlike the zygotic embryo, there is no vascular connection between the
embryoid and the leaf tissue. A new plantlet is formed directly from
the embryoid when it is separated from the leaf and begins to germinate
on a suitable substratum. In striking contrast to examples of embryoid
formation to be considered later in this chapter, the bog orchid is a
clear case in which the potentiality of vegetative cells to regenerate
embryoids is expressed within the organization of an intact plant.

B. Origin of Embryoids from Reproductive Cells

In angiosperms, the most common source for the spontaneous origin of embryo-like structures is the synergids, which become egg-like and develop into embryos with or without fertilization. Occasionally, embryo-like structures doubtless do originate from antipodals or endosperm nuclei through irregularities in the fertilization process. There is little or no evidence that embryos formed from the accessory cells of the embryo sac ever reach maturity. Embryonic abnormality such as the formation of multiple embryos by cleavage of the zygote or suspensor is a common feature of gymnosperms, but is less frequent in angiosperms. Regardless of their origin, the occurrence of more than one embryo in the ovule denotes polyembryony; developmental aspects of the different types of polyembryony in angiosperms have been reviewed in depth by P. Maheshwari and Sachar (1963).

The cells of the integument and nucellus of some plants (for example, species of *Citrus, Mangifera, Eugenia*, etc.) form perfectly normal embryos routinely as a part of their developmental cycle. In *Citrus*, the number of embryos formed per seed is a function of the age of the plant, nucelli of adult plants tending to be more embryogenic than those of young plants (Furusato *et al.*, 1957). The conversion of the diploid cells of the ovular tissue into a diploid embryo does not necessarily involve the complex sequence of events that occur with fixed chronology in the formation of the embryo following the fusion of the egg and sperm. Two features characterize the formation of accessory embryos from the sporophytic tissues of the ovule. First, cells destined to form embryos have a dense cytoplasm and they actively divide to form small groups of embryonic cells. Second, as the cells divide, they gradually push their way into the embryo sac cavity where they compete with each other and complete their development in a chaotic environment (Fig. 14.2). Yet a sharp demarcation of adventive embryos from zygotic embryos cannot be drawn except by their lateral position and lack of a well defined suspensor.

Depending upon the species, the appearance of adventive embryos may be abrupt, or may be conditional upon successful pollination and fertilization. In the former case, the basic stimulus to division of the nucellar cells to form embryoids is probably furnished by substances released by the surrounding degenerating cells, much in the way that a wound hormone acts (Haberlandt, 1922). In support of this thesis, Haberlandt (1921) tried to induce adventive embryos in *Oenothera lamarckiana* by pricking ovules with a needle or squeezing them gently. The results of this promising approach did not gain much acceptance in view of the limited number of ovules which positively responded to the treatment and a subsequent failure to reproduce the results (Beth, 1938). Activation of nucellar cells by discharge from the pollen tubes has been claimed to induce the formation of adventive embryos in fertilized ovules of some plants (see Maheshwari, P. and Rangaswamy, 1965), but there is no hard evidence to support this.

Although attempts have been made to achieve artificial induction of adventive embryony by application of hormones, the results have

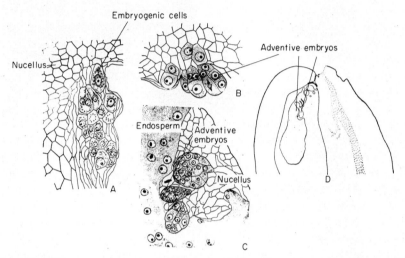

FIG. 14.2. Different stages in the development of adventive embryos from the nucellus of *Mangifera*. **A**, *M. indica* var. 'Olour', portion of the nucellus showing embryogenic cells. **B**, *M. indica* var. 'Higgins' showing young adventive embryos. **C**, *M. odorata*, portion of the ovule showing nucellus, adventive embryos and endosperm nuclei. **D**, *M. indica* var. 'Olour', upper part of the ovule with many adventive embryos. (After Maheshwari, P. and Rangaswamy, 1958)

been uniformly disappointing. Injection of ovaries of *Datura stramonium* (van Overbeek *et al.*, 1941b) and *D. fatuosa* (Chopra and Sachar, 1957) with solutions of IBA and NAA has been shown to result in the formation of tumorous masses of richly cytoplasmic cells which fail to differentiate into embryoids. According to Fagerlind (1946), smearing decapitated pistils of unfertilized ovules of *Hosta* with lanolin paste containing auxin led to production of young adventive embryos whose growth, after a transient increase, came almost to a standstill. In contrast, the use of hormones was found to be relatively effective in eliminating adventive embryos from ovules of *Citrus natsudaidai* and *C. unshiu* (Furusato, 1953).

The idea that nucellar tissues of normally polyembryonic species of *Citrus* can yield a continuous supply of genetically uniform adventive embryos has gained credence from the work of Rangaswamy (1958a, b, 1961; Maheshwari, P. and Rangaswamy, 1958). When a piece of nucellus excised from fertilized ovules of *C. microcarpa* was cultured in a medium containing casein hydrolyzate, it proliferated and formed a callus. The callus apparently contained all the elements of a highly

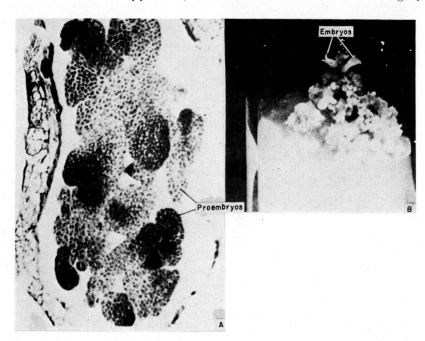

Fɪɢ. 14.3. Formation of pseudobulbils and plantlets in explanted nucelli of *Citrus reticulata*. **A**, Section of the ovule 15 days after inoculation showing proliferation of adventive embryos from the nucellus. **B**, Pseudobulbils appearing on the callus. (**A** Sabharwal, unpublished; **B** after Sabharwal, 1962; photographs courtesy of P. S. Sabharwal)

organized seedling, since it formed a large number of embryo-like regenerants termed pseudobulbils which eventually formed plantlets. The well documented behavior of explanted nucelli of *C. aurantifolia* and *C. reticulata* (Sabharwal, 1962, 1963) also conforms to the pattern described in *C. microcarpa* (Fig. 14.3). In both *C. aurantifolia* and *C. reticulata*, it has been claimed that the callus is intrinsically embryogenic, since it is formed from cells of the nucellus destined to form embryos. More recently, adventive embryos have been successfully induced in explanted nucelli originating from unpollinated ovules and monoembryonic and embryoless varieties of *Citrus* and some other genera of the Rutaceae (Singh, 1963; Rangan *et al.*, 1968; Button and Bornman, 1971; Mitra and Chaturvedi, 1972; Kochba and Spiegel-Roy, 1972; Kochba *et al.*, 1972; Button *et al.*, 1974; Murashige, 1974). The embryoids arose either directly from the nucellus or from a callus which originated from the nucellus. On the basis of these results, the prospects seem bright for the use of nucellus culture as a standard procedure for clonal propagation of *Citrus*, especially to establish virus-free clones (Bitters *et al.*, 1972).

The above examples, although restricted to a single family, nevertheless show that excision and culture of the nucellus in an enriched medium alter its normal growth pattern to achieve an unlimited type of growth. This was indeed part of the background which led to the realization that in the absence of correlative influences from their neighboring cells, somatic cells of angiosperms enter a regenerative phase and behave like embryos.

II. REGENERATION OF EMBRYOIDS BY CHEMICALS

The selective use of chemicals to inhibit growth in certain parts of the embryo and to promote growth in other parts has been exploited to induce adventive embryony. The ability of embryos to respond to chemicals is conditioned by their developmental stage: they are particularly sensitive during certain periods of early embryogeny. Seeds of *Eranthis hiemalis* have attracted considerable attention for this study because at the time of shedding they contain undifferentiated embryos of no more than a few hundred cells. With suitable care, test solutions can be injected into seeds which are then planted in the soil. Alternatively, seeds are soaked in the test solutions and planted or solutions are applied to seeds after planting. Irrespective of the mode of treatment, modifications in the morphology of embryos are observed as they mature in the soil. When seeds are treated with NAA, 2,4-D or 2,4,5-trichlorophenoxyacetic acid (Haccius, 1955a, b), the commonest type of adventive growth is the appearance of multiple cotyledons in the embryo. Less obvious types of variations clearly do exist and one that may represent a stepping stone in the transition to polyembryony is the formation of twin embryos. These are often fused at the base and have presumably originated by regeneration of the undifferentiated embryo. Haccius (1955b) has attempted to project these findings into a more general format by analogy to the well known phenomenon of apical dominance where the main shoot inhibits the growth of the side branches. Since the chemicals damage the apical

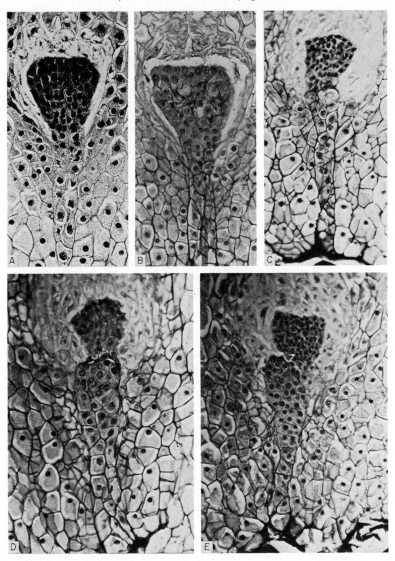

Fig. 14.4. When embryos of *Eranthis hiemalis* are damaged by exposure to low pH (3·4–4·5), the cells of the suspensor become active and regenerate new embryos. **A**, Undifferentiated embryo of the mature seed. **B–E**, Acidity-damaged embryos, 8, 14, 18 and 22 days, respectively, after beginning of treatment. **F–I**, Regeneration of the new embryo from the suspensor cells which have survived the acid treatment, 35, 42, 56 and 70 days, respectively, after beginning of treatment. (After Haccius, 1963; photographs courtesy of B. Haccius)

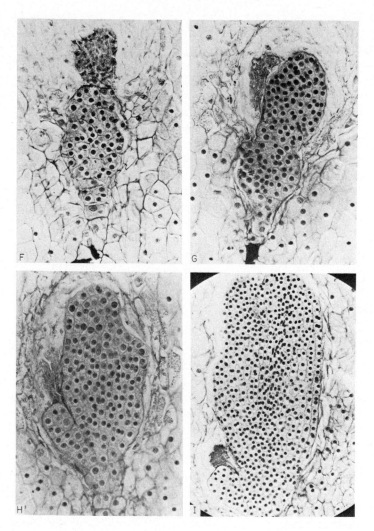

meristem of the undifferentiated embryo, the cells from which the future cotyledons arise grow into new embryos. The degeneration of the meristematic cells of the embryo is then but the first step in an unfolding sequence of events that result in multiple embryos.

With some treatments deep-seated degenerative changes occur in the embryo with corresponding variations in adventive growth. For example, embryos of freshly harvested seeds were found to be damaged and arrested in growth by colchicine (Haccius, 1957b), maleic hydrazide (Haccius, 1957a, 1959b), isopropyl-phenylcarbamate (Haccius,

1959b), low pH (Haccius, 1963, 1965b), and X-rays (Haccius and Reichert, 1964). The degenerative effects were manifest in the conspicuously enlarged cells, the nuclei of which exhibited chromosomal abnormalities. However, the suspensor or small groups of cells or single cells of the damaged embryo proliferated to form viable embryos (Fig. 14.4). That similar faculties reside within the embryonic tissues of other plants was demonstrated by Miettinen and Waris (1958) in a study on the effect of glycine on the growth of seedlings of *Oenanthe aquatica*. Here, seedlings which became morbid as a result of prolonged growth in a medium containing glycine generally sloughed off cell aggregates into the medium, which eventually regenerated neomorphic plants. From these experiments the inference seems clear that the regenerative powers of the embryonal cells are expressed only when the original meristems are rendered nonfunctional.

III. INDUCTION OF EMBRYOIDS IN ORGAN, TISSUE AND CELL CULTURES

A. Carrot Tissue Cultures

The culture of the storage tissue of domestic carrot provided the starting point for the spectacular demonstration of the ability of somatic cells to regenerate whole plants. This concept, known as totipotency, implies that each and every cell of a multicellular organism is potentially immortal and contains within it all the essential instructions required to produce the whole organism in full multicellularity and structure. By the simple expedient of growing slabs of secondary phloem of carrot in a medium containing mineral salts, sucrose and coconut milk, Reinert (1958, 1959) obtained prolific growth of an amorphous callus. It is a great general feature of this strain of carrot tissue that it could be subcultured repeatedly in an undifferentiated state in a medium containing coconut milk (7%) and IAA (10^{-5} mg/l). The first indication of morphogenetic potency of the callus was observed upon its transfer to a completely synthetic medium enriched with an elaborate mixture of amino acids, amides, vitamins, a purine and an auxin. The tissue eventually became granular in this medium and formed nodule-like structures on its exposed surface. If the partly differentiated tissue was transferred to a medium lacking auxin, young plants with normal shoot and root appeared (Fig. 14.5). The clinching evidence was provided by histological studies which showed that plantlets arose from perfectly normal bipolar embryos which had their origin in certain marginal cells of the callus. These experiments thus exploited withdrawal of certain constituents from the medium, first coconut milk and then auxin, which promoted particular stages of an embryogenic sequence in an unorganized callus.

Although the first demonstration of the formation of adventive embryos from carrot tissue explants can thus be traced to the work of

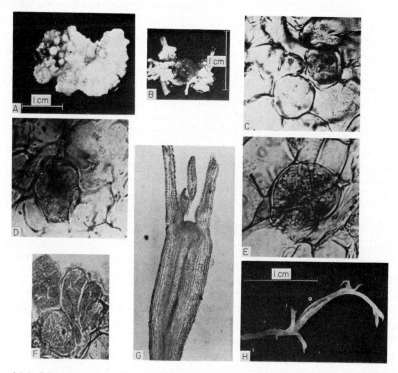

Fig. 14.5. Morphogenesis of carrot root tissue culture. **A**, Tissue originally grown in a medium containing coconut milk, 3 months after transfer to a synthetic medium containing auxin, showing nodules. **B**, Similar tissue after transfer from a synthetic medium containing auxin to one without auxin, showing shoot initiation. **C–F**, Stages in the formation of proembryos. **G**, Longitudinal section of an adventive embryo. **H**, Plantlet formed from an adventive embryo. (After Reinert, 1959)

Reinert, the classical experiments upon which our concepts of totipotency and adventive embryogeny are founded have come from the laboratory of Steward, who provided the major stimulus for work in the field, beginning with the innovation of methods for large-scale cultivation of carrot tissue explants. The stage was set for these investigations by the discovery that callus derived from carrot phloem grown with gentle agitation in a liquid medium containing coconut milk dissociated itself to form suspensions of single cells showing a bewildering array of sizes and shapes. The vitality of these cells was revealed by their intense cytoplasmic streaming. These cells multiplied by divisions and wall formation which took various forms such as normal equatorial division in densely cytoplasmic isodiametric cells, or unequal division in some highly vacuolate cells, or tubular filamentous growth followed by cell wall formation, or infrequently, formation of bud-like outgrowths in large spherical cells. Cells formed by repeated divisions remained attached to one another to form aggregate masses. When the aggregates reached a certain size, the inner cells became lignified and enclosed by a sheath of cambium-like cells. Eventually, lateral root primordia appeared in the cambial region. A normal carrot plant with a characteristically thickened primary root, leaves and stem which flowered and set seeds is assembled in a culture flask when cell aggregates with lateral roots are transferred to a stationary semi-solid medium, the first sign of morphogenesis in the new environment being the formation of a shoot diametrically opposite the root on the common axis (Steward et al., 1958a, b). By this means it was possible to recapitulate the life cycle of a carrot plant repeatedly in culture through a pathway involving disassembly of secondary phloem cells and their subsequent assembly. In this work, Steward (1963b) has emphasized the occurrence of many multicellular structures in cell suspension cultures which to a surprising degree resemble stages in the normal carrot embryogeny. In suspension cultures of tobacco cells, Bergmann (1959, 1960) found that cells allowed to settle on agar plates gave rise to filamentous structures which were organized like proembryos.

The most important milestone in this series of investigations occurred a few years later when it was shown that free cells of carrot in culture formed an enormous number of adventive embryos which faithfully reproduced the different stages of embryogeny as if they were exact replicas of zygotic embryos. This discovery was made when embryos isolated from immature seeds of domestic carrot grown from single cells, or from seeds of wild carrot (Queen Anne's Lace), were allowed to grow in a medium containing coconut milk. The free cells that sloughed off from the "germinated embryos" when plated on a semi-solid agar medium formed in orderly succession literally thousands of embryoids including proembryos, the globular, heart-shaped, torpedo-shaped and mature stages which later organized into plantlets (Fig. 14.6). The essential parts of this work which will illustrate the value of bold and unorthodox exploration of a field have been widely published and publicized in a number of articles (Steward, 1963a, 1967, 1970a, b; Steward et al., 1964a, b, 1969, 1970).

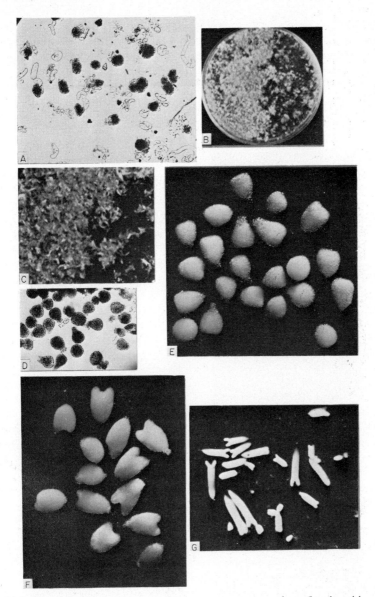

Fig. 14.6. Free cells of carrot embryo form an enormous number of embryoids which faithfully reproduce the different stages in zygotic embryogeny. **A**, Proembryogenic mass in a filtered suspension. **B**, Crop of embryoids in a petri dish culture plated with a cell suspension. **C**, Higher magnification of **B**. (After Steward *et al.*, 1964b; photographs courtesy of F. C. Steward) **D–G**, Whole mounts of embryoids of different stages obtained from a callus of carrot petiole: **D**, globular proembryoids; **E**, heart-shaped embryoids; **F**, embryoids at the stage of cotyledonary initiation; **G**, mature embryoids. (After Halperin, 1966a; photographs courtesy of W. Halperin)

The unusual plasticity of carrot for the induction of embryoids was further heightened by the demonstration that callus cultures originating from roots, petioles, umbellate peduncles and epidermal strips from hypocotyls of both wild and domestic carrot formed a large number of embryoids in liquid or solid culture (Wetherell and Halperin, 1963; Halperin and Wetherell, 1964; Nakajima and Yamaguchi, 1967; Kato,

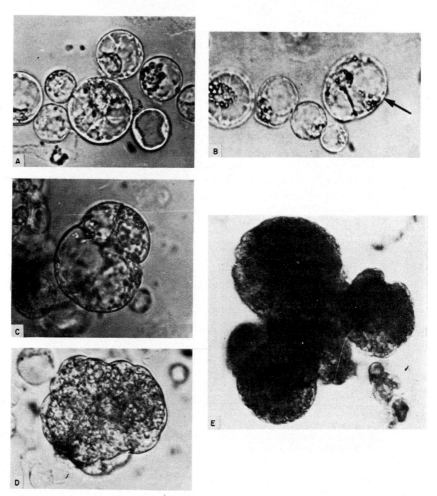

FIG. 14.7. Cells regenerating from isolated carrot cell protoplasts form embryoids which simulate stages of zygotic embryos. **A**, Isolated protoplasts. **B**, Wall formation (arrow), 2 days after isolation. **C**, One week old embryoidal cell group. **D**, Embryoid, 2–3 weeks old. **E**, Embryoid, 3–4 weeks old. (**B**, **C**, **E** after Grambow et al., 1972; photographs **A** (unpublished), **C**, **D** (unpublished) courtesy of O. L. Gamborg)

1968; Linser and Neumann, 1968). In some strains of carrot, seedlings originating from embryoids are able to form a second generation of embryoids directly on the hypocotyl and leaves (Homès, 1967a; Homès and Guillaume, 1967; Vermylen-Guillaume, 1969). Other proof of the plasticity of carrot was also forthcoming. A very elegant one is the isolation of protoplasts from cells separated directly from carrot root (Kameya and Uchiyama, 1972) or from suspension cultures (Grambow *et al.*, 1972) and their subsequent culture. The basic technique used here is simple. Small pieces of the tissue or cell aggregates are held in a medium of high osmolarity secured by the addition of mannitol and treated with a mixture of highly potent cellulase and pectinase solutions. After incubation in the enzyme mixture for 12–15 h, protoplasts are isolated, washed and cultured in an osmotically adjusted mineral salt medium. The isolated protoplasts formed walls, divided and formed clusters of cells which regenerated embryoids directly or indirectly from a callus (Fig. 14.7). The convergence of experiments and observations from many laboratories was finally capped by the unequivocal demonstration of the transformation of isolated single cells of carrot into embryoids (Backs-Hüsemann and Reinert, 1970; Reinert *et al.*, 1971). Serial microscopic observations have shown that the isolated cell does not directly give rise to a polarized bicellular structure analogous to the product of first division of the zygote, but to a conglomerate of embryogenic and parenchymatous cells from which embryoids finally emerge (Fig. 14.8). Other workers (Nakajima and Yamaguchi, 1967; Benbadis, 1973; Haccius, 1973; McWilliam *et al.*, 1974) have also noted differences between the early developmental pathways of embryoids and zygotic embryos. The most explicit comparative analysis of the early development of an embryoid and a zygotic embryo of carrot is provided by the last-named group of workers, who have shown that whereas the first longitudinal division in the embryoid takes place at the two- or three-celled stage, the zygotic proembryo produces a uniseriate filament of eight cells before the first longitudinal division.

The extensive literature dealing with adventive embryony that has burgeoned during the past few years attests to the widening recognition of the potential importance of the method for clonal multiplication of plants and for an understanding of the critical steps involved in the initiation of embryogenic behavior in the somatic cells of plants. As this field continues to expand, two different objectives can be recognized in the resulting flow of investigation. One of these has been to extend the method to a number of species of plants belonging to different taxa. A second focus, and a principal theme to be discussed later,

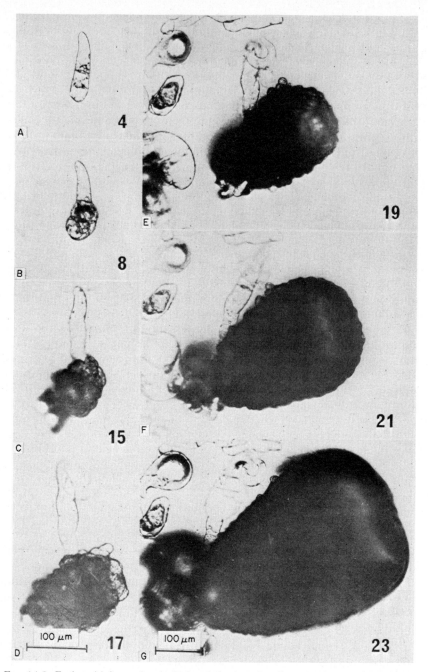

Fig. 14.8. Embryoid formation in isolated single cells of carrot tissue culture begins with an unequal division, followed by the formation of a complex of parenchymatous and embryogenic cells. Embryoid is initiated from the latter group of cells. **A–D**, Formation of the complex of cells. **E–G**, Initiation and growth of a heart-shaped embryo. Numbers in the right-hand corner indicate days after isolation of the cell. (After Backs-Hüsemann and Reinert, 1970)

is to define in precise chemical terms the requirements for induction of embryoids in the differentiated somatic cells.

B. Other Plants

Great success in embryoid induction in a number of plants has been obtained when cultures are started with whole embryos or parts of embryos. Work with tropical stem parasites is revealing in the profound variations in the propensity of embryos of different genera and of different ages for embryoid induction when they are grown in suitable media. Proembryos of the mistletoe *Dendrophthoe falcata*, grown in a medium supplemented with IAA and casein hydrolyzate, proliferated a callus which later differentiated embryoids, while mature embryos of this species never formed embryoids (Johri and Bajaj, 1962, 1963, 1965). Mature embryos of another mistletoe, *Scurrula pulverulenta*, possess a pronounced capacity for proliferation and a limited capacity for differentiation and differ principally from the *D. falcata* type in that embryoids fail to differentiate beyond the heart-shaped stage (Johri and Bhojwani, 1970). Intermediate behavior is characteristic of *Cuscuta reflexa* (Maheshwari, P. and Baldev, 1961, 1962) and *Amyema pendula* (Johri and Bajaj, 1964). Immature embryos of the former exhibited full embryogenic capacity when they were grown in a medium containing IAA and casein hydrolyzate while embryogenic behavior of mature embryos was restricted to their radicular ends (Fig. 14.9). In the latter species, in the absence of organized callus, papillate outgrowths appeared on the radicular end of the embryo which differentiated into late embryonic stages. Embryoids were induced even on mature embryos of *Nuytsia floribunda* when they were grown in a medium containing IBA, kinetin and casein hydrolyzate. However, the callus which differentiated embryoids could not be satisfactorily maintained on the same medium without loss of capacity to form embryoids. A sustained potentiality for embryoid differentiation was found to be maintained upon substitution of 2,4-D for kinetin in the medium (Nag and Johri, 1969). Although the hormonal requirements of embryos of stem parasites are somewhat fastidious, the presence of hormones undoubtedly determines whether cells will simply continue to divide or organize as they divide.

Under certain conditions of culture, embryoids are also induced on cultured embryos of other plants including several dicots [*Santalum album* (Rao, 1965b; Rao and Rangaswamy, 1971); *Coriandrum sativum*, *Arabidopsis thaliana* (Steward et al., 1966); *Solanum melongena* (Yamada et al., 1967); *Sium suave* (Ammirato and Steward, 1971); *Ilex aquifolium*

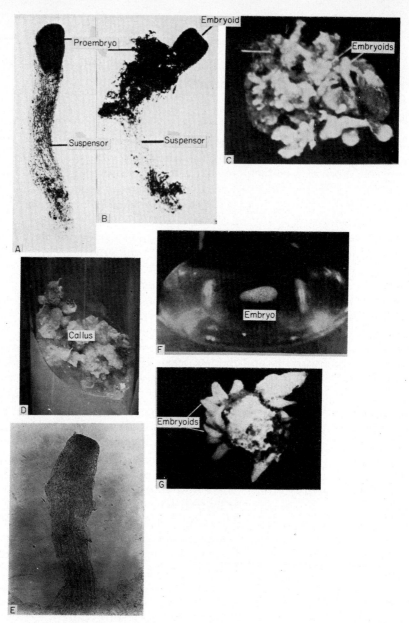

Fig. 14.9. Embryoid formation in cultured embryos of stem parasites. **A–C**, *Dendrophthoe falcata*: **A**, globular embryo at the time of culture; **B**, portion of a 3 week old callus with an accessory embryo; **C**, portion of a 20 week old callus with a large number of accessory embryos. (After Johri and Bajaj, 1965) **D, E**, *Scurrula pulverulenta*: **D**, a culture, about 20 weeks old, of a mature embryo in a medium containing casein hydrolyzate, IAA and kinetin; **E**, whole mount of a heart-shaped embryoid. (After Johri and Bhojwani, 1970; photographs courtesy of B. M. Johri and S. S. Bhojwani) **F, G**, Embryoid formation in immature embryos of *Cuscuta reflexa*: **F**, embryo at the time of culture in a medium containing casein hydrolyzate and IAA; **G**, a 40 day old culture showing accessory embryos. (After Maheshwari and Baldev, 1962)

(Hu and Sessex, 1971)], monocots [barley (Norstog, 1970); oil palm (Rabéchault *et al.*, 1970)] and a few gymnosperms [*Biota orientalis* (Konar and Oberoi, 1965); *Ephedra foliata* (Sankhla *et al.*, 1967c); *Zamia integrifolia* (Norstog, 1965b; Norstog and Rhamstine, 1967); pine (Sommer and Brown, 1974)]. In barley, embryoids arise in response to kinetin treatment and their occurrence is restricted to a particular region of the embryo opposite the scutellum (Norstog, 1970). The significance of this positional relationship is not entirely clear. In addition, freely suspended cells derived from cultured embryos of endive (*Cichorium endivia*) have been shown to form plantlets by embryoid

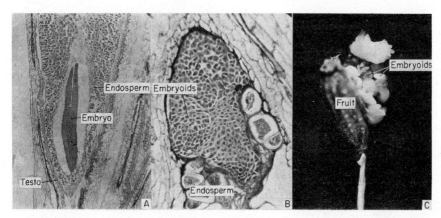

Fig. 14.10. Proliferation of the contained embryo and its differentiation into embryoids are seen in ovule cultures of *Anethum graveolens*. **A**, Section of the ovule at the time of culture. (After Johri and Sehgal, 1963b) **B**, Section of the ovule 5 weeks after culture in a medium containing casein hydrolyzate, showing the mature embryo and accessory proembryos. **C**, Ruptured fruit showing embryoids on a medium containing yeast extract. (After Johri and Sehgal, 1963a; photographs courtesy of B. M. Johri)

formation, although the latter did not simulate typical stages in true embryogenesis (Vasil *et al.*, 1964; Vasil and Hildebrandt, 1966a).

In ovary and ovule culture of some plants, proliferation of the contained embryo and its eventual differentiation into embryoids are seen. This was the case when ovaries of *Anethum graveolens* were cultured in media containing casein hydrolyzate or yeast extract. While formation of normal monoembryonic seeds was the rule, in a small percentage of ovaries the embryonal mass cleaved and produced accessory embryos comparable to the zygotic embryos (Johri and Sehgal, 1963a, b, 1965, 1966) (Fig. 14.10). Similar results were obtained when ovaries of *Ranunculus sceleratus* (Sachar and Guha, 1962) and *Foeniculum vulgare* (Sehgal, 1964; Johri and Sehgal, 1966) containing proembryo-bearing ovules and of *Ammi majus* (Sehgal, 1972) containing zygote-bearing ovules were cultured. Cleavage of the embryonal mass suggestive of an early stage of embryoid formation was observed when ovules of cotton were cultured in a medium containing an extract of pollinated cotton ovules, or casein hydrolyzate and kinetin (Joshi and Johri, 1972). Endosperm cultures of some plants have yielded facsimiles of embryoids which fail to undergo further development (Satsangi and Mohan Ram, 1965; Bhojwani, 1966) (Fig. 14.11). Since leaf buds, roots and whole plants have been induced in endosperm cultures of several plants by subtle

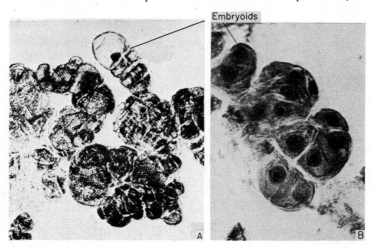

Fig. 14.11. Squash preparations of the endosperm of *Ricinus communis* (**A**) and *Croton bonplandianum* (**B**) yield cell aggregates resembling early stages of embryoid development. (**A** after Satsangi and Mohan Ram, 1965; **B** after Bhojwani, 1966)

alterations in medium composition (see Chapter 6), it is perhaps a matter of time before complete development of embryoids will be achieved.

The success that has attended embryoid induction in the vegetative parts of carrot has encouraged attempts to study the potentiality of the vegetative parts of other plants to regenerate embryoids. As a result, there is now a long list of plants in which latent embryogenic tendencies of somatic cells have been demonstrated and in which clones of embryogenic tissue have been established. Included in this list are stem segments of *Foeniculum vulgare* (Maheshwari, S. C. and Gupta, 1965), tobacco (Haccius and Lakshmanan, 1965), *Didiscus coerulea* (Ball and Joshi, 1966), *Tylophora indica* (Rao *et al.*, 1970), *Coffea canephora* (Staritsky, 1970), *Asparagus officinalis* (Steward and Mapes, 1971b), *Euphorbia pulcherrima* (Nataraja, 1971a), *Pergularia minor, Asclepias curassavica* (Prabhudesai and Narayanaswamy, 1974), *Petunia inflata, P. hybrida* (Handro *et al.*, 1972; Rao *et al.*, 1973a, b) and *Antirrhinum majus* (Poirier-Hamon *et al.*, 1974), hypocotyls of *Apium graveolens* (Reinert *et al.*, 1966), *Asparagus officinalis* (Wilmar and Hellendoorn, 1968), *Sinapis alba* (Bajaj and Bopp, 1972), *Cucurbita pepo* (Jelaska, 1972, 1974), *Conium maculatum* (Nétien and Raynaud, 1972), tobacco (Prabhudesai and Narayanaswamy, 1973) and *Anethum graveolens* (Ratnamba and Chopra, 1974), leaf petiole of parsley *(Petroselinum hortense)* (Vasil and Hildebrandt, 1966b) and tobacco (Prabhudesai and Narayanaswamy, 1973), leaf mesophyll of *Kalanchoe pinnata* (Wadhi and Mohan Ram, 1964; Mohan Ram and Wadhi, 1965), *Macleaya cordata* (Kohlenbach, 1965), *Rauvolfia serpentina* (Mitra and Chaturvedi, 1970), *Mesembryanthemum floribundum* (Mehra and Mehra, 1972), *Petunia inflata, P. hybrida* (Handro *et al.*, 1972; Rao *et al.*, 1973a, b) and *Asparagus officinalis* (Jullien, 1974), shoot apex of *Cymbidium* (Steward and Mapes, 1971a), mesocotyl of *Bromus inermis* (Gamborg *et al.*, 1970; Constabel *et al.*, 1971), pedicels and floral buds of *Nigella damascena* (Raman and Greyson, 1974), fruit pericarp of *Cucurbita pepo* (Schroeder, 1968), nucellus of *Citrus aurantifolia* and *C. sinensis* (Mitra and Chaturvedi, 1972; Button *et al.*, 1974), root of *Atropa belladonna* (Thomas and Street, 1970, 1972) and seedlings of *Cheiranthus cheiri* (Khanna and Staba, 1970). Figure 14.12 shows the stages in embryoid formation in *Petroselinum hortense* petioles (Vasil and Hildebrandt, 1966b), which can be stated to be typical of those found in many other species listed here.

Mention was made earlier of the remarkable plasticity of carrot enabling it to regenerate embryoids from virtually any part of the plant. Among other equally versatile plants is *Ranunculus sceleratus*, a herbaceous annual which has served as the subject of a series of studies by Konar and Nataraja (1964, 1965a–d, 1969; Nataraja and Konar, 1970). The initial culture procedure involved placing flower buds at the stage of differentiation of floral organs on a medium containing coconut milk and IAA. A callus regenerating from the explant gave rise to numerous embryoids without any further treatment. Embryoids set free in the medium instead of completing normal development germinated precociously into seedlings which bore spontaneously a second generation of embryoids all along their stem and hypocotyl by division of the epidermis (Konar and Nataraja, 1964, 1965a, b, 1969). It was also found that the friable callus dissociating into free cells and cell clumps on a shake culture developed into embryoids which were capable of producing new embryoids by budding (Konar and Nataraja, 1965c). These embryoids had their origin in the peripheral cells of the aggregate which were in immediate contact with the medium. Under

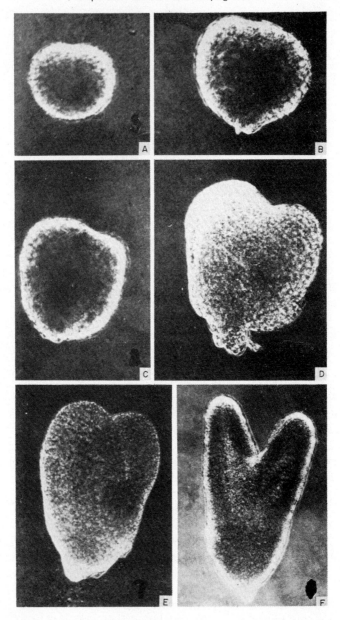

FIG. 14.12. Various stages in embryoid development such as globular (**A, B**), heart-shaped (**C, D**) and early cotyledonary stages (**E, F**) are obtained from the petiole callus of *Petroselinum hortense* grown in a high salt medium containing adenine. (After Vasil and Hildebrandt, 1966b)

all cultural conditions embryoids faithfully displayed the usual stages in embryogeny, including two- to four-celled, globular, heart-shaped and torpedo-shaped stages (Fig. 14.13). In later studies (Konar and Nataraja, 1965d; Nataraja and Konar, 1970), isolated sepals, petals,

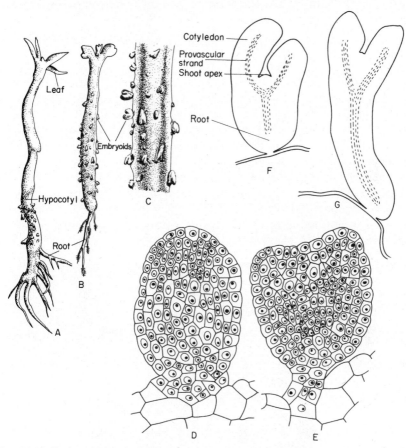

FIG. 14.13. Embryoid formation in *Ranunculus sceleratus*. **A–C**, One month old seedlings bearing accessory embryos. Globular (**D**), heart-shaped (**E**) and developing cotyle-donary stage embryoids (**F, G**). (After Konar and Nataraja, 1965b)

anthers as well as shoot tips, stem, petiole and lamina of adult plants were found to yield plentiful crops of embryoids upon culture in a medium containing coconut milk and 2,4-D. Callus tissues were also established from organs of other members of the Ranunculaceae, but only one originating from the floral bud of *Consolida orientalis* revealed structures which bore some resemblance to embryos (Nataraja, 1971b).

Different parts of ginseng *(Panax ginseng)* (Butenko *et al.*, 1968) and *Pterotheca falconeri* (Mehra and Mehra, 1971) have also been shown to express high morphogenetic potency upon culture by the formation of embryoids. In view of the active interest in this field in several laboratories, this list cannot be considered as being closed and other interesting species may yet be added.

C. Origin of Embryoids

A subject of considerable interest relates to the identification of the specific cell types from which embryoids arise. From anatomical observations of embryo explants of *Cuscuta reflexa*, P. Maheshwari and Baldev (1961) have implicated epidermal cells as progenitors of embryoids. These cells divide to form globular aggregates which elongate and eventually organize the characteristic shoot apex flanked by embryonic leaves. Analysis of adventive embryogeny in hypocotyls of carrot (Kato and Takeuchi, 1966; Haccius and Lakshmanan, 1969) and *Ranunculus sceleratus* (Konar and Nataraja, 1965b; Konar *et al.*, 1972a) have also demonstrated unequivocally the transformation of epidermal cells into embryoids (Fig. 14.14).

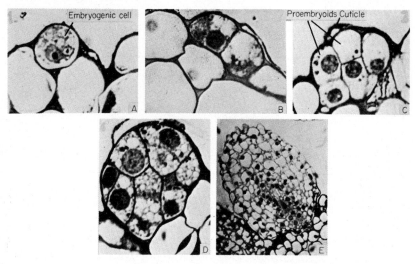

Fig. 14.14. Embryoids are initiated from highly cytoplasmic cells of the stem epidermis of plantlets of *Ranunculus sceleratus*. **A, B**, A single and a pair of potential embryogenic epidermal cells, respectively. **C**, Two two-celled proembryoids and a potential embryogenic cell, respectively. **D**, A young embryoid, showing darkly stained amyloplasts. **E**, A globular embryoid showing a central core of meristematic cells. (After Konar *et al.*, 1972a)

In tissue cultures any cell within the undifferentiated aggregate mass is potentially capable of producing an embryoid. The embryogenic cells are generally characterized by a dense cytoplasm and large nuclei and nucleoli. These cells are also distinct from the nonembryogenic cells by their higher ribosome density and profiles of rough ER (Halperin and Jensen, 1967; Thomas *et al.*, 1972). The possibility that these properties are associated with an intense synthesis of RNA as the first step towards embryogeny has been suggested (Kohlenbach, 1970; Danilina, 1972; Sussex, 1972) but warrants further examination. The transition of the callus cells to an embryogenic state has been associated by inference with a change in the distribution of microtubules from a random orientation to one arranged in linear parallel arrays (Wochok, 1973a).

In carrot tissue cultures, the embryogenic cells divide repeatedly to form masses of meristematic tissue designated as the proembryogenic mass that initiates the production of embryoids. The proembryogenic clusters grow initially without tissue and organ differentiation and embryoids subsequently develop on this dense tissue mass (Sussex, 1972; Jones, L. H., 1974).

In the stem callus of *Tylophora indica*, embryoids have been traced to certain cells embedded deep inside or in its periphery (Rao and Narayanaswami, 1972). A diagnostic feature of the deep-seated embryoid is its precision in cell lineage, cell division and cell disposition which closely corresponds to the segmentation pattern of the zygotic embryo. This distinguishes it from the peripheral embryoid which does not follow the established segmentation pattern. These facts are compatible with the view that a normal pattern of development observed in the deep-seated embryoid is presumably dictated by its very position, surrounded as it is by many layers of cells which simulate to some extent the physical environment of hydrostatic pressure around a zygote.

In *Ranunculus sceleratus*, embryoid development appears to proceed initially by an unequal transverse division in the epidermal cell which gives rise to an embryonal cell and a suspensor cell (Konar *et al.*, 1972a). Since ontogenetically the embryoid suspensor is not comparable with a similar organ of the zygotic embryo, the term as used here implies reference to a group of cells of the proembryoid which are not incorporated into the embryoid proper and which thus constitute a narrow band of cells which connects the proembryoid to its mother tissue. Structures answering this description have been recorded in embryoids originating in tissue cultures of several plants (Halperin and Wetherell, 1965b; Hu and Sussex, 1971; Khanna and Staba, 1970; McWilliam *et al.*, 1974). No firm ideas regarding the function of the embryoid suspensor have evolved and it is even doubtful whether it plays the same role here as in the zygotic embryo. Since embryoids presumably absorb nutrients through their cotyledons (Ammirato and Steward, 1971), the suspensor, if present, appears to be a redundant organ. Embryoid suspensors of *Atropa belladonna* are unique in showing a pronounced ability to initiate a new category of adventive embryos by budding (Konar *et al.*, 1972b). On the basis of the limited information presently available, we can agree with Haccius (1965a) that interpretation of the role of suspensor in the embryoid remains an open question.

IV. FACTORS CONTROLLING EMBRYOID FORMATION

A central theme of the foregoing account is that cells must be exposed to some chemical environment if they are to become embryogenic. The importance of this theme cannot be overemphasized. It is now necessary to give some consideration to the role of these and other specific stimuli as they are known to operate in the expression of cellular totipotency.

A. Isolation of Cells

Some of the rigorous and detailed analyses of the requirements for embryoid induction are those that have been carried out on carrot tissue cultures. Steward *et al.* (1964a, b) have pressed into service the argument that isolation of cells from the influence of the neighboring cells in an organized tissue is necessary to persuade them to express their inherent embryogenic tendencies. An isolated cell comes close to satisfying the likeness of a zygote, namely, it is free from the influence exerted by other cells and so develops like a zygote into an embryo. This argument is eminently reasonable and persuasive in a qualitative sense, but has not been supported by results of other studies in which typical embryoids arise on the surface of, or deep inside, cultured organs and well organized callus tissues by transformation of cells which are in organic connection with their immediate neighbors. These studies have been alluded to earlier and will not be taken up here again. Suffice it to say at this stage that the view that a cell should be separated from its neighbors before it can become embryogenic needs reinforcement.

B. Role of Coconut Milk

Our understanding of the chemical factors controlling embryoid induction has unfortunately been somewhat bedeviled by the strong arguments and counterarguments on the role of coconut milk in the process. We have mentioned the fact that Steward and his associates accomplished successful induction of embryoids when a cell suspension of the seed embryo of wild or domestic carrot which was established in a liquid medium containing coconut milk was plated on a semi-solid medium of the same composition. Long before these results were obtained, the importance of coconut milk for the growth and proliferation of carrot explants was established. In fact, the promotion of growth induced by coconut milk in the otherwise nongrowing cells of the secondary phloem parenchyma of carrot was so dramatic that this strangely assorted pair of plants—carrots and coconuts—has figured in popular articles and public lectures as classical examples of induced growth in plants (Steward, 1960, 1963c). Since coconut milk and similar substances of endospermic origin normally nourish the zygote, it has been succinctly argued that to cause somatic cells to develop like zygotes, they should be given the nutrients and stimuli that normally nourish the zygote in the ovule (Steward et al., 1970).

Acceptance of the above viewpoint presents unexpected difficulties. Some investigators have found that coconut milk is a superfluous constituent of the medium for inducing embryoids and that in certain cases its presence may even be inhibitory. According to Halperin and Wetherell (1964; Halperin, 1966b), cultures derived from the vegetative parts of wild carrot formed embryoids just as readily in a simple medium supplemented with adenine and 2,4-D or kinetin and 2,4-D as in coconut milk. Differentiation of very small embryoids in the tissues was found to be inhibited when they were plated on a medium containing coconut milk (Halperin, 1964). From these results it is doubtful whether coconut milk contributes any unique substances for embryogeny. Any idea that wild carrot was somewhat exceptional was also dispelled when it was shown that tissue cultures derived from the domestic carrot formed embryoids in a defined medium consisting of major salts, trace elements and organic addenda according to the formula of Murashige and Skoog or White, and 2,4-D (Reinert, 1963, 1967; Reinert et al., 1966). It is also significant that a well documented report on embryoid formation in tissue cultures of domestic carrot (Kato and Takeuchi, 1963) did not include coconut milk in the medium. Sussex and Frei (1968) found that for long-term tissue cultures of domestic carrot the only requirement for adventive embryogeny was addition of an auxin, preferably IAA, to the basal medium which alone would support neither embryogeny nor continued proliferation of the tissue. Coconut milk alone led to rapid proliferation of the tissue without inducing embryoids, while in conjunction with IAA, it inhibited embryogeny. It has been suggested that in those instances where coconut milk has been reported to be mandatory for embryogeny, the additive may in reality be affecting the subsequent growth of preformed embryoids rather than their initiation. That coconut milk plays a role in the growth of embryoids is thus unquestionable. Yet this is not to imply that it is the controlling agent primarily responsible for inducing embryoid formation. This reservation simply points to one of the many difficulties in identifying the causal agent when complex natural extracts are used in tissue culture media. It may be that in looking for a common denominator for embryoid induction the focus on the role of coconut milk was misplaced.

C. Role of Auxins

Since auxins such as IAA and 2,4-D have found their way into culture media, it would appear to be of prime importance to learn whether or not they are critical for induction of embryoids. In a system that has been as closely examined as carrot, the available data present a confusing story. Although the earlier work of Halperin and Wetherell (1964) showed a requirement for both auxin and cytokinin for embryogeny, in later work (Halperin, 1966a, 1970) efficiency of embryoid formation was found to be higher in the presence of 2,4-D alone. A key factor in the use of 2,4-D seems to be its concentration, since addition of the auxin at a concentration >0.1 mg/l did not permit development of embryoids to maturity. Petrů (1970) was able to induce embryoids in a tissue which was subcultured for over 6 years in a medium containing 1.0–5.0 mg/l IAA by supplementing the medium additionally with 0.5 mg/l 2,4-D. Norreel and Nitsch (1968, 1970) have claimed a greater effectiveness for NAA (1.0 mg/l) over 2,4-D in inducing embryoid formation in tissue cultures of carrot petiole. Other workers, on the contrary, have obtained embryoid formation in tissues which have never been exposed to auxins (Steward et al., 1964b) or by lowering the effective auxin concentration of the medium by transferring tissues from a higher auxin concentration to a lower concentration or to an auxin-free medium (Reinert, 1959; Reinert and Backs, 1968; Pilet, 1961; Homès, 1967a; Yamada et al., 1967) or to a medium containing an antiauxin (Newcomb and Wetherell, 1970). In retrospect, part of the confusion on the role of auxin in embryogenesis is due to the nature of the tissues used, namely, freshly isolated versus habituated after repeated subculture, size of the inoculum, concentration and type of auxin. Sussex and Frei (1968), who have considered these questions, suggest that auxin is indeed a limiting factor in adventive embryogeny. Results such as those of Reinert (1959) and Pilet (1961) can perhaps be explained by assuming that there is a carry-over of a

sufficient amount of auxin for embryogenesis by large inocula transferred from a high level of auxin to an auxin-free medium. Similarly,
Halperin's (1967) observation that embryologically competent cells
fail to develop embryoids in dilute suspensions of the inoculum can
be readily explained on the basis of a lack of an adequate amount of
auxin in the dilute inoculum rather than a loss of critical metabolites.

The inhibitory effects of high auxin concentrations on somatic cell
embryogenesis may be influenced by an array of complex factors. For
example, it has been shown that the ethylene-producing chemical 2-
chloroethylphosphonic acid mimics the effects of 2,4-D on carrot cell
embryogenesis, suggesting that the suppression of organization resulting from the addition of auxin might be in part due to the auxin-in-
duced production of ethylene and the subsequent action of this hormone (Wochok and Wetherell, 1971). Another tenuous lead to explain
the action of auxin on embryogenesis is the observation that auxin
prevents the induction of isoperoxidases which are necessary for
embryoid maturation (Wochok and Burleson, 1974). Since it is custo-
mary to maintain embryogenic clones of cells in the unorganized
state simply by culturing them in relatively high concentrations of
auxin, the analysis of the mechanism of auxin action in somatic cell
embryogenesis is a pressing problem of both theoretical and practical
importance.

Differentiation of embryoids from diverse plant organs, most of
which follow a seemingly general sequence in which the organ first
forms a callus, has been achieved by subtle alterations in the hormonal
constituents supplied in the medium. Despite the fact that special pre-
treatment of the explants, such as exposure to γ-rays (Norreel and Rao,
1974), increases their embryogenic potential, a requirement for hormones is a *sine qua non* for the process. Perhaps the most remarkable
single feature of the hormonal control of embryoid induction is that
it is evoked by an auxin, generally 2,4-D alone or in combination with
a cytokinin such as kinetin. Given the right hormonal mixture, environmental parameters such as light and temperature might appear to be
of secondary importance, although in some particularly recalcitrant
tissues, such as the wound callus of *Nicotiana* stem, hormonal environment is effective only under conditions of high intensity light (Haccius
and Lakshmanan, 1965). Growth inhibitors such as abscisic acid might
also play a role particularly in processes concerned with embryo maturation, by inhibiting uncontrolled growth (Ammirato, 1974). In a few
cases, the view has emerged prominently that a complex series of
sequential treatments in different media containing synergistic combinations of growth hormones are the determining factors in embryoid

formation. In the stem segments of *Asparagus officinalis*, these treatments consisted of induction of callus in White's medium supplemented with coconut milk and NAA, separation of cells and cell clumps in a medium in which NAA is replaced by 2,4-D, induction of embryoids in a high salt medium (Murashige and Skoog medium) supplemented with NAA, and formation of root and shoot primordia by transfer of embryoids to a medium containing coconut milk and IAA (Steward and Mapes, 1971b). The reconstitution of cells into embryoids can also be observed when leaf mesophyll callus of this same species is transferred in sequence from a medium enriched with an auxin and a cytokinin to one without these hormones (Jullien, 1974). These results are in keeping with the fact that "it is not enough to understand the digits in the combination lock to the door of cell growth and cell division, for one needs also to know the correct sequence in which to apply them and perhaps also the amount of time that ought to lapse between the different stimuli to which the cells are subjected" (Steward *et al.*, 1969).

D. Role of Nitrogen Compounds

A study of the effect of nitrogen compounds on embryogenesis has led to another set of perplexing observations. Halperin and Wetherell (1965a) suggested that embryoid initiation in wild carrot tissue was specifically induced by the presence of ammonium ion in the medium. The case for this hypothesis was weakened by the completely conflicting data of Reinert and coworkers (Reinert, 1968; Reinert *et al.*, 1967; Reinert and Tazawa, 1969), who found that embryoids were induced by nitrogen equivalent amounts of NH_4NO_3 or KNO_3 or by organic nitrogen compounds such as amino acids and amides. Moreover, any stimulation of embryogenesis induced by ammonium ion has been shown to be solely due to its serving as a substrate for the synthesis of amino acids (Tazawa and Reinert, 1969). The fact that the medium used by Halperin and Wetherell included kinetin, which was later shown to be inhibitory for embryoid formation, also speaks strongly against any specific effect of ammonium ion.

Depending upon the type of nitrogen supplied in the medium, two distinct forms of plantlet regeneration can be identified in carrot tissue cultures. Transfer of cells from a high to a low auxin-containing medium induced embryoids in high proportions when the high auxin medium contained reduced nitrogen. In the absence of reduced nitrogen in the medium, rhizogenesis—a morphogenetic event characterized by the appearance of root-bearing clumps of tissue—occurred (Halperin, 1966a). A correlation between nitrogen supply in the medium and morphogenesis is also borne out by the experiments of Steward *et al.* (1970), who found a preferential induction of roots in cultures grown in a medium poor in reduced nitrogen. Recently, it has been reported that carrot tissues grown in O_2-limited media have high embryogenic potential (Kessell and Carr, 1972); despite the parallelism between the effects of high amounts of reduced nitrogen and partial anaerobiosis, it is not clear whether reduced nitrogen induces embryogenesis or conditions the nutrient milieu for selection of cells with embryogenic potential. Some caution is warranted in making deductions on the effects of inorganic additives of the medium since the work of Butenko *et al.* (1967) has indicated the possible importance of osmotic and toxic effects of high salt concentrations of the medium on initiation and growth of embryoids.

E. Role of Other Nutrients

The role of other nutrient additives of the medium in inducing embryogenesis has been barely touched upon, and it is not known whether there are other constituents in the medium that may specifically interfere with the full expression of embryogenic potential of a given strain of tissue. The fact that embryoids are not formed in the absence of sugar in the medium, or in its presence in exceedingly low concentrations, must be regarded as a strong indication of the necessity of sugar for embryogenesis (Homès, 1967b, 1968; Neumann and deGarcia, 1974). At high concentrations of sugar, embryoids do not differentiate beyond the proembryo stage; this remarkable effect stems indirectly from the osmotic effect of high sugar in inhibiting normal growth processes.

In summary, this review of selected examples presented above illustrates very well the fact that embryoid initiation is a complex process. Even in a versatile system such as carrot tissue culture, it is doubtful whether any one substance can be singled out as the controlling factor. It would nevertheless appear that under normal circumstances the prime factors that switch the cultured cells to an embryogenic pathway are the presence of a cell division factor and sufficient nutrients for the normal synthetic activities of the cells. The different supplements to the medium such as auxins, cytokinins, coconut milk and ammonium ion serve one or both of these functions. More complicated media serve only to obscure the nature of the real morphogenetic influences at work in embryoid formation and may even lead to the development of abnormal structures as found by some workers (Hill, 1967; Vermylen-Guillaume, 1969) in carrot tissue cultures. Attempts to explain the chemical control of embryogenesis in tissue cultures must be flexible enough to accommodate the specific roles of the different constituents of the medium in terms of their known metabolic effects.

F. Loss of Embryogenic Potency

Several workers (Steward and Mapes, 1963; Halperin, 1966a; Reinert and Backs, 1968; Sussex and Frei, 1968; Thomas and Street, 1970, 1972) have observed that the embryogenic ability of cells of carrot attenuates in serial transfers. Cells of certain strains of carrot even lose their ability to produce embryoids after a period of prolonged cultivation. Obviously there is a relation between the rate of growth of cells in culture and their capacity for embryogenesis. Generally, slow growing cells retain their embryogenic capacity longer than fast growing cells, and treatments which decelerate growth such as transferring tissues to a less enriched medium or removing the hormonal components of the medium prolong this capacity (Meyer-Teuter and Reinert, 1973; Mouras and Lutz, 1973). A systematic depletion of medium components such as iron and EDTA has also been reported to induce the loss of embryogenic power with passage of time (Mestre et al., 1972, 1973).

A more promising interpretation of the loss of embryogenic potency is based on the concept that cytological changes in the cultured cells give rise to non-totipotent cell lines, which by competitive elimination become the only cell types in a culture (Smith and Street, 1974). Since extensive changes in nuclear cytology are known to occur in serially cultured cells, it is not unreasonable to expect a subtle effect of this on the embryogenic potency of cells in a population. Supporting this view is the observation that a mixed culture containing initially equal numbers of diploid and tetraploid cells became predominantly tetraploid with a concomitant loss of embryogenic potency. The exact mechanism conferring selective advantage on the non-totipotent cell types in a culture remains to be determined.

Few papers have evaluated the difficulties to be overcome to achieve continued embryogenesis in long-term cell cultures. A study by Nag and Street (1973) has shown that embryogenic strains of carrot cells can be preserved by freeze-storage techniques without any appreciable impairment and loss of morphogenetic potential. In nucellar cultures of Citrus sinensis, prolonged starvation and aging might lead to rejuvenation of embryogenic potential, resulting in dramatic increases in the production of embryoids upon subsequent subcultures (Kochba and Button, 1974). The development of these techniques is of considerable interest as possible models to establish banks of particular cell types.

V. COMMENTS

Studies described in this chapter have provided evidence beyond any reasonable doubt that mature plant cells which acquire overt functional and structural specialization have the capacity to regenerate whole plants of the type of which they were integral parts. The significant aspect of this study is the development from cells in culture of embryoids similar in structure to embryos that develop within the embryo sac. The documentation of the number of species in which initiation of a zygotic subprogram in the differentiated cells is achieved provides a broad confirmation necessary to establish the validity of the phenomenon as a secure generalization. The apparent ability of free cells to simulate the stages in the normal embryogeny of the species in question and form new plants poses a fundamental question. Is it due to the release of cells from the restrictive influence of their neighboring cells or is it due to the presence of specific nutritional substances in the medium? From the discussion in earlier pages, it is quite clear that removal of cells from the restrictive influence as such is not necessary to induce an embryogenic type of growth in all species thus far investigated; the requirement for a particular set of nutrient substances cannot also be assured for free cells of plants to direct them to an embryogenic pathway. These statements plainly suggest that more work is needed to achieve a deeper insight into the factors controlling somatic cell embryogenesis.

While we can visualize the potential use of cell culture techniques for propagation of a large number of plants of identical genome, the recent review of Murashige (1974) on plant propagation through tissue culture is a poignant reminder as to how much is yet to be done before the principles established in these studies are applied for large-scale propagation of plants.

15. Adventive Embryogenesis: Induction of Haploid Embryoids

The development of the egg or the accessory cells of the embryo sac into embryos without actual fusion with the sperm nucleus, or the development of the cells of the male gametophyte into embryos in the absence of fusion with the egg, introduces into the sporophyte cell types that differ in their chromosome number from those originating from a normally fertilized egg. Embryos that develop from the reduced gametophytic cells in the absence of fertilization are haploids at the cellular level, that is, consist of a single set, the gametic number of chromosomes in their somatic cells. Closely following the demonstration of totipotency and successful induction of embryoids from the somatic cells of plants, investigations on the induction of haploid embryoids provide a focal point for examining the regenerative potencies of germ cells. Research on haploids in general, and on the methods of inducing haploid embryoids and plants in angiosperms in particular, has been the subject of a recent international symposium (Kasha, 1974).

I. EMBRYO SAC-BASED HAPLOIDS

Haploid embryos may appear in the embryo sac spontaneously or may be induced by various physical and chemical agents. Spontaneous origin of embryos from some haploid component of the embryo sac is a sporadic event and has so far been recorded in about 100 species (Kimber and Riley, 1963; Magoon and Khanna, 1963). In a few plants, embryos have been traced to the egg cell or the synergids (Maheshwari, P., 1950), but in the majority of cases, in the absence of cytologically substantiated accounts, the exact origin of haploid embryos in the embryo sac remains uncertain. Origin of haploids which appear in large numbers in normal interspecific crosses in barley has been attributed to the loss of the male chromosomes from the zygote during its early division phase (Kasha and Kao, 1970). Other genetic abnormalities are by no means uncommon in the formation of haploids and have been documented on occasions during developmental stages. Isolated mutations or chromosome changes within the cells of the embryo sometimes occur and may give rise to abnormal plants with less than the diploid number of chromosomes.

Several procedures such as delayed pollination, use of abortive pollen, distant hybridization, *in vitro* pollination, and application of temperature shocks, radiation and chemicals have formed the basis for investigations to induce haploid embryos in plants (Lacadena, 1974). However, these methods are too laborious for routine work and the extremely low frequency of production of haploids from any given stock has prevented more critical examination of their potential use. Owing to such limitations, haploids originating from the embryo sac have had relatively little impact on contemporary studies and were regarded more as abnormalities than as basic tools for research. Technical considerations of isolating and culturing the delicate female gametophytes of angiosperms have thus far precluded attempts to study *in vitro* embryoid induction on them, although sporadic experiments to secure embryoid formation on the female gametophyte of a gymnosperm have been promising (Norstog, 1965b).

A few instances of formation of haploid embryos from the male gametophyte have been detected through genetic analysis; embryos and plants so formed have come to be referred to as androgenic haploids, and the process itself is known as androgenesis. Interspecific crosses between different species of *Nicotiana* (Clausen and Lammerts, 1929; Kostoff, 1929, 1934; Kehr, 1951; Burk and Gerstel, 1961) and *Hordeum* (Davies, 1958) have yielded plants possessing paternal characteristics and hence thought to be androgenic in origin. Gerassimowa (1936) described a haploid plant with features typical of the male parent when X-rayed female flowers of *Crepis tectorum* were pollinated with pollen of a marked male parent having recessive characters. Using similar procedures a few androgenic haploids have also been recovered in *Antirrhinum majus* (Maly, 1958) and *Arabidopsis thaliana* (Gerlach-Cruse, 1970). In both instances, it appeared that X-irradiation killed the egg nucleus and that the male gamete alone developed into the embryo. Occasional appearance of androgenic haploids in normal populations of some plants is somewhat more fortuitous than those described above and seems to be essentially independent of any specific breeding programs or outside agencies (Campos and Morgan, 1958, 1960; Goodsell, 1961; Haustein, 1961). In no case has it been definitely established whether embryos originated from the vegetative cell or the generative cell of the male gametophyte. There have been few other studies on androgenic haploids and, by and large, interest in their production languished until methods could be evolved to produce them reproducibly in quantity.

II. ANTHER-DERIVED HAPLOIDS

If, as seen above, genetic techniques contributed information indicating the possibility of obtaining haploid plants, it is the tissue culture technique of Guha and Maheshwari (1964) that is widely recognized as the principal cornerstone in the induction of haploid embryoids and plants in great frequency. These workers found that when excised anthers of *Datura innoxia* (Solanaceae) were cultured in a mineral salt medium in conjunction with coconut milk and other complex organic substances and growth hormones, embryo-like outgrowths appeared from the sides of the anther in about 6–7 weeks (Fig. 15.1). In later studies they confirmed the haploid nature of embryoids and their origin from microspores (Guha and Maheshwari, 1966, 1967). It was found that in the cultured anthers, a variable but substantial proportion of microspores began to divide repeatedly to produce a multicellular pollen grain. Later, the exine ruptured, liberating the shapeless mass of cells which organized structures not unlike zygotic embryos. Thus, the emergence from cultured anthers of plantlets having their origin in microspores or pollen grains is one of the hallmarks of anther-derived haploids. The term "anther androgenesis" has been suggested for this phenomenon to distinguish it from "ovule androgenesis" involving the development of the male gametophyte in the female cytoplasm described in the previous section (Pandey, 1973). Comprehensive treatments of the production of embryoids from anthers are to be found in the reviews by Sunderland (1971, 1973), Norreel (1973a) and Pandey (1973).

The discovery of Guha and Maheshwari served to launch an extensive research on the induction of embryoids from anthers of other plants. According to a recent compilation (Smith, H. H., 1974) production of haploids by anther culture has been reported in 17 genera, 23 species and four interspecific hybrids representing five angiosperm families. Among those investigated, it is the peculiar virtue of anthers from flowers of plants belonging to the Solanaceae to respond to excision

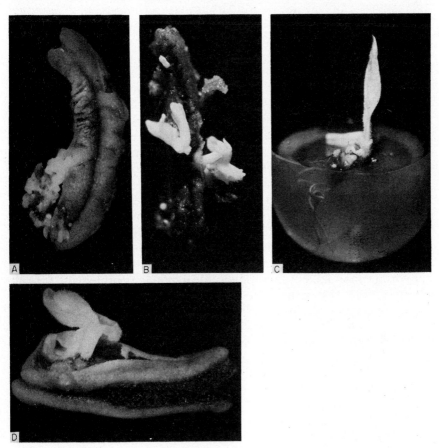

FIG. 15.1. Culture of excised anthers of *Datura innoxia* on complex media containing coconut milk and other additives results in the appearance of embryoids. **A, B,** Anthers showing embryoids 7 and 9 weeks, respectively, after culture in a medium containing indoleacetic acid and kinetin. **C, D,** Formation of seedlings from cultured anthers. (After Guha and Maheshwari, 1967; photographs courtesy of S. C. Maheshwari)

and culture by production of embryoids. Convincing demonstration of the direct transformation of microspores into embryoids has been provided in cultured anthers of *Datura stramonium* (Guha and Maheshwari, 1967), *D. metel* (Narayanaswamy and Chandy, 1971), *D. meteloides*, *D. muricata* (Nitsch, 1972), *D. wrightii* (Kohlenbach and Geier,

1972), *Nicotiana tabacum*, *N. sylvestris* (Bourgin and Nitsch, 1967), *N. alata*, *N. langsdorffii*, *N. glutinosa* 2N, 4N (Nitsch, 1969), *N. affinis*, *N. rustica* (Nitsch and Nitsch, 1969), *N. otophora* (Collins *et al.*, 1972), *N. debneyi*, *N. paniculata*, *N. plumbaginifolia*, *N. longiflora*, *N. undulata* (Nakamura and Itagaki, 1973), *N. knightiana*, *N. raimondii*, *N. attenuata* (Collins and Sunderland, 1974), *N. sanderae*, *N. clevelandii* (Vyskot and Novák, 1974), *Atropa belladonna* (Zenkteler, 1971), *Lycium halimifolium* (Zenkteler, 1972), *Petunia hybrida*, *P. hybrida* × *P. axillaris* (Raquin and Pilet, 1972), *Solanum dulcamara* (Zenkteler, 1973), *S. tuberosum* (Dunwell and Sunderland, 1973), *Capsicum annuum* (George and Narayanaswamy, 1973; Wang, Sun, Wang and Chien, 1973), *Hyoscyamus niger*, *H. albus* and *H. pusillus* (Raghavan, 1975b). In various species of *Nicotiana* (Nitsch, 1970; Nitsch and Nitsch, 1970; Niizeki, 1973a), and *Hyoscyamus* (Raghavan, 1975b), *Atropa belladonna* (Zenkteler, 1971), *Datura metel* (Narayanaswamy and Chandy, 1971; Iyer and Raina, 1972), *Capsicum annuum* (George and Narayanaswamy, 1973), embryoids go through the globular, heart-shaped and torpedo-shaped stages typical of the ontogeny of normal embryos before they elongate and form shoot and root meristems (Fig. 15.2). It may be significant that at least in *Nicotiana tabacum* androgenic embryoids have a higher cell number per unit area than corresponding stages of zygotic embryos, suggesting that the former may have a more favorable genomic complement for cell division than the latter (Norreel, 1973b). That the pathway of embryoid induction through the direct transformation of the microspore might be widely prevalent in other angiosperm families is indicated by the success obtained in producing embryoids from cultured anthers of certain cereals (Gramineae) (Guha *et al.*, 1970; Thomas and Wenzel, 1975).

Nitsch and Nitsch (1969) reported rearing a large number of haploid plants from cultured anthers of *Nicotiana tabacum*, *N. sylvestris*, *N. affinis* and *N. rustica*. Although the plants flowered profusely, flowers were smaller than those formed on diploid plants, and further, did not set seeds. Varying degrees of success have been recorded in rearing haploid plants from cultured anthers of other species listed above. From these well documented studies, the capacity of microspores in cultured anthers to form haploid embryoids and plants can now scarcely be questioned.

In the formation of anther-derived haploids, a useful distinction may be drawn between species in which haploids arise directly from the microspores and those in which they arise through the intervention of a callus. This latter process is exemplified in the androgenesis of *Oryza sativa* var. *japonica* and *O. sativa* × *O. sativa* var. *indica*, described by

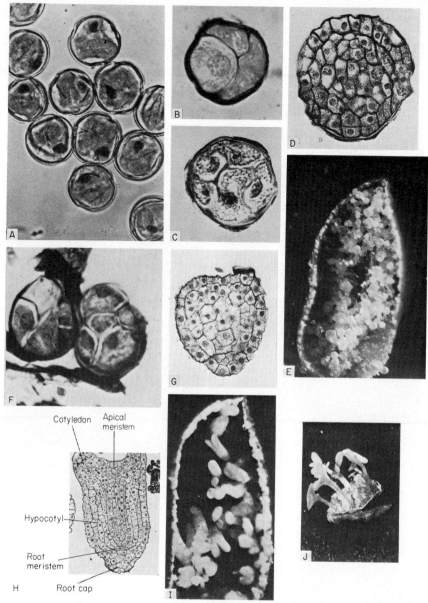

Fig. 15.2. Pollen embryoids of *Nicotiana tabacum* go through stages typical of zygotic embryos. **A**, Microspores at an early stage before division of the nucleus. **B**, Three-celled stage. **C**, **D**, Globular embryoids. **E**, Globular embryoids contained within an anther. **F**, Young proembryoid. **G**, Heart-shaped embryoid. **H**, Torpedo-shaped embryoid. **I**, Torpedo-shaped and cotyledonary stage embryoids contained within an anther. **J**, Androgenic plantlet. (**A, B, E, F, I** after Nitsch, 1969; **C, D, G, H** after Norreel, 1970; **J** after Nitsch and Nitsch, 1969; copyright 1969, American Association for the Advancement of Science; photographs courtesy of C. Nitsch and B. Norreel)

Niizeki and Oono (1968, 1971). Starting with anthers containing uninucleate microspores cultured in a medium supplemented with 2,4-D
or NAA, it was found that in about 2 weeks the microspores developed
into multicellular bodies which later gave rise to an exceedingly dense
callus. Subculture of the callus in an embryogenesis-inducing medium

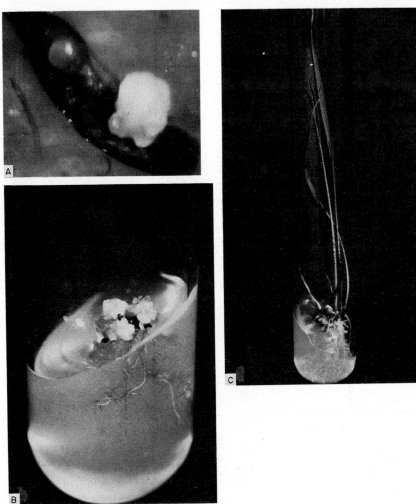

Fig. 15.3. Culture of excised anthers of rice results in the formation of a callus from
the microspore which later regenerates haploid plants. **A**, Callus emerging from the
anther after 35 days in culture. **B**, Rooting of callus 5 days after subculture. **C**, Differentiation of seedlings 26 days after subculture. (After Iyer and Raina, 1972; photographs courtesy of R. D. Iyer)

containing specific concentrations of auxins and a cytokinin led to initiation of haploid seedlings. Studies with a number of other plants,
including other subspecies of *O. sativa* (Harn, 1969; Harn and Hwang,
1970; Iyer and Raina, 1972), *O. sativa* var. *indica* × *O. sativa* var. *japonica*
hybrid (Woo and Tung, 1972; Woo and Su, 1975), *Brassica oleracea*
(Kameya and Hinata, 1970), *Lolium* × *Festuca* hybrid (Nitzsche, 1970),
Lolium multiflorum, Hordeum vulgare (Clapham, 1971), *Pelargonium hortorum* (Abo El-Nil and Hildebrandt, 1971, 1973), *Agropyron* (Kimata
and Sakamoto, 1971), *Aegilops* (Kimata and Sakamoto, 1971, 1972),
Setaria italica (Ban *et al.*, 1971), *Lycopersicon esculentum* (Sharp *et al.*, 1971a;
Gresshoff and Doy, 1972a), *Solanum nigrum* (Harn, 1971, 1972a, b), *S.
verrucosum* (Irikura and Sakaguchi, 1972), *Arabidopsis thaliana* (Gresshoff
and Doy, 1972b), *Lilium longiflorum* (Sharp *et al.*, 1971b), *Asparagus officinalis* (Pelletier *et al.*, 1972), *Prunus armeniaca* (Harn and Kim, 1972),
Triticum species (Kimata and Sakamoto, 1971; Chu *et al.*, 1973; Picard,
1973; Ouyang *et al.*, 1973; Wang, Chu, Sun, Wu, Yin and Hsü, 1973),
Triticale (Wang, Sun, Wang and Chien, 1973) *Coffea arabica* (Sharp *et
al.*, 1973) and *Populus* species (Sato, 1974), soon showed that the microspore → callus → plantlet route was a successful alternate pathway for
haploidy (Fig. 15.3). Production of albino plants in high proportions
is a general but as yet unexplained phenomenon observed in the formation of pollen plantlets in cereal grains (Niizeki and Oono, 1971; Clapham, 1971, 1973; Kimata and Sakamoto, 1972; Ouyang *et al.*, 1973
Wang, Chu, Sun, Wu, Yin and Hsü, 1973; Wang, Sun, Wang and
Chien, 1973); in rice nearly 50% of the haploids formed are reported
to be albinos (Sun *et al.*, 1974b).

In *Datura metel* (Iyer and Raina, 1972), *D. meteloides, D. innoxia* (Geier
and Kohlenbach, 1973) and *Capsicum annuum* (Wang, Sun, Wang and
Chien, 1973), microspores may give rise to embryoids either directly or
through the intermediary of a callus. According to Geier and Kohlenbach (1973) in both *D. meteloides* and *D. innoxia*, cell division in the
microspore unaccompanied by cell enlargement results directly in embryoids, whereas cell division associated with cell enlargement fore-
shadows callus growth (Fig. 15.4). This does not, however, explain why
certain microspores choose one or the other pathway of embryoid
formation. The control system favoring callus growth and embryoid
formation, difficult to decipher in *D. meteloides* and *D. innoxia*, is equally
ill-defined in *Festuca arundinacea* and *Phleum pratense*, where the microspores, after producing typical early stage embryoids, differentiate into
callus (Niizeki and Kita, 1974). In all the three species of *Datura* referred
to above, a plantlet is formed from the haploid callus through stages
reminiscent of developing zygotic embryos; in other plants, callus

tissues in culture regenerate roots, shoots and whole plants without recourse to a pathway involving embryo-like stages. Finally, the failure of
haploid callus grown from pollen grains to regenerate plantlets also
seems to be well established, despite the fact that the period of growth

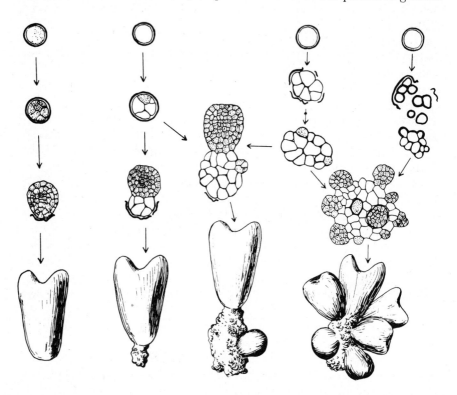

Fig. 15.4. The pathways of embryoid formation from microspores of *Datura meteloides*
and *D. innoxia* are illustrated schematically. The dotted cells represent embryogenic
components. (After Geier and Kohlenbach, 1973)

in some cases may extend for several months (Tulecke, 1953, 1959;
Tulecke and Sehgal, 1963; Konar, 1963; Yamada *et al.*, 1963; Bonga
and Fowler, 1970; Razmologov, 1973; Gresshoff and Doy, 1974;
Novák, 1974; Jordan, 1974; Bonga, 1974a; Bonga and McInnis, 1975).
It thus seems that in the induced morphogenesis of microspores and
pollen grains there is a gradation of morphogenetic pathways of which
the least complex is the formation of an undifferentiated callus.

A. Requirements for Androgenesis

Anther Age

The ease with which embryoids are produced in cultured anthers of several species has occasioned great interest in the precise age of the anther and the stage of development of the microspore at which induction is accomplished. Do anthers of all ages respond to induction or do they respond only during a specified period in their developmental cycle? Can embryoid formation be ascribed to any particular cell type in the pollen grain or is this associated with the pollen grain as a whole? Some aspects of these problems have been usefully clarified in *Nicotiana* and *Datura* anthers, while investigations are still proceeding with other plants. Anthers of *N. tabacum* cultured at the stage of mature pollen grain in which starch accumulation had commenced and the large central vacuole had disappeared generally failed to form embryoids (Nitsch *et al.*, 1968; Nitsch and Nitsch, 1969). Formation of embryoids occurred with high frequency when anthers containing fully individualized microspores undergoing the first mitotic division were cultured. Embryoid formation also occurred in anthers cultured immediately after the first microspore mitosis was complete (Norreel, 1970; Nöth and Abel, 1971; Sunderland and Wicks, 1971; Niizeki and Grant, 1971; Niizeki, 1973a; Engvild, 1974). Some authors (Nakata and Tanaka, 1968; Carlson, 1970), using different cultivars of *N. tabacum*, have successfully raised plants from anthers containing pollen tetrads. Flexibility of anther age is a feature that *N. tabacum* has in common with *Datura innoxia* in which anthers are at the height of sensitivity when cultured just prior to, during or immediately after the first microspore mitosis (Norreel, 1970; Engvild *et al.*, 1972; Sunderland *et al.*, 1974) (Table 15.I). In *D. innoxia*, anthers with tetrads and fully bicellular pollen grains are also productive (Sopory and Maheshwari, 1972). The critical phase for induction in both *N. tabacum* and *D. innoxia* probably corresponds to the S or G_2 phase of the mitotic cycle of the microspore nucleus (Sunderland and Wicks, 1969; Sunderland *et al.*, 1974). The uninucleate stage of the microspore has been found to be favorable for *Oryza sativa* (Guha *et al.*, 1970; Niizeki and Oono, 1971), *Atropa belladonna* (Zenkteler, 1971; Narayanaswamy and George, 1972; Rashid and Street, 1973), *Datura metel* (Iyer and Raina, 1972), *Asparagus officinalis* (Pelletier *et al.*, 1972), *Lycium halimifolium* (Zenkteler, 1972), *Triticum aestivum* (Wang, Chu, Sun, Wu, Yin and Hsü, 1973; Ouyang *et al.*, 1973), *Pelargonium hortorum* (Abo El-Nil and Hildebrandt, 1973) and *Capsicum annuum* (Wang, Sun, Wang and Chien, 1973; Kuo *et al.*, 1973). In *Brassica oleracea* (Kameya and Hinata, 1970) and *Lolium* × *Festuca* hybrid (Nitzsche, 1970), only anthers as far advanced in microsporogenesis as those containing nearly mature pollen grains give rise to callus and plantlets. In *Arabidopsis thaliana* (Gresshoff and Doy, 1972b) and *Lycopersicon esculentum* (Gresshoff and Doy, 1972a), callus initiation

TABLE 15.I. Effect of pollen stage on plantlet formation in *Datura innoxia* anther cultures. (From Sunderland *et al.*, 1974)

Pollen stage	Total number of anthers cultured	Number of anthers producing plantlets	% of anthers producing plantlets
1	35	13	37
2	27	17	63
3	113	96	85
4	20	19	95
5	122	84	69
6	116	16	14

Pollen stages: 1, tetrads and young microspores in G_1 of the cell cycle; 2, midphase vacuolate microspores; 3, late phase microspores with nuclei completing DNA synthesis or in G_2; 4, first pollen mitosis; 5, young pollen grains with generative and vegetative nuclei separated by a wall, but with the microspore vacuole still intact; 6, slightly older pollen grains lacking a vacuole, and generally free of starch.

occurs when anthers are cultured at the microspore meiosis stage; in the last-named species anthers cultured at the stage of the first microspore mitosis are also responsive to callus induction (Sharp *et al.*, 1971a). However, it should be kept in mind that although the majority of microspores in a given anther will conform to a specific developmental stage, the possibility is not precluded that embryoids are initiated from microscopes of other developmental stages which also occur to lesser degrees in the same anther.

Nutritional Requirements

Despite the fact that experimental androgenesis has been achieved in several plants, it is difficult to make generalizations about the exact nutritional requirements for embryoid formation. This is due to the fact that the basic nutrients required vary from one species to another and may also be related to the experimental techniques and cultural conditions. In the widely exploited species, *Nicotiana tabacum*, the minimal medium for androgenesis is as simple as a 2% sucrose solidified with agar. However, only a small percentage of cultured anthers supplied with this limited diet formed embryoids which remained arrested at the globular stage (Nitsch, 1971). The optimal medium devised by Bourgin and Nitsch (1967) consists, in addition to sucrose, of several major salts, minor salts, and organic addenda such as inositol, glycine, nicotinic acid, pyridoxine, thiamine, folic acid and biotin. A great weight of evidence indicates that among the mineral elements, iron is especially important for embryogenesis, best results being obtained when the element is supplied in a chelated form (Nitsch, 1969; Sopory and Maheshwari, 1973; Rashid and Street, 1973). The organic substances appear to be optional since embryoid induction has been achieved in anthers of several species of *Nicotiana* cultured in a sucrose-containing mineral salt medium with or without these compounds (Nitsch, 1969, 1972; Nitsch and Nitsch, 1969). According to Sharp *et al.* (1971a), anthers of *N. tabacum* cultured in White's inorganic salt medium in which iron is supplied as FeEDTA, and in which nicotinic acid, pyridoxine, thiamine, Ca-pantothenate and inositol are also present, form embryoids even in the absence of added sucrose, while relatively high concentrations of sucrose lead to initiation of albino shoots.

Several investigators have shown that enrichment of the medium with growth hormones or complex organic additives such as yeast extract, casein hydrolyzate or coconut milk in conjunction with auxins or cytokinins is essential for callus or embryoid formation in cultured anthers of some species (Guha and Maheshwari, 1964; Nakata and Tanaka, 1968; Guha *et al.*, 1970; Zenkteler, 1971; Clapham, 1971; Iyer and Raina, 1972; Kohlenbach and Geier, 1972; Narayanaswamy and Chandy, 1972; George and Narayanaswamy, 1973; Sun *et al.* 1973; Wang, Chu, Sun, Wu, Yin and Hsü, 1973; Wang, Sun, Wang, Chien, 1973). In relatively older flower buds of tobacco, addition of activated charcoal to the medium has been reported to enhance anther response (Nakamura and Itagaki, 1973; Anagnostakis, 1974). However, as the field continues to expand it has become clear that in some cases the additives can be eliminated from the culture medium without any reduction in the embryogenic capacity of the anthers. Although Guha and Maheshwari (1964) supplemented Nitsch's or White's medium with coconut milk, grape juice, plum juice, and casein hydrolyzate along with IAA and kinetin in their pioneer study on the anthers of *Datura innoxia*, later work (Sunderland *et al.*, 1974) has shown that the use of Bourgin and Nitsch's (1967) medium is in no way inferior for embryoid formation in this species. A lack of effect of organic additives cannot be entirely ruled out since embryoid induction occurs in anthers containing tetrads and bicellular pollen grains only in a medium containing coconut milk or kinetin (Sopory and Maheshwari, 1972). Rashid and Street (1973) obtained embryoids from anthers of *Atropa belladonna* with high frequency when they were cultured in a mineral salt medium in which iron was supplied as the ferric salt of ethylenediamine-di-*O*-hydroxyphenylacetic acid (FeEDDHA); addition of auxin and cytokinin used to induce embryoids in an earlier work on the same species (Zenkteler, 1971; Narayanaswamy and George, 1972) resulted in callus growth. Other studies have also drawn attention to the possible importance of growth hormones in determining the pathway of differentiation of microspores into embryoids or callus. For example, embryoids originating from anthers of *indica* subspecies of rice cultured in a medium enriched with auxin resumed further growth only upon transfer to an auxin-free medium, whereas subculture in the original medium led to differentiation of a continuously growing friable callus (Guha *et al.*, 1970; Guha-Mukherjee, 1973). A further observation of interest is that anthers which habitually differentiate a callus from microspores do so in a medium enriched with growth hormones or organic additives (Nakata and Tanaka, 1968; Kameya and Hinata, 1970; Nitzsche, 1970; Gresshoff and Doy, 1972a, b; Abo El-Nil and Hildebrandt, 1973; Ouyang *et al.*, 1973; Wenzel and Thomas, 1974). Where microspores give rise to a haploid callus, transfer of the tissue to a medium containing a different mixture of the same or different hormones or organic additives is necessary to obtain root and shoot initiation and plantlet formation (Niizeki and Oono, 1968, 1971; Harn and Hwang, 1970; Kameya and Hinata, 1970; Nitzsche, 1970; Gresshoff and Doy, 1972a, b; Narayanaswamy and Chandy, 1972; George and Narayanaswamy, 1973; Clapham, 1973; Picard and de Buyser, 1973; Wang, Chu, Sun, Wu, Yin and Hsü, 1973; Wang, Sun, Wang and Chien, 1973).

It was stated earlier that androgenesis occurs when intact anthers containing microspores at an appropriate stage of development are cultured. The question posed initially was whether association of the somatic tissues of the anther with the embryogenic microspores represents a causal relationship. A recent experiment has provided an affirmative answer to this question. The experiment in question involves administering a cold shock at 3°C to flower buds of *Datura innoxia* and culturing the free pollen grains in a medium habituated by the addition of an anther extract of the same species (Nitsch, C. and Norreel, 1973). The controlling effect of somatic tissues of the anther is apparently nonspecific since *Datura* anther extract is also beneficial for embryogenesis in microspores isolated from cold-treated anthers of *Lycopersicon pimpinellifolium* and *L. esculentum* (Debergh and Nitsch,

1973). A similar conclusion was derived from the successful culture of pollen grains of *Nicotiana tabacum* in the anthers of *Petunia hybrida* or in the callus tissue originating from explanted *Petunia* corolla (Pelletier and Durran, 1972; Pelletier and Ilami, 1972; Pelletier, 1973). It has been suggested that the somatic tissues of the anther provide a conditioning effect for the microspores during the early stages of culture, presumably when the crucial decision favoring an embryogenic pathway is made. Whether the influence of the somatic tissues is based on the action of some overt growth hormones on the microspores, or on the removal of some inhibitors from the latter, has not yet been determined.

In a major extension of this work, C. Nitsch (1974) was able to induce embryogenesis in free microspores of tobacco grown in a fully defined medium. In brief, the procedure consisted of administering a cold shock to flower buds for 48 h at 3°C and conditioning the anthers by allowing them to grow for 4 days in the regular anther culture medium. The microspores were squeezed out of the anthers and transferred to the embryogenesis-inducing medium consisting of glutamine, serine, inositol, in addition to mineral salts and sucrose. The possibility of contamination with somatic tissues of the anther was vigorously excluded by filtration and centrifugation of the pollen suspension. Under such conditions, the free microspores showed typical stages of embryogenesis similar to those occurring within cultured anthers. To some extent the results reinforce the notion that those crucial events signaling the change in the pattern of growth of the microspores were initiated during the conditioning phase, probably by drawing upon the materials of the anther wall. It would be a great step forward if free microspores could be induced to form a large number of embryoids by single step culture in a defined medium.

Other Factors

An overriding and obvious factor regulating embryoid induction is the trauma resulting from separation of the anther from the flower bud and its culture in an artificial medium. According to Picard (1973), mechanical stimulation of the intact anthers may also divert microspores from their normal division pattern to one resembling the early stage of an embryoid. Some experiments show that a switch in the normal gametophytic pattern of development of the microspore can be associated with internal metabolic changes. Bennett and Hughes (1972) have described a study in which spraying wheat inflorescence with ethrel (2-chlorethylphosphonic acid) induced additional mitotic divisions in the microspores of the intact anther very much like early division phase in embryogenic microspores of cultured anthers. Wang *et al.* (1974) have reported an increased androgenetic response in anther cultures of rice sprayed with this same chemical. Thus, additional means of induction of androgenesis presumably exist, and it may be that one of these is a change in the endogenous hormone level of the microspores.

Among the environmental factors studied, temperature in particular is suspected to affect androgenesis. Much of the work in this area is confined to *Nicotiana tabacum* in which optimum temperature for embryogenesis has been shown to be around 25°C. Lower temperatures tend to shift the balance toward the production of abortive embryoids and at temperatures below 15°C growth of embryoids is terminated prematurely (Sunderland, 1971, 1973). Specific light regimes, or lack of light thereof, do not seem to have any effect on embryoid induction; however, light is necessary to prevent etiolation and promote normal growth of plantlets (Devreux, 1970). Further, anther responses vary from one plant to another and are also dependent upon the age of the plant.

B. Mechanism of Embryoid Induction

Cytological Changes

In the ontogeny of the seed plant, formation of the microspore following meiosis in the microspore mother cell is the beginning of a short-lived male gametophytic phase. The microspore matures into the pollen grain, which germinates and forms a pollen tube. The latter serves to transport the sperms to the vicinity of the female gametophyte. The microspore nucleus undergoes one or two mitotic divisions to form a mature pollen grain. The first division is characteristically asymmetric and results in the formation of a small generative cell and large vegetative cell. The generative cell divides again to produce two sperm cells. The second division generally takes place after the pollen grains are discharged from the anther. Interestingly enough, the vegetative cell either degenerates soon after its formation or survives as a vestigial unit in the pollen tube.

Androgenesis is an event of subtle complexity in which a microspore destined to form pollen tube and gametes is turned on to form a sporophytic type of growth. In the anthers of *Nicotiana tabacum* (Bernard, 1971; Sunderland and Wicks, 1971) and *Datura metel* (Iyer and Raina, 1972) cultured at the uninucleate microspore stage embryoids are initiated from the larger vegetative cell arising from an initial asymmetric division. This cell loses its morphogenetic individuality, and after a short inductive period of 6–12 days begins to divide repeatedly by a

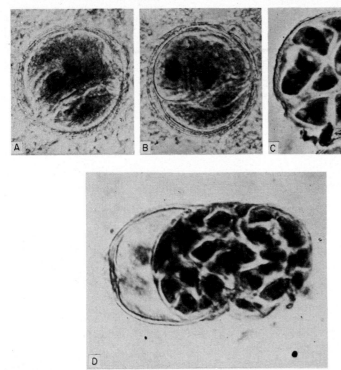

FIG. 15.5. Early ontogeny of embryoids from microspores of *Datura metel*. **A**, Two-celled pollen from an anther showing the generative cell and vegetative cell; the nucleus of the latter has divided into two. **B**, Wall formation between two vegetative daughter nuclei. **C**, A multicellular embryoid. **D**, Embryoid emerging from the exine. (After Iyer and Raina, 1972; photographs courtesy of R. D. Iyer)

series of internal divisions without intervening growth until a compact multicellular mass of cells typical of somatic cell size is produced (Fig. 15.5). Eventually the embryoid is liberated from the exine and finds its way into the anther locule where it completes further development. In *Hordeum vulgare* in which the microspore gives rise to a callus rather than to an embryoid, the cell aggregate formed from the vegative cell fails to form an organized structure upon its release from the pollen (Clapham, 1971) (Fig. 15.6). Continued divisions of the vegetative cell provoke the disintegration of the generative cell which at best divides only once or twice before disappearing completely. In a slight deviation, some embryogenic microspores of *Triticale* appear to form two seemingly identical nuclei after the first division. One of the nuclei answering the staining reactions of a typical generative nucleus

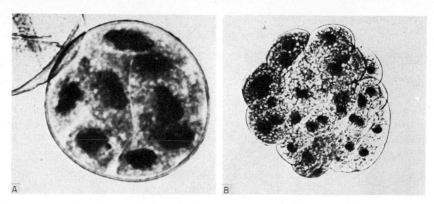

FIG. 15.6. Failure of cell aggregates formed in *Hordeum vulgare* to form embryoids. **A**, Multicellular pollen grain still enclosed in the exine. **B**, Pollen callus after bursting of the exine. (After Clapham, 1971; photographs courtesy of D. Clapham)

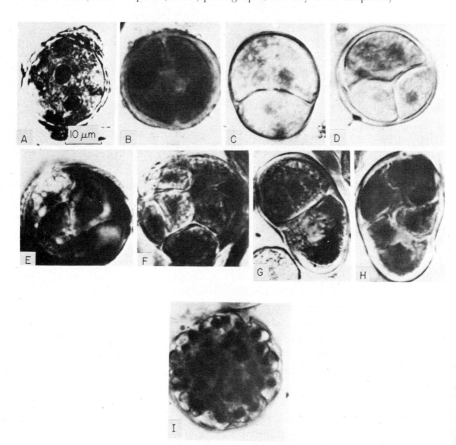

FIG. 15.7. In *Nicotiana sylvestris*, both vegetative and generative cells formed from the first microspore mitosis contribute to embryoid formation. **A**, Microspore with two nuclei. **B, C**, Bicellular microspores. **D, E**, Four-celled microspores. **F**, Microspore with seven or eight cells. **G, H**, Elongated embryoids. **I**, Late globular embryoid. (After Rashid and Street, 1974b)

produces several free nuclei, although the haploid callus itself is formed from the second, presumably the vegetative nucleus (Sun *et al.*, 1973, 1974a). Formation of an embryoid from the generative cell is a rare event, but has been reported (Devreux *et al.*, 1971).

In *Nicotiana tabacum*, *N. sylvestris* (Rashid and Street, 1974b), *Datura innoxia* (Norreel, 1970; Sopory and Maheshwari, 1972; Sunderland *et al.*, 1974), *Atropa belladonna* (Rashid and Street, 1973) and some *indica* subspecies of rice (Guha-Mukherjee, 1973), microspores give rise to two equal cells by a symmetric division and they jointly contribute to the formation of the embryoid. In *N. sylvestris*, where detailed cytological studies have been made, embryoids are formed almost exclusively by an equal division in the uninucleate microspore, whereas microspores which divide unequally turn nonembryogenic (Rashid and Street, 1974b) (Fig. 15.7). An increase in the proportion of microspores with equal cells secured by the administration of a cold shock to anthers at the time of the first pollen mitosis is believed to be the key to an enhancement of embryoid formation in pollen cultures of *Datura innoxia* (Nitsch, C. and Norreel, 1973).

In yet another embryogenic pathway described in *Datura innoxia* (Sunderland *et al.*, 1974), embryoids are formed by the participation of both vegetative and generative nuclei which undergo endoreduplication followed by fusion between chromosome complements. Less obvious types of chromosomal rearrangements clearly do take place during androgenesis, but it is difficult in most cases to document these occurrences. What determines the particular course of development to be followed by microspores in an anther is not understood, although the stage in the cell cycle of the microspore nucleus at the time of culture is an important consideration. In any event, by whatever pathway they are formed, the end products are not always easy to distinguish.

Cytochemical Changes

In discussions on the expression of cell characteristics in the male gametophytic generation of angiosperms, considerable emphasis has been given to the role of nucleic acids and their derivatives. During development of a pollen grain, the vegetative cell, which retains the great bulk of the cytoplasm, shows a rapid synthesis of RNA and protein and a negligible synthesis of DNA. The generative cell, which is cut off with only a small amount of cytoplasm, undergoes DNA synthesis to 2C level but exhibits little or no RNA and protein synthesis. Differentiation of the pollen grain therefore involves switching off the DNA synthesizing system in the vegetative cell, while RNA synthesis (transcription) proceeds; in the generative cell, transcription is slowed down without impairing DNA synthesis (Woodard, 1958; Sauter, 1969).

Transformation of the microspore into an embryoid is basically a problem in which a cell conditioned for terminal differentiation is switched onto a pathway of unlimited growth. This morphogenetic switch should clearly involve changes in biosynthetic processes, going from the precisely regulated metabolism of the male gametophyte to those concerned with the increased synthesis of macromolecules necessary for rapid cell growth and division. On the basis of the established role of nucleic acid and protein synthetic activity during differentiation of the microspore into a pollen grain, it follows that probably no events are closer to the mechanism of induction of androgenic embryoids than those concerned with the patterns of synthesis of DNA, RNA and proteins.

Unfortunately, specific information concerning the mechanism of androgenesis is scarce and the underlying events at the molecular level—the signal for division of the vegetative nucleus, the fate of the gametophytic program, and the role of the sporophytic program—all remain to be established. A recent cytochemical study of nucleic acid and protein changes in the embryogenic microspores of *Nicotiana tabacum* (Bhojwani *et al.*, 1973) has indicated a decrease in the amount of cytoplasmic RNA and protein in the vegetative cell prior to its division. It has been suggested that suppression of information concerned with gametophytic growth (that is, pollen tube and gamete formation) already established in the vegetative cell is the primary biochemical event in the embryogenic microspore. This is perhaps an oversimplification of a spectrum of changes that occur during the inductive period. For example, assessment of the fate of the gametophytic program of the vegetative cell involves the question whether this is transmitted to the adjacent generative cell; this might explain how the gamete-producing function of the generative cell is suppressed during androgenesis.

The essential role of histones in gene regulation has been documented by research programs of recent years. A general statement of histones being active in the generative nucleus of the embryogenic pollen in a transcription-suppressing role has been formulated by Sunderland (1971) on the basis of the reported presence of DNA-bound, lysine-rich histones in the generative nucleus of the microspores of *Paeonia* (Sauter, 1969). The assumption is also made that histones are absent in the vegetative nucleus or are present in a form that does not hinder RNA transcription. Decisive experiments to test the assumption have not been carried out.

The directive signals for embryoid initiation must also involve information which directs specific sections of the genetic apparatus of the vegetative cell to differential transcription of RNA and synthesis of proteins to effect cellular differentiation. Mascarenhas' (1971) study of RNA metabolism of germinating pollen grains of *Tradescantia* suggests a possible role for rRNA and tRNA which are synthesized about 24 h before and after the first microspore mitosis, after which their synthesis stops. Pollen grains cultured after termination of rRNA and tRNA synthesis germinate normally and form pollen tubes. In order to shift the microspores to an androgenic pattern, it seems reasonable to suggest that anthers should be cultured before the genes for rRNA and tRNA are turned off, but no evidence has been sought to substantiate this.

The transcriptional changes as well as changes in the control system governing DNA synthesis in the embryogenic microspores could be as diverse as the pathways involved in embryogenesis. Their elucidation will, it is hoped, provide clues to how androgenesis may be controlled to our advantage.

Ultrastructural Changes

Induction of pollen embryoids in general reflects a breakdown of the normal functional machinery of the microspore, and its acquisition of new morphogenetic capabilities. Is this transformation based on stable changes in the ultrastructural morphology of the cell? Since microspores from cultured anthers of only one species have been examined in the electron microscope, a definite answer to this question cannot be given. Moreover, there is no single dependable criterion to identify an embryogenic microspore at its very early stage of induction in a heterogeneous collection of pollen grains. Considering the distribution of organelles in the vegetative cell of a normally germinating pollen grain as an essential base line, the vegetative cell of the embryogenic pollen of *Nicotiana tabacum* has been shown to pass through a short phase characterized by a subcellular morphology typical of the male gametophyte, after which its structural profile remains unchanged (Vazart, B., 1971; Dunwell and Sunderland, 1974a). Characterized in this way, the vegetative cell had a dense assemblage of ribosomes, mitochondria, plastids, and ER interspersed with numerous vacuoles. The ribosomes were found to occur attached to ER, or in polysomal configurations or inside the plastids. The earliest change seen in the electron microscope that can be specifically related to embryoid induction is the suppression of formation of starch grains in the cytoplasm of the vegetative cell. This is followed by the complete degeneration of the cytoplasmic organelles and the appearance of multivesiculate bodies resembling lysosomes (Dunwell and Sunderland, 1974b). The basis for this change is unknown, but it is presumably linked to the operation of accessory mechanisms that prevent the continuation of the gametophytic program. Another aspect of the ultrastructure of the vegetative cell, described by B. Vazart (1973a, b) is a renewed synthesis of ribosomes and the elaboration of the plastids and mitochondria which occur coincident with the first sporophytic mitosis, heralding the embryogenic pathway. From these observations it seems that at least for tobacco microspores certain changes at the subcellular level signify their embryogenic potential.

C. Production of Homozygous Lines

Ploidy Level of Plants

Chromosome numbers offer tangible evidence for the ploidy level of cells, and as revealed by cytogenetic studies, phenotypic anomalies in plants can be traced to karyotypic alterations. Cytological examinations of anther-derived haploids have been made in some cases to correlate cytogenetic variations to chromosome patterns and to screen putative haploids for genetic and breeding work. In anther-derived populations of *Nicotiana tabacum* and *N. otophora*, the mode is precisely determined at the haploid level with no departures other than small proportions of diploid or aneuploid cells (Collins *et al.*, 1972). Where a more extensive degree of diploidy exists, it is not unusual for the plant to bear both haploid and diploid flowers (Sunderland, 1970). In *Datura innoxia*, 2N, 3N and 4N plants are produced in addition to the regular haploid type (Nitsch and Nitsch, 1970; Engvild *et al.*, 1972) while cells with 2N and 3N chromosomes have been recorded in anther-derived haploids of *Atropa belladonna* (Zenkteler, 1971). Since plants of higher ploidy level originate directly from microspores when anthers from progressively older flower buds are cultured, there is a possible correlation between ploidy level of regenerated plants and flower bud stage at culture (Engvild *et al.*, 1972; Engvild, 1973, 1974).

The presence of plants of different ploidy levels is frequently observed in anther-derived populations arising through the intervention of a callus. For example, *Datura metel* plants emerging directly from microspores appeared to be haploids, whereas the karyotype profile of plants regenerating from the callus shifted over to a grouping consisting predominantly of diploids and triploids (Narayanaswamy and Chandy, 1971; Iyer and Raina, 1972). Similarly, haploid to pentaploid plants have been recovered from anther cultures of *Oryza sativa*; elevation of chromosome numbers by ploidy shifts tended to be expressed by some distinguishing vegetative feature in the regenerated plants (Nishi and

Mitsuoka, 1969; Niizeki and Oono, 1971; Watanabe, 1974). Anther-derived diploids and triploids have been reported in *Petunia* (Raquin and Pilet, 1972; Engvild, 1973; Wagner and Hess, 1974), and diploids, triploids and aneuploids in *Solanum nigrum* (Harn, 1971, 1972a, b). From another cultures of different species of *Populus* only diploid plants have been thus far obtained (Sato, 1974). In rye anther cultures, plant-lets arising either directly from the microspores or through the inter-mediary of a callus are invariably diploid, with an occasional triploid or tetraploid form (Wenzel and Thomas, 1974; Thomas and Wenzel, 1975).

Several mechanisms might account for the shift in ploidy levels of plants originating from microspores. One obvious possibility is that of endomitosis (doubling of chromosomes without cell or nuclear division). It is well known that plant tissues cultured in media containing complex organic additives show a high frequency of natural chromosome duplication which might account for the formation of diploids and tetraploids. Indeed, in anthers of *Nicotiana suaveolens* × *N. langsdorffii* hybrid (Guo, 1972) and *N. tabacum* (Niizeki, 1973b, 1974b) grown in media containing growth hormones, cells with aberrant chromosome numbers are frequently observed. Other mitotic abnormalities such as failure of the spindle mechanism and formation of a restitution nucleus might also contribute to a multiplication of chromosome numbers. The origin of diploid plants can also be traced to the somatic tissues of the anther such as the tapetal cells, anther wall or the connective tissue which proliferate as a callus, like a pollen callus. Some investigators have suggested that nuclear fusions between the generative and vegetative nuclei which have already undergone one or two successive cycles of DNA synthesis might account for 3N and 5N plants (Narayanaswamy and Chandy, 1971; Engvild *et al.*, 1972; Sunderland *et al.*, 1974). Finally, in plants such as *Datura* in which meiotic irregularities during microsporogenesis are common, microspores with diploid, triploid and tetraploid numbers of chromosomes have been documented. The consistent occurrence of irregular microspores with a nonhaploid number of chromosomes agrees with the ploidy level of plants regenerating from anthers to suggest yet another route to the formation of nonhaploid plants (Collins, Dunwell and Sunderland, 1974).

It follows that uniformly haploid cell populations from pollen callus can be obtained only if the natural endomitosis of the cells and the undesirable growth of the diploid tissues are controlled. Although it has not been possible to control spontaneous chromosome doubling in cell populations by subtle changes in medium composition or by altering the physical conditions of culture, wide differences are to be expected between cell types in regard to their endomitotic potential.

Consistent with this statement is the observation of Niizeki and Oono (1971), who found that haploid callus derived from certain varieties of *Oryza sativa* could be subcultured for several years without shift in ploidy level of cells, while callus from some other varieties turned polyploid. The technical hurdle of selectively eliminating diploid cells in a population can, however, be overcome by the use of specific inhibitors. For example, some workers (Gupta and Carlson, 1972; Niizeki, 1974a) have found that *p*-fluorophenylalanine inhibits the growth of diploid cells of tobacco without affecting the growth of haploid cells. This discovery has an important application in maintaining stable populations of haploid cells in culture and in preferentially selecting haploid cells from mixed populations.

The culture of callus, exclusively of pollen origin, is plainly marked by genetic variability since such callus is formed by the division of a number of pollen grains which grow together in the same anther. Clearly, maintenance of uniform lines of haploid cells in culture for use in the development of mutants requires the rigid exclusion of any cross contamination. With this aim in mind, some workers have developed techniques to raise callus from single pollen grains. Formation of cell clusters from isolated pollen grains of *Brassica oleracea* grown in microcultures in a medium containing coconut milk has been reported (Kameya and Hinata, 1970). Starting with a preparation in which pollen grains of *Lycopersicon esculentum* spread on a filter paper disc were placed on top of a horizontally placed cultured anther of the same species in a way analogous to a nurse tissue culture, Sharp *et al.* (1972) isolated numerous clones of haploid callus originating from single pollen grains. Binding (1972) obtained masses of tissues from individual pollen grains of *Petunia hybrida* by suspending a large population of the pollen in a nutrient medium containing 2,4-D, kinetin and a high concentration of sucrose. As shown in *Atropa belladonna* (Rashid and Street, 1974a), callus outgrowths from explanted segments of haploid plants can serve as a source of suspension of haploid cells. Omission of a single essential constituent such as auxin from the medium will induce latent embryogenic potential in these cells during a limited number of subcultures. The interesting possibilities raised by these results suggest that a broad series of studies is obviously needed to determine whether haploid cell lines of single cell origin can in general be obtained and maintained in continuous culture.

Diploidization

Since haploid plants are usually sterile, diploidization of haploids to obtain sexually fertile populations of homozygous diploids becomes an important operation. From a practical standpoint, the most productive approach to diploidization is the spontaneous doubling of chromosomes that occurs in populations of haploid plants (Sunderland, 1970; Collins et al., 1972). A second, and perhaps the most common means of diploidization exploits the tendency of haploid callus to undergo endomitosis in culture. Callus tissues originating from pollen grains, after prolonged growth in the same medium, generally become homozygous diploid cell lines due to endomitosis during growth (Nishi and Mitsuoka, 1969; Kameya and Hinata, 1970). When pollen grains give rise directly to embryoids and plantlets, stem segments or other parts of the haploid plant are induced to form a callus in vitro. Repeated divisions of the cells of the callus accompanied by chromosome doubling result in the formation of a diploid tissue which is then transferred to a medium supporting regeneration of roots and buds. Subtle alterations in the hormonal components of the medium are necessary to induce callus growth from explants and the induction of roots and shoots from the former (Kadotani, 1969; Nitsch et al., 1969a; Kochhar et al., 1971; Harn, 1972c; Kasperbauer and Collins, 1972). A simple sequence of treatments consists of culture of explants in a medium containing auxin to induce callus growth, followed by a combination of auxin and kinetin to induce root and shoot initiation (Kochhar et al., 1971). Nitsch et al. (1969a) induced callus growth in explants of Nicotiana tabacum and N. sylvestris by growing them in a medium containing 2,4-D and benzyladenine which were replaced by IAA, adenine and zeatin to induce differentiation. The method of Kasperbauer and Collins (1972) uses a single formula, and starting with leaf mid-veins of N. tabacum, recognizable leafy shoots are formed in about 2–3 weeks in culture.

Colchicine treatment of haploid seedlings has also been successfully employed to produce diploids in tobacco (Tanaka and Nakata, 1969; Harn and Kim, 1971; Burk et al., 1972). Rashid and Street (1974a) obtained regeneration of diploid embryoids in high proportions when a haploid cell suspension of Atropa belladonna growing in a medium containing NAA and kinetin was exposed to colchicine, and subsequently transferred to an auxin-free medium. With some systems, colchicine-induced diploids may be inferior and in such cases spontaneous diploidy of callus tissues is the preferred method to regenerate diploid plants. The meiotic stability of doubled haploids of tobacco arising by endomitosis or colchicine treatment has been confirmed by cytological studies (Collins and Sadasivaiah, 1972) and the relatively low frequency of nuclear aberrations observed in such diploids qualifies them for use in breeding programs. The concept of genetic purity of doubled haploids of rice has received some support from an analysis of the peroxidase isozyme patterns of their leaves (Woo and Su, 1975). Since potentially embryogenic pollen grains of haploid anthers have been shown to regenerate diploid plants by endoduplication in culture (Chandy and Narayanaswamy, 1971), androgenesis of haploids may also serve as an alternative method for the production of completely homozygous diploids.

III. APPLICATIONS

The use of haploids and their homozygous derivatives as a tool in plant breeding, a hopeless pursuit several years ago, is now a realistic practical goal for geneticists and plant breeders (Melchers and Labib, 1970; Melchers, 1972). Because haploid plants can be obtained in quantity by anther culture techniques and transformed into true-breeding isogenic diploid lines by simple procedures, experimental systems are becoming available for planning breeding programs for crop improvement and the development of new varieties. Screening of haploid progenies from heterozygous F_1 individuals has already been applied in tobacco genetics with respect to leaf color (Burk, 1970; Nakata, 1971), leaf shape (Melchers, 1971) and resistance to black shank (Collins et al., 1971), tobacco mosaic virus and wildfire diseases (Nakata and Kurihara, 1972). The isolation of nullihaploid plants from cultured anthers of monosomic tobacco plants and the regeneration of nullisomics provide a tool for determining linkage relationships and the development of genetic maps in this plant (Mattingly and Collins, 1974). The feasibility of producing breeding lines of haploid tobacco plants differing in alkaloid contents has recently been demonstrated (Collins, Legg and Kasperbauer, 1974). In some cases, androgenesis has opened the way to the evolution of disease-resistant selections, as seen in the formation of *Geranium* plants free of virus-like leaf symptoms by anther culture (Abo El-Nil and Hildebrandt, 1971).

The selection of haploid mutants from populations of androgenic haploids by irradiation techniques has been pursued by some workers. Nitsch et al. (1969b) reported the appearance of tobacco mutants with variant patterns of leaves and flowers by irradiating plantlets at the time of emergence with γ-rays. Devreux and Saccardo (1971) X-irradiated anthers of tobacco at the time of the first microspore mitosis and obtained upon culture from such anthers several aberrant phenotypes with different plant height, leaf shape and colors.

A great potential of haploid cell lines is that they provide a foothold within higher plants for applying the powerful techniques of molecular biology and microbial genetics in uncovering and selectively recovering defined biochemical and temperature-sensitive mutants (Zenk, 1974). Tulecke (1960) isolated naturally occurring arginine-requiring strains of *Ginkgo* pollen by culturing them in media containing high concentrations of arginine. Experiments of Carlson (1970) have indicated that haploid cell lines of tobacco are useful in the induction and isolation of auxotrophic mutants. By the use of ethyl-methyl-sulfonate (EMS), six auxotrophs having impaired ability to synthesize hypoxanthine, biotin, *p*-aminobenzoic acid, arginine, lysine and proline have been isolated. Plants regenerated from the mutant cells showed different leaf shapes and growth characteristics very much like morphological mutants (Carlson, 1973a). The search for other mutagenic agents has inevitably led to the use of streptomycin and 5-bromodeoxyuridine (BUdR) and has resulted in the recovery of haploid cell lines of certain plants resistant to these drugs (Binding *et al.*, 1970; Maliga *et al.*, 1973a, b). The likelihood of emergence of agriculturally significant haploid mutants in tobacco has been enhanced by the discovery that plants regenerated from mutant cell cultures resistant to methionine sulfoximine, which is a doubtful structural analog of the toxin produced by the wildfire disease pathogen (*Pseudomonas tabaci*), showed milder disease symptoms when inoculated with the pathogen (Carlson, 1973b). So far there is only minimal evidence that haploid cell lines can be selected for extreme temperature minima and maxima (Melchers and Bergmann, 1959).

At a still more complex level, pollen-derived haploid cells of *Arabidopsis thaliana* and *Lycopersicon esculentum* have been successfully used for the transfer and subsequent expression of three systems of genes from the bacterium *Escherichia coli* (Doy *et al.*, 1973). The genes involved are the galactose and lactose operons coding for information for the utilization of galactose and lactose and a mutant suppressor gene which can correct amber nonsense codons in mRNA. It appears that such phage-mediated phenomena of gene transfer and associated functions of transcription and translation (termed "transgenosis") can result in inheritance within the cytoplasm in addition to nuclear inheritance. This study is of strategic importance since it strengthens the case for the use of genetic engineering techniques for information transfer between plant cells to produce desirable strains. Such experiments should also reveal to what extent we can control or predict the subsequent development of an individual cell by input of information at specific stages during its development and maturation.

IV. COMMENTS

During the brief span of slightly over 10 years since the discovery of pollen embryogenesis, considerable progress has been made in our understanding of the morphological and cytological aspects of embryoid induction, although many gaps in our knowledge still remain. Information is especially needed on the molecular mechanisms by which the microspores switch their assigned developmental route and form embryoids instead of pollen tubes and gametes. Because the genome is the ultimate seat of the informational blueprint of the cell, there is some important relationship between pollen embryogenesis and gene activity, but this remains to be elucidated.

One of the legitimate goals of anther culture is to increase the yield of haploids significantly, but this has seldom been achieved. It is a significant fact that generally less than 5% of the total microspore population of an anther gives rise to embryoids, although in some cases slightly higher percentages have been achieved. A knowledge of the mechanism of pollen embryoid induction should strengthen future attempts to bring a higher percentage of microspores into the embryogenic pathway.

Section III

From Seed to Seedling

16. Seed Dormancy: Developmental Block in the Embryo

The mature, full-grown embryo constitutes the structural link between the completed gametophytic and the ongoing sporophytic generations of the plant, and the attention it attracts from morphologists, physiologists, ecologists and biochemists is a measure of its importance as a biological unit which represents a crucial phase in the life cycle of the plant. Two fundamentally distinct but interdependent phenomena characterize the mature embryo. First, while still enclosed in the dry seed, it remains quiescent or dormant, retaining a capacity to stay metabolically inactive almost indefinitely, or for a period of time spanning a sizable segment of an unfavorable season, or until such time as an essential environmental condition is supplied. Second, this temporary halt of growth rate finds its counterpart in the reactivation of growth during germination which stirs up things quite effectively, and with amazing efficiency, leading to sweeping and permanent structural changes resulting in a new sporophytic plant. In both these respects, embryos are unmatched by any other part of the adult plant save the winter buds and underground stems and are not greatly different from spores of lower plants.

The state of suspended animation of the contained embryo and its later reactivation confront plant biologists with two major but closely related problems: (1) What are the special biochemical and physiological attributes of a contained embryo which enable it to remain in a depressed metabolic state and resume growth under favorable conditions? (2) What are the physiological processes initiated in the embryo which restore metabolic normality after a period of temporary suspension of growth? A survey of the literature on seed dormancy and germination leaves little doubt that the mature seed with its enclosed embryo has served as an eminently useful subject for a series of pioneering investigations on the processes and phenomena involved in the adaptation of the plant to different developmental phases during its life cycle.

The discussion of this chapter will focus on seed dormancy with special emphasis on the developmental block imposed on the embryo. As we mentioned incidentally in the introduction, the seed with its contained embryo may remain quiescent or dormant for a period of time after harvest. The period of arrested growth in the embryo is fundamental to further development as a protection against adverse environmental conditions. Beyond this there is the notion that dormancy or quiescence mechanisms in the seed provide a selective advantage for the survival of the species. At this point, a basic distinction must be made between quiescence and dormancy as applied to seeds. A quiescent seed is in an interrupted state of development which can be easily overcome by confronting it with appropriate environmental conditions for germination such as water, oxygen and a suitable temperature. On the other hand, seeds of most wild and cultivated plants of the temperate climate which contain well developed embryos do not germinate immediately after being shed, although external conditions are conducive to germination. Such seeds are dormant and will sprout only when they are given one of the specific pretreatments which are recognized as being necessary to break dormancy. A number of distinct forms of dormancy have been recognized, depending upon the pretreatments applied (Villiers, 1972c). These include, most importantly, mechanical restriction of the seed coat investments requiring their disruption to overcome dormancy; presence of chemical inhibitors which are removed by leaching the seeds; internal blocks to germination processes which are overcome by light or temperature treatments; and immaturity of the embryo which has to attain normal physiological and morphological growth. To summarize: we can envisage dormancy of the seed in terms of restraint imposed by external factors, so that while still encapsulated in the seed, the embryo is able to express a mechanism which prevents its own growth. A major impediment along the difficult road of understanding the physiology of dormancy concerns the nature of the mechanism imposing developmental arrest, and that involved in its termination.

In the following sections, attention will be confined to seeds in which dormancy is broken by light or temperature treatments, since such seeds have become especially amenable to experimental analysis and the results have immediate relevance to the developmental physiology of the embryo. However, it would lead us too far from the goal of this book to discuss here in any extensive detail all of the investigations that have been undertaken on breakage of seed dormancy by specific light or temperature treatments. Rather, greater emphasis will be given to key concepts which have been discovered in recent years, and which fit closely into the general terms of reference governing the subject of this book. For further information on the physiology of seed dormancy, the reader is directed to the comprehensive articles by Amen (1968) and Villiers (1972c), while authoritative presentations of the literature on the involvement of light and temperature in seed dormancy are provided in the reviews of Toole et al. (1956), Koller et al. (1962) and Stokes (1965).

I. LIGHT AND SEED DORMANCY

A. Phytochrome Action

Seeds of several species show a type of dormancy which is overcome by exposure to light (positively photoblastic seeds, for example, achene of lettuce) while in a few others dormancy is induced by light (negatively photoblastic seeds, for example, *Phacelia tenacetifolia*). Light may act directly on the embryo, or its action may be indirect on the extraembryonal tissues, such as the endosperm. Spectral requirements of positively photoblastic seeds have been studied in considerable detail for many years, and especially during the past few years there has been evolving an intriguing picture of light effects on seed germination. Suffice it to state here that it was the classical study of Borthwick et al. (1952) that led to the discovery of the red–far-red reaction system in light-sensitive seeds, namely, potentiation of germination by red light and its inhibition by far-red light. The effects of red and far-red light are mutually reversible, and if they are given in succession, the effect of the type of radiation given last will prevail. The mutual photoreversibility is due to the presence of a pigment, phytochrome, which can exist in two photoreversible states, one absorbing in the red region (P_r phytochrome) and the other in the far-red (P_{fr}

phytochrome) region of the spectrum. This physiological investigation has recently been reviewed (Rollin, 1972) and need not be detailed here.

Since seeds of a considerable number of other species also respond to red and far-red irradiances, it would appear that phytochrome is of wide-spread occurrence in positively photoblastic seeds and acts as a photoreceptor. What is the actual site of the photoreceptor system in the seed? In lettuce, it is probably localized in the tip of the hypocotyl itself (Ikuma and Thimann, 1959). The role of the ensuing photochemical reaction in stimulating germination is not entirely clear, but there is reason to believe that it leads to the secretion of cytolytic enzymes which decompose the endosperm and pericarp and overcome the mechanical resistance to embryo expansion (Ikuma and Thimann, 1963; Pavlista and Haber, 1970). Supporting this view is the observation that dormancy can be reimposed in naked embryos which germinate freely in the dark by incubating them in an osmoticum such as mannitol (Scheibe and Lang, 1965; Bewley and Fountain, 1972). The data of Ikuma and Thimann (1963) illustrate the possible function of light even more explicitly. These workers found that intact achenes of lettuce injected with polysaccharidases germinated in substantial numbers even in complete darkness. Thus the picture of light action on the embryo enclosed in the seed is more complex than was anticipated.

Perhaps in a similar way, by controlling the mechanical restraint of the seed coat, light prevents the growth of the embryo and inhibits germination of negatively photoblastic seeds. In the case of *Phacelia tenacetifolia*, isolation of the embryo from the mature seed, or removal of that part of the seed coat and endosperm which directly covers the radicle, allowed for full germination in light whereas incubation of isolated embryos or operated seeds in osmotica reinstated light sensitivity. That inhibition of seed germination in light could be ascribed to light-induced inhibition of the generation of expansive force by the embryo was shown by keeping isolated embryos in an osmoticum in the dark, when they still germinated (Chen and Thimann, 1966; Chen, 1970). While this analysis holds true to some degree in other cases (for example, *Nemophila insignis*; Chen, 1968), it is not sufficient to provide an adequate explanation for light inhibition of germination of negatively photoblastic seeds of *Citrullus colocynthis* in which the testa is lined with a thin inner membrane (Koller *et al.*, 1963). In this species light sensitivity is greatly reduced in excised embryos. If the testa is removed leaving only the inner membrane of the embryo, or if the excised embryo is covered with a filter paper, light inhibition is reinstated. Since mechanical restraint imposed by the inner membrane or by the filter paper does not appear to be significant, light may be assumed to inhibit germination by interfering with gaseous exchange.

In the germination of negatively photoblastic seeds, there is no convincing evidence for the involvement of phytochrome. On the other hand, detection of the pigment by *in vivo* spectrophotometry in excised embryos of positively photoblastic seeds (Tobin and Briggs, 1969; McArthur and Briggs, 1970; Boisard and Malcoste, 1970) has strengthened its postulated role in germination. According to Tobin and Briggs (1969), embryos excised from dry seeds of *Pinus sylvestris* show low levels of phytochrome which increase substantially upon hydration. It is clearly desirable that a closer look be taken at the actual locus of synthesis of phytochrome along the embryonic axis to see if there is any relation between the initiation of growth in the embryo and the site of phytochrome synthesis.

B. Light and Growth Hormones

Stimulation of germination of light-sensitive seeds by GA and to some extent by cytokinins has been widely documented in the literature. The interactions of light quality and hormones in seed germination are not yet clear in detail, but the analysis of Ikuma and Thimann (1960) on GA-induced germination of lettuce seeds has contributed much useful information. A possible model of germination of lettuce seeds confers upon red light the ability to induce the synthesis of GA, which in turn initiates the processes of germination. Favoring this view is the fact that GA often substitutes completely for red light when supplied to dark-imbibed seeds, but the validity of the model was generally discounted when no clear-cut increase in GA content of seeds occurred following irradiation with a saturating dose of red light. More recently, qualitative and quantitative changes in the levels of cytokinins have been reported in light-stimulated seeds to suggest a possible role for these hormones in germination (van Staden and Wareing, 1972; van Staden, 1973). The picture of light and hormone interaction has unfolded further through the use of isolated embryos. It appears that germination of embryos isolated from lettuce seeds is inhibited by abscisic acid and that this inhibition is relieved by cytokinins but not by GA. It is likely that since hormones have multiple effects extending to several sites, GA perhaps acts at a site independent of the other two hormones. Reasoning that GA or red light provokes the synthesis of cytokinins in the embryo has not been supported by experiments using isolated embryos (Bewley and Fountain, 1972; Black *et al.*, 1974).

Within the scope of this chapter, these examples must suffice to indicate the variability of light and hormone action on embryos of intact seeds. In a sense, light effects on seeds reveal a lot about the sensitivity of their contained embryos, since a loss of photosensitivity invariably occurs in the embryos of light-sensitive seeds upon removal of the seed coat (Toole *et al.*, 1956). To sum up, the study of the effects of light has contributed in some measure to our understanding of the physiology of seed germination, but it remains insufficient to account for the control of morphogenesis and initiation of growth in the dormant embryo.

II. TEMPERATURE AND SEED DORMANCY

Effects of temperature in overcoming dormancy are seen most dramatically in seeds of early-ripening species of fruit trees which fail to germinate until they are exposed in the presence of oxygen for weeks or months to low temperature, a treatment known as after-ripening. In a few rare cases, similar effects may be observed by exposing seeds to high temperatures (Stokes, 1965). In horticultural practice, after-ripening of seeds is achieved by layering them during winter in moist flats by a process known as stratification.

In the study of after-ripening of seeds of a great many species, it has been repeatedly emphasized that the problem is basically concerned with the physiological growth of the embryo. A fully after-ripened seed may in fact be considered to possess a morphologically and physiologically mature embryo which will germinate under conditions of appropriate warmth, moisture and aeration. All indications suggest that changes during after-ripening involve a radical reorientation of the metabolic status of the cells of the embryo which enables them to embark upon a phase of division and differentiation under conditions conducive to germination. It would appear that characterization of the physiological changes in the embryo responsible for the switch in the biosynthetic capacities of the cells might throw light on the basis of after-ripening.

Several problems relating to after-ripening in the seed, especially with reference to the kinetics of temperature action, interaction of temperature and light, and the influence of the seed coat, have been studied extensively and much of the pertinent work done in this area has been reviewed by Stokes (1965). In the following account, attention will be devoted to the action of low temperature on the development of the dormant embryo in seeds of several species which have been subjected to detailed analysis.

A. After-ripening of Seeds with Mature Embryos

A stringent test for evaluating the effect of chilling treatment is the ability of the embryo excised from the after-ripened seed to attain normal growth in culture. This is clearly demonstrated by the work of Flemion (1931), who has provided an accurate account of the after-ripening process in the seeds of *Sorbus aucuparia* which germinate fully after chilling at 1°C for 2–4 months. Embryos removed from non-after-ripened seeds do not grow in culture. When embryos are excised from seeds stratified at temperatures of $-5°, 1°, 5°, 10°$ and $15°$C and cultured at room temperature, best growth occurs in embryos of seeds pretreated

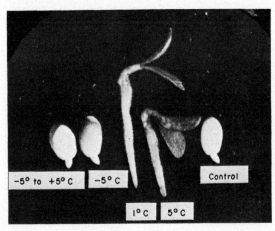

Fig. 16.1. Best growth of excised embryos of *Sorbus aucuparia* occurs when seeds have been stratified at 1° and 5°C. (After Flemion, 1931)

at 1°C and 5°C (Fig. 16.1). Growth of the excised embryo also depends upon the duration of the chilling treatment. If seeds are stratified for periods of 2, 4, 6 and 8 weeks, optimum growth occurs in embryos excised from seeds chilled for 6–8 weeks. Likewise, less than 10% of embryos excised from mature non-after-ripened seeds of *Polygonum scandens* grow in culture, but there is a several-fold increase in growth when embryos are excised from stratified seeds or if cultured embryos are subjected to low temperatures (Justice, 1940). In apple, attention has been drawn to the fact that irrespective of the duration of cold treatment of seeds, germination of excised embryos occurs only if they are planted with their cotyledons fully in contact with the nutrient medium (Côme *et al.*, 1968; Thévenot and Côme, 1971b, 1973b).

Seeds of *Symphoricarpos racemosus* respond to chilling only after seed coats are softened either by treatment at 25°C for 3 months in moist acid peat moss, or by a combination of a brief treatment in sulfuric acid and several weeks at 25°C (Flemion, 1934a). Embryos excised from dry seeds or from seeds which have been maintained at 5°C or at 20°C do not grow; on the other hand, considerable measure of success is

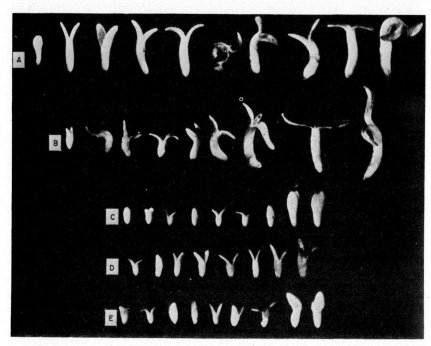

Fig. 16.2. Embryos of *Symphoricarpos racemosus* respond to after-ripening after seed coats have been softened in sulfuric acid or peat moss. Growth of embryos excised from seeds after various treatments are compared in this figure. **A**, Soaked for 75 min in concentrated sulfuric acid followed by 70 days at 5°C in moist peat moss. **B**, At 25°C in moist peat moss for 3 months, then transferred to 5°C for 70 days. **C**, At 25°C for 4 months in moist peat moss. **D**, At 5°C for 70 days in moist peat moss. **E**, Embryos from dry seeds. All cultures were aerated. (After Flemion, 1934a)

achieved with embryos excised from seeds in which the seed coats are softened by the methods referred to above and then chilled (Fig. 16.2).

Observations on the culture of embryos of cold-requiring seeds of *Rhodotypos kerrioides* are also instructive. When embryos from non-after-ripened seeds are excised and placed under conditions suitable for growth, they exhibit normal or near-normal growth up to a certain stage and eventually produce plants commonly referred to as physiological dwarfs (Flemion, 1933a, b). Dwarfness is a characteristic feature of seedlings obtained from non-after-ripened seeds of several species (Flemion, 1934b; Davidson, 1933; von Veh, 1936, 1939), and persists for years unless the seedlings are chilled or given some other suitable treatments. It has been suggested that dwarfism of seedlings originating from nonstratified apple seeds may be due to the high activity of peroxidases which apparently destroy auxin (Rychter and Lewak, 1971). From a comparative study of the rates of cell division and cell elongation in normal and dwarf seedlings of *Rhodotypos kerrioides* and *Prunus persica*, it has been concluded that physiological dwarfness is due to a lag in the rate of mitosis in the internodes (Ledbetter, 1960). The final appearance of seedlings obtained from embryos given varying periods of chilling can serve as a direct expression of dormancy; for example, embryos excised from fully after-ripened seeds form normal seedlings and the degree of dwarfness increases with decreasing periods of chilling (Tukey and Carlson, 1945a, b; Visser, 1956). Pollock and Olney (1959) have shown that irrespective of the other environmental parameters, when excised embryos of *P. persica* var. 'Elberta' were grown at different temperatures, dwarfness or nondwarfness was determined within a narrow range of temperatures between 23° and 27°C; at temperatures below 23°C the embryos developed into normal plants and over 27°C seedlings were severely dwarfed. By comparison of the growth in culture of embryos excised from seeds of *Agrostemma githago* at various times during ripening, Borriss and Arndt (1956a) showed that dormancy developed in different parts of the embryo at different times and to varying degrees. While embryos from unripe seeds developed uniformly in culture, growth inhibition was strongly manifest in the hypocotyl, weak in the cotyledons and intermediate in the radicle of mature embryos.

One must add to these experimental findings the fact that exposure of seeds or excised embryos of a few species to photoperiodic regimes or to a specific quality of light substitutes for their low temperature requirement. Cultured embryos of several varieties of cold-requiring seeds are remarkably sensitive to photoperiod and grow rapidly when exposed to continuous illumination at a moderate temperature (Lammerts, 1943; Flemion, 1956; Taylor, 1957). When embryos excised from non-after-ripened seeds of *Heracleum sphondylium* were maintained in culture, illumination in the early stages enhanced their rate of development in comparison to those grown in the dark (Stokes, 1953b). The light requirement for growth of excised embryos of non-after-ripened apple was met by short exposures to red light and the red light effect was repeatedly reversible by far-red light, thus implicating phytochrome in overall control of embryo growth (Lewak *et al.*, 1970; Smoleńska and Lewak, 1971). Although the evidence is circumstantial, perhaps, it is not unreasonable to infer that one of the factors that prevents the growth of the embryo in the non-after-ripened seed is lack of illumination, the requirement for which is lost after it has been after-ripened. Dormancy of cold-requiring apple seeds is overcome by incubating them in an atmosphere of nitrogen, although the basis for this effect is unknown (Tissaoui and Côme, 1973).

B. After-ripening of Seeds with Immature Embryos

In the examples described above, the embryo is fully developed in the seed at the time of shedding and after-ripening involves physiological changes in the embryo as a result of which germination is improved. There is evidence to indicate that even immature embryos of some early-ripening fruits possess a physiological system for dormancy which can be overcome by subjecting cultured embryos to low temperatures (Zagaja, 1961, 1962; Zagaja *et al.*, 1960; Abou-Zeid and Gruppe, 1972). Embryo dormancy of immature seeds of *Acer pseudoplatanus* is quite different in that at no time during maturation do embryos respond to a chilling treatment, although the same treatment of mature seeds elicits a positive response in the embryo (Thomas *et al.*, 1973). The mechanisms that implement cold-sensitivity or its lack thereof may be expressed in terms of altered metabolic sequences, particularly those involved in switching cells from accumulation of storage materials during maturation to the synthesis of metabolites needed for initiating germination. There are a few cases where the embryo is rudimentary and underdeveloped in the mature seed, and during after-ripening, in addition to physiological changes, we also observe changes in the morphology of the embryo. In these seeds, the morphological development of the embryo during after-ripening is undoubtedly of overshadowing importance for its subsequent growth. Examples of seeds with rudimentary embryos which require stratification for growth are *Heracleum sphondylium, Fraxinus nigra, F. excelsior, Pinus sylvestris, Picea abies, Anona crassiflora*, etc. (Stokes, 1965; Rizzini, 1973).

The potential value of seeds with dimunitive embryos for studies of the after-ripening process is indicated by the work of Stokes (1952b, c, 1953a, b) on *Heracleum sphondylium*. During after-ripening, the embryo develops from a spherical mass of cells comprising 0·4% of the dry weight of the seed to a fully developed organ weighing about 30% of the seed dry weight. The effect of low temperature in this species is seen from a comparison of the growth of the embryo at 2°C and 15°C (Fig. 16.3). At 2°C, the embryo reaches the mature size in about 9 weeks, while at the elevated temperature, in spite of an initial rapid development, there is virtually no growth after 6 weeks. It has been suggested that the initial growth of the embryo at both temperatures is sustained by nutrients present in'and around it, while the accelerated growth at low temperature is due to mobilization and utilization of nutrients from the surrounding endosperm (Stokes, 1952b).

Embryos of fully mature black ash (*Fraxinus nigra*) and European ash (*F. excelsior*) seeds are morphologically developed, but are somewhat smaller than the length they attain before germination. Growth of the embryo to its normal size takes place upon imbibition of the seed

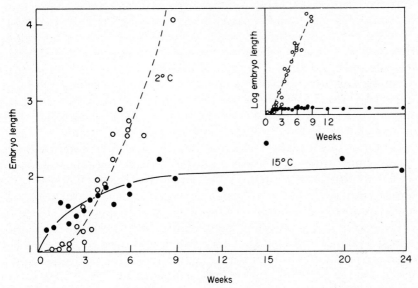

FIG. 16.3. Effect of low and high temperatures on the growth of embryos of *Heracleum sphondylium*. Embryo lengths are plotted as multiples of the original mean size. (After Stokes, 1952b)

at a warm temperature (22–25°C) for about 3 months (Steinbauer, 1937; Villiers and Wareing, 1964). Cells of the embryonic root meristem sampled during this period were characterized by marked alterations in ultrastructural pattern such as the development of ER into longer and more interconnected units, differentiation of plastids into starch-containing amyloplasts, appearance of golgi vesicles, microbodies and multivesicular bodies and elaboration of mitochondria. Lipid bodies which were passively massed in the cell along the plasma membrane and protein body membrane showed a general tendency for breakdown and dissolution until few, if any, remained in the cytoplasm (Villiers, 1971). Seeds can be maintained in a state of enforced dormancy for periods of 6–7 years without germination and apparent loss of viability by keeping them fully imbibed at 22–25°C. Cytological changes in the cells of the root meristem during this period have been followed by periodic electron microscopic examination of samples (Villiers, 1972a). By the end of the first year of imbibition, ER became dilated and sac-like in some areas of the cytoplasm, while in other areas they coiled around protein bodies and cytoplasmic organelles in close association. At the end of the second year, the organelles engulfed by ER were found to be in a state of degeneration. In this state they appeared to possess acid hydrolase activity and were similar in structure to the cytolysomes found in pathological animal cells (Villiers, 1967a). In many cells cytoplasm had become devoid of organelles after imbibition for 4 years and by this time some of the remaining mitochondria became enlarged into abnormal structures. By the fifth year other pathological changes involving membrane damage, swelling and bursting of plastids and mitochondria had occurred. However, even at this stage seeds showed a tendency for resumption of embryonic capabilities and germinate after appropriate after-ripening treatment.

Embryos of *F. excelsior* which attain full size after imbibition at a warm temperature also require a low temperature treatment at 5–6°C for 5–6 months for germination. Certain of the changes found to occur during the early stage of prolonged dormancy characterize the embryos during imbibition of seeds at the low temperature. These include association of ER and protein bodies, and appearance of cytolysomes. The only cytological change which appeared to be dependent upon the chilling treatment was an activation of the nucleolus, resulting in a large increase in its volume (Villiers, 1972b).

C. Mechanism of the Low Temperature Effect

Physiological Mechanism

An attractive explanation to account for the physiological basis of after-ripening is that a hormonal factor necessary for cell division is being synthesized at the low temperature. A crucial test of this hypothesis is to apply hormones to excised embryos, whole seeds or seedlings and see whether they can substitute for the chilling treatment. When GA was applied to embryos isolated from non-after-ripened seeds of *Malus arnoldiana* (Barton, 1956), this was found to be the case. Similar results have also been obtained by GA treatment of non-after-ripened seeds and embryos of other species (Monin, 1964; Stokes, 1965; Frankland and Wareing, 1966; Bradbeer and Pinfield, 1967; Jarvis *et al.*, 1968a; Baskin and Baskin, 1970; Côme and Durand, 1971; Le Page-Degivry,

1973b; Frost-Christensen, 1974). In some seeds, cytokinins can substitute for cold treatment although they appear to be less effective than GA (Frankland, 1961; Tzou et al., 1973; Pinfield et al., 1974). Significant increases in the amount of endogenous gibberellins (Frankland and Wareing, 1962; Ross and Bradbeer, 1968, 1971; Sińska and Lewak, 1970) and cytokinins (van Staden et al., 1972b; Brown and van Staden, 1973) found in seeds during chilling might substantiate a role for these hormones in dormancy breaking. There is some evidence to indicate that hormones synthesized during cold treatment may cause the mobilization of reserve materials of the embryo necessary for initiation of growth (Smoleńska and Lewak, 1974).

Fujii (1969) found that although embryos excised from non-after-ripened seeds of *Eragrostis ferruginea* developed normally in culture, they lost their ability to do so when seeds were after-ripened. Addition of yeast extract to the medium reinstated growth in the after-ripened embryos, suggesting that the synthesis by the endosperm of a substance similar to yeast extract might be involved in the after-ripening process.

A major alternative theory to account for the after-ripening process is that chilling treatment inactivates growth inhibitors and releases the embryo to express its full growth potential. For example, excision of the cotyledons, presumed to be the site of production of inhibitors, has been shown to result in normal growth of non-after-ripened embryos in culture (Flemion and Prober, 1960; Bulard and Monin, 1963; Côme and Thévenot, 1968; Thévenot and Côme, 1971a, 1973a; Abou-Zeid and Neumann, 1973). If embryos of non-after-ripened peach were allowed to grow initially with cotyledons intact and then decotylated, they showed features typical of non-after-ripened embryos (Flemion and Prober, 1960). This indicates that the inhibitor was probably active during the early stages of growth of the embryo and leaves a permanent impression thereof, perhaps by inactivating a hormone-synthesizing system. In another line of evidence, presence of the growth inhibitor, abscisic acid, and a decrease in its concentration during cold treatment have been demonstrated in embryos or whole seeds of several species (Lipe and Crane, 1966; Sondheimer et al., 1968; Martin et al., 1969; Rudnicki, 1969; Le Page-Degivry, 1970b, 1973a; Brown and van Staden, 1971; Diaz and Martin, 1972; van Staden and Brown, 1972; Paul et al., 1973). In some instances, application of abscisic acid has been shown to inhibit germination of after-ripened embryos and impose secondary dormancy in them (Durand et al., 1973; Le Page-Degivry, 1973c). If abscisic acid acts *in vivo* as it does *in vitro*, it seems a reasonable supposition that the growth inhibitor prevents growth of the embryo by inhibiting the synthesis of GA (Rudnicki et al., 1972).

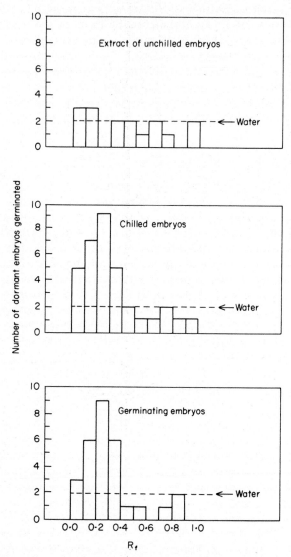

Fig. 16.4. Chilling of the seeds of *Fraxinus excelsior* causes the production of a germination promoter, as shown here by a bioassay based on the effects of embryo extracts of *F. excelsior* on the germination of dormant embryos of the same species. Extracts were separated by paper chromatography. Number of embryos germinating in water is shown by the dotted horizontal line. (After Villiers and Wareing, 1965a)

Using a bioassay based on growth promotion of dormant excised embryos, or growth inhibition of excised nondormant embryos of *Fraxinus excelsior*, Villiers and Wareing (1960, 1965a, b; Kentzer, 1966) have detected naturally occurring inhibitors and stimulators of germination in chilled seeds of this species. Figure 16.4 illustrates the results of a bioassay for detecting germination stimulators. Since GA can substitute for the chilling requirement of embryos and abscisic acid can reimpose dormancy in otherwise nondormant embryos (Villiers, 1968b), the identity of the germination-stimulating substance to the former and of the inhibitor to the latter appears likely.

Factors regulating dormancy of seeds of *Euonymus europaeus* have been the subject of continuing investigations, and hence merit a brief account. The seeds require stratification for a period of about 3–4 months at 4°C before embryos excised from them will grow in culture (Monin, 1966). Among a number of substances tested, only GA was found to substitute effectively for cold treatment and induce growth of excised embryos in high percentage (Monin, 1959; 1967b, c; Béranger-Novat and Monin, 1971a, b; Gambade, 1972). The key reaction during stratification would thus appear to be the synthesis of GA in the embryos. It is consistent with this suggestion that embryo extracts of stratified seeds have revealed, by chromatography and bioassay, the presence of GA-like substances (Monin, 1967a, 1968); in another line of evidence, inhibition of growth of isolated embryos by AMO-1618, an inhibitor of GA biosynthesis, has also been noted (Béranger-Novat and Monin, 1974). Although auxins have also been detected in the embryos of stratified seeds, there was no evidence for an overall regulation of dormancy by an auxin–gibberellin balance (Monin, 1967g). In order to account for the cytoplasmic metabolic activities of GA, it has been proposed that the final step in the growth of the embryo is the production of enzymes whose action enables the tip of the radicle to penetrate through the endosperm and seed coat. Some support for this hypothesis comes from the observation that cold requirement of seeds can be overcome by cellulase or by surgically removing the endosperm and seed coat at the site of the radicle or plumule. Studies on other cold-requiring seeds also permit assertions which indicate that dormancy is due to the mechanical restriction of the growth potential of the embryo by the seed coat (Black and Wareing, 1959; Fujii, 1969; Baskin and Baskin, 1971; Pinfield and Stobart, 1972; Pinfield et al., 1972) or by the inability of the embryo to hydrolyze endosperm reserves (Curtis and Cantlon, 1965, 1968).

Dormancy regulation in *Euonymus* also involves the active participation of an inhibitor. The effect of GA described above relates to release from dormancy if the hormone is absorbed by the root primordium. As Fig. 16.5 shows, GA does not substitute for cold treatment if it is imbibed through the cotyledons (Bulard and Monin, 1960a, b). As a result, the concept has developed that GA treatment of the cotyledons leads to the liberation of inhibitors, although it is not clear why the inhibitors are inactive when the hormone is translocated through the root. Using a bioassay based on GA-induced growth of dwarf pea, an inhibitor of GA has been identified in embryos of *Euonymus* (Monin, 1967d, e). Since *para*-coumaric acid was found to inhibit GA-induced growth of embryos, there is suggestive evidence for the identity of the inhibitor to this compound (Monin, 1967f). Despite the fact that we have here a confusing picture of the dormancy mechanism, these results may mean no more than that natural dormancy in seeds is geared to a balance and interaction of endogenous hormones and inhibitors (Khan, 1971; Webb and Wareing, 1972). In summary, we can state that the most fruitful generalization from the last few years' work is that it is not definitely established whether after-ripening leads to the synthesis of growth hormones or results in the destruction of inhibitors. The experiments in support of both alternatives are suggestive but not sufficiently conclusive to clinch the issue. No doubt, the promotor–inhibitor theme is a simple and attractive model of embryo dormancy in which temperature and other factors play a role.

Molecular Mechanism

The major question of concern to physiologists is, of course, the molecular mechanism whereby synthesis of growth promoters or growth inhibitors during cold treatment of seeds potentiates their germination during a subsequent period at an elevated temperature. The existence of a correlation between the progress of after-ripening of seeds and changes in the photosynthetic and respiratory activities, reducing sugar content, enzyme activity, and lipid and sugar metabolism of embryos, has been described by several workers (Borriss and Schneider, 1955a, b; Borriss and Arndt, 1956b; Borriss and Baum, 1957; Stokes, 1965; Bradbeer and Colman, 1967; LaCroix and Jaswal, 1967, 1973; Maciejewska *et al.*, 1974). But how much of this information will enable us to formulate an attack on the mechanism of low temperature action is a moot question.

Large, low-temperature-dependent increases in the synthesis and accumulation of cellular RNA in embryos of cold-treated seeds have been noted by some workers (Wood and Bradbeer, 1967; Khan *et al.*, 1968; Villiers, 1972b). Wood and Bradbeer (1967) have documented an increase in all RNA fractions during chilling of cotyledons of *Corylus*

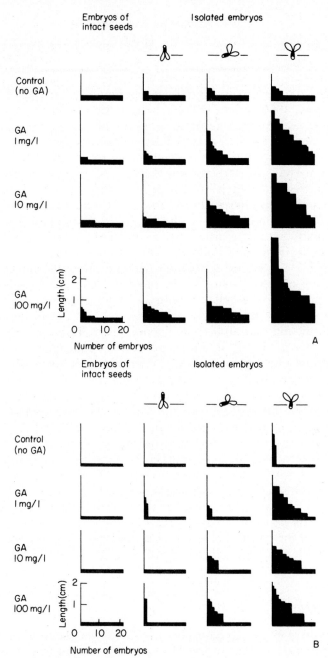

FIG. 16.5. Effect of different concentrations of gibberellic acid on the growth of the hypocotyl (**A**) and root (**B**) of isolated embryos of *Euonymus europaeus*, planted with the cotyledons, hypocotyl or root in contact with the medium. Effects of gibberellic acid on the growth of embryos of the intact seed are also shown in the figure. (After Bulard and Monin, 1960b; print courtesy of C. Bulard)

avellana, including a steep rise in rRNA during the later stages of chilling. Stratification of seeds of Bartlett pear is accompanied by an increase in nucleic acid synthesizing capacity as well as by an increase in the chromatin-bound RNA polymerase activity of embryos (Khan *et al.,* 1968; Khan, 1972). A dramatic rise in the activities of aminoacyl tRNA synthetases in embryos of cold-treated seeds discovered more recently (Tao and Khan, 1974) suggests that increased RNA synthesis is critically involved in the basic dormancy-breaking reaction, perhaps by accelerating the rate of synthesis of aminoacyl-tRNA synthetases, which in turn provide the signal for release from dormancy. In sour cherry (*Prunus cerasus*), during after-ripening there is an accumulation of nitrogen and phosphorus in the nucleotide- and nucleic-acid-containing fractions of the embryo axis (Pollock and Olney, 1959; Olney and Pollock, 1960). Other investigators (Bradbeer and Floyd, 1964; Ching and Ching, 1972; Simmonds and Dumbroff, 1974) have observed increases in the adenylate energy supply and energy charge in the cells of embryos of after-ripened seeds. These results apparently indicate that dormancy is due to a block in phosphate metabolism, and that an increased availability of phosphate acceptors may be associated with its breakage. A little understood facet of this hypothesis is whether the increased energy availability is the primary reaction of the dormancy-breaking mechanism, or is just one of the many secondary reactions that follow the activation of cells.

These experiments have put the case for an involvement of some phase of nucleic acid metabolism in dormancy breakage of embryos in a much sharper focus. In view of the fact that certain hormones can substitute for a cold requirement, there is obviously much to be learned from comparative experiments on the effects of cold treatment and hormones on the nucleic acid metabolism of embryos before a clear conclusion can be drawn. One problem to be faced in drawing a close analogy between breakage of dormancy and the observed changes in the nucleic acid metabolism of embryos is that these changes may be produced as a consequence, rather than being the cause, of relieving the state of dormancy.

III. COMMENTS

Completion of embryogenesis does not signal the cessation of developmental processes in the embryo. In fact, they continue in many subtle and varied ways during seed dormancy and germination. Work described in this chapter indicates that the dormant state of the embryo is an aspect of the problem of development of a plant organ which paradoxically expresses a mechanism for preventing its own growth. If that concept is correct, then the factors that regulate normal growth and development of the embryo should also apply to the induction of dormancy in seeds and their release from it. The view that dormancy results from a disturbance in the hormonal control mechanism of the embryo seems reasonable; whether such a disturbance is due to a change in the concentration of a highly specific hormone, or due to trivial changes in the total hormonal balance, is not established. Moreover, there appears to be little in the literature to indicate whether hormones are synthesized in the embryo itself or translocated from the surrounding tissues.

The factors, such as light, temperature and hormones, which regulate growth of the developing embryo are also involved in breaking its dormancy. The important fact is that in all cases the stimuli perceived by the cells are translated into chemical messengers which trigger off cellular activities and germination. While there is little precise information on the mode of action of the chemical stimuli in the embryos of seeds which germinate in response to specific dormancy-breaking treatments, this question has been explored more thoroughly in the embryos of quiescent seeds. We shall examine in the final chapter some of these investigations.

17. Seed Germination: Initiation of Growth Processes in the Embryo

Germination, as commonly understood, is the beginning of growth in a seed, spore or other similar metabolically inactive structures. Consequently, germination essentially involves the elimination of the quiescent or dormant state and its replacement with another phase during which developmental potencies of the system are activated. In the embryo, the physiological processes of germination may be viewed as the inverse of those occurring during initiation of the dormant state and study of one is likely to shed light on the other.

Seed germination is a vast area of study with ramifications extending into morphology, ultrastructure, developmental botany, physiology, biochemistry and molecular biology. In this final chapter of the book we shall restrict ourselves to a consideration of the biochemical aspects of germination particularly with reference to the molecular events triggered during germination. Information on the ultrastructure and physiology of germination given here is intended to serve as a background for the molecular biological aspects to be discussed later. Since seed germination has all the earmarks of an activation process, the principles evolved, if not the details, ought to be applicable to other systems where development is initiated after a period of metabolic standstill. For a general picture of the early literature on seed germination, the reader may turn to the book by Mayer and Poljakoff-Mayber (1975) or to the review by Brown (1965).

I. STRUCTURAL CHANGES IN THE EMBRYO DURING GERMINATION

As we have seen in previous chapters, during seed development, the embryo and surrounding tissues such as the endosperm and perisperm rapidly synthesize a growing reservoir of RNA, proteins, carbohydrates, fats and other storage products and potential sources of energy which are stockpiled for future use. In exalbuminous seeds, these substances are generally stored in the embryo, principally in the cotyledons, which consequently become bulky and prominent in the mature seed. In albuminous seeds, reserve materials are stored mainly in the endosperm and perisperm, and much less in the cotyledons and embryo axis.

Electron microscopic evidence now takes us a long way toward a unified picture of the structural changes taking place in the embryo during germination and supplements the array of biochemical studies to be reviewed later. Fine structure studies of the cells of dry embryos are, however, limited, because the latter pose a virtually impenetrable barrier to conventional fixatives and embedding media. Although the cell contents of the cotyledons of dry, desiccated cotton embryo appear to be ultrastructurally similar in morphology to those of the fully hydrated embryo (Yatsu, 1965), the results obtained by using an aqueous fixative leave much to be desired. Thanks to the power of improved fixation methods, especially the new aldehyde fixatives combined with osmium, our minimal image of the ultrastructure of the cell of a dry quiescent embryo includes mitochondria, plastids and ER, besides varying kinds of storage materials (Perner, 1965; Klein and Ben-Shaul, 1966; Paulson and Srivastava, 1968; Yoo, 1970; Durzan et al., 1971; Hallam, 1972; Vozzo, 1973). In contrast to the highly evolved organelles of the developing embryo, those of the mature dehydrated embryo are of the simple and attenuated type, usually present in cells which have low metabolic activity, or which exhibit a pathological condition (Fig. 17.1). There has not always been agreement as to whether cells of the quiescent embryo contain dictyosomes and ribosomes. Some workers have found them in cells of the dry embryo, while others have found them only after a period of soaking. The effects of quiescence pervade even the nucleoli which appear to be devoid of granular components (Bal and Payne, 1972). Despite the fact that organelles of the cells of the quiescent embryo appear to be relatively unspecialized from a functional point of view, if the basic requirement for cellular metabolism demands an assembly of essential organelles, then it may be misleading to describe the lack of metabolic activity as a cause for quiescence. Rather, subcellular simplification might be

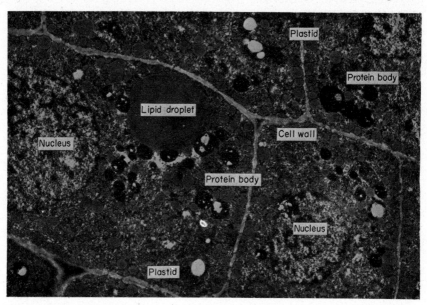

FIG. 17.1. The presence of electron-dense storage materials, and the simple and attenuated organelles in the root tip cells of dry lettuce embryo, give them the appearance of cells in a pathological state. (After Paulson and Srivastava, 1968; photograph courtesy of L. M. Srivastava)

viewed more appropriately as an expression of the quiescent state, the basic cause for which should be sought elsewhere.

In recent years, germinating embryos of pea (Setterfield *et al.*, 1959; Varner and Schidlovsky, 1963; Bain and Mercer, 1966b; Yoo, 1970), wheat (Setterfield *et al.*, 1959), peanut (Bagley *et al.*, 1963), *Tropaeolum* (Nougarède, 1963b), lima bean (Klein and Ben-Shaul, 1966), French bean (Öpik, 1968), *Yucca* (Horner and Arnott, 1966), lettuce (Srivastava and Paulson, 1968), maize (Berjak and Villiers, 1970; Deltour and Bronchart, 1971), pine (Durzan *et al.*, 1971), rye (Hallam *et al.*, 1972; Siwecka and Szarkowski, 1974) and *Setaria* (Rost, 1972) have been examined at the fine structural level of scrutiny. Although methods of tissue preparation employed in these studies differ from one another, in general, most of the findings concerning ultrastructural changes in the cells harmonize. In summary, we can associate the period beginning a few hours after imbibition with a progressive loss of reserve materials, an increased resolution of membranes and appearance of additional organelles and membrane systems not found in the dry embryo (Fig. 17.2). As seen in the electron microscope, the sequence of breakdown

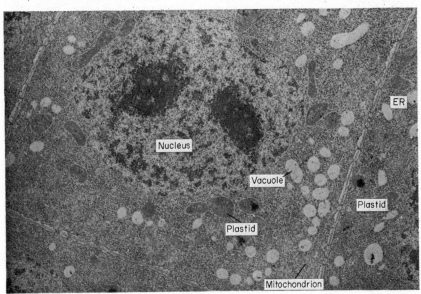

Fig. 17.2. Progressive loss of reserve materials and increased resolution of organelles and membrane systems in the embryo characterize the early stage of germination of lettuce achene; root tip cell, after 36 h germination. (After Srivastava and Paulson, 1968; photograph courtesy of L. M. Srivastava)

of reserve carbohydrate, protein and fat bodies takes various forms, and exhibits developmental continuity to a number of vacuoles which subsequently appear in the cells. When we consider the increased clarity of the organelles, we are inclined to think first of the mitochondria. This first thought may well be justified on functional grounds, for the notion that imbibition of the embryo is accompanied by a tremendous increase in respiratory activity is deeply entrenched in the literature. This impression is substantiated by a gradual definition and elaboration of mitochondrial substructure that one observes in the imbibed seed, such as formation of a clear double membrane, increased number of cristae and decrease in the electron opaque area. Elaboration of the plastid structure, especially the formation of an extensive lamellar system and stacks of grana and the rapid proliferation of the ER, are common but not universal features of the imbibition period. In some species, dictyosomes and microtubules are encountered after imbibition is under way. Finally the initial period of imbibition is followed closely by the redistribution of ribosomes that brings about a change from a high concentration of free ribosomes in the cells of the quiescent embryo to their fabrication into polysomes (Chapman and Rieber, 1967; Srivastava and Paulson, 1968; Villiers, 1968b; Berjak and Villiers, 1970; Durzan *et al.*, 1971; Deltour and Bronchart, 1971; Siwecka and Szarkowski, 1974). This links the contact of the embryo with water to an array of molecular events connected with the synthesis of the first proteins of germination. Fine structural changes of the type described above are reported to occur in the dormant, fully imbibed embryos of *Fraxinus excelsior* during the period of maturation when actual germination does not occur (Villiers, 1971). This seems entirely reasonable in view of the fact that during the maturation period the embryo undergoes extension growth and morphogenesis, although theoretically it will not germinate unless exposed to a chilling treatment. Many other aspects of ultrastructure, perhaps some subtle variations in chromatin, nucleoplasm, nucleolus and membrane structure at a still finer level, may be expected to be changed in the cells of embryos as they pass from a quiescent to an actively growing state, but an analysis of these changes must wait further work.

II. BIOCHEMISTRY OF GERMINATION

An important development during maturation of the embryo is the gradual desiccation and loss of water from cells. Although the dry embryo has a water content of the order of 5–10% of its weight, most of this water is actively bound to colloids and is unavailable for hydrolytic reactions. Similarly, the dry embryo may contain a complement of enzymes, coenzymes and substrates, which, of course, do not participate in any metabolic reactions. Thus, the biochemical state of the quiescent embryo in a seed may be described as one of expectant inactivity. Yet, it would seem obvious that some respiratory or oxidative metabolism which appears to be the common base for all cellular activities must be going on in the embryo, albeit at a negligible rate, to justify its capacity for growth and development. This has been found to be the case. Dry embryo respires at a low rate, while other metabolic activities are virtually at a standstill. The mechanism by which the embryo maintains a relatively low level of metabolism in the quiescent state, and yet remains capable of activation later, is one of the main current mysteries of the biochemistry of germination, an analysis of which has been gaining momentum in recent years. As is well known, a fundamental property of many seeds is that they require only contact with water at a favorable temperature for initiation of growth. This is clearly a universal and essential feature of germination, involving many structural and biochemical transformations in the cells.

A. Respiration

The general metabolic background of germinating seeds is indicated by high rates of respiration in the embryo and in the storage tissues of the endosperm and perisperm, beginning with the onset of imbibition and reaching a maximum a few hours later. In isolated lupin embryos, respiration which commences within 15 min after imbibition increases continuously until about 40 h (Czosnowski, 1962). Similar results have been reported in isolated bean embryo axes involving a considerably shorter period of imbibition (Oota, 1958). Since the total number of cells in the embryo does not increase during imbibition, the increased respiratory activity may well reflect an increase in the relative area of the dehydrated cell surface available for overall gaseous exchange.

In barley, and presumably in all cereals, most of the respiratory activity is due to the embryo, while the endosperm which constitutes the bulk of the grain is relatively inert (Barnell, 1937). Subsequently, as growth is initiated in the primordial organs of the embryo, respiratory activity of each organ assumes a characteristic pattern. While respiratory gas exchange increases during early germination, the respiratory quotient (RQ) decreases from an initially high value to a value near unity. However, at least during the initial phase of imbibition, changes in RQ with time are not necessarily harmonious with changes in substrate availability, but are probably due to low oxygen uptake restricted by hard and impermeable seed coats. In fact, comparative studies of oxygen uptake by attached and isolated embryos of barley (Brown, 1943) have confirmed that the latter have a much lower RQ than the former. Eventually, as the permeability barrier is overcome, changes in RQ may be explicable in terms of the oxidation of specific substrates, initially the soluble sugars, and subsequently the fats and proteins.

A number of investigators, working with mitochondria isolated from embryos and storage tissues, have demonstrated a pattern of rise and fall of mitochondrial bulk and their oxidative and phosphorylative activities during the progress of germination (Akazawa and Beevers, 1957; Hanson *et al.*, 1959; Howell, 1961; Cherry, 1963; Öpik, 1965; Malhotra and Spencer, 1970). It is probable that these changes reflect the highly complex pattern of mitochondrial biogenesis in the germinating seed involving the utilization of mitochondria pre-existing from the maturation phase of the seed, their disintegration and renewed synthesis. Electron microscopic studies have already established with great clarity an increase in mitochondrial number, structure and organization with the progress of germination.

From the survey of seeds of a number of species, it seems certain that many oxidative enzymes are present in the embryo of the dry seed at quite high levels, and that their concentrations change with the progress of germination, but it is doubtful whether all of them are related to respiratory metabolism. The presence of cytochromes, dehydro-genases, catalases, peroxidases, lipoxidases, phenolases, among others, has been demonstrated. Regarding the actual pathway of breakdown of substrates during respiration, the existence of glycolysis seems to be reasonably assured in many seeds, while hard evidence for the existence of the pentose phosphate pathway is apparent only in a few cases. On the basis of the ability of isolated mitochondria to negotiate the oxidation of tricarboxylic acid cycle intermediates, it seems that partial re-actions of the Krebs cycle are of widespread occurrence. A study with slices of castor bean endosperm has established that the fate of ^{14}C-labeled pyruvic acid follows the main course of the Krebs cycle (Neal and Beevers, 1960). All of this gives a picture of a normal respiratory pathway in the germinating seed. However, it is only when one analyzes the information available on individual species that one realizes how few facts there are on which to build detailed concepts of the biosynthetic steps involved.

B. Mobilization of Food Reserves during Germination

The most obvious changes in the metabolism of seeds during germination are the hydrolysis of reserve substances of the storage tissues and their utilization by the growing embryo. Underlying these reactions may be changes in enzyme levels, cellular organization, and as we have already seen, respiration.

In an early, informative study on the germinating seeds of *Vigna sesquipedalis*, Oota *et al.* (1953) made interesting comparisons between the changes in dry weight, water soluble matter, soluble sugar, insoluble polysaccharides, soluble and protein nitrogen and RNA in the cotyledons on the one hand, and the hypocotyl, radicle and epicotyl on the other hand. All compounds studied were found to register a net loss from the cotyledons and a net gain in the other organs of the embryo during the progress of germination. Figure 17.3, which shows the changes in the content of protein nitrogen, is fairly representative of the changes in the other parameters studied. Germination in lupin

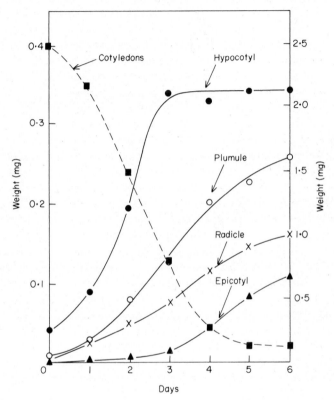

FIG. 17.3. Changes in the protein nitrogen content of embryonic organs of *Vigna sesquipedalis* during germination. Scale on the right-hand ordinate for cotyledons; on the left-hand ordinate for other embryonic parts. (After Oota *et al.*, 1953)

(McRary, 1940; Rogozińska, 1960), pea (Barker and Hollinshead, 1964; Beevers and Guernsey, 1966), French bean (Simon and Meany, 1965) and mung bean (*Phaseolus aureus*) (Paul and Mukherji, 1973) is also accompanied by a rapid loss of various nitrogenous components from the cotyledons and their accompanying increase in the embryo axis. Comparable results have been obtained in the germination of some albuminous dicotyledonous seeds and many monocotyledonous seeds

(Brown, 1946; Fukui and Nikuni, 1956; Ingle *et al.*, 1964; Kriedemann and Beevers, 1967a; Park and Chen, 1974) where growth of the embryo takes place at the expense of the endosperm reserves. The form in which the storage products of the endosperm are transmitted to the embryo cannot be assured from the nature of the reserves and in some seeds glucose from the storage tissue is first converted to sucrose and subsequently absorbed by the embryo axis (Edelman *et al.*, 1959; Kriedemann and Beevers, 1967b; Chen and Varner, 1969; Park and Chen, 1974). In Gramineae, early growth of the embryo axis is probably initiated by the breakdown of the storage reserves of the scutellum, coleorhiza or coleoptile, before translocation of carbohydrates from the endosperm begins (Dure, 1960; Price and Murray, 1969).

In gymnosperms, growth of embryos is related to the reserve materials of the female gametophyte in much the same way as growth of angiosperm embryos is related to the attached cotyledons and surrounding endosperm. Decreases in dry weight of the gametophyte due to loss of fats, proteins, nitrogen and phosphorus and increases in dry weight of the embryo due to utilization of these compounds observed during pine seed germination are consistent with this view (Sasaki and Kozlowski, 1969). Carbohydrates, structural components and soluble compounds of embryos of germinating Douglas fir (*Pseudotsuga menziesii*) seeds were apparently synthesized from raw materials of lipids, proteins, reserve phosphorus compounds, and perhaps nucleotides which disappeared from the gametophyte (Ching, 1966). In red pine (*Pinus resinosa*), initiation of embryonic growth, including differentiation of needle primordia, stomata and primary and secondary vascular tissues, appeared to be independent of the food reserves of the gametophyte. Nevertheless, as shown in Fig. 17.4, overall growth of the component parts of the embryo clearly depended upon the gametophyte, which was utilized only after completion of an initial phase of growth (Sasaki and Kozlowski, 1969).

One envisages the process of hydrolysis of reserve materials in terms of the action of specific enzymes. In germinating seeds of diverse plants, enzymes negotiating the breakdown of primary classes of cellular reserves such as carbohydrates, proteins, fats and phytins, nucleic acids, and cell wall materials such as cellulose and hemicellulose, have been identified. Whereas oxidative enzymes are present even in the dry seed, there is no significant activity of hydrolytic enzymes, which begin to appear as the seed imbibes water. This is illustrated in germinating pea seeds where the activity of amylase increases approximately 10-fold and that of phosphatase approximately 20-fold during a 6 day germination period (Young and Varner, 1959). More dramatic was the activity of 3'-nucleotidase, which registered an almost 80-fold increase during 40 h of germination of isolated wheat embryo (Shuster and Gifford, 1962). In dormant, but fully imbibed seeds of *Avena fatua* and *A. ludoviciana*, amylase activity was virtually absent, but it increased

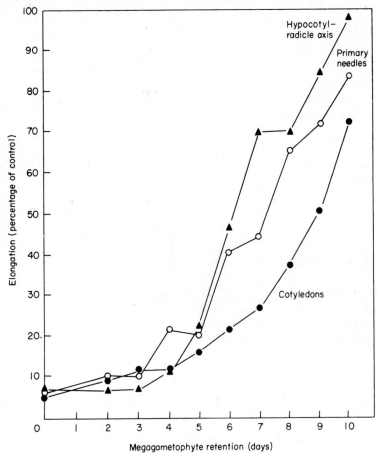

FIG. 17.4. Elongation of the hypocotyl–radicle axis, cotyledons and primary needles of red pine (*Pinus resinosa*) is dependent upon the number of days the megagametophyte is attached to the embryo. Data taken 35 days after beginning of germination, and expressed as percentage of the control. (After Sasaki and Kozlowski, 1969)

considerably during germination of nondormant seeds (Drennan and Berrie, 1962). This fact discourages any attempt to relate enzyme appearance to the state of seed proteins as they become hydrated, and implies a close relationship between physiological changes in the embryo and enzyme activation.

In germinating seeds, either activation of enzyme molecules already present or synthesis of new molecules may account for increases in enzyme activity. For examples of both these situations, attention is conveniently turned to castor beans and cereal grains, where this has been studied in considerable detail. In the endosperm of castor bean, activities of several key enzymes involved in the conversion of lipids to carbohydrates are very low during seed maturation, but increase rapidly during germination. Significant increases of aldolase and glucose-6-phosphate dehydrogenase are observed when the endosperm is allowed to imbibe at 0°C, when no conceivable synthesis of RNA and protein occurs. The proposition from these results is that enzymes appear by a reversible activation of their molecules which are inactivated during seed maturation. On the other hand, a pattern to which other hydrolytic enzymes of the endosperm conform is one of little or no activity at low temperatures, suggesting their *de novo* synthesis. The inhibition of both RNA and enzyme synthesis by actinomycin D, an inhibitor of mRNA synthesis, and the correlated decrease in enzyme activity and polysome level in the endosperm suggest that control mechanisms for *de novo* synthesis operate at the transcription level (Marrè *et al.*, 1965: Marrè, 1967).

In cereal grains starch is hydrolyzed to glucose by α-amylase, β-amylase and maltase. About 80% of the β-amylase of the ungerminated wheat grain exists in an inactive form probably bound to the protein glutenin by disulphide linkages. Its release and activation are thought to be accomplished during germination by the secretion of substances that reduce the disulphide linkage (Rowsell and Goad, 1962a, b). In contrast, the activity of α-amylase and maltase apparently results from *de novo* synthesis either in the aleurone cells or the scutellum of the grain. Much of the evidence in favor of this hypothesis has come from the work of Varner (Varner and Chandra, 1964; Filner and Varner, 1967) on the induction of α-amylase in barley grains, but in general the findings seem to be applicable to other cereal grains, and perhaps to the induction of other hydrolytic enzymes.

The first link in the chain of evidence that related *de novo* α-amylase synthesis to the hydrolysis of starch grains was the discovery of a close interaction between the embryo, the starchy endosperm and the proteinaceous aleurone layer of the barley grain. Although it was first thought that the aleurone layer supplied the enzyme for starch hydrolysis, the absence of any starch breakdown in embryo-less seeds incubated normally indicated that the embryo was instrumental in the process.

In many ways one of the most revealing experiments to clarify the role of the embryo in this system was that of Yomo (1958), who showed that the embryo factor causing α-amylase synthesis was a gibberellin-like substance. Subsequently, Yomo (1960) and Paleg (1960) independently showed that in embryo-less endosperm treated with GA as low as 10^{-11}M, starch hydrolysis occurred just as rapidly as when the embryo was present. This raised the hope of α-amylase induction by endogenous GA in the normal seed, where only such low concentrations of the hormone are ostensibly available. Considering the fact that the only living cells of the endosperm are those of the aleurone layer, it is perhaps not surprising that exogenous GA initiates α-amylase production in these cells. The picture that has unfolded is that in a normally germinating seed, the embryo produces GA, which diffuses into the aleurone layers and causes the synthesis of α-amylase. The enzyme is channeled to the starchy endosperm, resulting in the mobilization of starch. Conclusive evidence that all of the α-amylase induced by GA is the result of *de novo* synthesis, rather than of release of an active form from an inactive precursor, was obtained by labeling the newly synthesized enzyme with ^{18}O and separating it from the old enzyme labeled with ^{16}O by density gradient equilibrium centrifugation.

There are other examples such as wheat (Moro *et al.*, 1963) and wild oat (Naylor, 1966) where GA controls α-amylase activity, and pea (Young and Varner, 1959; Swain and Dekker, 1969) where increase in enzyme activity presumably results from *de novo* synthesis. In addition to α-amylase, following GA treatment, barley endosperm also produces a number of other hydrolytic enzymes *de novo* in response to hormone application (Jacobsen and Varner, 1967). In cotton, lipase activity is induced by GA (Black and Altschul, 1965), while in wheat, this enzyme is activated by hydroxylamine and glutamate (Tavener and Laidman, 1969).

The possiblity raised by these studies that the utilization of storage reserves of the seed during germination is dependent upon the embryo is also underlined in the work with seeds where such reserves are deposited predominantly in the cotyledons. For example, in the cotyledons of pea separated from the embryo axis, there is an arrest of RNA and protein degradation and a decrease in the level of hydrolytic enzymes (Oota and Osawa, 1954; Varner *et al.*, 1963; Bain and Mercer, 1966c; Guardiola and Sutcliffe, 1971; Chin *et al.*, 1972; Yomo and Varner, 1973). If lack of an attached embryo axis permits a prolongation of the life of the cotyledons, it may signify that the former provides directive signals which trigger the synthesis of degradative enzymes. In leguminous seeds, where the reserve materials of the cotyledons are metabolized during germination, an embryonal control over the mobilization of the endosperm reserves may not be apparent (Reid and Meier, 1972).

In recent years, a welcome trend has developed in using germinating seeds to study the metabolic pathways involved in the breakdown of storage substances, and a voluminous literature has accumulated on the pathways of utilization of starch, fat, protein, cellulose, hemicellulose, phytin, sterols and other compounds. In general, it is of little use to consider this literature in detachment from basic biochemistry of plant metabolism. Although details of the degradative routes followed by the compounds in seeds of different species vary, it can be said that in the presence of appropriate enzymes with given biocatalytic functions, conversion of these compounds to simple substances follows the pathways familiar to biochemists that need no repetition here. For an up-to-date review of this work, see Ching (1972).

III.   MOLECULAR ASPECTS OF GERMINATION

A. Control of Protein Synthesis

It is not enough, however, to emphasize the catabolic activity of the storage tissues of the seed, for it may give the impression that mobilization of food materials as a consequence of self-digestion of storage cell contents is the only sustained event during germination. Although it is a well established fact that germinating seeds possess potentialities for complex and diverse energy-dependent synthetic processes, the significance of this often passes unnoticed. It is important to realize that in contrast to the massive endosperm and cotyledons of angiosperms and the megagametophyte of gymnosperms where catabolic activities take place, the anabolic reactions are restricted to the relatively insignificant embryonic axis. Since the embryo is the growing entity of the seed, it will not be difficult to visualize the reasons for this.

In the previous section we have seen that germination is accompanied by an increased respiratory activity. Respiratory gas exchange results in the production of ATP, which is the driving force for initiating the synthesis of proteins (enzymes) and substrates. It is significant that when the embryo axis is imbibed in tritiated water, the first compounds which become radioactive after as little as 3 min are those amino acids which are related to the Krebs cycle and ATP output (Collins and Wilson, 1972). Respiratory metabolism precedes protein synthesis perhaps by a few hours. A sustained synthesis of proteins is clearly an indispensable requirement for embryonic growth, as shown by several workers (Waters and Dure, 1966; Walton, 1966; Walton and Soofi, 1969; Fujisawa, 1966; Abdul-Baki, 1969; Hallam et al., 1972; Stoddart et al., 1973; Yadav and Das, 1974; Subramanian, 1974). On the basis of simple incorporation of labeled amino acids into proteins, these workers demonstrated that isolated embryos synthesized proteins during the early hours of imbibition (Fig. 17.5). By a similar experimental approach, increased synthesis of RNA, mostly 23S and 16S plastid ribosomal RNAs, during germination has also been detected. RNA synthesis may be triggered off immediately following the onset of imbibition (Rejman and Buchowicz, 1971, 1973; Dobrzańska *et al.*, 1973) or

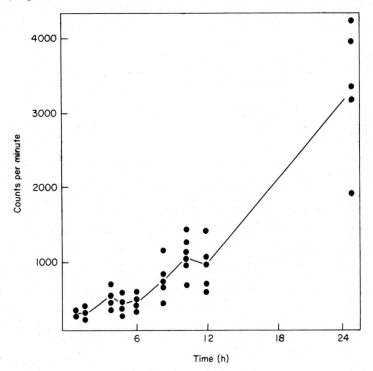

Fig. 17.5. Incorporation of ^{14}C-amino acid mixture into trichloroacetic-acid-precipitable material of rye embryo during imbibition. Values expressed as cpm per embryo, five embryos analyzed at each time. (After Hallam *et al.*, 1972)

a few hours thereafter (Sasaki and Brown, 1969; van de Walle and Bernier, 1969; Deltour, 1970; Vedel and D'Aoust, 1970; Frankland *et al.*, 1971; Walbot, 1973b).

In the context of this chapter, we are concerned mainly with the sequential transcription and translation of informational RNA in the synthesis of the first proteins of germination. It is an accepted fact that genetic information for the synthesis of proteins is coded in the nuclear DNA of the cell. The uncoding of this information in protein synthesis involves both a transcription step (DNA to mRNA) and a translation step (mRNA to proteins). Assuming that the embryo contained within the seed has its own blueprint for the adult organization, the intriguing question is: how is this information uncoded to provide the first proteins of germination?

The Concept of Stored mRNA

The picture that has emerged from an analysis of the germination of seeds of several species is that acceleration of protein synthesis following imbibition does not depend upon newly synthesized mRNA, and that the genome of the dry seed is equipped with a long-lived mRNA which directs the synthesis of the first proteins of germination. The stored mRNA is transcribed during embryogenesis, but is not translated until imbibition of the seed is under way. This concept seems to provide a reasonable framework for the analysis of the molecular aspects of germination, as the following review of the work from a few laboratories will show.

Dure and Waters' work (1965; see also Waters and Dure, 1966) on the embryo of cotton provided the first clear evidence for the presence of a long-lived mRNA active in seed germination. That there is a requirement for sustained protein synthesis in the germination of the seed was confirmed by the addition of the protein synthesis inhibitor, cycloheximide, to the germinating medium; the drug caused a virtual shutdown of protein synthesis and arrest of embryo development. To test whether proteins synthesized during germination were coded on mRNA synthesized simultaneously, the effects of actimomycin D were examined. Although the drug did not affect the growth of the embryo or protein synthesis during the first 36 h of germination, sucrose-gradient centrifugation profiles showed that the synthesis of mRNA and rRNA was inhibited. In addition, the ratio of polysome to monosome found in the dry seed was not changed during the course of the first few days of germination, or by actinomycin D treatment. Since protein synthesis and germination did not suffer rate reduction when transcription was presumably inhibited by actinomycin D, there are reasonable grounds for the view that proteins synthesized during this period are directed by stored mRNA and ribosomes preformed in the mature seed. This view has been strengthened by the isolation and characterization of a ribonucleoprotein particle believed to contain specific mRNAs from cotton embryos (Walbot et al., 1974).

In a series of studies, Marcus and associates (Marcus, 1969) have explored the mechanism by which stored mRNA is unmasked during the initial stages of germination of the wheat embryo. To the extent that this is a measure of competence in protein synthesis, ribosomal preparations from dry desiccated embryos were found to be less active than similar preparations from partially soaked embryos; however, any apparent ineffectiveness of the ribosomes at synthesizing polypeptides cannot be due to their inferior quality, for, in the presence of the synthetic messenger, poly-uridylic acid, inactive particles regained their full activity and readily incorporated amino acids into proteins (Table 17.I). This led to the idea that protein synthesis in the dry seed is limited

TABLE 17.I. ^{14}C-Phenylalanine incorporation into wheat embryo ribosomes. (From Marcus and Feeley, 1964)

Conditions	Ribosomes (mg RNA)	Counts per minute—in protein	
		0 day	1 day
No poly-uridylic acid	0·17	2	120
	0·85	9	881
With poly-uridylic acid	0·02	240	408
	0·16	3681	3522

The complete incubation medium contained 50 μmol tris buffer (pH 7·9), 5 μmol $MgCl_2$, 50 μmol KCl, 20 μmol 2-mercaptoethanol, 1 μmol ATP, 10 μmol creatine phosphate, 50 μg creatine phosphate kinase, 0·5 μmol GTP, 200 μg yeast tRNA, 2·1 m μmol ^{14}C-phenylalanine (101 000 cpm), and 0·27 ml 0 day, dialyzed 100000 g supernatant in a total volume of 1 ml. Incubation was for 40 min at 30°C and 100 mg poly-uridylic acid were added where indicated. The non-poly-uridylic acid tubes contained 62·5 mμmol of tyrosine, aspartic acid, glutamic acid, leucine, isoleucine, alanine, arginine, lysine, cysteine, histidine, proline, threonine, valine, glycine, serine, tryptophan, asparagine, glutamine and methionine. 0 day and 1 day refer, respectively, to dry, and 16 h germinated embryos.

by the complete absence of an active mRNA or by the low functional status of stored mRNA. More convincing evidence for this thesis, shown in Fig. 17.6, was the demonstration that in contrast to the soaked embryo, the dry embryo had hardly any polysomes (Marcus and Feeley, 1964, 1965; Allende and Bravo, 1966). Under in vivo conditions, protein synthesizing capacity and polysome formation rose rapidly after the embryo was placed in water and followed the fresh weight increase by about a 19 min lag. The rapidity with which protein synthesis and polysome formation are activated upon imbibition substantiates the presence of a functional informational and translational system in the

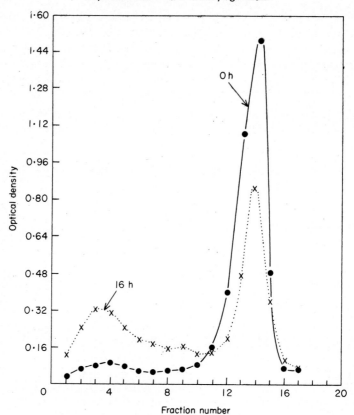

Fig. 17.6. Sucrose-gradient density centrifugation profile of ribosomes from dry (0 h) and imbibed (16 h) wheat embryos, showing polysome formation upon imbibition. The ribosomes (0·3 mg RNA) were centrifuged through a standard sucrose gradient and 15-drop fractions were collected. (After Marcus and Feeley, 1965)

dry embryo and rules out the possibility of synthesis of new mRNA (Marcus *et al.*, 1966).

Proof of the correctness of the concept of conserved mRNA in the quiescent wheat embryo has been supplied by the demonstration that a homogenate from the dry embryo could convert ribosomes to polysomes (Marcus and Feeley, 1966). Interestingly enough, this process is dependent upon the presence of exogenous ATP in the incubation medium. One polysome peak is illustrated in Fig. 17.7, which is a sucrose gradient centrifugation profile of ribosomes isolated from the incubated homogenate. In all likelihood the mechanism of unmasking stored messengers and attaching them to ribosomes involves the syn-

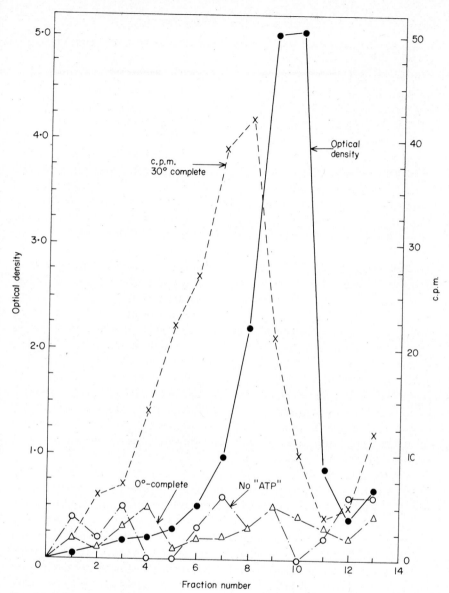

FIG. 17.7. Polysome formation *in vitro* by ribosomes of wheat embryos incubated for 12 min at 30°C with ATP and an ATP-generating system; ^{14}C-leucine was included during the last 2 min. Ribosomes isolated from the incubation mixture were centrifuged through a sucrose gradient. After determining absorbance, radioactivity of trichloroacetic acid-insoluble material was determined. In one of the controls incubation was at 0°C, and the 2 min pulse of radioactivity was at 30°C; in the second control, both manipulations were at 30°C, with ATP omitted during the initial incubation period and added just prior to the pulse. (After Marcus and Feeley, 1966)

thesis of endogenous ATP following imbibition (Obendorf and Marcus, 1974).

Mention should be made of three additional lines of evidence originating from another laboratory (Chen *et al.*, 1968a) which have provided some conception of the functional independence of the conserved mRNA and the newly synthesized mRNA in the wheat embryo. First, methylated-albumin-Kieselguhr-column analysis of ^{32}P incorporation into RNA showed that up to 24 h germinating embryos synthesized very little RNA, whereas from 24 to 48 h they synthesized typical short-lived mRNA. Yet protein synthesis begins by 16 h, at the most conservative estimate. Second, competition for DNA hybridization sites indicated that very few "new" species of RNA not present in the dry embryo were formed during the first 48 h of imbibition. Third, use of a cell-free amino acid incorporating system totally stripped of endogenous message showed that RNA extracted from ungerminated embryos was template-active. However, the genome of the embryo is not entirely at rest during the initial stages of germination, dominated by translation of stored mRNA, since transcription of ribosomal RNA starts as early as 2 h from imbibition (Chen *et al.*, 1971). Weeks and Marcus (1971) have actually isolated an active messenger fraction from the dry wheat embryo capable of stimulating amino acid incorporation and polysome formation in an *in vitro* system with an excess of all components except mRNA required for protein synthesis. Changes reported in the ratio of individual tRNAs during germination of the wheat embryo (Vold and Sypherd, 1968a, b) suggest that translation control of mRNA could be accomplished by the synthesis or modification of aminoacyl tRNAs or synthetases.

A substantial body of research has purported to demonstrate the absence of template-active messengers, and the presence of masked messengers in latent stage embryos of other species. The inhibition of RNA synthesis in nondormant embryos of *Fraxinus excelsior* by abscisic acid when protein synthesis proceeds normally is compatible with the idea of stable mRNA (Villiers, 1968b). In germinating embryos of *Vicia faba*, RNA synthesis did not appear to be a prerequisite for protein synthesis since the latter preceded the former by several hours, perhaps using stored mRNA (Jakob and Bovey, 1969). Dry embryos of peanut, *Pisum arvense*, *P. sativum*, *Oryza sativa* and *Pinus resinosa* appeared to be analogous to wheat embryo in the absence of polysomes and in the inability of ribosomes to incorporate amino acids into proteins (Marcus and Feeley, 1964, 1965; Barker and Rieber, 1967; Jachymczyk and Cherry, 1968; McCarthy *et al.*, 1971; App *et al.*, 1971; Sasaki and Brown, 1971; Beevers and Poulson, 1972). Polysome formation and protein synthesis by ribosome preparations which become pronounced with progress of imbibition have been suggested as indicators of the synthesis of new mRNA.

Since seeds thus far considered are quiescent and will germinate upon contact with water at a suitable temperature, it would be wrong to associate polysome formation with imbibition process *per se*. For example, dormant achenes of lettuce and seeds of pine, which require a light stimulus for germination, do not form polysomes even after prolonged periods of imbibition in the dark, unless they are exposed to light (Mitchell and Villiers, 1972; Yamamoto *et al.*, 1974). Protein synthesis obtained in homogenates of dry lettuce achenes incubated with artificial messengers seems to indicate that mRNA synthesis is a limiting factor in the dry embryo (Efron *et al.*, 1971). Entirely conflicting results were obtained by Fountain and Bewley (1973), who ran polysome profiles of dark-imbibed lettuce achenes in a separate investigation and found that polysome formation did indeed occur in the absence of the light stimulus. The reasons for the discrepancy between the results of different groups of workers remain unexplained.

A temporal sequence of synthesis and elimination of RNA species is apparently operating in the early phase of growth of the embryo of germinating mung bean seed. Evidence for this statement comes from competition experiments between labeled and unlabeled RNA isolated from the embryo at specific times after soaking. As the soaking period of the embryo was increased, RNA extracted from it became progressively less efficient as a competitor for high specific activity RNA synthesized during the early hours of soaking. Since mRNA is apparently limiting in the later stages of germination, production of new mRNA is the first prerequisite for continued growth of the embryo (Biswas, 1969). In the embryo of black eye pea (*Vigna unguiculata*) about half of the proteins synthesized during the initial stages of germination were coded on newly synthesized mRNA, as shown by actinomycin D effects. Possibly, the synthesis of new mRNA begins long before the stored mRNA is effectively used (Chakravorty, 1969). During the initial stages of germination, embryos of *Phaseolus angularis* and pea have been shown to synthesize DNA-like RNA with messenger properties and it is conceivable that the first proteins of germination are coded on templates of both new and old mRNA (Tanifuji *et al.*, 1969; Watanabe *et al.*, 1973). In rye embryos, synthesis of major classes of RNA can be detected in less than an hour after imbibition, and this has given rise to the speculation that early protein synthesis is coded for by both newly synthesized and long-lived mRNA (Sen and Osborne, 1974).

The experiments described above do not in themselves tell us whether RNA synthesized during embryogenesis is preserved during the quiescent period of the seed to serve as templates for the first proteins of germination. An ingenious labeling technique in which ^{32}P was supplied through the vascular tissue of the leaflet allowed Walbot (1971) to trace the fate of RNA synthesized in *Phaseolus vulgaris* during embryogenesis, rest period, and germination. In seeds which normally matured in about 35 days, RNA synthesis was rapid during the period

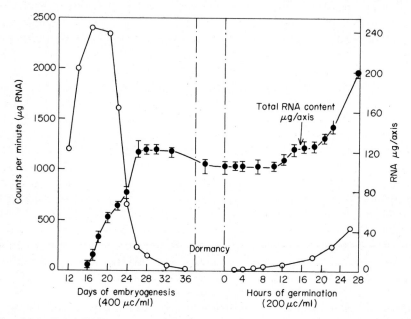

FIG. 17.8. Total RNA content of the embryos of *Phaseolus vulgaris* during embryogenesis, rest period and germination (closed circles). Open circles, specific activity of RNA after 2 h pulses of ^{32}P. (After Walbot, 1971)

from the sixteenth to twenty-fourth day of embryogenesis and then declined (Fig. 17.8). Most of this RNA was preserved during seed maturation and the subsequent period of quiescence, and remained essentially intact until about 6 h after imbibition, when the specific activity of the label began to decline. Moreover, synthesis of measurable amounts of new RNA did not begin until 12 h after imbibition, at a near-normal water content of the embryo, while protein synthesis was well under way within 15–20 min after imbibition at about 30% water content. Since RNA synthesis does not take place at a low water content, one might suspect that mRNA stored in the embryo is set aside during seed development and called into operation during germination to direct the synthesis of early proteins (Walbot, 1972). Continued growth of the embryo is dependent upon the synthesis of proteins coded on new mRNA transcribed under the direction of the total embryonic genome, although a transitional stage in seed germination at which preformed mRNA is exhausted and new messengers are activated is difficult to delineate (Walbot, 1973a).

Synthesis of Germination Enzymes on Stored mRNA Templates

A stringent test to implicate stored mRNA in protein synthesis during the initial stages of germination will be to follow the fate of particular proteins which are synthesized *de novo* during the period and demonstrate their independence upon simultaneous RNA synthesis. In the cotton embryo, a specific proteolytic enzyme, carboxypeptidase, is synthesized *de novo* in about 24 h after imbibition. The appearance of the enzyme is inhibited by cycloheximide, but not by actinomycin D at concentrations strong enough to stop all RNA synthesis and cause lethality in the embryo. This suggests that carboxypeptidase synthesis in the mature embryo is directed by mRNA which is a superfluous holdover from a stage in embryogenesis. In the relatively immature embryo, when presumably no conserved mRNA is present, enzyme synthesis is sensitive to actinomycin D. The critical stage in embryogenesis when transcription of preformed mRNA for carboxypeptidase occurs is said to be attained when the embryo weighs about 100 mg (Ihle and Dure, 1969). Isocitrase, which converts the stored lipids into carbohydrates, is another enzyme in cotton embryo which exhibits exactly the same syndrome (Ihle and Dure, 1972). These enzymes are apparently representatives of a group of germination enzymes for which transcription occurs more than halfway through embryogenesis, but translation is delayed until germination.

Other enzymes may be directly concerned with DNA synthesis in the cells of the embryo. One such enzyme is DNA polymerase, which catalyzes the polymerization of nucleotide triphosphates into DNA molecules. Using blasticidin, an inhibitor of protein synthesis, Mory *et al.* (1972, 1975) showed that in wheat embryo cellular DNA replication occurring about 15 h after imbibition was dependent upon proteins synthesized about 6 h earlier. This relationship was further strengthened by the demonstration that the level of DNA polymerase was higher in the germinated embryo than in the ungerminated one. Since wheat embryo synthesizes little or no RNA during the first 24 h following imbibition (Chen *et al.*, 1968a), a subtle chain of command is discernible between stored mRNA and synthesis of DNA polymerase.

B. "Masking" and "Unmasking" of mRNA

The validity of the conserved mRNA hypothesis, plausible and suggestive as it is, will be strengthened if we know the mechanism by which
messengers remain masked in the dry embryo and that by which they
are unmasked at germination. Nothing in the literature contradicts the
idea that stored mRNA is masked by a protein component and is
physically separated from the ribosomes. From an experiment designed
to study the effect of water stress at different stages of germination on
mRNA activity in wheat embryo, it appears that stable mRNA present
in the dry embryo is fundamentally different from mRNA produced
later during germination (Chen *et al.*, 1968b). Embryos imbibed for
24 h could be dehydrated without any adverse effects on growth upon
rehydration, whereas normal growth was impaired if they were dehydrated after imbibition for 72 h. Dehydration stopped all mRNA and
protein synthesis in both systems without affecting ribosome activity
per se. Hybridization experiments for the same complementary sites
between RNA from normally germinating embryos and dehydrated
embryos have shown that virtually the entire mRNA population transcribed during 24 h of germination was displaced by mRNA derived
from dehydrated 24 h embryos, whereas only less than 30% of mRNA
present in 72 h old embryos competed effectively with mRNA of dehydrated 72 h embryos. Since these results clearly indicate that mRNA
present in 24 h embryos is conserved through dehydration while that
of 72 h embryos disappears during dehydration (Fig. 17.9), the differences observed between embryos made artificially drought-resistant
and drought-sensitive may reflect differences in the transcription
mechanism. In simplified terms, it seems that in wheat embryo, the
control mechanism favoring a quiescent state involves a protection
against messenger destruction under conditions halting metabolic
activity and imposing secondary dormancy.

A hormonal control mechanism is operational in the masking of preformed mRNA concerned in carboxypeptidase synthesis in the cotton
embryo (Ihle and Dure, 1970). The hormone concerned is abscisic acid,
which inhibits precocious development of the enzyme in washed,
cultured, immature embryos. The effect of abscisic acid is reproducible
by culturing unwashed embryos or by treating washed embryos with
an extract of the ovular material. Since embryos excised from ovules
undergoing seed formation do not respond to abscisic acid or to ovular
extract, it appears that the hormone, known to be synthesized in the
ovules of cotton at the same time in ontogeny as the mRNA for carboxypeptidase is synthesized, inhibits translation of the latter until such time

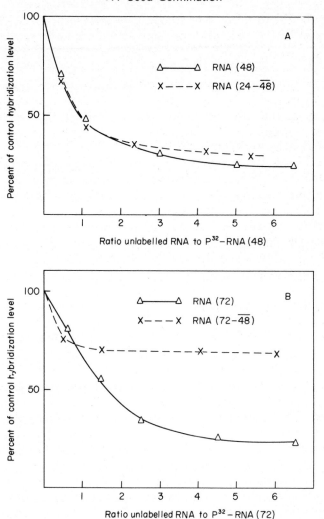

FIG. 17.9. Changes in mRNA during dehydration of 24 h germinated (**A**) and 72 h germinated (**B**) wheat embryos. **A,** Hybridization competition experiment contained in each sample 40 μg DNA, 134 μg ^{32}P RNA (48 h germinated) (8100 cpm/μg) transcribed in the interval of 24–48 h of germination and increasing amounts of unlabeled RNA derived from 24 h germinated, 48 h dehydrated (24–48) embryos. Self-competition with unlabeled RNA from 48 h germinated embryos is included. **B,** The hybridization competition experiment contained in each sample 10 μg DNA, 125 μg ^{32}P RNA (72 h germinated) (11 500 cpm/μg) transcribed in the interval of 48–72 h of germination and increasing amounts of unlabeled RNA derived from embryos germinated for 72 h and dehyrated for 48 h (72–48). Self-competition with unlabeled RNA from 72 h germinated embryos is included. (After Chen *et al.*, 1968b)

as the integuments become sclerified as in the mature seed. This last-mentioned event apparently removes the source of presumptive abscisic acid, and releases the mRNA for translation of carboxypeptidase for normal germination. The appearance of isocitrase in cotton embryo is under the same developmental regulation by abscisic acid as that of carboxypeptidase (Ihle and Dure, 1972). At the molecular level, a block in polyadenylation of the transcribed mRNA by abscisic acid is probably responsible for the delayed translation of germination enzymes (Harris and Dure, 1972).

The center of attention in these experiments was the synthesis of the first proteins of germination, and despite the differences in the approach of the various investigators, the results clearly show that embryos of germinating seeds are able to synthesize some of their early proteins independently of genomic transcription. These results may go a long way toward explaining the molecular mechanisms of seed germination, although the work is too limited to permit formulation of principles that are more generally applicable.

IV. LOSS OF SEED VIABILITY

Seeds of many plants are capable of remaining viable for long periods of time after harvest. Yet, under optimum storage conditions, the embryo gradually loses its viability until, after a time which varies from species to species, the germination capacity of seeds is completely lost. From a practical standpoint, it is desirable to prolong the life of seeds not only under conditions of optimum storage, but also in environments adverse to seed longevity. This in turn has stimulated considerable interest in the manifestations of seed deterioration and in the mechanisms by which seeds lose their vigor and deteriorate. Among the physiological and biochemical attributes, delayed germination, changes in respiratory rate, enzyme complements, food reserves, membrane physiology and macromolecule synthesis in the embryos have been associated, if not directly then at least by implication, with deterioration and loss of vigor of seeds. Roberts (1972) has recently written a book devoted directly to the subject of seed viability in which manifestations of the loss of viability are considered in great detail.

Among the cytological changes characteristic of aging embryos is damage to the nucleus and to the membrane systems of the cells. According to Berjak and Villiers (1970, 1972a, b, c, d), degenerative changes in maize embryos subjected to an accelerated aging treatment are probably caused by release into the cytoplasm of hydrolases normally enclosed within the lysosomes. Viability loss of seeds was low when they were afforded cathodic protection through a negatively charged conductor (Pammenter *et al.*, 1974) Since this implies a reduction in free radical damage to cell membranes, lysosome disintegration in the aging embryos could conceivably be associated with free radical accumulation. One remarkable feature of the cells of embryos of aging seeds is a temporary increase in the rate of synthesis of RNA and proteins, and a stimulation of DNA replication and cell division before the release of hydrolytic enzymes. This suggests that the appearance of enzymes is regulated by control mechanisms progressing through a route perhaps involving the activation of newly derepressed genes. Later on, as recorded in aging seeds of other plants (van Onckelen *et al.*, 1974), there is probably a decrease in the synthetic activities of the cells of the aging embryo.

Little can be said about the actual mechanisms by which seeds lose their viability and deteriorate. Embryo transplantation studies have shown that growth of young wheat embryos is inhibited when they are implanted onto aged endosperm of the same cultivar (Floris, 1966, 1968, 1970). Moreover, under natural conditions of storage, minute chromosomal aberrations were present in increasing frequency in progressively nonviable seeds. The various structural novelties of chromosomes proved remarkably stable and occurred consistently in root tip cells of embryos sampled over a period of more than 3 years (Nuti Ronchi and Martini, 1962; Innocenti and Avanzi, 1971). One can see a possible explanation of this effect if the impairment of growth of the embryo is due to the accumulation of toxic or mutagenic substances in the endosperm during aging. A fraction separated from the methanol extract of aged wheat endosperm has been shown to duplicate the *in vivo* effect of the latter with regard to inhibition of embryo growth, and induction of chromosome aberrations (Floris *et al.*, 1972). Although it is not established that toxic compounds of the endosperm act on the embryo during storage, these results suggest that the same basic mechanism may be responsible for both chromosome breakage and low seed viability.

In rye seeds slowly deteriorating under normal storage conditions, a progressive loss of the ability of embryos to synthesize proteins could contribute to loss of viability (Hallam *et al.*, 1973). It is difficult to conceive of this without involving a loss of integrity of the components of the protein-synthesizing system. By determining the ability of mixtures of ribosomes and supernatant from either viable or nonviable embryos to synthesize proteins *in vitro*, it was shown that loss of viability was due to the fragmentation of the ribosomes and the low functional status of the supernatant (Roberts *et al.*, 1973). The suggestion that a lesion in the supernatant could result from the lability of the transferase enzyme involved in the binding of aminoacyl-tRNA to ribosomes (Roberts and Osborne, 1973) is within the explanatory capacity of the currently known concepts in protein biosynthesis.

The use of the embryo culture method in testing seeds for viability was briefly described in Chapter 13; in seed-testing practices, biochemical approaches to the determination of seed viability and vigor have gone hand-in-hand with culture methods. Among such approaches, one which has commanded close attention of seed testers is the classical reduction of tetrazolium salts by embryos. Interesting relationships between deterioration of seed samples and changes in O_2 uptake, respiratory enzymes, starch content, glucose utilization, and extent of chromosomal damage in isolated embryos have been described. The potential value of these methods for measuring seed vigor has recently been reviewed (Abdul-Baki and Anderson, 1972) and in general it can be said that although they are somewhat tedious to perform, they appear to yield promising indices of viability of samples in seed testing trials.

V. COMMENTS

The treatment of germination presented in this chapter is admittedly brief relative to the monumental amount of information available. Much of the difficulty in understanding the control mechanisms of germination as a unified concept lies in the fact that an array of different phenomena, each with its own independent control, is involved in the germination process. These include a latent period of quiescence and dormancy, digestion of storage reserves, mobilization of food materials, growth of the primordial meristems and finally their emergence. The experiments reviewed here have focused on some potentially fruitful approaches to the analysis of the contributions of each of these control mechanisms to germination as a whole. The increasing use of molecular and biophysical approaches to a problem which has been traditionally attacked from morphological and physiological points of view promises a future of considerable advances in our understanding of germination.

Appendix

Compiled below is the mineral salt composition of nutrient media used for successful culture of embryos of certain plants. These media are referred to in the book after the investigators who formulated them for general plant tissue culture work. This is not a thorough list of the media used in embryo culture work, but includes recipes of some of the most commonly used ones. In addition to major salts and microelements, invariably, in all cases, the medium also contains vitamins and other organic additives, and a carbon energy source such as sucrose. Reference to the original papers cited in the text and at the end of the appendix is recommended for further details, and for the constituents of nutrient media used in the culture of embryos of other plants discussed in the book.

Unless otherwise stated, amounts are given in milligrams per liter.

1. Bourgin and Nitsch's Medium

Major salts		Microelements	
$Ca(NO_3)_2 \cdot 4H_2O$	500	$MnSO_4 \cdot 4H_2O$	25
KNO_3	125	H_3BO_3	10
$MgSO_4 \cdot 7H_2O$	125	$ZnSO_4 \cdot 4H_2O$	10
KH_2PO_4	125	$Na_2MoO_4 \cdot 2H_2O$	0·25
		$CuSO_4 \cdot 5H_2O$	0·025
		Na_2EDTA	37·3
		$FeSO_4 \cdot 7H_2O$	27·8

2. Halperin's Medium

Major salts		Microelements	
$MgSO_4 \cdot 7H_2O$	185	$MnSO_4 \cdot H_2O$	7
$CaCl_2$	166	$ZnSO_4 \cdot 7H_2O$	4·05
KH_2PO_4	68	H_3BO_3	2·4
		KI	0·375
		$(NH_4)_6Mo_7O_{24} \cdot 4H_2O$	0·0925
		$CuSO_4 \cdot 5H_2O$	0·01
		Na_2EDTA	18·6
		$FeSO_4 \cdot 7H_2O$	13·9

In addition to the above, the medium contains nitrogen, supplied in various forms, but mainly as $(NH_4)Cl$, or as KNO_3.

3. Heller's Medium

Major salts		*Microelements*	
KCl	750	$ZnSO_4 \cdot 7H_2O$	1
$NaNO_3$	600	H_3BO_3	1
$MgSO_4 \cdot 7H_2O$	250	$MnSO_4 \cdot 4H_2O$	0·1
$NaH_2PO_4 \cdot H_2O$	125	$CuSO_4 \cdot 5H_2O$	0·03
$CaCl_2 \cdot 2H_2O$	75	$AlCl_3$	0·03
		$NiCl_2 \cdot 6H_2O$	0.03
		KI	0·01
		$FeCl_3 \cdot 6H_2O$	1

4. Knop's Medium

Major salts		*Microelements*	
$Ca(NO_3)_2$	500	KH_2PO_4	125
KNO_3	125	$FeCl_3 \cdot 6H_2O$	1
$MgSO_4$	125		

5. Knudson's Medium

Major salts		*Microelement*	
$Ca(NO_3)_2$	1000	$FePO_4$	50
K_2HPO_4	250		
$MgSO_4 \cdot nH_2O$	250		
$(NH_4)_2SO_4$	500		

6. Murashige–Skoog Medium

Major salts		*Microelements*	
NH_4NO_3	1650	H_3BO_3	6·2
KNO_3	1900	$MnSO_4 \cdot 4H_2O$	22·3
$CaCl_2 \cdot 2H_2O$	440	$ZnSO_4 \cdot 4H_2O$	8·6
$MgSO_4 \cdot 7H_2O$	370	KI	0·83
KH_2PO_4	170	$Na_2MoO_4 \cdot 2H_2O$	0·25
		$CuSO_4 \cdot 5H_2O$	0·025
		$CoCl_2 \cdot 6H_2O$	0.025
		Na_2EDTA	37·3
		$FeSO_4 \cdot 7H_2O$	27·8

7. Modified Murashige–Skoog Medium (Monnier, 1973)

Major salts		*Microelements*	
NH_4NO_3	825	H_3BO_3	12·4
KNO_3	1900	$MnSO_4 \cdot H_2O$	33·6
$CaCl_2 \cdot 2H_2O$	880	$ZnSO_4 \cdot 7H_2O$	21
$MgSO_4 \cdot 7H_2O$	370	KI	1·66
KH_2PO_4	170	$Na_2MoO_4 \cdot 2H_2O$	0·5
KCl	350	$CuSO_4 \cdot 5H_2O$	0·05
		$CoCl_2 \cdot 6H_2O$	0·05
		Na_2EDTA	14·9
		$FeSO_4 \cdot 7H_2O$	11·1

8. Nitsch's Medium

Major salts		Microelements	
$Ca(NO_3)_2 \cdot 4H_2O$	500	$MnSO_4 \cdot 4H_2O$	3
KNO_3	125	$CuSO_4 \cdot 5H_2O$	0·025
$MgSO_4 \cdot 7H_2O$	125	H_3BO_3	0·5
KH_2PO_4	125	$ZnSO_4 \cdot 7H_2O$	0·5
		H_2SO_4	0·0005 ml
		$Na_2MoO_4 \cdot 2H_2O$	0·025
		Ferric citrate	10

9. Modified Olsen's Medium (Rijven, 1952)

Major salts		Microelements	
$Ca(NO_3)_2 \cdot 4H_2O$	168	$MnSO_4 \cdot 4H_2O$	0·4
KNO_3	149	H_3BO_3	0·4
$MgSO_4 \cdot 7H_2O$	101	$ZnSO_4 \cdot 7H_2O$	0·2
		$CuSO_4 \cdot 5H_2O$	0·1
		$(NH_4)_2MoO_4$	0·05
		Ferric citrate	50

0·01 M phosphate buffer to one liter

10. Modified Pfeffer's Medium (Loo and Wang, 1943)

Major salts		Microelement	
$Ca(NO_3)_2$	328	$FeCl_3$	16
KNO_3	80		
$MgSO_4 \cdot 7H_2O$	98		
KH_2PO_4	13		
KCl	74		

11. Randolph and Cox's Medium

Major salts		Microelement	
$Ca(NO_3)_2 \cdot 4H_2O$	236·8	$FeSO_4 \cdot 7H_2O$	2
KNO_3	85		
KCl	65		
$Na(PO_3)n$	10		
$MgSO_4 \cdot 7H_2O$	36		

Modification of the medium involves an addition of 0·4 mg/l $MnSO_4$

12. Rappaport's Medium

Major salts		Microelements	
$Ca(NO_3)_2 \cdot 4H_2O$	236·8	$MnSO_4 \cdot 4H_2O$	0·5
KNO_3	85	Ferric citrate	30
KCl	65		
$Na(PO_3)n$	10		
$Mg(SO_4) \cdot 7H_2O$	36		

13.   Modified Robbins and Schmidt's Medium
(Raghavan and Torrey, 1963)

Major salts		*Microelements*	
$Ca(NO_3)_2 \cdot 4H_2O$	480	H_3BO_3	0·56
$MgSO_4 \cdot 7H_2O$	63	$MnCl_2 \cdot 4H_2O$	0·36
KNO_3	63	$ZnCl_2$	0·42
KCl	42	$CuCl_2 \cdot 2H_2O$	0·27
KH_2PO_4	50	$(NH_4)_6Mo_7O_{24} \cdot 4H_2O$	1·55
		Ferric tartrate	3·08

14.   Street's Medium

Major salts		*Microelements*	
$Ca(NO_3)_2 \cdot 4H_2O$	288	H_3BO_3	1·5
KNO_3	80	$CuSO_4 \cdot 5H_2O$	0·02
KCl	65	$ZnSO_4 \cdot 7H_2O$	2·65
$MgSO_4 \cdot 7H_2O$	740	KI	0·75
NaH_2PO_4	21·5	H_2MoO_4	0·0017
Na_2SO_4	453·4	$MnCl_2 \cdot 4H_2O$	6·0
		FeEDTA	1

15.   Tukey's Medium

Major salts		*Microelement*	
KNO_3	135	$Fe_3(PO_4)_2$	185
KCl	680		
$CaSO_4$	185		
$Ca_3(PO_4)_2$	185		
$MgSO_4 \cdot 7H_2O$	185		

16.   White's Medium

Major salts		*Microelements*	
$Ca(NO_3)_2 \cdot 4H_2O$	300	$MnSO_4 \cdot 4H_2O$	7
$MgSO_4 \cdot 7H_2O$	720	$ZnSO_4 \cdot 7H_2O$	3
Na_2SO_4	200	H_3BO_3	1·5
$NaH_2PO_4 \cdot H_2O$	16·5	$CuSO_4 \cdot 5H_2O$	0·001
KNO_3	80	MoO_3	0·0001
KCl	65	KI	0·75
		$Fe_2(SO_4)_3$	2·5

17.   Modified White's Medium (Matsubara, 1962)

Major salts		Microelements	
$Ca(NO_3)_2$	197·2	$MnSO_4$	4·5
$MgSO_4$	361·2	$ZnSO_4$	1·52
Na_2SO_4	199·6	H_3BO_3	1·44
NaH_2PO_4	14·4	KI	0·75
KNO_3	80	Ferric citrate	4·0
KCl	67·2		

18.   Modified White's Medium (Rangaswamy, 1961)

Major salts		Microelements	
$Ca(NO_3)_2 \cdot 4H_2O$	260	$MnSO_4 \cdot 4H_2O$	3
$MgSO_4 \cdot 7H_2O$	360	$ZnSO_4 \cdot 7H_2O$	0·5
Na_2SO_4	200	H_3BO_3	0·5
NaH_2PO_4	165	$CuSO_4 \cdot 5H_2O$	0·025
KNO_3	80	Na_2MoO_4	0·025
KCl	65	$CoCl_2$	0·025
		Ferric citrate	10

REFERENCES TO MEDIUM COMPOSITION

Journal references for medium composition are given in the same order as the media listed above.

1. *Ann Physiol. Végét.* **9**, 377–382 (1967).
2. *Lloydia* **24**, 139–145 (1961); *Amer. Jour. Bot.* **53**, 443–453 (1966).
3. *Ann Biol.* **30**, 261–281 (1954).
4. *Landw. Versuchs-Stat.* **30**, 292–294 (1884).
5. *Bot. Gaz.* **79**, 345–379 (1925).
6. *Physiol. Plantarum* **15**, 473–497 (1962).
7. *Soc. Bot. Fr. Mémoires, Colloq. Morphologie* 179–194 (1973).
8. *Amer. Jour. Bot.* **38**, 566–577 (1951).
9. *Acta Bot. Neerl.* **1**, 157–200 (1952).
10. *Science* **98**, 544 (1943).
11. *Proc. Amer. Soc. Hort. Sci.* **43**, 284–300 (1943); *Amer. Jour. Bot.* **40**, 538–545 (1953).
12. *Bot. Rev.* **20**, 201–225 (1954).
13. *Amer. Jour. Bot.* **50**, 540–551 (1963).
14. *Discovery N. S.* **15**, 286–292 (1954).
15. *Proc. Amer. Soc. Hort. Sci.* **32**, 313–322 (1934); *Bot. Gaz.* 99, 630–665 (1938).
16. P. R. White, "The Cultivation of Animal and Plant Cells". New York, Ronald Press (1963).
17. *Bot. Mag. Tokyo* **75**, 10–18 (1962).
18. *Phytomorphology* **11**, 109–127 (1961).

References

ABBE, E. C. and STEIN, O. L. (1954). The growth of the shoot apex in maize: Embryogeny. *Amer. Jour. Bot.* **41**, 285–293.

ABDUL-BAKI, A. A. (1969). Metabolism of barley seed during early hours of germination. *Plant Physiol.* **44**, 733–738.

ABDUL-BAKI, A. A. and ANDERSON, J. D. (1972). Physiological and biochemical deterioration of seeds. *In* "Seed Biology" (T. T. Kozlowski, ed.), Vol. 2, pp. 283–315. Academic Press, New York and London.

ABO EL-NIL, M. M. and HILDEBRANDT, A. C. (1971). Differentiation of virus-symptomless *Geranium* plants from anther callus. *Plant Disease Report.* **55**, 1017–1020.

ABO EL-NIL, M. M. and HILDEBRANDT, A. C. (1973). Origin of androgenetic callus and haploid *Geranium* plants. *Canad. Jour. Bot.* **51**, 2107–2109.

ABOU-ZEID, A. (1972). Embryoachsenkultur von Kirschen in flüssiger Nährlösung. *Gartenbauwiss.* **37**, 273-280.

ABOU-ZEID, A. and GRUPPE, W. (1972). Das Wachstum von Kirschenembryonen verschiedener Sorten in Abhängigkeit von dem Grad der Embryoentwicklung, Temperaturbehandlung und Nährmedium. *Gartenbauwiss.* **37**, 225–238.

ABOU-ZEID, A. and NEUMANN, K.-H. (1973). Preliminary investigations on the influence of cotyledons on the development of cherry embryos (*Prunus avium* L.). *Zeitschr. Pflanzenphysiol.* **69**, 299–305.

ABRAHAM, A. and RAMACHANDRAN, K. (1960). Growing *Colocasia* embryos in culture. *Curr. Sci.* **29**, 342–343.

ABRAHAM, A. and THOMAS, K. J. (1962). A note on the *in vitro* culture of excised coconut embryos. *Indian Cocon. Jour.* **15**, 84–88.

AGARWAL, S. (1962). Embryology of *Quinchamalium chilense* Lam. *In* "Plant Embryology—A Symposium", pp. 162–169. Council of Scientific and Industrial Research, New Delhi.

AKAZAWA, T. and BEEVERS, H. (1957). Mitochondria in the endosperm of the germinating castor bean: A developmental study. *Biochem. Jour.* **67**, 115–118.

ALBAUM, H. G. (1938). Normal growth, regeneration, and adventitious outgrowth formation in fern prothalli. *Amer. Jour. Bot.* **25**, 37–44.

ALEXANDER, L. J. (1956). Embryo culture of tomato interspecific hybrids. *Phytopathology* **46**, 6 (abstract).

ALEXANDER, M. P. and RAO, T. C. R. (1968). *In vitro* culture of bamboo embryos. *Curr. Sci.* **37**, 415–416.

ALLENDE, J. E. and BRAVO, M. (1966). Amino acid incorporation and aminoacyl transfer in a wheat embryo system. *Jour. Biol. Chem.* **241**, 5813–5818.

ALTSCHUL, A. M., SNOWDEN, J. E., JR., MANCHON, D. D., JR. and DECHARY, J. M. (1961). Intracellular distribution of seed proteins. *Arch. Biochem. Biophys.* **95**, 402–404.

ALVAREZ, M. R. (1968). Quantitative changes in nuclear DNA accompanying postgermination embryonic development in *Vanda* (Orchidaceae). *Amer. Jour. Bot.* **55**, 1036–1041.

ALVAREZ, M. R. and KING, D. O. (1969). Peroxidase localization, activity, and isozyme patterns in the developing seedling of *Vanda* (Orchidaceae). *Amer. Jour. Bot.* **56**, 180–186.

ALVAREZ, M. R. and SAGAWA, Y. (1965a). A histochemical study of embryo sac development in *Vanda* (Orchidaceae). *Caryologia* **18**, 241–249.

ALVAREZ, M. R. and SAGAWA, Y. (1965b). A histochemical study of embryo development in *Vanda* (Orchidaceae). *Caryologia* **18**, 251–261.

AMEMIYA, A. (1964). Effect of peptone on growth of rice embryo (Studies on the embryo culture in rice plant. 4). *Bull. Natl Inst. Agric. Sci. Tokyo Ser. D. (Plant Physiol.)* **11**, 151–210.

AMEMIYA, A., AKEMINE, H. and TORIYAMA, K. (1956a). Cultural conditions and growth of immature embryo in rice plant (Studies in the embryo culture in rice plant. 1). *Bull. Natl Inst. Agric. Sci. Tokyo Ser. D. (Plant Physiol.)* **6**, 1–40.

AMEMIYA, A., AKEMINE, H. and TORIYAMA, K. (1956b). The first germinative stage and varietal differences in growth response of cultured embryo of rice plant (Studies on the embryo culture rice plant. 2). *Bull. Natl Inst. Agric. Sci. Tokyo Ser. D. (Plant Physiol.)* **6**, 41–60.

AMEN, R. D. (1968). A model of seed dormancy. *Bot. Rev.* **34**, 1–31.

AMMIRATO, P. V. (1974). The effects of abscisic acid on the development of somatic embryos from cells of caraway (*Carum carvi* L.). *Bot. Gaz.* **135**, 328–337.

AMMIRATO, P. V. and STEWARD, F. C. (1969). Indirect effects of irradiation: Morphogenetic effects of irradiated sucrose. *Develop. Biol.* **19**, 87–106.

AMMIRATO, P. V. and STEWARD, F. C. (1971). Some effects of environment on the development of embryos from cultured free cells. *Bot. Gaz.* **132**, 149–158.

ANAGNOSTAKIS, S. L. (1974). Haploid plants from anthers of tobacco: Enhancement with charcoal. *Planta* **115**, 281–283.

ANDREWS, C. J. and SIMPSON, G. M. (1969). Dormancy studies in seed of *Avena fatua*. 6. Germinability of the immature caryopsis. *Canad. Jour. Bot.* **47**, 1841–1849.

ANDRONESCU, D. I. (1919). Germination and further development of the embryo of *Zea mays* separated from the endosperm. *Amer. Jour. Bot.* **6**, 443–452.

APP, A. A., BULIS, M. G. and McCARTHY, W. J. (1971). Dissociation of ribosomes and seed germination. *Plant Physiol.* **47**, 81–86.

ARDITTI, J. (1966a). The effect of tomato juice and its fractions on the germination of orchid seeds and on seedling growth. *Amer. Orch. Soc. Bull.* **35**, 175–182.

ARDITTI, J. (1966b). The effects of niacin, adenine, ribose and niacinamide coenzymes on germinating orchid seeds and young seedlings. *Amer. Orch. Soc. Bull.* **35**, 892–898.

ARDITTI, J. (1967a). Factors affecting the germination of orchid seeds. *Bot. Rev.* **33**, 1–97.

ARDITTI, J. (1967b). Niacin biosynthesis in germinating × *Laeliocattleya* orchid embryos and young seedlings. *Amer. Jour. Bot.* **54**, 291–298.

AREKAL, G. D. (1963). Embryological studies in Canadian representatives of the tribe Rhinantheae, Scrophulariaceae. *Canad. Jour. Bot.* **41**, 267–302.

ARNOLD, C.-G. and CRUSE, D. (1967). Die Wirkung von Thalidomid auf die Embryoentwicklung von *Arabidopsis thaliana*. *Zeitschr. Pflanzenphysiol.* **56**, 292–294.

ASAKAWA, S. (1961). A preliminary study concerning the relationship between an embryo and endosperm during stratification. *Jour. Jap. Forest Soc.* **43**, 195–197.

ASEN, S. (1948). Embryo culture of rose seeds. *Amer. Rose Ann.* 151–152.

ASHLEY, T. (1972). Zygote shrinkage and subsequent development in some *Hibiscus* hybrids. *Planta* **108**, 303–317.

ASPINALL, D. and PALEG, L. G. (1971). The deterioration of wheat embryo and endosperm function with age. *Jour. Exptl Bot.* **22**, 925–935.

AUGUSTEN, H. (1956). Wachstumsversuche mit isolierten Weizen-Embryonen. *Planta* **48**, 24–46.

AVANZI, S., BRUNORI, A. and D'AMATO, F. (1969). Sequential development of meristems in the embryo of *Triticum durum*. A DNA autoradiographic and cytophotometric analysis. *Develop. Biol.* **20**, 368–377.

AVANZI, S., CIONINI, P. G. and D'AMATO, F. (1970). Cytochemical and autoradiographic analyses on the embryo suspensor cells of *Phaseolus coccineus*. *Caryologia* **23**, 605–638.

AVANZI, S., BUONGIORNO-NARDELLI, M., CIONINI, P. G. and D'AMATO, F. (1971). Cytological localization of molecular hybrids between rRNA and DNA in the embryo suspensor cells of *Phaseolus coccineus*. A preliminary note. *Rend. Classe Sci. Fisiche Matemat. Natural. Ser. 8,* **50**, 357–361.

AVANZI, S., CORSI, G., D'AMATO, F., FLORIS, C. and MELETTI, P. (1967). The chromosome breaking effect of the irradiated endosperm in water-soaked seeds of *durum* wheat. *Mutation Res.* **4**, 704–707.

AVERY, G. S., JR. (1930). Comparative anatomy and morphology of embryos and seedlings of maize, oats, and wheat. *Bot. Gaz.* **89**, 1–39.

AVERY, G. S., JR., BERGER, J. and SHALUCHA, B. (1941). The total extraction of free auxin and auxin precursor from plant tissue. *Amer. Jour. Bot.* **28**, 596–607.

AVERY, G. S., JR., BERGER, J. and SHALUCHA, B. (1942a). Total auxin extraction from wheat. *Amer. Jour. Bot.* **29**, 612–616.

AVERY, G. S., JR., BERGER, J. and SHALUCHA, B. (1942b). Auxin content of maize kernels during ontogeny, from plants of varying heterotic vigor. *Amer. Jour. Bot.* **29**, 765–772.

BACKS-HÜSEMANN, D. and REINERT, J. (1970). Embryobildung durch isolierte Einzelzellen aus Gewebekulturen von *Daucus carota*. *Protoplasma* **70**, 49–60.

BAGLEY, B. W., CHERRY, J. H., ROLLINS, M. L. and ALTSCHUL, A. M. (1963). A study of protein bodies during germination of peanut (*Arachis hypogaea*) seed. *Amer. Jour. Bot.* **50**, 523–532.

BAHME, R. B. (1949). Nicotinic acid as a growth factor for certain orchid embryos. *Science* **109**, 522–523.

BAIN, J. M. and MERCER, F. V. (1966a). Subcellular organization of the developing cotyledons of *Pisum sativum* L. *Austral. Jour. Biol. Sci.* **19**, 49–67.

BAIN, J. M. and MERCER, F. V. (1966b). Subcellular organization of the cotyledons in germinating seeds and seedlings of *Pisum sativum* L. *Austral. Jour. Biol. Sci.* **19**, 69–84.

BAIN, J. M. and MERCER, F. V. (1966c). The relationship of the axis and the cotyledons in germinating seeds and seedlings of *Pisum sativum* L. *Austral. Jour. Biol. Sci.* **19**, 85–96.

BAJAJ, Y. P. S. (1964). Development of ovules of *Abelmoschus esculentus* L. var. *Pusa Sawani in vitro*. *Proc. Natl Inst. Sci. India* **30B**, 175–185.

BAJAJ, Y. P. S. (1966a). Growth of *Hyoscyamus niger* ovaries in culture. *Phyton* **23**, 57–62.

BAJAJ, Y. P. S. (1966b). Behavior of embryo segments of *Dendrophthoe falcata* (L.f.) Ettings. *in vitro*. *Canad. Jour. Bot.* **44**, 1127–1131.

BAJAJ, Y. P. S. (1967). *In vitro* studies on the embryos of two mistletoes, *Amyema pendula* and *Amyema miquelii*. *N. Zeal. Jour. Bot.* **5**, 49–56.

BAJAJ, Y. P. S. (1968). Some factors affecting growth of embryos of *Dendrophthoe falcata* in cultures. *Canad. Jour. Bot.* **46**, 429–433.

BAJAJ, Y. P. S. (1970). Growth responses of excised embryos of some mistletoes. *Zeitschr. Pflanzenphysiol.* **63**, 408–415.

BAJAJ, Y. P. S. and BOPP, M. (1972). Growth and organ formation in *Sinapis alba* tissue cultures. *Zeitschr. Pflanzenphysiol.* **66**, 378–381.

BAL, A. K. and PAYNE, J. F. (1972). Nucleolar ultrastructure in the quiescent embryonic cells of the dry seed of *Allium cepa* L. *Experientia* **28**, 680–682.

BALANSARD, J. and PELLESSIER, F. (1943). Action sur les germes de Blé isolés de la saponine du Quillaya. *Compt. Rend. Soc. Biol.* **137**, 461–462.

BALATKOVÁ, V. and TUPÝ, J. (1968). Test-tube fertilization in *Nicotiana tabacum* by means of an artificial pollen tube culture. *Biol. Plantarum* **10**, 266–270.

BALATKOVÁ, V. and TUPÝ, J. (1972a). Some factors affecting the seed set after *in vitro* pollination of excised placentae of *Nicotiana tabacum* L. *Biol. Plantarum* **14**, 82–88.

BALATKOVÁ, V. and TUPÝ, J. (1972b). The stimulatory effect of uracil and 5-bromouracil on the seed set in *Papaver somniferum* L. *Biol. Plantarum* **14**, 140–145.

BALATKOVÁ, V. and TUPÝ, J. (1973). The significance of the methods of stigmatal and placental pollination *in vitro* in *Antirrhinum majus* L.; seed and callus formation on placentae. *Biol. Plantarum* **15**, 102–106.

BALL, E. (1946). Development in sterile culture of stem tips and subjacent regions of *Tropaeolum majus* L. and of *Lupinus albus* L. *Amer. Jour. Bot.* **33**, 301–318.

BALL, E. (1956a). Growth of the embryo of *Ginkgo biloba* under experimental conditions. I. Origin of the first root of the seedling *in vitro*. *Amer. Jour. Bot.* **43**, 488–495.

BALL, E. (1956b). Growth of the embryo of *Ginkgo biloba* under experimental conditions. II. Effects of a longitudinal split in the tip of the hypocotyl. *Amer. Jour. Bot.* **43**, 802–810.

BALL, E. (1959). Growth of the embryo of *Ginkgo biloba* under experimental conditions. III. Growth rates of root and shoot upon media absorbed through the cotyledons. *Amer. Jour. Bot.* **46**, 130–139.

BALL, E. and JOSHI, P. C. (1966). Adventive embryos in a callus culture of *Didiscus coerulea*. *Amer. Jour. Bot.* **53**, 612 (abstract).

BAN, Y., KOKUBU, T. and MIYAJI, Y. (1971). Production of haploid plant by anther-culture of *Setaria italica*. *Bull. Fac. Agric. Kagoshima Univ.* **21**, 77–81.

BANERJEE, S. N. (1968). Effect of ionizing radiation on developing embryo-gametophyte complex of *Pinus resinosa* Ait. *Bot. Mag. Tokyo* **81**, 11–21.

BANERJEE, S. N. and RADFORTH, N. W. (1967). Gibberellin-like substances in the fruits of *Ginkgo biloba* L. *Plant Cell Physiol.* **8**, 207–209.

BAPTIST, N. G. (1956). γ-Aminobutyric acid and other free amino acids in the coconut. *Nature* **178**, 1403–1404.

BAPTIST, N. G. (1963). Free amino acids in the endosperm of the developing coconut (*Cocos nucifera*). *Jour. Exptl Bot.* **14**, 29–41.

BARKER, G. R. and HOLLINSHEAD, J. A. (1964). Nucleotide metabolism in germinating seeds. The ribonucleic acid of *Pisum arvense*. *Biochem. Jour.* **93**, 78–83.

BARKER, G. R. and RIEBER, M. (1967). The development of polysomes in the seed of *Pisum arvense*. *Biochem. Jour.* **105**, 1195–1201.

BARNELL, H. R. (1937). Analytic studies in plant respiration. VII. Aerobic respiration in barley seedlings and its relation to growth and carbohydrate supply. *Proc. Roy. Soc. Lond.* **B123**, 321–342.

BARNES, R. L. (1961). Adenine catabolism in pine embryos and tissue cultures. *Bot. Gaz.* **123**, 141–143.

BARNES, R. L. and NAYLOR, A. W. (1962). Formation of β-alanine by pine tissues supplied with intermediates in uracil and orotic acid metabolism. *Plant Physiol.* **37**, 171–175.

BARON, F. J. (1962). Seed components in relation to embryo growth and germination of sugar pine (*Pinus lambertiana* Dougl.). *Plant Physiol.* **37** (Suppl.), lv.

BARTELS, H. (1957a). Kultur isolierter Koniferenembryonen. *Naturwiss.* **44**, 290–291.

BARTELS, H. (1957b). Kultur isolierter Koniferenembryonen. II. *Naturwiss.* **44**, 595–596.

BARTON, L. V. (1956). Growth response of physiologic dwarfs of *Malus arnoldiana* Sarg. to gibberellic acid. *Contrib. Boyce Thompson Inst.* **18**, 311–317.

BASKIN, J. M. and BASKIN, C. C. (1970). Replacement of chilling requirement in seeds of *Ruellia humilis* by gibberellic acid. *Planta* **94**, 250–252.

BASKIN, J. M. and BASKIN, C. C. (1971). Effect of chilling and gibberellic acid on growth potential of excised embryos of *Ruellia humilis*. *Planta* **100**, 365–369.

BEAMISH, K. I. (1955). Seed failure following hybridization between the hexaploid *Solanum demissum* and four diploid *Solanum* species. *Amer. Jour. Bot.* **42**, 297–304.

BEASLEY, C. A. (1971). *In vitro* culture of fertilized cotton ovules. *BioScience* **21**, 906–907.

BEASLEY, C. A. (1973). Hormonal regulation of growth in unfertilized cotton ovules. *Science* **179**, 1003–1005.

BEASLEY, C. A. and TING, I. P. (1973). The effects of plant growth substances on *in vitro* fiber development from fertilized cotton ovules. *Amer. Jour. Bot.* **60**, 130–139.

BEASLEY, C. A. and TING, I. P. (1974). Effects of plant growth substances on *in vitro* fiber development from unfertilized cotton ovules. *Amer. Jour. Bot.* **61**, 188–194.

BEASLEY, C. A., TING, I. P. and FEIGEN, L. A. (1971). Test tube cotton. *California Agric.* **25** (No. 10), 6–8.

BEASLEY, J. O. (1940). Hybridization of American 26-chromosome and Asiatic 13-chromosome species of *Gossypium*. *Jour. Agric. Res.* **60**, 175–181.

BEAUDRY, J. R. (1951). Seed development following the mating *Elymus virginicus* L. × *Agropyron repens* (L.) Beauv. *Genetics* **36**, 109–133.

BEEVERS, L. and GUERNSEY, F. S. (1966). Changes in some nitrogenous components during the germination of pea seeds. *Plant Physiol.* **41**, 1455–1458.

BEEVERS, L. and POULSON, R. (1972). Protein synthesis in cotyledons of *Pisum sativum* L. I. Changes in cell-free amino acid incorporation capacity during seed development and maturation. *Plant Physiol.* **49**, 476–481.

BELL, P. R. (1960). Interaction of nucleus and cytoplasm during oogenesis in *Pteridium aquilinum* (L.) Kuhn. *Proc. Roy. Soc. Lond.* **B153**, 421–432.

BELL, P. R. (1961). Failure of nucleotides to diffuse freely into the embryo of *Pteridium aquilinum*. *Nature* **191**, 91–92.

BELL, P. R. (1963). The cytochemical and ultrastructural peculiarities of the fern egg. *Jour. Linn. Soc. (Bot.)* **58**, 353–359.

BELL, P. R. (1966). Organelle in der Eizelle von *Dryopteris filix-mas* (L.) Schott. *Naturwiss.* **53**, 232.

BELL, P. R. (1969). The cytoplasmic vesicles of the female reproductive cells of *Pteridium aquilinum*. *Zeitschr. Zellforsch.* **96**, 49–62.

BELL, P. R. (1970). The archegoniate revolution. *Sci. Progr. Oxford* **58**, 27–45.

BELL, P. R. (1972). Nucleocytoplasmic interaction in the eggs of *Pteridium aquilinum* maturing in the presence of thiouracil. *Jour. Cell. Sci.* **11**, 739–755.

BELL, P. R. (1974). Nuclear sheets in the egg of a fern, *Dryopteris filix-mas*. *Jour. Cell. Sci.* **14**, 69–83.

BELL, P. R. and MÜHLETHALER, K. (1962a). A membrane peculiar to the egg in the gametophyte of *Pteridium aquilinum*. *Nature* **195**, 198.

BELL, P. R. and MÜHLETHALER, K. (1962b). The fine structure of the cells taking part in oogenesis in *Pteridium aquilinum* (L.) Kuhn. *Jour. Ultrastr. Res.* **7**, 452–466.

BELL, P. R. and MÜHLETHALER, K. (1964a). The degeneration and reappearance of mitochondria in the egg cells of a plant. *Jour. Cell Biol.* **20**, 235–248.

BELL, P. R. and MÜHLETHALER, K. (1964b). Evidence for the presence of deoxyribonucleic acid in the organelles of the egg cells of *Pteridium aquilinum*. *Jour. Mol. Biol.* **8**, 853–862.

BELL, P. R., FREY-WYSSLING, A. and MÜHLETHALER, K. (1966). Evidence for the discontinuity of plastids in the sexual reproduction of a plant. *Jour. Ultrastr. Res.* **15**, 108–121.

BENBADIS, A. (1973). Analyse des aptitudes embryogènes des divers types de cellules isolées de *Daucus carota* L. *Soc. Bot. Fr. Mémoires, Colloq. Morphologie*, p. 223–234.

BENNETT, M. D. and HUGHES, W. G. (1972). Additional mitosis in wheat pollen induced by ethrel. *Nature* **240**, 566–568.

BÉRANGER-NOVAT, N. and MONIN, J. (1971a). A propos de la levée de dormance des embryons d'*Evonymus europaeus* L. par l'acide gibbérellique. *Compt. Rend. Acad. Sci. Paris* **272**, 1368–1371.

BÉRANGER-NOVAT, N. and MONIN, J. (1971b). Étude de l'intensité respiratoire d'embryons dormants d'*Evonymus europaeus* L. cultivés aseptiquement en présence d'acide gibbérellique. *Compt. Rend. Acad. Sci. Paris* **272**, 1865–1868.

BÉRANGER-NOVAT, N. and MONIN, J. (1974). A propos de la levée de dormance des embryons d'*Evonymus europaeus* L. par le froid. *Compt. Rend. Acad. Sci. Paris* **278**, 881–884.

BERGER, C. and ERDELSKÁ, O. (1973). Ultrastructural aspects of the embryo sac of *Jasione montana* L.: Cell walls. *Caryologia* **25** (Suppl.), 109–120.

BERGMANN, L. (1959). A new technique for isolating and cloning cells of higher plants. *Nature* **184**, 648–649.

BERGMANN, L. (1960). Growth and division of single cells of higher plants *in vitro*. *Jour. Gen. Physiol.* **43**, 841–851.

BERJAK, P. and VILLIERS, T. A. (1970). Ageing in plant embryos. I. The establishment of the sequence of development and senescence in the root cap during germination. *New Phytol.* **69**, 929–938.

BERJAK, P. and VILLIERS, T. A. (1972a). Ageing in plant embryos. II. Age-induced damage and its repair during early germination. *New Phytol.* **71**, 135–144.

BERJAK, P. and VILLIERS, T. A. (1972b). Ageing in plant embryos. III. Acceleration of senescence following artificial ageing treatment. *New Phytol.* **71**, 513–518.

BERJAK, P. and VILLIERS, T. A. (1972c). Ageing in plant embryos. IV. Loss of regulatory control in aged embryos. *New Phytol.* **71**, 1069–1074.

BERJAK, P. and VILLIERS, T. A. (1972d). Ageing in plant embryos. V. Lysis of the cytoplasm in non-viable embryos. *New Phytol.* **71**, 1075–1079.

BERLYN, G. P. (1962). Developmental patterns in pine polyembryony. *Amer. Jour. Bot.* **49**, 327–333.

BERLYN, G. P. and MIKSCHE, J. P. (1965). Growth of excised pine embryos and the role of the cotyledons during germination *in vitro. Amer. Jour. Bot.* **52**, 730–736.

BERNARD, S. (1971). Développement d'embryons haploides à partir d'anthères cultivées *in vitro.* Étude cytologique comparée chez le tabac et le pétunia. *Rev. Cytol. Biol. Végét.* **34**, 165–188.

BETH, K. (1938). Untersuchungen über die Auslösung von Adventivembryonie durch Wundreiz. *Planta* **28**, 296–343.

BEWLEY, J. D. and FOUNTAIN, D. W. (1972). A distinction between the actions of abscisic acid, gibberellic acid and cytokinins in light-sensitive lettuce seed. *Planta* **102**, 368–371.

BHALLA, P. R. (1971). Gibberellin-like substances in developing watermelon seeds. *Physiol. Plantarum* **24**, 106–111.

BHATNAGAR, S. P. and SABHARWAL, G. (1966). Female gametophyte and endosperm of *Iodina rhombifolia* Hook. & Arn. *Phytomorphology* **16**, 588–591.

BHOJWANI, S. S. (1966). Morphogenetic behavior of mature endosperm of *Croton bonplandianum* Baill. in culture. *Phytomorphology* **16**, 349–353.

BHOJWANI, S. S. (1969). Differentiation of haustoria in the germinating embryos of mistletoe without host stimulus. *Experientia* **25**, 543.

BHOJWANI, S. S. and JOHRI, B. M. (1970). Cytokinin-induced shoot bud differentiation in mature endosperm of *Scurrula pulverulenta. Zeitschr. Pflanzenphysiol.* **63**, 269–275.

BHOJWANI, S. S. and JOHRI, B. M. (1971). Morphogenetic studies on cultured mature endosperm of *Croton bonplandianum. New Phytol.* **70**, 761–766.

BHOJWANI, S. S., DUNWELL, J. M. and SUNDERLAND, N. (1973). Nucleic acid and protein contents of embryogenic tobacco pollen. *Jour. Exptl Bot.* **24**, 863–871.

BILS, R. F. and HOWELL, R. W. (1963). Biochemical and cytological changes in developing soybean cotyledons. *Crop Sci.* **3**, 304–308.

BINDING, H. (1972). Nuclear and cell divisions in isolated pollen of *Petunia hybrida* in agar suspension cultures. *Nature New Biol.* **237**, 283–285.

BINDING, H., BINDING, K. and STRAUB, J. (1970). Selektion in Gewebekulturen mit haploiden Zellen. *Naturwiss.* **57**, 138–139.

BIRNBAUM, E. H., BEASLEY, C. A. and DUGGER, W. M. (1974). Boron deficiency in unfertilized cotton (*Gossypium hirsutum*) ovules grown *in vitro. Plant Physiol.* **54**, 931–935.

BISALPUTRA, T. and ESAU, K. (1964). Polarized light study of phloem differentiation in embryo of *Chenopodium album. Bot. Gaz.* **125**, 1–7.

BISSON, C. S. and JONES, H. A. (1932). Changes accompanying fruit development in the garden pea. *Plant Physiol.* **7**, 91–105.

BISWAS, B. B. (1969). Ribosomes in cotyledons of mungbean seeds at different stages of germination. *Arch. Biochem. Biophys.* **132**, 198–204.

BITTERS, W. P., MURASHIGE, T., RANGAN, T. S. and NAUER, E. (1972). Investigations on establishing virus-free *Citrus* plants through tissue culture. *Proceed-*

ings of the Fifth Conference of the International Organization of Citrus Virologists (W. C. Price, ed.), pp. 267–271, University of Florida Press, Gainesville.

BLACK, H. S. and ALTSCHUL, A. M. (1965). Gibberellic acid-induced lipase and α-amylase formation and their inhibition by aflatoxin, *Biochem. Biophys. Res. Comm.* **19**, 661–664.

BLACK, M. and WAREING, P. F. (1959). The role of germination inhibitors and oxygen in the dormancy of the light-sensitive seed of *Betula* spp. *Jour. Exptl Bot.* **10**, 134–145.

BLACK, M., BEWLEY, J. D. and FOUNTAIN, D. (1974). Lettuce seed germination and cytokinins: Their entry and formation. *Planta* **117**, 145–152.

BLAKE, M. A. (1939). Some results of crosses of early ripening varieties of peaches. *Proc. Amer. Soc. Hort. Sci.* **37**, 232–241.

BLAKESLEE, A. F. and SATINA, S. (1944). New hybrids from incompatible crosses in *Datura* through culture of excised embryos on malt media. *Science* **99**, 331–334.

BLOCISZEWSKI, T. (1876). Physiologische Untersuchungen über die Keimung und weitere Entwickelung einiger Samentheile bedecktsamiger Pflanzen. *Landw. Jahrb.* **5**, 145–161.

BLOWERS, J. W. (1958). Gibberellin for orchids. *Orch. Rev.* **66**, 128–130.

BOGORAD, L. (1950). Factors associated with the synthesis of chlorophyll in the dark in seedlings of *Pinus jeffreyi*. *Bot. Gaz.* **111**, 221–241.

BOISARD, J. and MALCOSTE, R. (1970). Analyse spectrophotométrique du phytochrome dans l'embryon de courge (*Cucurbita pepo*) et de potiron (*Cucurbita maxima*). *Planta* **91**, 54–67.

BONGA, J. M. (1965). *Arceuthobium pusillum* Peck: Collection of seeds and *in vitro* culture of the early seedling stage. *Canad. Jour. Bot.* **43**, 1307–1308.

BONGA, J. M. (1968). The effect of growth substances on the development of *Arceuthobium pusillum* Peck cultured *in vitro*. *In* "Biochemistry and Physiology of Plant Growth Substances" (F. Wightman and G. Setterfield, eds), pp. 867–874. Runge Press, Ottawa.

BONGA, J. M. (1969). The morphology and anatomy of holdfasts and branching radicles of *Arceuthobium pusillum* cultured *in vitro*. *Canad. Jour. Bot.* **47**, 1935–1938.

BONGA, J. M. (1971). Formation of holdfasts, callus, embryoids and haustorial cells in the *in vitro* cultures of dwarf mistletoe *Arceuthobium pusillum*. *Phytomorphology* **21**, 140–153.

BONGA, J. M. (1974a). *In vitro* culture of microsporophylls and megagametophyte tissue of *Pinus*. *In Vitro* **9**, 270–277.

BONGA, J. M. (1974b). The formation of branching strands at the apex of radicles of *Arceuthobium pusillum* cultured *in vitro*. *Canad. Jour. Bot.* **52**, 2113–2115.

BONGA, J. M. and CHAKRABORTY, C. (1968). *In vitro* culture of a dwarf mistletoe, *Arceuthobium pusillum*. *Canad. Jour. Bot.* **46**, 161–164.

BONGA, J. M. and FOWLER, D. P. (1970). Growth and differentiation in gametophytes of *Pinus resinosa* cultured *in vitro*. *Canad. Jour. Bot.* **48**, 2205–2207.

BONGA, J. M. and McINNIS, A. H. (1975). Stimulation of callus development from immature pollen of *Pinus resinosa* by centrifugation. *Plant Sci. Lett.* **4**, 199–203.

BONNER, J. (1938). Nicotinic acid and the growth of isolated pea embryos. *Plant Physiol.* **13**, 865–868.

BONNER, J. and AXTMAN, G. (1937). The growth of plant embryos *in vitro*. Preliminary experiments on the role of accessory substances. *Proc. Natl Acad. Sci. U.S.A.* **23**, 453–457.

BONNER, J. and BONNER, D. (1938). Ascorbic acid and the growth of plant embryos. *Proc. Natl Acad. Sci. U.S.A.* **24**, 70–75.

BONNET, C. (1754). "Recherches sur l'usage des feuilles dans les plantes, et sur quelques autres sujets relatifs à l'histoire de la végétation". Elie Luzac, Gottingue and Leide.

BORCHERT, R. (1968). Spontane Diploidisierung in Gewebekulturen des Megagametophyten von *Pinus lambertiana*. *Zeitschr. Pflanzenphysiol.* **59**, 389–392.

BORRISS, H. (1967). Untersuchungen über die Steuerung der Enzymaktivität in pflanzlichen Embryonen durch Cytokinine. *Wiss. Z. Univ. Rostock Math.-Naturwiss. Reihe* **16**, 629–639.

BORRISS, H. and ARNDT, M. (1956a). Die Entwicklung isolierter *Agrostemma*-Embryonen in Abhängigkeit vom Reife- und Nachreifezustand der Samen. *Flora* **143**, 492–498.

BORRISS, H. and ARNDT, M. (1956b). Die Temperaturabhängigkeit der Stärke-bildung in *Agrostemma*-Embryonen und ihre Beziehung zur Keimungs-bereitschaft der Samen. *Naturwiss.* **43**, 255–256.

BORRISS, H. and BAUM, F. (1957). Der Aminosäuregehalt von *Agrostemma*-Samen verschiedener Keimungsbereitschaft. *Naturwiss.* **44**, 404–405.

BORRISS, H. and SCHNEIDER, G. (1955a). Die Beziehungen zwischen N-Haus-halt und keimungsphysiologischem Verhalten von *Agrostemma*-Samen. *Naturwiss.* **42**, 102–103.

BORRISS, H. and SCHNEIDER, G. (1955b). Über den Nachweis eines Peptids in den Samen von *Agrostemma githago* L. und seine keimungsphysiologische Bedeutung. *Naturwiss.* **42**, 103.

BORTHWICK, H. A., HENDRICKS, S. B., PARKER, M. W., TOOLE, E. H. and TOOLE, V. K. (1952). A reversible photoreaction controlling seed germina-tion. *Proc. Natl Acad. Sci. U.S.A.* **38**, 662–666.

BOUHARMONT, J. (1961). Embryo culture of rice on sterile medium. *Euphytica* **10**, 283–293.

BOURGIN, J.-P. and NITSCH, J. P. (1967). Obtention de *Nicotiana* haploïdes à partir d'étamines cultivées *in vitro*. *Ann. Physiol. Végét.* **9**, 377–382.

BOUVINET, J. and RABÉCHAULT, H. (1965a). Recherches sur la culture "*in vitro*" des embryons de palmier à huile (*Elaeis guineensis* Jacq.). II. Effets de l'acide gibbérellique. *Oléagineux* **20**, 79–87.

BOUVINET, J. and RABÉCHAULT, H. (1965b). Effets de l'acide gibbérellique sur les embryons du palmier à huile (*Elaeis guineensis* Jacq. var. *dura*) en culture *in vitro*. *Compt. Rend. Acad. Sci. Paris* **260**, 5336–5338.

Bower, F. O. (1926). "The Ferns (Filicales)", Vol. II. Cambridge University Press.

Bower, F. O. (1928). "The Ferns (Filicales)", Vol. III. Cambridge University Press.

Bower, F. O. (1935). "Primitive Land Plants". MacMillan, London.

Boyes, J. W. and Thompson, W. P. (1937). The development of the endosperm and embryo in reciprocal interspecific crosses in cereals. *Jour. Genet.* **34**, 203–227.

Bradbeer, J. W. (1968). Studies in seed dormancy. IV. The role of endogenous inhibitors and gibberellin in the dormancy and germination of *Corylus avellana* L. seeds. *Planta* **78**, 266–276.

Bradbeer, J. W. and Colman, B. (1967). Studies in seed dormancy. I. The metabolism of $[2-{}^{14}C]$ acetate by chilled seeds of *Corylus avellana* L. *New Phytol.* **66**, 5–15.

Bradbeer, J. W. and Floyd, V. M. (1964). Nucleotide synthesis in hazel seeds during after-ripening. *Nature* **201**, 99–100.

Bradbeer, J. W. and Pinfield, N. J. (1967). Studies in seed dormancy. III. The effects of gibberellin on dormant seeds of *Corylus avellana* L. *New Phytol.* **66**, 515–523.

Bradbury, D. (1929). A comparative study of the developing and aborting fruits of *Prunus cerasus*. *Amer. Jour. Bot.* **16**, 525–542.

Brady, T. (1970). Activities of polytene chromosomes in *Phaseolus*. *Jour. Cell Biol.* **47**, 23a (abstract).

Brady, T. (1973a). Feulgen cytophotometric determination of the DNA content of the embryo proper and suspensor cells of *Phaseolus coccineus*. *Cell Diffn* **2**, 65–75.

Brady, T. (1973b). Cytological studies on the suspensor polytene chromosomes of *Phaseolus:* DNA content and synthesis, and the ribosomal cistrons. *Caryologia* **25** (Suppl.), 233–259.

Breddy, N. C. (1953). Observations on the raising of orchids by asymbiotic cultures. *Amer. Orch. Soc. Bull.* **22**, 12–17.

Brewbaker, J. L. and Emery, G. C. (1962). Pollen radiobotany. *Radiation Bot.* **1**, 101–154.

Briarty, L. G., Coult, D. A. and Boulter, D. (1969). Protein bodies of developing seeds of *Vicia faba*. *Jour. Exptl Bot.* **20**, 358–372.

Brink, R. A. and Cooper, D. C. (1939). Somatoplastic sterility in *Medicago sativa*. *Science* **90**, 545–546.

Brink, R. A. and Cooper, D. C. (1940). Double fertilization and development of the seed in angiosperms. *Bot. Gaz.* **102**, 1–25.

Brink, R. A. and Cooper, D. C. (1941). Incomplete seed failure as a result of somatoplastic sterility. *Genetics* **26**, 487–505.

Brink, R. A. and Cooper, D. C. (1944). The antipodals in relation to abnormal endosperm behavior in *Hordeum jubatum* × *Secale cereale* hybrid seeds. *Genetics* **29**, 391–406.

Brink, R. A. and Cooper, D. C. (1947). The endosperm in seed development. *Bot. Rev.* **13**, 423–541.

BRINK, R. A., COOPER, D. C. and AUSHERMAN, L. E. (1944). A hybrid between *Hordeum jubatum* and *Secale cereale*. *Jour. Hered.* **35**, 67–75.

BROCK, R. D. (1954). Spontaneous chromosome breakage in *Lilium* endosperm. *Ann. Bot. N.S.* **18**, 7–14.

BROCK, R. D. (1955). Chromosome balance and endosperm failure in hyacinths. *Heredity* **9**, 199–222.

BROOKS, H. J. and HOUGH, L. F. (1958). Vernalization studies with peach embryos. *Proc. Amer. Soc. Hort. Sci.* **71**, 95–102.

BROWN, C. L. and GIFFORD, E. M., JR. (1958). The relation of the cotyledons to root development of pine embryos grown *in vitro*. *Plant Physiol.* **33**, 57–64.

BROWN, H. T. (1906). On the culture of the excised embryos of barley on nutrient solutions containing nitrogen in different forms. *Trans. Guinness Res. Lab.* **1**, 288–299.

BROWN, H. T. and MORRIS, G. H. (1890). Researches on the germination of some of the Gramineae. *Jour. Chem. Soc.* **57**, 458–528.

BROWN, L. C. and KURTZ, E. B., JR. (1959). The *in vitro* synthesis of fats in cottonseed. *Agron. Jour.* **51**, 49–50.

BROWN, N. A. C. and VAN STADEN, J. (1971). Germination inhibitors in aqueous seed extracts of four South African Proteaceae. *Jour. S. Afr. Bot.* **37**, 305–315.

BROWN, N. A. C. and VAN STADEN, J. (1973). The effect of stratification on the endogenous cytokinin levels of seed of *Protea compacta* and *Leucadendron daphnoides*. *Physiol. Plantarum* **28**, 388–392.

BROWN, R. (1943). Studies in germination and seedling growth. I. The water content, gaseous exchange, and dry weight of attached and isolated embryos of barley. *Ann. Bot. N.S.* **7**, 93–113.

BROWN, R. (1946). Studies on germination and seedling growth. III. Early growth in relation to certain aspects of nitrogen metabolism in the seedling of barley. *Ann. Bot. N.S.* **10**, 73–96.

BROWN, R. (1965). Physiology of seed germination. *In* "Encyclopedia of Plant Physiology" (W. Ruhland, ed.), Vol. 15/2, pp. 894–908. Springer-Verlag, Berlin.

BROWN, R. C. and MOGENSEN, H. L. (1972). Late ovule and early embryo development in *Quercus gambelii*. *Amer. Jour. Bot.* **59**, 311–316.

BRUCHMANN, H. (1910). Die Keimung der Sporen und die Entwicklung der Prothallien von *Lycopodium clavatum* L., *L. annotinum* L. und *L. selago* L. *Flora* **101**, 220–267.

BRUCHMANN, H. (1912). Zur Embryologie der Selaginellaceen. *Flora* **104**, 180–224.

BRUCHMANN, H. (1913). Zur Reduktion des Embryoträgers bei Selaginellen. *Flora* **105**, 337–346.

BRUNORI, A. (1967). Relationship between DNA synthesis and water content during ripening of *Vicia faba* seed. *Caryologia* **20**, 333–338.

BRUNORI, A. and D'AMATO, F. (1967). The DNA content of nuclei in the embryo of dry seeds of *Pinus pinea* and *Lactuca sativa*. *Caryologia* **20**, 153–161.

BRYAN, G. S. and EVANS, R. I. (1956). Chromatin behavior in the development and maturation of the egg nucleus of *Zamia umbrosa*. *Amer. Jour. Bot.* **43**, 640–646.

BRYAN, G. S. and EVANS, R. I. (1957). Types of development from the central nucleus of *Zamia umbrosa*. *Amer. Jour. Bot.* **44**, 404–415.

BRYANT, L. R. (1935). A study of the factors affecting the development of the embryo sac and the embryo in the McIntosh apple. *Univ. N.H. Agric. Expt. Stat. Bull.* **61**, 1–40.

BUCHHOLZ, J. T. (1940). The embryogeny of *Torreya*, with a note on *Austrotaxus*. *Bull. Torrey Bot. Club* **67**, 731–754.

BUCHHOLZ, J. T. (1945). Embryological aspects of hybrid vigor in pines. *Science* **102**, 135–142.

BUCHHOLZ, J. T. (1946). Volumetric studies of seeds, endosperms, and embryos in *Pinus ponderosa* during embryonic differentiation. *Bot. Gaz.* **108**, 232–244.

BUCHHOLZ, J. T. and STIEMERT, M. L. (1945). Development of seeds, and embryos in *Pinus ponderosa*, with special reference to seed size. *Trans. Illinois State Acad. Sci.* **38**, 27–50.

BUCKNER, G. D. and KASTLE, J. H. (1917). The growth of isolated plant embryos. *Jour. Biol. Chem.* **29**, 209–213.

BUELL, K. M. (1952). Developmental morphology in *Dianthus*. II. Starch accumulation in ovule and seed. *Amer. Jour. Bot.* **39**, 458–467.

BUELL, K. M. (1953). Developmental morphology in *Dianthus*. III. Seed failure following interspecific crosses. *Amer. Jour. Bot.* **40**, 116–123.

BUFFARD-MOREL, J. (1968). Effets du glucose, du lévulose, du maltose et du saccharose sur le développement des embryons de palmier à huile (*Elaeis guineensis* Jacq. var. *Dura* Bec.) en culture *in vitro*. *Compt. Rend. Acad. Sci. Paris* **267**, 185–188.

BULARD, C. (1952). Culture aseptique d'embryons de *Ginkgo biloba*: Rôle des cotylédons dans l'absorption du sucre et la croissance de la tige. *Compt. Rend. Acad. Sci. Paris* **235**, 739–741.

BULARD, C. (1954). Essais de culture aseptique de gemmules isolées de *Ginkgo biloba*. *Ann. Univ. Saraviensis Naturwiss.-Sci.* **3**, 202–209.

BULARD, C. (1960). Cultures aseptiques de plantules de blé separées de l'albumen: Action de l'acide gibbérellique. *Compt. Rend. Acad. Sci. Paris* **250**, 3716–3718.

BULARD, C. (1967a). Modifications provoquees par divers régulateurs de croissance sur des plantules de *Ginkgo biloba* L. cultivées *in vitro*. *Soc. Bot. Fr., Colloq. Morphol. Exper.*, 119–130.

BULARD, C. (1967b). Un cas d'inhibition de croissance de l'épicotyle chez *Ginkgo biloba* L. obtenu sous l'influence de gibberellines. *Compt. Rend. Acad. Sci. Paris* **265**, 1301–1304.

BULARD, C. (1968). Acquisitions récentes sur la dormance embryonnaire. *Bull. Soc. Fr. Physiol. Végét.* **14**, 11–23.

BULARD, C. and DEGIVRY, M.-T. (1965). La dormance des graines de *Pinus coulteri* Don.: Mise en évidence et essai de localisation. *Phyton* **22**, 55–60.

BULARD, C. and LE PAGE-DEGIVRY, M.-T. (1968). Quelques précisions sur les conditions d'obtention d'une inhibition de la croissance épicotylaire chez

Ginkgo biloba L. sous l'effet de l'acide gibbérellique. *Compt. Rend. Acad. Sci. Paris* **266**, 356–359.

BULARD, C. and MONIN, J. (1960a). Action de l'acide gibbérellique sur des embryons dormants d'*Evonymus europaeus* cultivés *in vitro*. *Compt. Rend. Acad. Sci. Paris* **250**, 2922–2924.

BULARD, C. and MONIN, J. (1960b). Graines et embryons dormants d'*Evonymus europaeus*: Différentes modalités dans l'éveil de leur dormance par l'acide gibbérellique. *Compt. Rend. Acad. Sci. Paris* **250**, 4197–4199.

BULARD, C. and MONIN, J. (1963). Étude du comportement d'embryons de *Fraxinus excelsior* L. prélevés dans des graines dormantes et cultivés *in vitro*. *Phyton* **20**, 115–125.

BURGEFF, H. (1934). Pflanzliche Avitaminose und ihre Behebung durch Vitaminzufuhr. *Ber. Deut. Bot. Ges.* **52**, 384–390.

BURGEFF, H. (1936). "Samenkeimung der Orchideen". G. Fischer, Jena.

BURGEFF, H. (1959). Mycorrhiza of orchids. *In* "The Orchids" (C. L. Withner, ed.), pp. 361–395. Ronald Press, New York.

BURK, L. G. (1970). Green and light-yellow haploid seedlings from anthers of sulfur tobacco. *Jour. Hered.* **61**, 279.

BURK, L. G. and GERSTEL, D. U. (1961). Haploid *Nicotiana* deficient for two chromosomes. *Jour. Hered.* **52**, 203–206.

BURK, L. G., GWYNN, G. R. and CHAPLIN, J. F. (1972). Diploidized haploids from aseptically cultured anthers of *Nicotiana tabacum*. *Jour. Hered.* **63**, 355–360.

BURROWS, W. J. and CARR, D. J. (1970). Cytokinin content of pea seeds during their growth and development. *Physiol. Plantarum* **23**, 1064–1070.

BUTANY, W. T. (1958). Value of embryo culture in rice breeding. *Rice News Teller* **6**, 10–12.

BUTENKO, R. G., GRUSHVITSKII, R. V. and SLEPYAN, L. I. (1968). Organogenesis and somatic embryogenesis in a tissue culture of ginseng (*Panax ginseng*) and other *Panax* L. species. *Bot. Zh.* **53**, 906–911 (*Biol. Abst.* 1968, **50**, 6883).

BUTENKO, R. G., STROGONOV, B. P. and BABAEVA, Z. A. (1967). Somatic embryogenesis in carrot tissue culture under conditions of high salt concentrations in the medium. *Dokl. Akad. Nauk. SSSR.* **175**, 1179–1181 (*Biol. Abst.* 1968, **49**, 5731).

BUTTON, J. and BORNMAN, C. H. (1971). Development of nucellar plants from unpollinated and unfertilized ovules of the Washington Navel orange *in vitro*. *Jour. S. Afr. Bot.* **37**, 127–133.

BUTTON, J., BORNMAN, C. H. and CARTER, M. (1971). *Welwitschia mirabilis*: Embryo and free-cell culture. *Jour. Exptl Bot.* **22**, 922–924.

BUTTON, J., KOCHBA, J. and BORNMAN, C. H. (1974). Fine structure of and embryoid development from embryogenic ovular callus of 'Shamouti' orange (*Citrus sinensis* Osb). *Jour. Exptl Bot.* **25**, 446–457.

BUTTROSE, M. S. (1960). Submicroscopic development and structure of starch granules in cereal endosperms. *Jour. Ultrastr. Res.* **4**, 231–257.

BUTTROSE, M. S. (1963a). Ultrastructure of the developing wheat endosperm. *Austral. Jour. Biol. Sci.* **16**, 305–317.

BUTTROSE, M. S. (1963b). Ultrastructure of the developing aleurone cells of wheat grain. *Austral. Jour. 'Biol. Sci.* **16**, 768–774.

CALVIN, C. L. (1966). Anatomy of mistletoe (*Phoradendron flavescens*) seedlings grown in culture. *Bot. Gaz.* **127**, 171–183.

CÃMARA, A. (1943). Transplantação de embriões. *Agron. Lusit.* **5**, 375–386.

CAMEFORT, H. (1958). Rôle du suc nucléaire et des nucléoles dans la formation du cytoplasme du proembryon chez le *Pinus laricio* (var. *austriaca*). *Compt. Rend. Acad. Sci. Paris* **246**, 2014–2017.

CAMEFORT, H. (1959). Sur la nature cytoplasmique des inclusions dites "vitellines" de l'oosphère du *Pinus laricio* (var. *austriaca*): Étude en microscopie électronique. *Compt. Rend. Acad. Sci. Paris* **248**, 1568–1570.

CAMEFORT, H. (1960). Evolution de l'organisation du cytoplasme dans la cellule centrale et l'oosphère du *Pinus laricio* Poir. (var. *austriaca*). *Compt. Rend. Acad. Sci. Paris* **250**, 3707–3709.

CAMEFORT, H. (1962). L'organisation du cytoplasme dans l'oosphère et la cellule du "*Pinus laricio*" Poir. (var. *austriaca*). *Ann. Sci. Nat. Bot. Biol. Végét.* **3**, 265–291.

CAMEFORT, H. (1963). L'évolution des plastes pendant la croissance et la diferénciation du gamète femelles des pins. *Jour. Microscopie* **2**, 26–27 (abstract).

CAMEFORT, H. (1965a). Une interprétation nouvelle de l'organisation du protoplasme de l'oosphère des pins. *In* "Travaux dédiés à Lucien Plantefol", pp. 407–436. Masson & Cie, Paris.

CAMEFORT, H. (1965b). L'organisation du protoplasme dans le gamète ferénciation du gamète femelles des pins. *Jour. Microscopie* **2**, 26–27 (abstract).

CAMEFORT, H. (1966a). Observations sur les mitochondries et les plastes d'origine pollinique après leur entrée dans une oosphère chez le pin noir (*Pinus laricio* Poir. var. *austriaca=Pinus nigra* Arn.). *Compt. Rend. Acad. Sci. Paris* **263**, 959–962.

CAMEFORT, H. (1966b). Étude en microscopie électronique du néocytoplasme des proembryons coenocytiques du *Pinus laricio* Poir. var *austriaca* (*Pinus nigra* Arn.) dont les noyaux ont émigré à la base de l'oosphère. *Compt. Rend. Acad. Sci. Paris* **263**, 1371–1374.

CAMEFORT, H. (1966c). Étude en microscopie électronique de la degénérescence du cytoplasme maternel dans les oosphères embryonnées du *Pinus laricio* Poir. var. *austriaca* (*P. nigra* Arn.). *Compt. Rend. Acad. Sci. Paris* **263**, 1443–1446.

CAMEFORT, H. (1967a). Origine et evolution structurale d'un cytoplasme propre a l'embryon ou néocytoplasme, chez les Pins. *Ann. Université A.R.E.R.S.* **5**, 75–88.

CAMEFORT, H. (1967b). Observations sur les mitochondries et les plastes de la cellule centrale et de l'oosphère du *Larix decidua* Mill. (*Larix europea* D.C.). *Compt. Rend. Acad. Sci. Paris* **265**, 1293–1296.

CAMEFORT, H. (1967c). Fécondation et formation d'un néocytoplasme chez le *Larix decidua* Mill. (*Larix europea* D.C.). *Compt. Rend. Acad. Sci. Paris* **265**, 1784–1787.

CAMEFORT, H. (1968a). Sur l'organisation du néocytoplasme dans les proembryons tétranucléés du *Larix decidua* Mill. (*Larix europea* D.C.) et l'origine des mitochondries et des plastes de l'embryon chez cette espèce. *Compt. Rend. Acad. Sci. Paris* **266**, 88–91.

CAMEFORT, H. (1968b). Cytologie de la fécondation et de la proembryogénèse chez quelques Gymnospermes. *Bull. Soc. Bot. Fr.* **115**, 137–160.

CAMEFORT, H. (1969). Fécondation et proembryogénèse chez les Abiétacées (notion de néocytoplasme). *Rev. Cytol. Biol. Végét.* **32**, 253–271.

CAMEFORT, H. (1970). Particularités structurales du gamète femelle chez le *Cryptomeria japonica* D. Don. Formation de complexes plastes-réticulum pendant la période de maturation du gamète. *Compt. Rend. Acad. Sci. Paris* **270**, 49–52.

CAMPBELL, D. H. (1930). "The Structure and Development of Mosses and Ferns". MacMillan, New York.

CAMPBELL, W. F. (1966). Irradiation in successive generations: Effects on developing barley (*Hordeum distichum* L.) embryos *in situ. Radiation Bot.* **6**, 525–534.

CAMPOS, F. F. and MORGAN, D. T., JR. (1958). Haploid pepper from a sperm. *Jour. Hered.* **49**, 134–137.

CAMPOS, F. F. and MORGAN, D. T., JR. (1960). Genetic control of haploidy in *Capsicum frutescens* L. following crosses with untreated and X-rayed pollen. *Cytologia* **25**, 362–372.

CAPPELLETTI, C. (1933). Osservasioni sulla germinazione asimbiotica dei semi di orchidee del genere *Cymbidium. Bull. Soc. Ital. Biol. Sperm.* **8**, 1–4.

CAREW, D. P. and SCHWARTING, A. E. (1958). Production of rye embryo callus. *Bot. Gaz.* **119**, 237–239.

CARLSON, P. S. (1970). Induction and isolation of auxotrophic mutants in somatic cell cultures of *Nicotiana tabacum. Science* **168**, 487–489.

CARLSON, P. S. (1973a). Somatic cell genetics of higher plants. *In* "Genetic Mechanisms of Development" (F. H. Ruddle, ed.), pp. 329–353. Academic Press, New York and London.

CARLSON, P. S. (1973b). Methionine-sulfoximine resistant mutants of tobacco. *Science* **180**, 1366–1368.

CARR, D. J. and SKENE, K. G. M. (1961). Diauxic growth curves of seeds, with special reference to French beans (*Phaseolus vulgaris* L.). *Austral. Jour. Biol. Sci.* **14**, 1–12.

CASS, D. D. (1972). Occurrence and development of a filiform apparatus in the egg of *Plumbago capensis. Amer. Jour. Bot.* **59**, 279–283.

CASS, D. D. and JENSEN, W. A. (1970). Fertilization in barley. *Amer. Jour. Bot.* **57**, 62–70.

CASS, D. D. and KARAS, I. (1974). Ultrastructural organization of the egg of *Plumbago zeylanica. Protoplasma* **81**, 49–62.

CAVE, C. F. and BELL, P. R. (1974a). The nature of the membrane around the egg of *Pteridium aquilinum* (L.) Kuhn. *Ann. Bot. N.S.* **38**, 17–21.

CAVE, C. F. and BELL, P. R. (1974b). The synthesis of ribonucleic acid and protein during oogenesis in *Pteridium aquilinum. Cytobiologie* **9**, 331–343.

CAVE, M. S. and BROWN, S. W. (1954). The detection and nature of dominant lethals in *Lilium*. II. Cytological abnormalities in ovules after pollen irradiation. *Amer. Jour. Bot.* **41**, 469–483.

CHAKRAVORTY, A. K. (1969). Ribosomal RNA synthesis in the germinating black eye pea (*Vigna unguiculata*). I. The effect of cycloheximide on RNA synthesis in the early stages of germination. *Biochim. Biophys. Acta* **179**, 67–82.

CHANDY, L. P. and NARAYANASWAMY, S. (1971). Diploid and haploid androgenic plantlets from haploid *Datura in vitro*. *Indian Jour. Exptl Biol.* **9**, 472–475.

CHANG, C. W. (1963a). Incorporation of phosphorus-32 into nucleic acids during embryonic development of barley. *Nature* **198**, 1167–1169.

CHANG, C. W. (1963b). Comparative growth of barley embryos *in vitro* and *in vivo*. *Bull. Torrey Bot. Club* **90**, 385–391.

CHANG, C. W. (1968). Effects of ionizing radiation on nucleic acids during embryonic development: Metabolism during embryogeny and at embryonic tissue level. *Canad. Jour. Bot.* **46**, 51–56.

CHANG, C. W. and MERICLE, L. W. (1964). Effects of ionizing radiation on nucleic acids during embryonic development. I. Quantitative and qualitative analyses at embryonic "maturity". *Radiation Bot.* **4**, 1–12.

CHAPMAN, J. A. and RIEBER, M. (1967). Distribution of ribosome in dormant and imbibed seeds of *Pisum arvense*: Electron microscopic observations. *Biochem. Jour.* **105**, 1201–1202.

CHATTERJI, U. N. and SANKHLA, N. (1965). Effect of growth substances on the mature embryo of *Merremia dissecta* (Jacq.) Hallier F. cultivated *in vitro*. *In* "Tissue Culture" (C. V. Ramakrishnan, ed.), pp. 389–397. Dr. W. Junk, The Hague.

CHEN, D. and OSBORNE, D. J. (1970). Hormones in the translational control of early germination in wheat embryos. *Nature* **226**, 1157–1160.

CHEN, D., SARID, S. and KATCHALSKI, E. (1968a). Studies on the nature of messenger RNA in germinating wheat embryos. *Proc. Natl Acad. Sci. U.S.A.* **60**, 902–909.

CHEN, D., SARID, S. and KATCHALSKI, E. (1968b). The role of water stress on the inactivation of messenger RNA of germinating wheat embryos. *Proc. Natl Acad. Sci. U.S.A.* **61**, 1378–1383.

CHEN, D., SCHULTZ, G. and KATCHALSKI, E. (1971). Early ribosomal RNA transcription and appearance of cytoplasmic ribosomes during germination of the wheat embryo. *Nature New Biol.* **231**, 69–72.

CHEN, S. S. C. (1968). Germination of light-inhibited seed of *Nemophila insignis*. *Amer. Jour. Bot.* **55**, 1177–1183.

CHEN, S. S. C. (1970). Action of light and gibberellic acid on the growth of excised embryos from *Phacelia tenacetifolia* seeds. *Planta* **95**, 336–340.

CHEN, S. S. C. and CHANG, J. L. L. (1972). Does gibberellic acid stimulate seed germination via amylase synthesis? *Plant Physiol.* **49**, 441–442.

CHEN, S. S. C. and PARK, W.-M. (1973). Early actions of gibberellic acid on the embryo and on the endosperm of *Avena fatua* seeds. *Plant Physiol.* **52**, 174–176.

CHEN, S. S. C. and THIMANN, K. V. (1966). Nature of seed dormancy in *Phacelia tenacetifolia*. *Science* **153**, 1537–1539.

CHEN, S. S. C. and VARNER, J. E. (1969). Metabolism of ^{14}C-maltose in *Avena fatua* seeds during germination. *Plant Physiol.* **44**, 770–774.

CHEN, S. S. C. and VARNER, J. E. (1970). Respiration and protein synthesis in dormant and after-ripened seeds of *Avena fatua*. *Plant Physiol.* **46**, 108–112.

CHERRY, J. H. (1963). Nucleic acid, mitochondria, and enzyme changes in cotyledons of peanut seeds during germination. *Plant Physiol.* **38**, 440–446.

CHESNOY, L. (1967). Nature et évolution des formations dites "asteroïdes" de la cellule centrale de l'archégone du *Juniperus communis* L. Étude en microscopie photonique et électronique. *Compt. Rend. Acad. Sci. Paris* **264**, 1016–1019.

CHESNOY, L. (1969a). Sur la participation du gamète mâle à la constitution du cytoplasme de l'embryon chez le *Biota orientalis* Endl. *Rev. Cytol. Biol. Végét.* **32**, 273–294.

CHESNOY, L. (1969b). Sur l'origine du cytoplasme des embryons chez le *Biota orientalis* Endl. (Cupressacées). *Compt. Rend. Acad. Sci. Paris* **268**, 1921–1924.

CHESNOY, L. (1971). Étude cytologique des gamètes, de la fécondation et de la proembryogénèse chez le *Biota orientalis* Endl. Observations en microscopie photonique et électronique. I. Le gamète femelle. *Rev. Cytol. Biol. Végét.* **34**, 257–304.

CHESNOY, L. (1973). Sur l'origine paternelle des organites du proembryon du *Chamaecyparis lawsoniana* A. Murr. (Cupressacées). *Caryologia* **25** (Suppl.), 223–232.

CHESNOY, L. and THOMAS, M.-J. (1969). Sur la présence de mitochondries Feulgen positives dans la zone périnucléaire du gamète femelle du *Pseudotsuga menziesii* (Mirb.) Franco. Étude cytochimique et ultrastructurale. *Compt. Rend. Acad. Sci. Paris* **268**, 55–58.

CHESNOY, L. and THOMAS, M.-J. (1971). Electron microscopy studies on gametogenesis and fertilization in gymnosperms. *Phytomorphology* **21**, 50–63.

CHIN, T. Y., POULSON, R. and BEEVERS, L. (1972). The influence of axis removal on protein metabolism in cotyledons of *Pisum sativum* L. *Plant Physiol.* **49**, 482–489.

CHING, T. M. (1966). Compositional changes of Douglas fir seeds during germination. *Plant Physiol.* **41**, 1313–1319.

CHING, T. M. (1972). Metabolism of germinating seeds. *In* "Seed Biology" (T. T. Kozlowski, ed.), Vol. 2, pp. 103–218. Academic Press, New York and London.

CHING, T. M. and CHING, K. K. (1972). Content of adenosine phosphates and adenylate energy in germinating ponderosa pine seeds. *Plant Physiol.* **50**, 536–540.

CHOPRA, R. N. (1958). *In vitro* culture of ovaries of *Althaea rosea* Cav. *In* "Modern Developments in Plant Physiology" (P. Maheshwari, ed.), pp. 87–89. University of Delhi.

CHOPRA, R. N. (1962). Effect of some growth substances and calyx on fruit and seed development of *Althaea rosea* Cav. *In* "Plant Embryology—A Symposium", pp. 170–181. Council of Scientific and Industrial Research, New Delhi.

CHOPRA, R. N. and AGARWAL, S. (1958). Some further observations on the endosperm haustoria in the Cucurbitaceae. *Phytomorphology* **8**, 194–201.

CHOPRA, R. N. and BASU, B. (1965). Female gametophyte and endosperm of some members of the Cucurbitaceae. *Phytomorphology* **15**, 217–223.

CHOPRA, R. N. and RAI, K. S. (1958). Response of ovules of *Argemone mexicana* L. to colchicine treatment *in vivo*. *Phytomorphology* **8**, 107–113.

CHOPRA, R. N. and SABHARWAL, P. S. (1963). *In vitro* culture of ovules of *Gynandropsis gynandra* (L.) Briq. and *Impatiens balsamina* L. *In* "Plant Tissue and Organ Culture—A Symposium" (P. Maheshwari and N. S. Rangaswamy, eds), pp. 257–264. International Society of Plant Morphologists, Delhi.

CHOPRA, R. N. and SACHAR, R. C. (1957). Effect of some growth substances on fruit development. *Phytomorphology* **7**, 387–397.

CHOPRA, R. N. and SACHAR, R. C. (1963). Endosperm. *In* "Recent Advances in the Embryology of Angiosperms" (P. Maheshwari, ed.), pp. 135–170. International Society of Plant Morphologists, Delhi.

CHOPRA, V. L. and SWAMINATHAN, M. S. (1963). Sprout inhibition and radiometric properties in irradiated potatoes. *Naturwiss.* **50**, 374–375.

CHOUDHURY, B. (1955a). Embryo culture technique—I. The growth of immature tomato embryo *in vitro*. *Indian Jour. Hort.* **12**, 143–151.

CHOUDHURY, B. (1955b). Embryo culture technique—II. "Embryo factors" and immature tomato embryo. *Indian Jour. Hort.* **12**, 152–154.

CHOUDHURY, B. (1955c). Embryo culture technique—III. Growth of hybrid embryos (*Lycopersicon esculentum* × *Lycopersicon peruvianum*) in culture medium. *Indian Jour. Hort.* **12**, 155–156.

CHU, Z.-C., WANG, C.-C., SUN, C.-S., CHIEN, N.-F., YIN, K.-C. and HSÜ, C. (1973). Investigations on the induction and morphogenesis of wheat (*Triticum vulgare*) pollen plants. *Acta Bot. Sinica* **15**, 1–11.

CLAPHAM, D. (1971). *In vitro* development of callus from the pollen of *Lolium* and *Hordeum*. *Zeitschr. Pflanzenzüchtg* **65**, 285–292.

CLAPHAM, D. (1973). Haploid *Hordeum* plants from anthers *in vitro*. *Zeitschr. Pflanzenzüchtg* **69**, 142–155.

CLAUSEN, R. E. and LAMMERTS, W. E. (1929). Interspecific hybridization in *Nicotiana*. X. Haploid and diploid merogony. *Amer. Naturl.* **63**, 279–282.

CLOWES, F. A. L. (1953). The cytogenerative centre in roots with broad columellas. *New Phytol.* **52**, 48–57.

CLUTTER, M., BRADY, T., WALBOT, V. and SUSSEX, I. (1974). Macromolecular synthesis during plant embryogeny. Cellular rates of RNA synthesis in diploid and polytene cells in bean embryos. *Jour. Cell Biol.* **63**, 1097–1102.

CLUTTER, M. E. and SUSSEX, I. M. (1968). Ultrastructural development of bean embryo cells containing polytene chromosomes. *Jour. Cell Biol.* **39**, 26a (abstract).

Cocucci, A. E. and Jensen, W. A. (1969a). Orchid embryology: The mature megagametophyte of *Epidendrum scutella*. *Kurtziana* **5**, 23–38.

Cocucci, A. and Jensen, W. A. (1969b). Orchid embryology: Megagametophyte of *Epidendrum scutella* following fertilization. *Amer. Jour. Bot.* **56**, 629–640.

Coe, G. E. (1954). Distribution of carbon 14 in ovules of *Zephyranthes drummondii*. *Bot. Gaz.* **115**, 342–346.

Collins, D. M. and Wilson, A. T. (1972). Metabolism of the axis and cotyledons of *Phaseolus vulgaris* seeds during early germination. *Phytochemistry* **11**, 1931–1935.

Collins, G. B. and Sadasivaiah, R. S. (1972). Meiotic analysis of haploid and doubled haploid forms of *Nicotiana otophora* and *N. tabacum*. *Chromosoma* **38**, 387–404.

Collins, G. B. and Sunderland, N. (1974). Pollen-derived haploids of *Nicotiana knightiana*, *N. raimondii*, and *N. attenuata*. *Jour. Exptl Bot.* **25**, 1030–1039.

Collins, G. B., Dunwell, J. M. and Sunderland, N. (1974). Irregular microspore formation in *Datura innoxia* and its relevance to anther culture. *Protoplasma* **82**, 365–378.

Collins, G. B., Legg, P. D. and Kasperbauer, M. J. (1972). Chromosome numbers in anther-derived haploids of two *Nicotiana* species. *N. tabacum* L. and *N. otophora* Gris. *Jour. Hered.* **63**, 113–118.

Collins, G. B., Legg, P. D. and Kasperbauer, M. J. (1974). Use of anther-derived haploids in *Nicotiana*. I. Isolation of breeding lines differing in total alkaloid content. *Crop Sci.* **14**, 77–80.

Collins, G. B., Legg, P. D., Litton, C. C. and Kasperbauer, M. J. (1971). Inheritance of resistance to black shank in *Nicotiana tabacum*. *Canad. Jour. Genet. Cytol.* **13**, 422–428.

Colonna, J.-P., Cas, G. and Rabéchault, H. (1971). Mise au point d'une méthode de culture *in vitro* d'embryons de caféiers. Application à deux variétés de caféiers cultivés. *Compt. Rend. Acad. Sci. Paris* **272**, 60–63.

Côme, D. and Durand, M. (1971). Influence de l'acide gibbérellique sur la livée de dormance des embryons de pommier (*Pirus malus* L.) par le froid. *Compt. Rend. Acad. Sci. Paris* **273**, 1937–1940.

Côme, D. and Thévenot, C. (1968). Influence de l'excision d'un cotylédon sur la germination des embryons de pommier (*Pirus malus* L.) dormants. *Compt. Rend. Acad. Sci. Paris* **267**, 1832–1834.

Côme, D., Thévenot, C. and Tissaoui, T. (1968). Influence de la température sur la germination, l'intensité respiratoire et l'imbibition des embryons de pommier non dormants (variété Golden Delicious) ensemencés a plat ou debout sur les cotylédons. *Rev. Gén. Bot.* **75**, 611–626.

Conger, A. D. and Randolph, M. L. (1959). Magnetic centers (free radicals) produced in cereal embryos by ionizing radiation. *Radiation Res.* **11**, 54–66.

Connors, C. H. (1919). Growth of fruits of peach. *N.J. Agric. Expt. Stat. Ann. Rep.* **40**, 82–88.

CONSTABEL, F., MILLER, R. A. and GAMBORG, O. L. (1971). Histological studies on embryos produced from cell cultures of *Bromus inermis. Canad. Jour. Bot.* **49**, 1415–1417.

COOPER, D. C. and BRINK, R. A. (1944). Collapse of the seed following the mating of *Hordeum jubatum* × *Secale cereale. Genetics* **29**, 370–390.

COOPER, D. C. and BRINK, R. A. (1945). Seed collapse following matings between diploid and tetraploid races of *Lycopersicon pimpinellifolium. Genetics* **30**, 376–401.

COOPER, D. C. and BRINK, R. A. (1949). The endosperm–embryo relationship in an autonomous apomict, *Taraxacum officinale. Bot. Gaz.* **111**, 139–153.

CORCORAN, M. R. and PHINNEY, B. O. (1962). Changes in amounts of gibberellin-like substances in developing seed of *Echinocystis, Lupinus* and *Phaseolus. Physiol. Plantarum* **15**, 252–262.

CORTI, E. F. and MAUGINI, E. (1964). Passagio di corpi figurati fra cellule del tappeto e cellula centrale nell'archegonio dei pini. *Caryologia* **17**, 1–39.

COULTER, J. M. and CHAMBERLAIN, C. J. (1903). The embryogeny of *Zamia. Bot. Gaz.* **35**, 184–194.

COULTER, J. M. and CHAMBERLAIN, C. J. (1917). "Morphology of Gymnosperms". University of Chicago Press.

COX, E. A., STOTZKY, G. and GOOS, R. D. (1960). *In vitro* culture of *Musa balbisiana* Colla embryos. *Nature* **185**, 403–404.

COX, L. G., MUNGER, H. M. and SMITH, E. A. (1945). A germination inhibitor in the seed coats of certain varieties of cabbage. *Plant Physiol.* **20**, 289–294.

CRÉTÉ, P. (1963). Embryo. *In* "Recent Advances in the Embryology of Angiosperms" (P. Maheshwari, ed.), pp. 171–220. International Society of Plant Morphologists, Delhi.

CROCKER, W. and BARTON, L. V. (1957). "Physiology of Seeds". The Chronica Botanica Co., Waltham, Mass.

CURTIS, E. J. C. and CANTLON, J. E. (1965). Studies of the germination process in *Melampyrum lineare. Amer. Jour. Bot.* **52**, 552–555.

CURTIS, E. J. C. and CANTLON, J. E. (1968). Seed dormancy and germination in *Melampyrum lineare. Amer. Jour. Bot.* **55**, 26–32.

CURTIS, J. T. (1947a). Undifferentiated growth of orchid embryos on media containing barbiturates. *Science* **105**, 128.

CURTIS, J. T. (1947b). Studies on the nitrogen nutrition of orchid embryos. I. Complex nitrogen sources. *Amer. Orch. Soc. Bull.* **16**, 654–660.

CURTIS, J. T. and NICHOL, M. A. (1948). Culture of proliferating orchid embryos *in vitro. Bull. Torrey Bot. Club* **75**, 358–373.

CURTIS, J. T. and SPOERL, E. (1948). Studies on the nitrogen nutrition of orchid embryos. II. Comparative utilization of nitrate and ammonium nitrogen. *Amer. Orch. Soc. Bull.* **17**, 111–114.

CUTTER, V. M., JR. and FREEMAN, B. (1954). Development of the syncytial endosperm of *Cocos nucifera. Nature* **173**, 827.

CUTTER, V. M., JR. and FREEMAN, B. (1955). Nuclear aberrations in the syncytial endosperm of *Cocos nucifera. Jour. Elisha Mitchell Sci. Soc.* **71**, 49–58.

CUTTER, V. M., JR. and WILSON, K. S. (1954). Effect of coconut endosperm and other growth stimulants upon the development *in vitro* of embryos of *Cocos nucifera*. *Bot. Gaz.* **115**, 234–240.

CUTTER, V. M., JR., WILSON, K. S. and DUBÉ, G. R. (1952a). The isolation of living nuclei from the endosperm of *Cocos nucifera*. *Science* **115**, 58–59.

CUTTER, V. M., JR., WILSON, K. S. and DUBÉ, J. F. (1952b). The endogenous oxygen uptake of tissues in the developing fruit of *Cocos nucifera*. *Amer. Jour. Bot.* **39**, 51–56.

CUTTER, V. M., JR., WILSON, K. S. and FREEMAN, B. (1955). Nuclear behavior and cell formation in the developing endosperm of *Cocos nucifera*. *Amer. Jour. Bot.* **42**, 109–115.

CZOSNOWSKI, J. (1962). Metabolism of excised embryos of *Lupinus luteus* L. I. Effect of metabolic inhibitors and growth substances on the water uptake. *Acta Soc. Bot. Polon.* **31**, 135–152.

DAHLGREN, K. V. O. (1934). Die Embryologie von *Impatiens roylei*. *Svensk. Bot. Tidskr.* **28**, 103–125.

DAHMEN, W. J. and MOCK, J. J. (1971). Sterilization techniques for seeds and excised embryos of corn (*Zea mays* L.). *Iowa Jour. Sci.* **46**, 7–11.

D'ALASCIO-DESCHAMPS, R. (1972). Le sac embryonnaire du lin après la fécondation. *Botaniste* **50**, 273–288.

D'ALASCIO-DESCHAMPS, R. (1973). Organisation du sac embryonnaire du *Linum catharticum* L., espèce récoltée en station naturelle; étude ultrastructurale. *Bull. Soc. Bot. Fr.* **120**, 189–200.

DALBY, A. and DAVIES, I. I. (1967). Ribonuclease activity in the developing seeds of normal and opaque-2 maize. *Science* **155**, 1573–1575.

DANIELSSON, B. (1950). Embryokulturer av stenfruktträd. *Sver. Pomol. Fören. Arsskr. Stockholm* **51**, 200–206.

DANILINA, A. N. (1972). Morphogenesis in tissue cultures of *Daucus carota*. *Phytomorphology* **22**, 160–164.

DATTA, M. (1955). The occurrence and division of free nuclei in the endospermal milk in some Palmae. *Trans. Bose Res. Inst. Calcutta* **19**, 117–125.

DAVIDSON, O. W. (1933). The germination of "non-viable" peach seeds. *Proc. Amer. Soc. Hort. Sci.* **30**, 129–132.

DAVIDSON, O. W. (1934). Growing trees from "non-viable" peach seeds. *Proc. Amer. Soc. Hort. Sci.* **32**, 308–312.

DAVIES, D. R. (1958). Male parthenogenesis in barley. *Heredity* **12**, 493–498.

DAVIES, D. R. (1960). The embryo culture of inter-specific hybrids of *Hordeum*. *New Phytol.* **59**, 9–14.

DAVIES, W. E. (1963). Herbage legume breeding. *Plant Breeding Abst.* **33**, 10–11 (abstract).

DAVIS, G. L. (1961a). The life history of *Podolepis jaceoides* (Sims) Voss—II. Megasporogenesis, female gametophyte and embryogeny. *Phytomorphology* **11**, 206–219.

DAVIS, G. L. (1961b). The occurrence of synergid haustoria in *Cotula australis* (Less.) Hook. f. (Compositae). *Austral. Jour. Sci.* **24**, 296–297.

DAVIS, G. L. (1962). Embryological studies in the Compositae. II. Sporogenesis, gametogenesis, and embryogeny in *Ammobium alatum* R. Br. *Austral. Jour. Bot.* **10**, 65–75.

DAVIS, G. L. (1963). Embryological studies in the Compositae. 3. Sporogenesis, gametogenesis and early embryogeny in *Minuria denticulata* (D.C.) Benth. (Astereae). *Proc. Linn. Soc. N.S.W.* **88**, 35–40.

DAVIS, G. L. (1966). "Systematic Embryology of the Angiosperms". Wiley, New York.

D'CRUZ, R. and KALE, V. R. (1957). Preliminary studies of the growth rate in the excised embryos of legumes. *Poona Agric. Coll. Mag.* **48**, 37–39.

DEBERGH, P. and NITSCH, C. (1973). Premiers résultats sur la culture *in vitro* de grains de pollen isolés chez le tomate. *Compt. Rend. Acad. Sci. Paris* **276**, 1281–1284.

DECAPITE, L. (1955). La coltura dei frutti *in vitro* da fiori recisi di *Fragaria chiloensis* Ehrh. × *F. virginia* Duch. var Marshali e di *Pisum sativum* L. var. Zekka. *Ric. Sci. Ital.* **25**, 532–538.

DEGIVRY, M.-T. (1966). Revue bibliographique sur les problèmes liés aux cultures *in vitro* d'embryons immatures. *Bull. Sci. Bourgogne* **24**, 57–87.

DEKRUIJFF, E. (1906). Composition of coconut water and presence of diastase in coconuts. *Bull. Dept. Agric. Indes Neerland.* No. 4, 1–8.

DELTOUR, R. (1970). Synthèse et translocation de RNA dans les cellules radiculaires de *Zea mays* au début de la germination. *Planta* **92**, 235–239.

DELTOUR, R. and BRONCHART, R. (1971). Changements de l'ultrastructure des cellules radiculaires de *Zea mays* au début de la germination. *Planta* **97**, 197–207.

DEMAGGIO, A. E. (1961a). Morphogenetic studies on the fern *Todea barbara* (L.) Moore—I. Life history. *Phytomorphology* **11**, 46–64.

DEMAGGIO, A. E. (1961b). Morphogenetic studies on the fern *Todea barbara* (L.) Moore—II. Development of the embryo. *Phytomorphology* **11**, 64–79.

DEMAGGIO, A. E. (1963). Morphogenetic factors influencing the development of fern embryos. *Jour. Linn. Soc. (Bot.)* **58**, 361–376.

DEMAGGIO, A. E. and WETMORE, R. H. (1961a). Growth of fern embryos in sterile culture. *Nature* **191**, 94–95.

DEMAGGIO, A. E. and WETMORE, R. H. (1961b). Morphogenetic studies on the fern *Todea barbara*. III. Experimental embryology. *Amer. Jour. Bot.* **48**, 551–565.

DESCHAMPS, R. (1969). Premiers stades du développement de l'embryon et de l'albumen du lin: Étude au microscope électronique. *Rev. Cytol. Biol. Végét.* **32**, 379–390.

DEVINE, V. (1948). Note on the culture of *Lychnis* embryos. *Proc. Iowa Acad. Sci.* **55**, 95–97.

DEVINE, V. (1950). Embryogeny of *Lychnis alba*. *Amer. Jour. Bot.* **37**, 197–208.

DEVREUX, M. (1963). Effets de l'irradiation gamma chronique sur l'embryogénèse de *Capsella bursa-pastoris* Moench. VI. Congr. Nucl. (Roma), *L'Energia Nucleare in Agricoltura* (CNEN Vallecchi), 199–217.

DEVREUX, M. (1970). New possibilities for the *in vitro* cultivation of plant cells. *Eurospectra* **9**, 105–110.

DEVREUX, M. and SACCARDO, F. (1971). Mutazioni sperimentali osservate su piante aploidi di tabacco ottenute per colture *in vitro* di antere irradiate. *Atti. Assoc. Genet. Ital.* **16**, 69–71 (abstract).

DEVREUX, M., SACCARDO, F. and BRUNORI, A. (1971). Plantes haploides et lignes isogeniques de *Nicotiana tabacum* obtenues par cultures d'antheres et de tiges *in vitro. Caryologia* **24**, 141–148.

DEVREUX, M. and SCARASCIA MUGNOZZA, G. T. (1962). Action des rayons gamma sur les premiers stades de développement de l'embryon de *Nicotiana rustica* L. *Caryologia* **15**, 279–291.

DEVREUX, M. and SCARASCIA MUGNOZZA, G. T. (1964). Effect of gamma radiation of the gametes, zygote and proembryo in *Nicotiana tabacum* L. *Radiation Bot.* **4**, 373–386.

DEVREUX, M. and SCARASCIA MUGNOZZA, G. T. (1965). Action of gamma radiation on the zygote in *Nicotiana. Radiation Bot. Suppl.* **5**, 283–292.

DEXHEIMER, J. (1973a). Quelques aspects ultrastructuraux du prothalle femelle alvéolaire du *Ginkgo biloba. Compt. Rend. Acad. Sci. Paris* **276**, 2789–2792.

DEXHEIMER, J. (1973b). Étude ultrastructurale du gamétophyte femelle de *Ginkgo biloba*. I. Les cellules a reserves. *Caryologia* **25** (Suppl.), 85–96.

DIAZ, D. H. and MARTIN, G. C. (1972). Peach seed dormancy in relation to endogenous inhibitors and applied growth substances. *Jour. Amer. Soc. Hort. Sci.* **97**, 651–654.

DIBOLL, A. G. (1967). Ultraviolet microscopy distinguishes the insoluble polysaccharides from the non-polysaccharides in periodic acid-Schiff-colored biological tissues. *Caryologia* **20**, 101–105.

DIBOLL, A. G. (1968). Fine structural·development of the megagametophyte of *Zea mays* following fertilization. *Amer. Jour. Bot.* **55**, 787–806.

DIBOLL, A. G. and LARSON, D. A. (1966). An electron microscopic study of the mature megagametophyte in *Zea mays. Amer. Jour. Bot.* **53**, 391–402.

DIERS, L. (1964). Bilden sich während der Oogenese bei Moosen und Farnen die Mitochondrien und Plastiden aus dem Kern? *Ber. Deut. Bot. Ges.* **77**, 369–371.

DIERS, L. (1965). Elektronmikroskopische Untersuchungen über die Eizellbildung und Eizellreifung des Lebermooses *Sphaerocarpus donnellii* Aust. *Zeitschr. Naturforsch.* **20b**, 795–801.

DIERS, L. (1966). On the plastids, mitochondria and other cell constituents during oögenesis of a plant. *Jour. Cell Biol.* **28**, 527–543.

DIETERICH, K. (1924). Über Kultur von Embryonen ausserhalb des Samens. *Flora* **117**, 379–417.

DOBRZAŇSKA, M., TOMASZEWSKI, M., GRZELCZAK, Z., REJMAN, E. and BUCHOWICZ, J. (1973). Cascade activation of genome transcription in wheat. *Nature* **244**, 507–509.

DOERPINGHAUS, S. L. (1947). Differences between species of *Datura* in utilization of five carbohydrates. *Amer. Jour. Bot.* **34**, 583 (abstract).

DONINI, B. (1963). Effetti in prima generazione dell'irraggiamento dei gameti in *Hordeum vulgare* L. *Atti. Assoc. Genet. Ital.* **8**, 268–278.

DONINI, B. and HUSSAIN, S. (1968). Development of the embryo in *Triticum durum* following irradiation of male or female gamete. *Radiation Bot.* **8**, 289–295.

DORE SWAMY, R. and MOHAN RAM, H. Y. (1967). Cultivation of embryos of *Drosophyllum lusitanicum* Link—An insectivorous plant. *Experientia* **23**, 675.

DOY, C. H., GRESSHOFF, P. M. and ROLFE, B. G. (1973). Biological and molecular evidence for the transgenosis of genes from bacteria to plant cells. *Proc. Natl Acad. Sci. U.S.A.* **70**, 723–726.

DRENNAN, D. S. H. and BERRIE, A. M. M. (1962). Physiological studies of germination in the genus *Avena* I. The development of amylase activity. *New Phytol.* **61**, 1–9.

DUBARD, M. and URBAIN, J.-A. (1913). De l'influence de l'albumen sur le développement de l'embryon. *Compt. Rend. Acad. Sci. Paris* **156**, 1086–1089.

DUFFUS, C. M. and ROSIE, R. (1975). Biochemical changes during embryogeny in *Hordeum distichum*. *Phytochemistry* **14**, 319–323.

DULIEU, H. L. (1963). Sur la fécondation *in vitro* chez le *Nicotiana tabacum* L. *Compt. Rend. Acad. Sci. Paris* **256**, 3344–3346.

DULIEU, H. L. (1966). Pollination of excised ovaries and culture of ovules of *Nicotiana tabacum* L. *Phytomorphology* **16**, 69–75.

DUNSTAN, W. R. (1906). Report on a sample of coconut "water" from Ceylon. *Trop. Agric. Ceylon* **26**, 377–378.

DUNWELL, J. M. and SUNDERLAND, N. (1973). Anther culture of *Solanum tuberosum* L. *Euphytica* **22**, 317–323.

DUNWELL, J. M. and SUNDERLAND, N. (1974a). Pollen ultrastructure in anther cultures of *Nicotiana tabacum*. I. Early stages of culture. *Jour. Exptl Bot.* **25**, 352–361.

DUNWELL, J. M. and SUNDERLAND, N. (1974b). Pollen ultrastructure in anther cultures of *Nicotiana tabacum*. II. Changes associated with embryogenesis. *Jour. Exptl Bot.* **25**, 363–373.

DURAND, M., THÉVENOT, C. and CÔME, D. (1973). Influences de l'acide abscissique sur la germination et la levée de dormance des embryons de pommier (*Pirus malus* L.). *Compt. Rend. Acad. Sci. Paris* **277**, 53–55.

DURE, L. S. (1960). Gross nutritional contributions of maize endosperm and scutellum to germination growth of maize axis. *Plant Physiol.* **35**, 919–925.

DURE, L. S. and JENSEN, W. A. (1957). The influence of gibberellic acid and indoleacetic acid on cotton embryos cultured *in vitro*. *Bot. Gaz.* **118**, 254–261.

DURE, L. and WATERS, L. (1965). Long-lived messenger RNA: Evidence from cotton seed germination. *Science* **147**, 410–412.

DURZAN, D. J. and CHALUPA, V. (1968). Free sugars, amino acids and soluble proteins in the embryo and female gametophyte of jack pine as related to climate at the seed source. *Canad. Jour. Bot.* **46**, 417–428.

DURZAN, D. J., MIA, A. J. and RAMAIAH, P. K. (1971). The metabolism and subcellular organization of the jack pine embryo (*Pinus banksiana*) during germination. *Canad. Jour. Bot.* **49**, 927–938.

DUTT, M. (1953). Dividing nuclei in coconut milk. *Nature* **171**, 799.

DUVICK, N. (1952). Free amino acids in the developing endosperm of maize. *Amer. Jour. Bot.* **39**, 656–661.

EAMES, A. J. (1936). "Morphology of Vascular Plants. Lower Group". McGraw-Hill, New York.

EDELMAN, J., SHIBKO, S. I. and KEYS, A. J. (1959). The role of the scutellum of cereal seedlings in the synthesi˙ and transport of sucrose. *Jour. Exptl Bot.* **10**, 178–189.

EFRON, D., EVENARI, M. and DE GROOT, N. (1971). Amino acid incorporation activity of lettuce seed ribosomes during germination. *Life Sci.* **10**, 1015–1019.

EHRENBERG, A., EHRENBERG, L. and STRÖM, G. (1969). Radiation-induced free radicals in embryo and endosperm of barley kernels. *Radiation Bot.* **9**, 151–158.

EID, A. A. H., DE LANGHE, E. and WATERKEYN, L. (1973). *In vitro* culture of fertilized cotton ovules. I–The growth of cotton embryos. *La Cellule* **69**, 359–371.

EMSWELLER, S. L., ASEN, S. and UHRING, J. (1962). *Lilium speciosum × L. auratum. Lily Yearbook N. Amer. Lily Soc.* **15**, 7–15.

EMSWELLER, S. L. and UHRING, J. (1962). Endosperm–embryo incompatibility in *Lilium* species hybrids. *In* "Advances in Horticultural Science and their Applications" (J.-C. Garnaud, ed.), Vol. 2, pp. 360–367. Pergamon Press, Oxford.

ENGLEMAN, E. M. (1966). Ontogeny of aleurone grains in cotton embryo. *Amer. Jour. Bot.* **53**, 231–237.

ENGVILD, K. C. (1964). Growth and chlorophyll formation of dark-grown pine embryos on different media. *Physiol. Plantarum* **17**, 866–874.

ENGVILD, K. C. (1973). Triploid petunias from anther cultures. *Hereditas* **74**, 144–147.

ENGVILD, K. C. (1974). Plantlet ploidy and flower-bud size in tobacco anther cultures. *Hereditas* **76**, 320–322.

ENGVILD, K. C., LINDE-LAURSEN, I. and LUNDQVIST, A. (1972). Anther cultures of *Datura innoxia*: Flower bud stage and embryoid level of ploidy. *Hereditas* **72**, 331–332.

ERICSON, M. C. and CHRISPEELS, M. J. (1973). Isolation and characterization of glucosamine-containing storage glycoproteins from the cotyledons of *Phaseolus aureus*. *Plant Physiol.* **52**, 98–104.

ERNST, R. (1967a). Effect of select organic nutrient additives on growth *in vitro* of *Phalaenopsis* seedlings. *Amer. Orch. Soc. Bull.* **36**, 694–704.

ERNST, R. (1967b). Effect of carbohydrate selection on the growth rate of freshly germinated *Palaenopsis* and *Dendrobium* seed. *Amer. Orch. Soc. Bull.* **36**, 1068–1073.

ERNST, R., ARDITTI, J. and HEALEY, P. L. (1971). Carbohydrate physiology of orchid seedlings. II. Hydrolysis and effects of oligosaccharides. *Amer. Jour. Bot.* **58**, 827–835.

ESEN, A. and SOOST, R. K. (1973). Seed development in *Citrus* with special reference to 2x × 4x crosses. *Amer. Jour. Bot.* **60**, 448–462.

ESENBECK, E. and SUESSENGUTH, K. (1925). Über die aseptische Kultur pflanzlicher embryonen, zugleich ein Beitrag zum Nachweis der Enzymausscheidung. *Arch. Exp. Zellforsch.* **1**, 547–586.

EUNUS, A. M. (1955). The effects of X-rays on the embryonal growth and development of *Hordeum vulgare* L. *Jour. Exptl Bot.* **6**, 409–421.

EVANS, A. M. (1962). Species hybridization in *Trifolium*. I. Methods of overcoming species incompatibility. *Euphytica* **11**, 164–176.

EYMÉ, J. (1965). Recherches sur la constitution cytoplasmique de l'archésporore et du sac embryonnaire de *"Lilium candidum"* L. *Botaniste* **48**, 99–155.

FAGERLIND, F. (1946). Hormonale Substanzen als Ursache der Frucht- und Embryobildung bei Pseudogamen *Hosta*-Biotypen. *Svensk. Bot. Tidskr.* **40**, 230–234.

FARQUHARSON, L. I. (1957). Hybridization of *Tripsacum* and *Zea*. *Jour. Hered.* **48**, 295–299.

FARRAR, K. R., BENTLEY, J. A., BRITTON, G. and HOUSLEY, S. (1958). Auxins in immature maize kernels. *Nature* **181**, 553–554.

FAVRE-DUCHARTRE, M. (1958). *Ginkgo*, an oviparous plant. *Phytomorphology* **8**, 377–390.

FILNER, P. and VARNER, J. E. (1967). A test for *de novo* synthesis of enzymes: Density labeling with H_2O^{18} of barley α-amylase induced by gibberellic acid. *Proc. Natl Acad. Sci. U.S.A.* **58**, 1520–1526.

FISHER, D. B. and JENSEN, W. A. (1972). Nuclear and cytoplasmic DNA synthesis in cotton embryos: A correlated light and electron microscope autoradiographic study. *Histochemie* **32**, 1–22.

FLEMION, F. (1931). After-ripening, germination, and vitality of seeds of *Sorbus aucuparia* L. *Contrib. Boyce Thompson Inst.* **3**, 413–440.

FLEMION, F. (1933a). Physiological and chemical studies of after-ripening of *Rhodotypos kerrioides* seeds. *Contrib. Boyce Thompson Inst.* **5**, 143–159.

FLEMION, F. (1933b). Dwarf seedlings from non-after-ripened embryos of *Rhodotypos kerrioides*. *Contrib. Boyce Thompson Inst.* **5**, 161–165.

FLEMION, F. (1934a). Physiological and chemical changes preceding and during the after-ripening of *Symphoricarpos racemosus* seeds. *Contrib. Boyce Thompson Inst.* **6**, 91–102.

FLEMION, F. (1934b). Dwarf seedlings from non-after-ripened embryos of peach, apple and hawthorn. *Contrib. Boyce Thompson Inst.* **6**, 205–209.

FLEMION, F. (1956). Effects of temperature, light and nutrients on physiological dwarfing in peach seedlings. *Plant Physiol.* **31**, (Suppl.), iii (abstract).

FLEMION, F. and PROBER, P. L. (1960). Production of peach seedlings from unchilled seeds. I. Effect of nutrients in the absence of the cotyledonary tissue. *Contrib. Boyce Thompson Inst.* **20**, 409–419.

FLEURY, P., DEYSSON, G. and DEYSSON, M. (1951). Action des divers inositols et de quelques composés voisins sur la croissance des plantules de *Pisum sativum* L. *Compt. Rend. Acad. Sci. Paris* **233**, 756–758.

FLINN, A. M. and PATE, J. S. (1968). Biochemical and physiological changes during maturation of fruit of the field pea (*Pisum arvense* L.). *Ann. Bot. N.S.* **32**, 479–495.

FLORIS, C. (1966). The possible role of the endosperm in the ageing of the embryo in the wheat seed. *Giorn. Bot. Ital.* **73**, 349–350 (abstract).

FLORIS, C. (1968). Comportamento di embrioni ed endospermi provenienti da semi di "*Triticum*" di età crescente. *Giorn. Bot. Ital.* **102**, 559–560 (abstract).

FLORIS, C. (1970). Ageing in *Triticum durum* seeds: Behaviour of embryos and endosperms from aged seeds as revealed by the embryo-transplantation technique. *Jour. Exptl Bot.* **21**, 462–468.

FLORIS, C., GIOVANNOZZI-SERMANNI, G. and MELETTI, P. (1972). Seed germination and growth in *Triticum*. I. Biological activity of extracts from *T. durum* endosperms. *Plant Cell Physiol.* **13**, 331–336.

FLORIS, C., MELETTI, P. and D'AMATO, F. (1970). Further observations on embryo–endosperm relations in irradiated water soaked seeds of *durum* wheat. *Mutation Res.* **10**, 253–255.

FONSHTEYN, L. M. (1961). Some data on the effects of irradiation of the endosperm of wheat seed upon the growth and development of plants. *Radiobiologia* (English translation) **1**, 446–451.

FORMAN, M. and JENSEN, W. A. (1965). Respiration and embryogenesis in cotton. *Plant Physiol.* **40**, 765–769.

FOUNTAIN, D. W. and BEWLEY, J. D. (1973). Polyribosome formation and protein synthesis in imbibed but dormant lettuce seeds. *Plant Physiol.* **52**, 604–607.

FRANKLAND, B. (1961). Effect of gibberellic acid, kinetin and other substances on seed dormancy. *Nature* **192**, 678–679.

FRANKLAND, B. and WAREING, P. F. (1962). Changes in endogenous gibberellins in relation to chilling of dormant seeds. *Nature* **194**, 313–314.

FRANKLAND, B. and WAREING, P. F. (1966). Hormonal regulation of seed dormancy in hazel (*Corylus avellana* L.) and beech (*Fagus sylvatica* L.). *Jour. Exptl Bot.* **17**, 596–611.

FRANKLAND, B., JARVIS, B. C. and CHERRY, J. H. (1971). RNA synthesis and the germination of light-sensitive lettuce seeds. *Planta* **97**, 39–49.

FRIDHANDLER, L. and QUASTEL, J. H. (1955). Absorption of amino acids from isolated surviving intestine. *Arch. Biochem. Biophys.* **56**, 424–440.

FRIDRIKSSON, S. and BOLTON, J. L. (1963). Preliminary report on the culture of alfalfa embryos. *Canad. Jour. Bot.* **41**, 439–440.

FROST-CHRISTENSEN, H. (1974). Embryo development in ripe seeds of *Eranthis hiemalis* and its relation to gibberellic acid. *Physiol. Plantarum* **30**, 200–205.

FUJII, T. (1969). Photocontrol of development of excised *Eragrostis* embryos. *Develop. Growth Diffn* **11**, 153–163.

FUJISAWA, H. (1966). Role of nucleic acid and protein metabolism in the initiation of growth at germination. *Plant Cell Physiol.* **7**, 185–197.

FUKUI, T. and NIKUNI, Z. (1956). Degradation of starch in the endosperms of rice seeds during germination. *Jour. Biochem. Japan* **43**, 33–40.

FURUSATO, K. (1953). Studies on polyembryony in *Citrus*. *Ann. Rep. Nat. Inst. Genetics Japan* No. 4, 56 (abstract).

FURUSATO, K., OHTA, Y. and ISHIBASHI, K. (1957). Studies on polyembryony in *Citrus*. *Seiken Zihô, Rep. Kihara Inst. Biol. Res.* No. 8, 40–48.

FURUYA, M. (1956). Effects of the vapour of methyl 2,4-dichlorophenoxy-acetate on growth and differentiation in *Phaseolus vulgaris* L. II. Behaviour of decotylated embryos in germination *in vitro* and the rôle of cotyledons in formative response. *Jap. Jour. Bot.* **15**, 270–284.

FURUYA, M. and OSAKI, S. (1955). Effects of the vapour of methyl 2,4-dichlorophenoxyacetate on growth and differentiation in *Phaseolus vulgaris* L. I. Formative effects induced in the seedling after various grades of application on dry seeds. *Jap. Jour. Bot.* **15**, 117–139.

FURUYA, M. and SOMA, K. (1957). The effects of auxins on the development of bean embryos cultivated *in vitro*. *Jour. Fac. Sci. Univ. Tokyo Sect. III. Bot.* **7**, 163–198.

GADWAL, V. R., JOSHI, A. B. and IYER, R. D. (1968). Interspecific hybrids in *Abelmoschus* through ovule and embryo culture. *Indian Jour. Genet. Plant Breed.* **28**, 269–274.

GALITZ, D. S. and HOWELL, R. W. (1965). Measurement of ribonucleic acids and total free nucleotides of developing soybean seeds. *Physiol. Plantarum* **18**, 1018–1021.

GALSTON, A. W. and DAVIES, P. J. (1969). Hormonal regulation in higher plants. *Science* **163**, 1288–1297.

GAMBADE, G. (1972). Étude cyto-histologique de la germination chez les espèces à embryons dormants. I. Influence d'un traitement par l'acide gib-bérellique sur le comportement des embryons d'*Euonymus europaeus* L. *Bull. Soc. Bot. Fr.* **119**, 151–166.

GAMBORG, O. L., CONSTABEL, F. and MILLER, R. A. (1970). Embryogenesis and production of albino plants from cell cultures of *Bromus inermis*. *Planta* **95**, 355–358.

GANESAN, A. T., SHAH, S. S. and SWAMINATHAN, M. S. (1957). Cause for the failure of seed-setting in the cross *Corchorus olitorius* × *C. capsularis*. *Curr. Sci.* **26**, 292–293.

GARCIA, V. (1962a). Embryological studies on the Loasaceae with special reference to the endosperm haustoria. *In* "Plant Embryology—A Symposium", pp. 157–161. Council of Scientific and Industrial Research, New Delhi.

GARCIA, V. (1962b). Embryological studies in the Loasaceae: Development of endosperm in *Blumenbachia hieronymi* Urb. *Phytomorphology* **12**, 307–312.

GARDINER, M. S. (1940). The effect of beta-indoleacetic acid upon isolated plant embryos. *Bull. Mount Desert Island Biol. Lab.*, 1–22.

GAUTHERET, R. J. (1935). "Recherches sur la culture des tissues végétaux: Essais de culture de quelques tissus méristèmatiques." Librairie E. le François, Paris.

GAUTHERET, R. J. (1937). Action de l'acide indol-β-acétique sur le développement de plantules et de fragments de plantules de *Phaseolus vulgaris*. *Compt. Rend. Séanc. Soc. Biol. Paris* **126**, 312–318.

GEIER, T. and KOHLENBACH, H. W. (1973). Entwicklung von Embryonen und embryogenem Kallus aus Pollenkörnern von *Datura meteloides* und *D. innoxia*. *Protoplasma* **78**, 381–396.

GEORGE, L. and NARAYANASWAMY, S. (1973). Haploid *Capsicum* through experimental androgenesis. *Protoplasma* **78**, 467–470.

GERASSIMOWA, H. (1936). Experimentell erhaltene haploide Pflanze von *Crepis tectorum* L. *Planta* **25**, 696–702.

GERLACH-CRUSE, D. (1969). Embryo- und Endospermentwicklung nach einer Röntgenbestrahlung der Fruchtknoten von *Arabidopsis thaliana* (L.) Heynh. *Radiation Bot.* **9**, 433–442.

GERLACH-CRUSE, D. (1970). Experimentelle Auslösung von Semigamie bei *Arabidopsis thaliana* (L.) Heynh. *Biol. Zentralbl.* **89**, 435–456.

GIANORDOLI, M. (1973). Étude ultrastructurale et cytochimique de l'oosphère mure chez le *Sciadopitys verticillata*. *Caryologia* **25** (Suppl.), 135–150.

GILMORE, A. E. (1950). A technique for embryo culture of peaches. *Hilgardia* **20**, 147–170.

GJØNNES, B. S. (1963). *Vitro* cultures of spruce. Some experiments on seedlings. *Medd. Norske Skogsforsoksv.* **18**, 85–105.

GODINEAU, J.-C. (1966). Ultrastructure du sac embryonnaire du *Crepis tectorum* L.: Les cellules du pôle micropylaire. *Compt. Rend. Acad. Sci. Paris* **263**, 852–855.

GODINEAU. J.-C. (1971). Ultrastructure du sac embryonnaire du *Crepis tectorum* L.: État apres la cellularisation et la fusion des noyaux polaires. *Ann. Université A.R.E.R.S.* **9**, 78–88.

GODINEAU, J.-C. (1973). Le sac embryonnaire des angiosperms. Morphogenèse et infrastructure. *Soc. Bot. Fr. Mémoires, Coll. Morphologie*, 25–54.

GONZALEZ Y SIOCO, B. M. (1914). The changes occurring in the ripening coconut. *Phillipine Agric. Forest.* **3**, 25–31.

GOODSELL, S. F. (1961). Male sterility in corn by androgenesis. *Crop. Sci.* **1**, 227–228.

GORTER, C. J. (1955). *In vitro* culture of *Cyclamen* embryos. *Koninkl. Nederl. Akad. Wetenschap. Proc.* **C58**, 377–385.

GRAHAM, J. S. D., MORTON, R. K. and RAISON, J. K. (1963). Isolation and characterization of protein bodies from developing wheat endosperm. *Austral. Jour. Biol. Sci.* **16**, 375–383.

GRAMBOW, H. J., KAO, K. N., MILLER, R. A. and GAMBORG, O. L. (1972). Cell division and plant development from protoplasts of carrot cell suspension cultures. *Planta* **103**, 348–355.

GRANT, W. F., BULLEN, M. R. and DENETTANCOURT, D. (1962). The cytogenetics of *Lotus*. I. Embryo-cultured interspecific diploid hybrids closely related to *L. corniculatus* L. *Canad. Jour. Genet. Cytol.* **4**, 105–128.

GREENSHIELDS, J. E. R. (1954). Embryology of interspecific crosses in *Melilotus*. *Canad. Jour. Bot.* **32**, 447–465.

GREENWOOD, M. S. and BERLYN, G. P. (1965). Regeneration of active root meristems *in vitro* by hypocotyl sections from dormant *Pinus lambertiana* embryos. *Canad. J. Bot.* **43**, 173–174.

GREENWOOD, M. S. and BERLYN, G. P. (1973). Sucrose-indole-3-acetic acid interactions on root regeneration by *Pinus lambertiana* embryo cuttings. *Amer. Jour. Bot.* **60**, 42–47.

GREENWOOD, M. S. and GOLDSMITH, M. H. M. (1970). Polar transport and accumulation of indole-3-acetic acid during root regeneration by *Pinus lambertiana* embryos. *Planta* **95**, 297–313.

GREGORY, F. G. and DE ROPP, R. S. (1938). Vernalization of excised embryos. *Nature* **142**, 481–482.

GREGORY, F. G. and PURVIS, O. N. (1936). Vernalization. *Nature* **138**, 249.

GREGORY, F. G. and PURVIS, O. N. (1938). Studies in vernalisation of cereals. II. The vernalisation of excised mature embryos, and of developing ears. *Ann. Bot. N.S.* **2**, 237–251.

GRESSHOFF, P. M. and DOY, C. H. (1972a). Development and differentiation of haploid *Lycopersicon esculentum* (tomato). *Planta* **107**, 161–170.

GRESSHOFF, P. M. and DOY, C. H. (1972b). Haploid *Arabidopsis thaliana* callus and plants from anther culture. *Austral. Jour. Biol. Sci.* **25**, 259–264.

GRESSHOFF, P. M. and DOY, C. H. (1974). Derivation of a haploid cell line from *Vitis vinifera* and the importance of the stage of meiotic development of anthers for haploid culture of this and other genera. *Zeitschr. Pflanzenphysiol.* **73**, 132–141.

GRIFFITH, E. and LINK, C. B. (1957). Germination and growth response of *Cattleya* seeds as influenced by the use of certain organic materials in the nutrient media. *Amer. Orch. Soc. Bull.* **26**, 184–192.

GRIS, A. (1864). Recherches anatomiques et physiologiques sur la germination. *Ann. Sci. Nat. (V. Bot.)* **2**, 1–123.

GUARDIOLA, J. L. and SUTCLIFFE, J. F. (1973). Mobilization of phosphorus in the cotyledons of young seedlings of the garden pea (*Pisum sativum* L.). *Ann. Bot. N.S.* **35**, 809–823.

GUHA, S. (1962). *In vitro* production of onion seeds. *In* "Plant Embryology—A Symposium", pp. 182–187. Council of Scientific and Industrial Research, New Delhi.

GUHA, S. and JOHRI, B. M. (1966). *In vitro* development of ovary and ovule of *Allium cepa* L. *Phytomorphology* **16**, 353–364.

GUHA, S. and MAHESHWARI, S. C. (1964). *In vitro* production of embryos from anthers of *Datura*. *Nature* **204**, 497.

GUHA, S. and MAHESHWARI, S. C. (1966). Cell division and differentiation of embryos in the pollen grains of *Datura in vitro*. *Nature* **212**, 97–98.

GUHA, S. and MAHESHWARI, S. C. (1967). Development of embryoids from pollen grains of *Datura in vitro*. *Phytomorphology* **17**, 454–461.

GUHA, S., IYER, R. D., GUPTA, N. and SWAMINATHAN, M. S. (1970). Totipotency of gametic cells and the production of haploids in rice. *Curr. Sci.* **39**, 174–176.

GUHA-MUKHERJEE, S. (1973). Genotypic differences in the *in vitro* formation of embryoids from rice pollen. *Jour. Exptl Bot.* **24**, 139–144.

GUNNING, B. E. S. and PATE, J. S. (1969). "Transfer cells". Plant cells with wall ingrowths, specialized in relation to short distance transport of solutes—Their occurrence, structure, and development. *Protoplasma* **68**, 107–133.

Guo, C. (1972). Effects of chemical and physical factors on the chromosome number in *Nicotiana* anther callus cultures. *In Vitro* **7**, 381–386.

Gupta, N. and Carlson, P. S. (1972). Preferential growth of haploid plant cells *in vitro*. *Nature New Biol.* **239**, 86.

Guttenberg, H. von and Lehle-Joerges, E. (1947). Über das Vorkommen von Auxin und Heteroauxin in ruhenden und keimenden Samen. *Planta* **35**, 281–296.

Guttenberg, H. von and Wiedow, H.-L. (1952). Über der Wirkstoffbedarf isolierter Haferembryonen. *Planta* **41**, 145–166.

Guzowska, I. and Zenkteler, M. (1969). *In vitro* development of immature embryos of *Cuscuta lupuliformis* Krock. *Bull. Soc. Sci. Lett. Poznan Ser. D* **9**, 37–48.

Györffy, P., Rédei, G. and Rédei, G. (1955). La substance de croissance du maïs laiteux. *Acta Bot. Acad. Sci. Hungar.* **2**, 57–76.

Haagen-Smit, A. J., Siu, R. and Wilson, G. (1945). A method for the culturing of excised, immature corn embryos *in vitro*. *Science* **101**, 234.

Haagen-Smit, A. J., Dandliker, W. B., Wittwer, S. H. and Murneek, A. E. (1946). Isolation of 3-indoleacetic acid from immature corn kernels. *Amer. Jour. Bot.* **33**, 118–120.

Haber, A. H. (1968). Ionizing radiations as research tools. *Ann. Rev. Plant Physiol.* **19**, 463–489.

Haber, A. H. and Luippold, H. J. (1960). Effects of gibberellin on gamma-irradiated wheat. *Amer. Jour. Bot.* **47**, 140–144.

Haberlandt, G. (1921). Über experimentelle Erzeugung von Adventivembryonen bei *Oenothera lamarckiana*. *Sitzungsber. Preussisch. Akad. Wiss. Berlin* **40**, 695–725.

Haberlandt, G. (1922). Über Zeilteilungshormone und ihre Beziehungen zur Wundheilung, Befruchtung, Parthenogenesis und Adventivembryonie. *Biol. Zentralbl.* **42**, 145–172.

Haccius, B. (1955a). Versuche zur somatischen Beeinflussung der Organbildung pflanzlicher Embryonen. *Experientia* **16**, 149.

Haccius, B. (1955b). Experimentally induced twinning in plants. *Nature* **176**, 355–357.

Haccius, B. (1956). Über Beeinflussung der Morphogenese pflanzlicher Embryonen durch Lithium-Ionen. *Ber. Deut. Bot. Ges.* **69**, 87–93.

Haccius, B. (1957a). Regenerationserscheinungen an pflanzlichen Embryonen nach Behandlung mit antimitotisch wirksamen Substanzen. *Beitr. Biol. Pfl.* **34**, 3–18.

Haccius, B. (1957b). Über die Regenerationsfähigkeit junger Embryonen von *Eranthis heimalis* nach Colchicin-Behandlung. *Naturwiss.* **44**, 18–19.

Haccius, B. (1959a). Morphoregulatorische Beeinflussung pflanzlicher Embryonen durch Phenylborsäure. *Naturwiss.* **46**, 153.

Haccius, B. (1959b). Über die unterschiedliche Antimitotica-Empfindlichkeit der Zellen noch undifferenzieter Embryonen von *Eranthis hiemalis*. *Zeitschr. Naturfors.* **14b**, 206–209.

Haccius, B. (1960). Experimentelle induzierte Einkeimblättrigkeit bei

Eranthis hiemalis. II. Monokotylie durch Phenylborsäure. *Planta* **54**, 482–497.

HACCIUS, B. (1963). Restitution in acidity-damaged plant embryos: Regeneration or regulation? *Phytomorphology* **13**, 107–115.

HACCIUS, B. (1965a). Haben "Gewebekultur-Embryonen" einen Suspensor? *Ber. Deut. Bot. Ges.* **78**, 11–21.

HACCIUS, B. (1965b). Weitere Untersuchungen über Somatogenese aus den Suspensorenzellen von *Eranthis hiemalis*-Embryonen. *Planta* **64**, 219–224.

HACCIUS, B. (1969). Anomalien der pflanzlichen Embryogenese nach Einwirkung von 2,4-D, TIBA und Morphaktinen, ein Vergleich. *Symposium über Morphaktine, Deut. Bot. Ges. Neue Folge* No. 3, 89–101.

HACCIUS, B. (1971). Zur derzeitigen Situation der Angiospermen-Embryologie. *Bot. Jahrb.* **91**, 309–329.

HACCIUS, B. (1973). Les premiers stades embryons végétaux zygotiques et somatiques sont-ils différents ou non? *Soc. Bot. Fr. Mémoires, Colloq. Morphologie*, 201–205.

HACCIUS, B. and LAKSHMANAN, K. K. (1965). Adventiv-Embryonen aus *Nicotiana*-Kallus, der bei hohen Lichtintensitäten kultiviert werde. *Planta* **65**, 102–104.

HACCIUS, B. and LAKSHMANAN, K. K. (1967). Experimental studies on monocotyledonous dicotyledons: Phenylboric acid-induced "dicotyledonous" embryos in *Cyclamen persicum*. *Phytomorphology* **17**, 488–494.

HACCIUS, B. and LAKSHMANAN, K. K. (1969). Adventiv-Embryonen-Embryoide-Adventiv-Knospen. Ein Beitrag zur Klärung der Begriffe. *Österr. Bot. Zeitschr.* **116**, 145–158.

HACCIUS, B. and REICHERT, H. (1964). Restitutionserscheinungen an pflanzlichen Meristemen nach Röntgenbestrahlung. II. Adventiv-Embryonie nach Samenbestrahlung von *Eranthis hiemalis*. *Planta* **62**, 355–372.

HACCIUS, B. and TROMPETER, G. (1960). Experimentelle induzierte Einkeimblättrigkeit bei *Eranthis hiemalis*. I. Synkotylie durch 2,4-dichlorophenoxyessigsäure. *Planta* **54**, 466–481.

HADDOCK, P. G. (1954). Sapling sugar pines grown from excised mature embryos. *Jour. Forest.* **52**, 434–437.

HAGMAN, M. and MIKKOLA, L. (1963). Observations on cross-, self-, and interspecific pollinations in *Pinus peuce* Griseb. *Silvae Genetica* **12**, 73–79.

HÅKANSSON, A. (1956). Seed development of *Picea abies* and *Pinus silvestris*. *Meddel. Skogsforsinst. Stockholm* **46**, 1–23.

HÅKANSSON, A. and ELLERSTRÖM, S. (1950). Seed development after reciprocal crosses between diploid and tetraploid rye. *Hereditas* **36**, 256–296.

HALL, C. B. (1948). Culture of *Solanum nigrum* embryos. *Proc. Amer. Soc. Hort. Sci.* **52**, 343–346.

HALL, O. L. (1954). Hybridization of wheat and rye after embryo transplantation. *Hereditas* **40**, 453–458.

HALL, O. L. (1956). Further experiments in embryo transplantation. *Hereditas* **42**, 261–262.

HALLAM, N. D. (1972). Embryogenesis and germination in rye (*Secale cereale* L.). I. Fine structure of the developing embryo. *Planta* **104**, 157–166.

HALLAM, N. D., ROBERTS, B. E. and OSBORNE, D. J. (1972). Embryogenesis and germination in rye (*Secale cereale* L.). II. Biochemical and fine structural changes during germination. *Planta* **105**, 293–309.

HALLAM, N. D., ROBERTS, B. E. and OSBORNE, D. J. (1973). Embryogenesis and germination in rye (*Secale cereale* L.). III. Fine structure and biochemistry of the non-viable embryo. *Planta* **110**, 279–290.

HALPERIN, W. (1964). Morphogenetic studies with partially synchronized cultures of carrot embryos. *Science* **146**, 408–410.

HALPERIN, W. (1966a). Alternative morphogenetic events in cell suspensions. *Amer. Jour. Bot.* **53**, 443–453.

HALPERIN, W. (1966b). Single cells, coconut milk, and embryogenesis *in vitro*. *Science* **153**, 1287–1288.

HALPERIN, W. (1967). Population density effects in embryogenesis in carrot cell cultures. *Exptl Cell Res.* **48**, 170–173.

HALPERIN, W. (1970). Embryos from somatic plant cells. *In* "Control Mechanisms of Cellular Phenotypes" (H. A. Padykula, ed.). Symposia of the International Society for Cell Biology, Vol. 9, pp. 169–191. Academic Press, New York and London.

HALPERIN, W. and JENSEN, W. A. (1967). Ultrastructural changes during growth and embryogenesis in carrot cell cultures. *Jour. Ultrastr. Res.* **18**, 428–443.

HALPERIN, W. and WETHERELL, D. F. (1964). Adventive embryony in tissue cultures of the wild carrot, *Daucus carota*. *Amer. Jour. Bot.* **51**, 274–283.

HALPERIN, W. and WETHERELL, D. F. (1965a). Ammonium requirement for embryogenesis *in vitro*. *Nature* **205**, 519–520.

HALPERIN, W. and WETHERELL, D. F. (1965b). Ontogeny of adventive embryos of wild carrot. *Science* **147**, 756–758.

HAMILTON, R. M. (1965). A soupçon of IBA. *Amer. Orch. Soc. Bull.* **34**, 429–430.

HANAWA, J. and ISHIZAKI, M. (1953). Malformation in *Sesamum indicum* L. caused by the operation on the embryo. *Sci. Rep. Fac. Lib. Arts Ed. Gifu Univ. (Nat. Sci.)* **1**, 55–61.

HANDRO, W., RAO, P. S. and HARADA, H. (1972). Contrôle hormonal de la formation de cals, bourgeons, racines et embryos sur des explantats de feuilles et de tiges de *Petunia* cultivés "*in vitro*". *Compt. Rend. Acad. Sci. Paris* **275**, 2861–2863.

HANNIG, E. (1904). Zur Physiologie pflanzlicher Embryonen. I. Ueber die Cultur von Cruciferen-Embryonen ausserhalb des Embryosacks. *Bot. Ztg* **62**, 45–80.

HANSON, J. B., VATTER, A. E., FISHER, M. E. and BILS, R. F. (1959). The development of mitochondria in the scutellum of germinating corn. *Agron. Jour.* **51**, 295–301.

HARBECK, M. (1963). Einige Beobachtungen bei der Aussaat verschiedener europaischer Erdorchideen auf sterilum Nahrboden. *Die Orchidee* **14**, 58–65.

HARN, C. (1969). Studies on the anther culture of rice. *Korean Jour. Breed.* **1**, 1–11.

HARN, C. (1971). Studies on anther culture in *Solanum nigrum*. *SABRAO Newsletter* **3**, 39–42.

HARN, C. (1972a). Studies on anther culture in *Solanum nigrum*. II. Cytological and histological observations. *SABRAO Newsletter* **4**, 27–32.

HARN, C. (1972b). Production of plants from anthers of *Solanum nigrum* cultured *in vitro*. *Caryologia* **25**, 429–437.

HARN, C. (1972c). Studies on the culture of haploid tobacco leaf. *Korean Jour. Bot.* **15**, 28–32.

HARN, C. and HWANG, J. (1970). Studies on the anther culture of rice. 2. Histological observation of haploid callus inoculated on differentiation medium. *Korean Jour. Bot.* **13**, 17–23.

HARN, C. and KIM, M. J. (1971). Studies on the anther culture of *Nicotiana tabacum* II. *Korean Jour. Bot.* **14**, 33–35.

HARN, C. and KIM, M. Z. (1972). Induction of callus from anthers of *Prunus armeniaca*. *Korean Jour. Breed.* **4**, 49–53.

HARRIS, B. and DURE, L., III (1974). Processing of stored mRNA in germinating cotton seeds. *Federation Proc.* **33**, 1343 (abstract).

HARRIS, C. J. (1935). The development of the flower and seed in *Galinsoga ciliata* (Raf.) Blake. *Univ. Pittsburgh Bull.* **32**, 131–137.

HARRIS, G. P. (1953). Amino acids and the growth of isolated oat embryos. *Nature* **172**, 1003.

HARRIS, G. P. (1956). Amino acids as sources of nitrogen for the growth of isolated oat embryos. *New Phytol.* **55**, 253–268.

HARWOOD, J. L., SODJA, A., STUMPF, P. K. and SPURR, A. R. (1971). On the origin of oil droplets in maturing castor bean seeds, *Ricinus communis*. *Lipids* **6**, 851–854.

HASHIMOTO, T. and RAPPAPORT, L. (1966). Variations in endogenous gibberellins in developing bean seeds. I. Occurrence of neutral and acidic substances. *Plant Physiol.* **41**, 623–628.

HASKELL, D. A. and POSTLETHWAIT, S. N. (1971). Structure and histogenesis of the embryo of *Acer saccharinum*. I. Embryo sac and proembryo. *Amer. Jour. Bot.* **58**, 595–603.

HATCHER, E. S. J. (1943). Auxin production during development of the grain in cereals. *Nature* **151**, 278–279.

HATCHER, E. S. J. (1945). Studies in the vernalisation of cereals. IX. Auxin production during development and ripening of the anther and carpel of spring and winter rye. *Ann. Bot. N.S.* **9**, 235–266.

HATCHER, E. S. J. and GREGORY, F. G. (1941). Auxin production during the development of the grain of cereals. *Nature* **148**, 626.

HAUSTEIN, E. (1961). Eine androgene haploide *Oenothera scabra*. *Planta* **56**, 475–478.

HAYNES, F. L. (1954). Potato embryo culture. *Amer. Potato Jour.* **31**, 282–288.

HEFTMANN, E. (1963). Biochemistry of plant steroids. *Ann. Rev. Plant Physiol.* **14**, 225–248.

HEGARTY, C. P. (1955). Observations on the germination of orchid seed. *Amer. Orch. Soc. Bull.* **24**, 457–464.

HEIT, C. E. (1955). The excised embryo method for testing germination quality of dormant seed. *Proc. Assoc. Off. Seed Analysts* **45**, 108–117.

HELMKAMP, G. and BONNER, J. (1953). Some relationships of sterols to plant growth. *Plant Physiol.* **28**, 428–436.

HENRY, M. P. (1956). Étude cytologique du lait de coco au cours du développement de la noix. *Compt. Rend. Acad. Sci. Paris* **245**, 401–404.

HESLOP-HARRISON, J. (1957). The physiology of reproduction in *Dactylorchis*. I. Auxin and the control of meiosis, ovule formation and ovary growth. *Bot. Notsier* **110**, 28–48.

HESSE, C. O. and KESTER, D. E. (1955). Germination of embryos of *Prunus* related to degree of embryo development and method of handling. *Proc. Amer. Soc. Hort. Sci.* **65**, 251–264

HILL, G. P. (1967). Multiple budding on carrot embryos arising in tissue culture. *Nature* **215**, 1098–1099.

HINTON, J. J. C. and MORAN, T. (1957). Methionine sulphoximine and the growth of the wheat embryo. *Brit. Jour. Nutrit.* **11**, 323–328.

HIRSCHBERG, K., HÜBNER, G. and BORRISS, H. (1972). Cytokinin-induzierte *de novo*-Synthese der Nitratreductase in Embryonen von *Agrostemma githago*. *Planta* **108**, 333–337.

HIRSH, D. H. (1959). Gibberellates: Stimulants for *Cattleya* seedlings in flasks. *Amer. Orch. Soc. Bull.* **28**, 342–344.

HOLLOWAY, J. E. (1917). The prothallus and young plant of *Tmesipteris*. *Trans. N. Zealand Inst.* **50**, 1–44.

HOLLOWAY, J. E. (1921). Further studies on the prothallus, embryo, and young sporophyte of *Tmesipteris*. *Trans. N. Zealand Inst.* **53**, 386–422.

HOLLOWAY, J. E. (1939). The gametophyte, embryo and young rhizome of *Psilotum triquetrum* Swartz. *Ann. Bot. N.S.* **3**, 313–336.

HOMÈS, J. (1967a). Induction de plantules dans des cultures *in vitro* de tissus de carotte. *Compt. Rend. Soc. Biol.* **161**, 730–732.

HOMÈS, J. L. A. (1967b). Action morphogénétique de glucose sur une souche embryogène de tissus de carotte cultivée *in vitro*. *Compt. Rend. Soc. Biol.* **161**, 1143–1145.

HOMÈS, J. L. A. (1968). Influence de la concentration en glucose sur le développement et la différenciation d'embryons formés dans des tissus de carotte cultivés *in vitro*. *In* "Les Cultures de Tissus de Plantes", pp. 49–60. Colloq. Nationaux du Centre National de la Recherche Scientifique, Paris, No. 920.

HOMÈS, J. L. A. and GUILLAUME, M. (1967). Phénomènes d'organogenèse dans des cultures *in vitro* de tissus de carotte (*Daucus carota* L.). *Bull. Soc. Roy. Bot. Belgique* **100**, 239–258.

HONMA, S. (1955). A technique for artificial culturing of bean embryos. *Proc. Amer. Soc. Hort. Sci.* **65**, 405–408.

HORNER, H. T., JR. and ARNOTT, H. J. (1965). A histochemical and ultrastructural study of *Yucca* seed proteins. *Amer. Jour. Bot.* **52**, 1027–1038.

HORNER, H. T., JR. and ARNOTT, H. J. (1966). A histochemical and ultrastructural study of pre- and post-germinated *Yucca* seeds. *Bot. Gaz.* **127**, 48–64.

HOTTA, Y. (1957). Roles of the cotyledon in the morphological differentiation of bean seedlings (Morphogenetical studies in *Vigna sesquipedalis*, II). *Bot. Mag. Tokyo* **70**, 383–390.

HOWELL, R. W. (1961). Changes in metabolic characteristics of mitochondria from soybean cotyledons during germination. *Physiol. Plantarum* **14**, 89–97.

HRISHI, N., MARIMUTHAMMAL, S. and DAS, L. D. V. (1969). Studies on the causes for seed failure in the intergeneric cross between sugarcane and maize. *Proc. Indian Acad. Sci.* **80B**, 169–177.

HU, C. Y. and SUSSEX, I. M. (1971). *In vitro* development of embryoids on cotyledons of *Ilex aquifolium*. *Phytomorphology* **21**, 103–107.

HUMPHREYS, J. L. (1958). Gibberellins for orchids. *Orch. Rev.* **66**, 176.

HYATT, D. W. (1965). Investigations in the orchid family. *Amer. Orch. Soc. Bull.* **34**, 787–790.

IHLE, J. N. and DURE, L., III (1969). Synthesis of a protease in germinating cotton cotyledons catalyzed by mRNA synthesized during embryogenesis. *Biochem. Biophys. Res. Comm.* **36**, 705–710.

IHLE, J. N. and DURE, L., III (1970). Hormonal regulation of translation inhibition requiring RNA synthesis. *Biochem. Biophys. Res. Comm.* **38**, 995–1001.

IHLE, J. N. and DURE, L. S., III (1972). The developmental biochemistry of cotton seed embryogenesis and germination. III. Regulation of the biosynthesis of enzymes utilized in germination. *Jour. Biol. Chem.* **247**, 5048–5055.

IKUMA, H. and THIMANN, K. V. (1959). Photosensitive site in lettuce seeds. *Science* **130**, 568–569.

IKUMA, H. and THIMANN, K. V. (1960). Action of gibberellic acid on lettuce seed germination. *Plant Physiol.* **35**, 557–566.

IKUMA, H. and THIMANN, K. V. (1963). The role of the seed-coats in germination of photosensitive lettuce seeds. *Plant Cell Physiol.* **4**, 169–185.

IMAI, I. and SIEGEL, S. M. (1973). A specific response to toxic cadmium levels in red kidney bean embryos. *Physiol. Plantarum* **29**, 118–120.

INGLE, J., BEEVERS, L. and HAGEMAN, R. H. (1964). Metabolic changes associated with the germination of corn. I. Changes in weight and metabolites and their redistribution in the embryo axis, scutellum, and endosperm. *Plant Physiol.* **39**, 735–740.

INGLE, J., BEITZ, D. and HAGEMAN, R. H. (1965). Changes in composition during development and maturation of maize seeds. *Plant Physiol.* **40**, 835–839.

INNOCENTI, A. M. and AVANZI, S. (1971). Seed aging and chromosome breakage in *Triticum durum* Desf. *Mutation Res.* **13**, 225–231.

IRIKURA, Y. and SAKAGUCHI, S. (1972). Induction of 12-chromosome plants from anther culture in a tuberous *Solanum*. *Potato Res.* **15**, 170–173.

ISLAM, A. S. (1964). A rare hybrid combination through application of hormone and embryo culture. *Nature* **201**, 320.

ISLAM, A. S. and RASHID, A. (1961). A new jute hybrid. *Jour. Hered.* **52**, 287–291.

ITO, K. (1951). Tomato pudding and orchid seed germination. *Na Pua Okika o Hawaii Nei* **1**, 37–38.

IVANOVSKAYA, E. V. (1946). Hybrid embryos of cereals grown on artificial nutrient medium. *Doklady Akad. Sci. SSSR* **54**, 445–448.

IVANOVSKAYA, E. V. (1962). The method of raising embryos on an artificial nutrient medium, and its application to wide hybridization. *In* "Wide Hybridization in Plants" (N. V. Tsitsin, ed.), pp. 134–142. Israel Program for Scientific Translations, Jerusalem.

IWAHORI, S. (1966). High temperature injuries in tomato. V. Fertilization and development of embryo with special reference to the abnormalities caused by high temperature. *Jap. Jour. Hort. Sci.* **35**, 379–386.

IWAHORI, S. (1967). Auxin of tomato fruit at different stages of its development with a special reference to high temperature injuries. *Plant Cell Physiol.* **8**, 15–22.

IYENGAR, C. V. K. (1942). Development of seed and its nutritional mechanism in Scrophulariaceae. Part I. *Rhamphicarpa longiflora* Benth., *Centranthera hispida* Br., and *Pedicularis zeylanica* Benth. *Proc. Natl Inst. Sci. India* **8**, 249–261.

IYER, R. D. and GOVILA, O. P. (1964). Embryo culture of interspecific hybrids in the genus *Oryza. Indian Jour. Genet. Plant Breed.* **24**, 116–121.

IYER, R. D. and RAINA, S. K. (1972). The early ontogeny of embryoids and callus from pollen and subsequent organogenesis in anther cultures of *Datura metel* and rice. *Planta* **104**, 146–156.

IYER, R. D., SULBHA, K. and RAMANUJAM, S. (1959). Embryo culture studies in jute and tomato. *Memoirs, Indian Bot. Soc.* **2**, 30–35.

JACHYMCZYK, W. J. and CHERRY, J. H. (1968). Studies on messenger RNA from peanut plants: *In vitro* polyribosome formation and protein synthesis. *Biochim. Biophys. Acta* **157**, 368–377.

JACOBSEN, J. V. and VARNER, J. E. (1967). Gibberellic acid-induced synthesis of protease by isolated aleurone layers of barley. *Plant Physiol.* **42**, 1596–1600.

JAFFE, L. (1956). Effect of polarized light on polarity of *Fucus. Science* **123**, 1081–1082.

JAFFE, L. F. (1968). Localization in the developing *Fucus* egg and the general role of localizing currents. *In* "Advances in Morphogenesis" (M. Abercrombie, J. Brachet and T. J. King, eds), Vol. 7, pp. 295–328. Academic Press, New York and London.

JAKOB, K. M. and BOVEY, F. (1969). Early nucleic acid and protein syntheses and mitoses in the primary root tips of germinating *Vicia faba. Exptl Cell Res.* **54**, 118–126.

JANSEN, L. L. and BONNER, J. (1949). Development of fruits from excised flowers in sterile culture. *Amer. Jour. Bot.* **36**, 826 (abstract).

JARVIS, B. C. and HUNTER, C. (1971). Changes in the capacity for protein synthesis in embryonic axes of hazel fruits during the breaking of dormancy by GA_3. *Planta* **101**, 174–179.

JARVIS, B. C., FRANKLAND, B. and CHERRY, J. H. (1968a). Increased nucleic acid synthesis in relation to the breaking of dormancy of hazel by gibberellic acid. *Planta* **83**, 257–266.

JARVIS, B. C., FRANKLAND, B. and CHERRY, J. H. (1968b). Increased DNA template and RNA polymerase associated with the breaking of seed dormancy. *Plant Physiol.* **43**, 1734–1736.

JAYASEKERA, R. D. E. and BELL, P. R. (1959). The effect of various experimental treatments on the development of the embryo of the fern *Thelypteris palustris*. *Planta* **54**, 1–14.

JAYASEKERA, R. D. E. and BELL, P. R. (1971). The synthesis and distribution of ribonucleic acid in developing archegonia of *Pteridium aquilinum*. *Planta* **101**, 76–87.

JAYASEKERA, R. D. E. and BELL, P. R. (1972). The effect of thiouracil on the viability of eggs and embryogeny in *Pteridium aquilinum*. *Planta* **102**, 206–214.

JAYASEKERA, R. D. E., CAVE, C. F. and BELL, P. R. (1972). The effect of thiouracil on the cytochemistry of the egg of *Pteridium aquilinum*. *Cytobiologie* **6**, 253–260.

JELASKA, S. (1972). Embryoid formation by fragments of cotyledons and hypocotyls in *Cucurbita pepo*. *Planta* **103**, 278–280.

JELASKA, S. (1974). Embryogenesis and organogenesis in pumpkin explants. *Physiol. Plantarum* **31**, 257–261.

JENNINGS, A. C. and MORTON, R. K. (1963). Changes in carbohydrate, protein, and non-protein nitrogenous compounds of developing wheat grain. *Austral. Jour. Biol. Sci.* **16**, 318–331.

JENNINGS, A. C., MORTON, R. K. and PALK, B. A. (1963). Cytological studies of protein bodies of developing wheat endosperm. *Austral. Jour. Biol. Sci.* **16**, 366–374.

JENSEN, C. O., SACKS, W. and BALDAUSKI, F. A. (1951). The reduction of triphenyltetrazolium chloride by dehydrogenases of corn embryos. *Science* **113**, 65–66.

JENSEN, W. A. (1962). "Botanical Histochemistry". Freeman, San Francisco.

JENSEN, W. A. (1964). Cell development during plant embryogenesis. *In* "Meristems and Differentiation". *Brookhaven Symp. Biol.* **16**, 179–202.

JENSEN, W. A. (1965a). The composition and ultrastructure of the nucellus in cotton. *Jour. Ultrastr. Res.* **13**, 112–128.

JENSEN, W. A. (1965b). The ultrastructure and histochemistry of the synergids of cotton. *Amer. Jour. Bot.* **52**, 238–256.

JENSEN, W. A. (1965c). The ultrastructure and composition of the egg and central cell of cotton. *Amer. Jour. Bot.* **52**, 781–797.

JENSEN, W. A. (1968a). Cotton embryogenesis: The tube-containing endoplasmic reticulum. *Jour. Ultrastr. Res.* **22**, 296–302.

JENSEN, W. A. (1968b). Cotton embryogenesis: The zygote. *Planta* **79**, 346–366.

JENSEN, W. A. (1968c). Cotton embryogenesis. Polysome formation in the zygote. *Jour. Cell Biol.* **36**, 403–406.

JENSEN, W. A. and FISHER, D. B. (1968). Cotton embryogenesis: The entrance and discharge of the pollen tube in the embryo sac. *Planta* **78**, 158–183.

Johansen, D. A. (1931a). Studies on the morphology of the Onagraceae—IV. *Stenosiphon linifolium. Bull. Torrey Bot. Club* **57**, 315–326.

Johansen, D. A. (1931b). Studies on the morphology of the Onagraceae—V. *Zauschneria latifolia*, typical of a genus characterized by irregular embryogeny. *Ann. N.Y. Acad. Sci.* **33**, 1–26.

Johansen, D. A. (1932). Studies on the morphology of the Onagraceae—VII. *Gayophytum ramosissimum. Bull. Torrey Bot. Club* **60**, 1–8.

Johansen, D. A. (1950). "Plant Embryology". Chronica Botanica Co., Waltham, Mass.

Johansen, E. L. and Smith, B. W. (1956). *Arachis hypogaea × A. diogoi*. Embryo and seed failure. *Amer. Jour. Bot.* **43**, 250–258.

Johnston, F. B. and Stern, H. (1957). Mass isolation of viable wheat embryos. *Nature* **179**, 160–161.

Johri, B. M. (1962a) Controlled growth of ovary and ovule. *In* "Proceedings of the Summer School of Botany" (P. Maheshwari, B. M. Johri and I. K. Vasil, eds), pp. 94–105. Ministry of Scientific Research and Cultural Affairs, New Delhi.

Johri, B. M. (1962b). Nutrition of the embryo sac. *In* "Proceedings of the Summer School of Botany" (P. Maheshwari, B. M. Johri and I. K. Vasil, eds), pp. 106–118. Ministry of Scientific Research and Cultural Affairs, New Delhi.

Johri, B. M. (1963a). Female gametophyte. *In* "Recent Advances in the Embryology of Angiosperms" (P. Maheshwari, ed.), pp. 69–103. International Society of Plant Morphologists, Delhi.

Johri, B. M. (1963b). Embryology and taxonomy. *In* "Recent Advances in the Embryology of Angiosperms" (P. Maheshwari, ed.), pp. 395–444. International Society of Plant Morphologists, Delhi.

Johri, B. M. and Bajaj, Y. P. S. (1962). Behaviour of mature embryo of *Dendrophthoe falcata* (L.f.) Ettings. *in vitro. Nature* **193**, 194–195.

Johri, B. M. and Bajaj, Y. P. S. (1963). *In vitro* response of the embryo of *Dendrophthoe falcata* (L.f.) Ettings. *In* "Plant Tissue and Organ Culture—A Symposium" (P. Maheshwari and N. S. Rangaswamy, eds), pp. 292–301. International Society of Plant Morphologists, Delhi.

Johri, B. M. and Bajaj, Y. P. S. (1964). Growth of embryos of *Amyema, Amylotheca*, and *Scurrula* on synthetic media. *Nature* **204**, 1220–1221.

Johri, B. M. and Bajaj, Y. P. S. (1965). Growth responses of globular proembryos of *Dendrophthoe falcata* (L.f.) Ettings. in culture. *Phytomorphology* **15**, 292–300.

Johri, B. M. and Bhojwani, S. S. (1965). Growth responses of mature endosperm in culture. *Nature* **208**, 1345–1347.

Johri, B. M. and Bhojwani, S. S. (1970). Embryo morphogenesis in the stem parasite *Scurrula pulverulenta. Ann. Bot. N.S.* **34**, 685–690.

Johri, B. M. and Garg, S. (1959). Development of endosperm haustoria in some Leguminosae. *Phytomorphology* **9**, 34–46.

Johri, B. M. and Guha, S. (1963a). The technique of *in vitro* culture in the study of physiology of reproduction. *Jour. Indian Bot. Soc.*, Maheshwari Comm. Vol. 42A, 58–73.

JOHRI, B. M. and GUHA, S. (1963b). *In vitro* development of onion plants from flowers. *In* "Plant Tissue and Organ Culture—A Symposium" (P. Maheshwari and N. S. Rangaswamy, eds), pp. 215–223. International Society of Plant Morphologists, Delhi.

JOHRI, B. M. and NAG, K. K. (1968). Experimental induction of triploid shoots *in vitro* from endosperm of *Dendrophthoe falcata* (L.f.) Ettings. *Curr. Sci.* **37**, 606–607.

JOHRI, B. M. and NAG, K. K. (1970). Endosperm of *Taxillus vestitus* Wall.: A system to study the effect of cytokinins *in vitro* in shoot bud formation. *Curr. Sci.* **39**, 177–179.

JOHRI, B. M. and NAG, K. K. (1974). Cytology and morphogenesis of embryo and endosperm tissues of *Dendrophthoe* and *Taxillus*. *Cytologia* **39**, 801–813.

JOHRI, B. M. and SEHGAL, C. B. (1963a). Chemical induction of polyembryony in *Anethum graveolens* L. *Naturwiss.* **50**, 47–48.

JOHRI, B. M. and SEHGAL, C. B. (1963b). Growth of ovaries of *Anethum graveolens* L. *In* "Plant Tissue and Organ Culture—A Symposium" (P. Maheshwari and N. S. Rangaswamy, eds), pp. 245–256. International Society of Plant Morphologists, Delhi.

JOHRI, B. M. and SEHGAL, C. B. (1965). *In vitro* production of neomorphs in *Anethum graveolens* L. *Nature* **205**, 1337.

JOHRI, B. M. and SEHGAL, C. B. (1966). Growth responses of ovaries of *Anethum*, *Foeniculum*·and *Trachyspermum*. *Phytomorphology* **16**, 364–378.

JOHRI, B. M. and SEHGAL, C. B. (1967). Plant cell, tissue and organ culture studies at the University of Delhi. *In* "Seminar on Plant Cell, Tissue and Organ Cultures, December 28–30, 1967", pp. 1–11. University of Delhi.

JOHRI, B. M. and SRIVASTAVA, P. S. (1973). Morphogenesis in endosperm cultures. *Zeitschr. Pflanzenphysiol.* **70**, 285–304.

JOHRI, B. M., AGRAWAL, J. S. and GARG, S. (1957). Morphological and embryological studies in the family Loranthaceae—I. *Helicanthes elastica* (Desr.) Daus. *Phytomorphology* **7**, 336–354.

JOHRI, M. M. and MAHESHWARI, S. C. (1965). Studies on respiration in developing poppy seeds. *Plant Cell Physiol.* **6**, 61–72.

JOHRI, M. M. and MAHESHWARI, S. C. (1966a). Changes in the carbohydrates, proteins and nucleic acids during seed development in opium poppy. *Plant Cell Physiol.* **7**, 35–47.

JOHRI, M. M. and MAHESHWARI, S. C. (1966b). Growth, development and respiration in the ovules of *Zephyranthes lancasteri* at different stages of maturation. *Plant Cell Physiol.* **7**, 49–58.

JONES, L. H. (1974). Factors influencing embryogenesis in carrot cultures (*Daucus carota* L.). *Ann. Bot. N.S.* **38**, 1077–1088.

JONES, R. L. (1969). The fine structure of barley aleurone cells. *Planta* **85**, 359–375.

JONES, R. L. (1974). The structure of the lettuce endosperm. *Planta* **121**, 133–146.

JORDAN, M. (1974). Multizelluläre Pollen bei *Prunus avium* nach *in vitro*-Kultur. *Zeitschr. Pflanzenzüchtg* **71**, 358–363.

JØRGENSEN, C. A. (1928). The experimental formation of heteroploid plants in the genus *Solanum*. *Jour. Genet.* **19**, 133–211.

Joshi, P. C. (1962). *In vitro* growth of cotton ovules. *In* "Plant Embryology—A Symposium", pp. 199–204. Council of Scientific and Industrial Research, New Delhi.

Joshi, P. C. and Johri, B. M. (1972). *In vitro* growth of ovules of *Gossypium hirsutum*. *Phytomorphology* **22**, 195–209.

Joshi, P. C. and Pundir, N. S. (1966). Growth of ovules in the cross *Gossypium arboreum* × *G. hirsutum* in vivo and *in vitro*. *Indian Cotton Jour.* **20**, 23–29.

Joy, K. W. and Folkes, B. F. (1965). The uptake of amino acids and their incorporation into the proteins of excised barley embryos. *Jour. Exptl Bot.* **16**, 646–666.

Jullien, M. (1974). La culture *in vitro* de cellules du tissu foliaire d'*Asparagus officinalis* L.: Obtention de souches à embryogenèse permanente et regénération de plantes entières. *Compt. Rend. Acad. Sci. Paris* **279**, 747–750.

Justice, O. L. (1940). Methods of breaking dormancy in isolated embryos of *Polygonum scandens*. *Amer. Jour. Bot.* **27**, 165 (abstract).

Kadotani, N. (1969). Studies on the haploid method of breeding by anther culture in tobacco. I. Production of diploid plants from haploid by pith and root tissue culture. *Iwata Tobacco Expt. Sta. Bull.* No. 2, 73–77.

Kameya, T. and Hinata, K. (1970). Production of haploid plants from pollen grains of *Brassica*. *Jap. Jour. Breed.* **20**, 82–87.

Kameya, T. and Uchiyama, H. (1972). Embryoids derived from isolated protoplasts of carrot. *Planta* **103**, 356–360.

Kandler, O. (1953). Über den "Synthetischen Wirkungsgrad" *in vitro* kultivierter Embryonen, Wurzeln und Sprosse. *Zeitschr. Naturforsch.* **8b**, 109–117.

Kandler, O. and Fink, E. (1955). Über die Abhängigkeit des synthetischen Wirkungsgrades *in vitro* kultivierter Embryonen von verschiedenen Aussenfaktoren. *Planta* **45**, 289–306.

Kaneko, K. (1957). Studies of the embryo culture on the interspecific hybridization of *Chrysanthemum*. *Jap. Jour. Genet.* **32**, 300–305.

Kano, K. (1965). Studies on the media for orchid seed germination. *Mem. Fac. Agric. Kagawa Univ.* No. 20, 1–68.

Kanta, K. (1960). Intraovarian pollination in *Papaver rhoeas* L. *Nature* **188**, 683–684.

Kanta, K. and Maheshwari, P. (1963a). Intraovarian pollination in some Papaveraceae. *Phytomorphology* **13**, 215–229.

Kanta, K. and Maheshwari, P. (1963b). Test-tube fertilization in some angiosperms. *Phytomorphology* **13**, 230–237.

Kanta, K. and Padmanabhan, D. (1964). *In vitro* culture of embryo segments of *Cajanus cajan* (L.) Millsp. *Curr. Sci.* **33**, 704–706.

Kanta, K., Rangaswamy, N. S. and Maheshwari, P. (1962). Test-tube fertilization in a flowering plant. *Nature* **194**, 1214–1217.

Kapil, R. N. and Jalan, S. (1962). Studies in the family Ranunculaceae: I. The embryology of *Caltha palustris* L. *In* "Plant Embryology—A Symposium", pp. 205–213. Council of Scientific and Industrial Research, New Delhi.

Kaplan, D. R. (1969). Seed development in *Downingia*. *Phytomorphology* **19**, 253–278.

KAPOOR, M. (1959). Influence of growth substances on the ovules of *Zephyranthes*. *Phytomorphology* **9**, 313–315.

KASHA, K. J. (ed.) (1974). "Haploids in Higher Plants. Advances and Potential". University of Guelph.

KASHA, K. J. and KAO, K. N. (1970). High frequency haploid production in barley (*Hordeum vulgare* L.). *Nature* **225**, 874–876.

KASPERBAUER, M. J. and COLLINS, G. B. (1972). Reconstitution of diploids from leaf tissue of anther-derived haploids in tobacco. *Crop Sci.* **12**, 98–101.

KATO, H. (1968). The serial observations of the adventive embryogenesis in the microculture of carrot tissue. *Sci. Papers Coll. Gen. Edu. Univ. Tokyo* **18**, 191–197.

KATO, H. and TAKEUCHI, M. (1963). Morphogenesis *in vitro* starting from single cells of carrot root. *Plant Cell Physiol.* **4**, 243–245.

KATO, H. and TAKEUCHI, M. (1966). Embryogenesis from the epidermal cells of carrot hypocotyl. *Sci. Papers Coll. Gen. Edu. Univ. Tokyo* **16**, 245–254.

KAUSIK, S. B. (1938). Studies in the Proteaceae. II. Floral anatomy and morphology of *Macadamia ternifolia* F. Muell. *Proc. Indian Acad. Sci.* **8B**, 45–62.

KAUSIK, S. B. (1942). Studies in Proteaceae. VII. The endosperm of *Grevillea robusta* Cunn. with special reference to the structure and development of the vermiform appendage. *Proc. Indian Acad. Sci.* **16B**, 121–140.

KEFFORD, N. P. and RIJVEN, A. H. G. C. (1966). Gibberellin and growth in isolated wheat embryos. *Science* **151**, 104–105.

KEHR, A. E. (1951). Monoploidy in *Nicotiana*. *Jour. Hered.* **42**, 107–112.

KEIM, W. F. (1953a). An embryo culture technique for forage legumes. *Agron. Jour.* **45**, 509–510.

KEIM, W. F. (1953b). Interspecific hybridization in *Trifolium* using embryo culture techniques. *Agron. Jour.* **45**, 601–606.

KENDE, H., HAHN, H. and KAYS, S. E. (1971). Enhancement of nitrate reductase activity by benzyladenine in *Agrostemma githago*. *Plant Physiol.* **48**, 702–706.

KENT, N. F. and BRINK, R. A. (1947). Growth *in vitro* of immature *Hordeum* embryos. *Science* **106**, 547–548.

KENTZER, T. (1966). The dynamics of gibberellin-like and growth-inhibiting substances during seed development of *Fraxinus excelsior* L. *Acta Soc. Bot. Polon.* **35**, 477–484.

KERR, T. and ANDERSON, D. B. (1944). Osmotic quantities in growing cotton bolls. *Plant Physiol.* **19**, 338–349.

KESSELL, R. H. J. and CARR, A. H. (1972). The effect of dissolved oxygen concentration on growth and differentiation of carrot (*Daucus carota*) tissue. *Jour. Exptl Bot.* **23**, 996–1007.

KESTER, D. E. (1953). Factors affecting the aseptic culture of Lovell peach seedlings. *Hilgardia* **22**, 335–365.

KESTER, D. E. and HESSE, C. O. (1955). Embryo culture of peach varieties in relation to season of ripening. *Proc. Amer. Soc. Hort. Sci.* **65**, 265–273.

KHAN, A. A. (1971). Cytokinins: Permissive role in seed germination. *Science* **171**, 853–859.

KHAN, A. A. (1972). ABA- and kinetin-induced changes in cell homogenates, chromatin-bound RNA polymerase and RNA composition. *In* "Plant Growth Substances 1970" (D. J. Carr, ed.), pp. 207–215. Springer-Verlag, Berlin.

KHAN, A. A. and ANOJULU, C. C. (1970). Abscisic acid changes in nucleotide composition of rapidly labelled RNA species of pear embryos. *Biochem. Biophys. Res. Comm.* **38**, 1069–1075.

KHAN, A. A. and HEIT, C. E. (1969). Selective effect of hormones on nucleic acid metabolism during germination of pear embryos. *Biochem. Jour.* **118**, 707–712.

KHAN, A. A., HEIT, C. E. and LIPPOLD, P. C. (1968). Increase in nucleic acid synthesizing capacity during cold treatment of dormant pear embryos. *Biochem. Biophys. Res. Comm.* **33**, 391–396.

KHAN, A. A., GASPAR, T., ROE, C. H., BOUCHET, M. and DUBUCQ, M. (1972). Synthesis of isoperoxidases in lentil embryonic axis. *Phytochemistry* **11**, 2963–2969.

KHAN, R. and RANDOLPH, L. F. (1960). Growth of excised embryos of wheat (*Triticum vulgare* var. Pb. 591) in different media with varying agar concentration. *Proc. Natl Acad. Sci. India* **30B**, 391–396.

KHANNA, P. and STABA, E. J. (1970). *In vitro* physiology and morphogenesis of *Cheiranthus cheiri* var. Clott of Gold and *C. cheiri* var. Goliath. *Bot. Gaz.* **131**, 1–5.

KIESSELBACH, T. A. (1949). The structure and reproduction of corn. *Univ. Nebraska Agric. Expt. Sta. Res. Bull.* No. 161, 1–96.

KIHARA, H. and NISHIYAMA, I. (1932). Different compatibility in reciprocal crosses of *Avena*, with special reference to tetraploid hybrids between hexaploid and diploid species. *Jap. Jour. Bot.* **6**, 245–305.

KIKUCHI, M. (1956). Studies on the decrepitude of seeds by the method of embryo transplanting. II. Effects of high temperature treatment of dry seeds upon their endosperms. *Proc. Crop Sci. Soc. Japan* **24**, 176.

KIMATA, M. and SAKAMOTO, S. (1971). Callus induction and organ rediferentiation of *Triticum*, *Aegilops* and *Agropyron* by anther culture. *Jap. Jour. Palynol.* **8**, 1–7.

KIMATA, M. and SAKAMOTO, S. (1972). Production of haploid albino plants of *Aegilops* by anther culture. *Jap. Jour. Genet.* **47**, 61–63.

KIMBER, G. and RILEY, R. (1963). Haploid angiosperms. *Bot. Rev.* **29**, 480–531.

KLEIN, S. and BEN-SHAUL, Y. (1966). Changes in cell fine structure of lima bean axes during early germination. *Canad. Jour. Bot.* **44**, 331–340.

KLYUCHAREVA, M. V. (1960). On the changes in deoxyribonucleic acid of the egg cell during fertilization in barley, wheat, and rye. *Doklady Akad. Nauk. SSSR* (English translation) **134**, 193–196.

KNUDSON, L. (1922). Nonsymbiotic germination of orchid seeds. *Bot. Gaz.* **73**, 1–25.

KNUDSON, L. (1951). Nutrient solutions for orchids. *Bot. Gaz.* **112**, 528–532.

KOCHBA, J. and BUTTON, J. (1974). The stimulation of embryogenesis and embryoid development in habituated ovular callus from the 'Shamouti' orange (*Citrus sinensis*) as affected by tissue age and sucrose concentration. *Zeitschr. Pflanzenphysiol.* **73**, 415–421.

KOCHBA, J. and SPIEGEL-ROY, P. (1972). Effect of culture media on embryoid formation from ovular callus of 'Shamouti' orange *Citrus sinensis* Zeitschr. *Pflanzenzüchtg* **69**, 156–162.

KOCHBA, J., SPIEGEL-ROY, P. and SAFRAN, H. (1972). Adventive plants from ovules and nucelli in *Citrus*. *Planta* **106**, 237–245.

KOCHBA, J., BUTTON, J., SPIEGEL-ROY, P., BORNMAN, C. H. and KOCHBA, M. (1974). Stimulation of rooting of *Citrus* embryoids by gibberellic acid and adenine sulphate. *Ann. Bot. N.S.* **38**, 795–802.

KOCHHAR, T., SABHARWAL, P. and ENGELBERG, J. (1971). Production of homozygous diploid plants by tissue culture technique. *Jour. Hered.* **62**, 59–61.

KOEHLER, L. D. and LINSKENS, H. F. (1967). Incorporation of protein and RNA precursors into fertilized *Fucus* eggs. *Protoplasma* **64**, 209–212.

KÖGL, F. and HAAGEN-SMIT, A. J. (1936). Biotin und Aneurin als Phytohormone. Ein Beitrag zur Physiologie der Keimung. 23. Mitteilung über pflanzliche Wachstumsstoffe. *Zeitschr. Physiol. Chem.* **243**, 209–226.

KOHLENBACH, H. W. (1965). Über organisierte Bildungen aus *Macleaya cordata* Kallus. *Planta* **64**, 37–40.

KOHLENBACH, H. W. (1970). Das durch Kinetin induzierte Nukléolus-Kernund Plasmawachstum isolierter Epidermiszellen von *Rhoeo spathacea*. *Zeitschr. Pflanzenphysiol.* **63**, 297–307.

KOHLENBACH, H. W. and GEIER, T. (1972). Embryonen aus *in vitro* kultivierten Antheren von *Datura meteloides* Dun., *D. wrightii* Regel und *Solanum tuberosum* L. *Zeitschr. Pflanzenphysiol.* **67**, 161–165.

KOLLER, D., MAYER, A. M., POLJAKOFF-MAYBER, A. and KLEIN, S. (1962). Seed germination. *Ann. Rev. Plant Physiol.* **13**, 437–464.

KOLLER, D., POLJAKOFF-MAYBER, A., BERG, A. and DISKIN, T. (1963). Germination-regulating mechanisms in *Citrullus colocynthis*. *Amer. Jour. Bot.* **50**, 597–603.

KONAR, R. N. (1958a). A qualitative survey of the free amino acids and sugars in the developing female gametophyte and embryo of *Pinus roxburghii* Sar. *Phytomorphology* **8**, 168–173.

KONAR, R. N. (1958b). A quantitative survey of some nitrogenous substances and fats in the developing embryos and gametophytes of *Pinus roxburghii* Sar. *Phytomorphology* **8**, 174–176.

KONAR, R. N. (1963). A haploid tissue from the pollen of *Ephedra foliata* Boiss. *Phytomorphology* **13**, 170–174.

KONAR, R. N. and NATARAJA, K. (1964). *In vitro* control of floral morphogenesis in *Ranunculus sceleratus* L. *Phytomorphology* **14**, 558–563.

KONAR, R. N. and NATARAJA, K. (1965a). Production of embryos on the stem of *Ranunculus sceleratus* L. *Experientia* **21**, 395.

KONAR, R. N. and NATARAJA, K. (1965b). Experimental studies in *Ranunculus sceleratus* L. Development of embryos from the stem epidermis. *Phytomorphology* **15**, 132–137.

KONAR, R. N. and NATARAJA, K. (1965c). Experimental studies in *Ranunculus sceleratus* L. Plantlets from freely suspended cells and cell groups. *Phytomorphology* **15**, 206–211.

KONAR, R. N. and NATARAJA, K. (1965d). Production of embryoids from the anthers of *Ranunculus sceleratus* L. *Phytomorphology* **15**, 245–248.

KONAR, R. N. and NATARAJA, K. (1969). Morphogenesis of isolated floral buds of *Ranunculus sceleratus* L. *in vitro*. *Acta. Bot. Neerl.* **18**, 680–699.

KONAR, R. N. and OBEROI, Y. P. (1965). *In vitro* development of embryoids on the cotyledons of *Biota orientalis*. *Phytomorphology* **15**, 137–140.

KONAR, R. N. and OBEROI, Y. P. (1969). Recent work on reproductive structures of living conifers and taxads—A review. *Bot. Rev.* **35**, 89–116.

KONAR, R. N. and RAMCHANDANI, S. (1958). The morphology and embryology of *Pinus wallichiana* Jack. *Phytomorphology* **8**, 328–346.

KONAR, R. N., THOMAS, E. and STREET, H. E. (1972a). Origin and structure of embryoids arising from epidermal cells of the stem of *Ranunculus sceleratus* L. *Jour. Cell. Sci.* **11**, 77–93.

KONAR, R. N., THOMAS, E. and STREET, H. E. (1972b). The diversity of morphogenesis in suspension cultures of *Atropa belladonna* L. *Ann. Bot. N.S.* **36**, 249–258.

KONZAK, C. F., RANDOLPH, L. F. and JENSEN, N. F. (1951). Embryo culture of barley species hybrids. Cytological studies of *Hordeum sativum* × *Hordeum bulbosum*. *Jour. Hered.* **42**, 125–134.

KOOIMAN, P. (1960). On the occurrence of amyloids in plant seeds. *Acta Bot. Neerl.* **9**, 208–219.

KOSTOFF, D. (1929). An androgenic *Nicotiana* haploid. *Zeitschr. Zellforsch.* **9**, 640–642.

KOSTOFF, D. (1930). Ontogeny, genetics, and cytology of *Nicotiana* hybrids. *Genetica* **12**, 33–139.

KOSTOFF, D. (1934). A haploid plant of *Nicotiana sylvestris*. *Nature* **133**, 949–950.

KOTOMORI, S. and MURASHIGE, T. (1965). Some aspects of aseptic propagation of orchids. *Amer. Orch. Soc. Bull.* **34**, 484–489.

KRAVTSOV, P. V. and KAS'YANOVA, V. G. (1968). Culture of isolated embryos as a method for prevention of sterility in distant hybrids of fruit plants. *Fiziol. Rastenii* (English translation) **15**, 784–786.

KRIEBEL, H. B. (1972). Embryo development and hybridity barriers in the white pines (section *Strobus*). *Silvae Genetica* **21**, 39–44.

KRIEDEMANN, P. and BEEVERS, H. (1967a). Sugar uptake and translocation in the castor bean seedling. I. Characteristics of transfer in intact and excised seedlings. *Plant Physiol.* **42**, 161–173.

KRIEDEMANN, P. and BEEVERS, H. (1967b). Sugar uptake and translocation in the castor bean seedling. II. Sugar transformations during uptake. *Plant Physiol.* **42**, 174–180.

KRUSE, A. (1973). *Hordeum* × *Triticum* hybrids. *Hereditas* **73**, 157–161.

KRUSE, A. (1974). An *in vivo/vitro* embryo culture technique. *Hereditas* **77**, 219–224.

KRUYT, W. (1952). Effects of some plant growth substances on early growth of pea embryos in sterile culture; a study in connection with the problem of hormonisation of seeds. *Koninkl. Nederl. Akad. Wetenschap. Proc.* **C55**, 503–513.

KRUYT, W. (1954). A study in connection with the problem of hormonization of seeds. *Acta Bot. Neerl.* **3**, 1–82.

KUO, J.-S., WANG, Y.-Y., CHIEN, N.-F., KU, S.-J., KUNG, M.-L. and HSU, H.-C. (1973). Investigations on the anther culture *in vitro* of *Nicotiana tabacum* L. and *Capsicum annuum* L. *Acta Bot. Sinica* **15**, 37–52.

KURAN, H. and MARCINIAK, K. (1969). Badania cytochemiczne woreczka zalążkowego *Lilium regale* w różnych fazach rozwoju. *Acta Soc. Bot. Polon.* **38**, 83–92.

KURTZ, E. B., JR. and MIRAMON, A. (1957). A system for the study of fat synthesis in flax embryos. *Plant Physiol.* **32** (Suppl.), xxxvii (abstract).

LABARCA, C. C., NICHOLLS, P. B. and BANDURSKI, R. S. (1965). A partial characterization of indoleacetylinositols from *Zea mays*. *Biochem. Biophys. Res. Comm.* **20**, 641–646.

LACADENA, J.-R. (1974). Spontaneous and induced parthenogenesis and androgenesis. *In* "Haploid Plants. Advances and Potential" (K. J. Kasha, ed.), pp. 13–32. University of Guelph.

LACROIX, L. J. and CANVIN, D. T. (1963). The role of light and other factors in the growth and differentiation of barley embryos. *Plant Physiol.* **38** (Suppl.), xxiii (abstract).

LACROIX, L. J. and JASWAL, A. S. (1967). Metabolic changes in after-ripening seed of *Prunus cerasus*. *Plant Physiol.* **42**, 479–480.

LACROIX, L. J. and JASWAL, A. S. (1973). Lipid and sugar metabolism during the after-ripening of sour cherry embryos. *Canad. Jour. Bot.* **51**, 1267–1270.

LACROIX, L. J., NAYLOR, J. and LARTER, E. N. (1962). Factors controlling embryo growth and development in barley (*Hordeum vulgare* L.). *Canad. Jour. Bot.* **40**, 1515–1523.

LAGARDE, R. V. (1929). Non-symbiotic germination of orchids. *Ann. Missouri Bot. Gard.* **16**, 499–514.

LAHILLE, A. (1920). Coconut water, its characteristics, composition and various uses. *Bull. Econ. Indochine*, *N.S.* **23**, 1–25.

LAIBACH, F. (1925). Das Taubwerden von Bastardsamen und die künstliche Aufzucht früh absterbender Bastardembryonen. *Zeitschr. Bot.* **17**, 417–459.

LAIBACH, F. (1929). Ectogenesis in plants. *Jour. Hered.* **20**, 201–208.

LAIBACH, F. (1930). Kreuzungsschwierigkeiten bei Pflanzen und die Möglichkeiten ihrer Behebung. *Ber. Deut. Bot. Ges.* **48**, 58–77.

LAKSHMANAN, M. and PADMANABHAN, D. (1968). Effect of ascochitine on the *in vitro* growth of embryos of *Clitoria ternatea* L. *Curr. Sci.* **37**, 321–322.

LAMI, R. (1927). Influence d'une peptone sur la germination de quelques Vandées. *Compt. Rend. Acad. Sci. Paris* **184**, 1579–1581.

LAMMERTS, W. E. (1942). Embryo culture an effective technique for shortening the breeding cycle of deciduous trees and increasing germination of hybrid seeds. *Amer. Jour. Bot.* **29**, 166–171.

. LAMMERTS, W. E. (1943). Effect of photoperiod and temperature on growth of embryo-cultured peach seedlings. *Amer. Jour. Bot.* **30**, 707–711.

LAMMERTS, W. E. (1946). Use of embryo culture in rose breeding. *Plants and Gardens* **2**, 111.

LANDES, M. (1939). The causes of self-sterility in rye. *Amer. Jour. Bot.* **26**, 567–571.

LANG, W. H. (1909). A theory of alternation of generations in archegoniate plants based upon the ontogeny. *New Phytol.* **8**, 1–12.

LaRue, C. D. (1936). The growth of plant embryos in culture. *Bull. Torrey Bot. Club* **63**, 365–382.

LaRue, C. D. (1942). The rooting of flower in sterile culture. *Bull. Torrey Bot. Club* **69**, 332–341.

LaRue, C. D. (1947). Growth and regeneration of the endosperm of maize in culture. *Amer. Jour. Bot.* **34**, 585–586 (abstract).

LaRue, C. D. (1948). Regeneration in the megagametophyte of *Zamia floridana*. *Bull. Torrey Bot. Club* **75**, 597–603.

LaRue, C. D. (1949). Cultures of the endosperm of maize. *Amer. Jour. Bot.* **36**, 798 (abstract).

LaRue, C. D. (1952). Growth of the scutellum of maize in culture. *Science* **115**, 315–316.

LaRue, C. D. (1954). Studies on growth and regeneration in gametophytes and sporophytes of gymnosperms. *In* "Abnormal and Pathological Plant Growth". *Brookhaven Symp. Biol.* **6**, 187–208.

LaRue, C. D. and AVERY, G. S., JR. (1938). The development of the embryo of *Zizania aquatica* in the seed and in artificial culture. *Bull. Torrey Bot. Club* **65**, 11–21.

LaRue, C. D. and MERRY, J. (1939). The development of excised maize embryos in an atmosphere of nitrogen. *Amer. Jour. Bot.* **26**, 18S (abstract).

LEDBETTER, M. C. (1960). Anatomical and morphological comparisons of normal and physiologically dwarfed seedlings of *Rhodotypos tetrapetala* and *Prunus persica*. *Contrib. Boyce Thompson Inst.* **20**, 437–458.

LEDINGHAM, G. F. (1940). Cytological and developmental studies of hybrids between *Medicago sativa* and a diploid form of *Medicago falcata*. *Genetics* **25**, 1–15.

LEE, A. E. (1952). The growth of excised immature sedge embryos in culture. *Bull. Torrey Bot. Club* **79**, 59–62.

LEE, A. E. (1955). Growth in culture of excised portions of lupine embryos. *Bot. Gaz.* **116**, 359–364.

LEE, F. A. and TUKEY, H. B. (1942). Chemical changes accompanying growth and development of seed and fruit of the Elberta peach. *Bot. Gaz.* **104**, 348–355.

LEE, J. H. and COOPER, D. C. (1958). Seed development following hybridization between diploid *Solanum* species from Mexico, Central and South America. *Amer. Jour. Bot.* **45**, 104–110.

LENZ, L. W. (1954). The endosperm as a barrier to intersectional hybridization in *Iris*. *Aliso* **3**, 57–58.

LENZ, L. W. (1955). Studies in *Iris* embryo culture. I. Germination of embryos of the subsection *Hexapogon* Benth. (sect. Regelia *sensu* Dykes). *Aliso* **3**, 173–182.

LENZ, L. W. (1956). Development of the embryo sac, endosperm and embryo in *Iris munzii* and the hybrid *I. munzii × I. sibirica* 'Caesar's Brother'. *Aliso* **3**, 329–343.

LEOPOLD, A. C. and SCOTT, F. I. (1952). Physiological factors in tomato fruit-set. *Amer. Jour. Bot.* **39**, 310–317.

LE PAGE, A. (1968). Mise en évidence d'une dormance associée à une immaturité de l'embryon chez *Taxus baccata* L. *Compt. Rend. Acad. Sci. Paris* **266**, 1028–1030.

LE PAGE-DEGIVRY, M.-T. (1970a). Dormance de graine associée à une immaturité de l'embryon: Étude en culture *in vitro* chez *Magnolia soulangeana* Soul. Bod. et *Magnolia grandiflora* L. *Planta* **90**, 267–271.

LE PAGE-DEGIVRY, M.-T. (1970b). Acide abscissique et dormance chez les embryons de *Taxus baccata* L. *Compt. Rend. Acad. Sci. Paris* **271**, 482–484.

LE PAGE-DEGIVRY, M.-T. (1973a). Intervention d'un inhibiteur lié dans la dormance embryonnaire de *Taxus baccata* L. *Compt. Rend. Acad. Sci. Paris* **277**, 177–180.

LE PAGE-DEGIVRY, M.-T. (1973b). Étude en culture *in vitro* de la dormance embryonnaire chez *Taxus baccata* L. *Biol. Plantarum* **15**, 264–269.

LE PAGE-DEGIVRY, M.-T. (1973c). Influence de l'acide abscissique sur le développement des embryons de *Taxus baccata* L. cultivés *in vitro*. *Zeitschr. Pflanzenphysiol.* **70**, 406–413.

LE PAGE-DEGIVRY, M.-T. and GARELLO, G. (1973). La dormance embryonnaire chez *Taxus baccata*: Influence de la composition du milieu liquide sur l'induction de la germination. *Physiol. Plantarum* **29**, 204–207.

LESLEY, J. W. and BONNER, J. (1952). The development of normal peach seedlings from seeds of early-maturing varieties. *Proc. Amer. Soc. Hort. Sci.* **60**, 238–242.

LETHAM, D. S. (1964). Isolation of a kinin from plum fruitlets and other tissues. *In* "Régulateurs Naturels de la Croissance Végétale", pp. 109–117. Colloq. Internationaux du Centre National de la Recherche Scientifique, Paris, No. 123.

LETHAM, D. S. (1968). A new cytokinin bioassay and the naturally occurring cytokinin complex. *In* "Biochemistry and Physiology of Plant Growth Substances" (F. Wightman and G. Setterfield, eds), pp. 19–31. Runge Press, Ottawa.

LETHAM, D. S. and MILLER, C. O. (1965). Identity of kinetin-like factors from *Zea mays*. *Plant Cell Physiol.* **6**, 355–359.

LEWAK, S., BIAŁEK, K. and SIŃSKA, I. (1970). Sensitivity of apple seed germination to light and some growth regulators. *Biol. Plantarum* **12**, 291–296.

LI, H.-W., WENG, T.-S., CHEN, C.-C. and WANG, W.-H. (1961). Cytogenetical studies of *Oryza sativa* L. and its related species. *Bot. Bull. Acad. Sinica* **2**, 79–86.

Li, T.-T. (1934). The development of embryo of *Ginkgo biloba in vitro. Sci. Rep. Natl Tsing Hua Univ. Ser. B* **2**, 29–35.

Li, T.-T. and Shen, T. (1934). The effect of "pantothenic acid" on the growth of the yeast and on the growth of the radicle of *Ginkgo* embryos in artificial media. *Sci. Rep. Natl Tsing Hua Univ. Ser. B* **2**, 53–60.

Lilleland, O. (1930). Growth study of the apricot fruit. *Proc. Amer. Soc. Hort. Sci.* **27**, 237–245.

Linser, H. and Neumann, K.-H. (1968). Untersuchungen über Beziehungen zwischen Zellteilung und Morphogenese bei Gewebekulturen von *Daucus carota*. I. Rhizogenese und Ausbildung ganzer Pflanzen. *Physiol. Plantarum* **21**, 487–499.

Linskens, H. F. (1969). Changes in the polysomal pattern after fertilization in *Fucus* eggs. *Planta* **85**, 175–182.

Lipe, W. N. and Crane, J. C. (1966). Dormancy regulation in peach seeds. *Science* **153**, 541–542.

List, A., Jr. and Steward, F. C. (1965). The nucellus, embryo sac, endosperm, and embryo of *Aesculus* and their interdependence during growth. *Ann Bot. N.S.* **29**, 1–15.

Lloyd, F. E. (1902). The comparative embryology of the Rubiaceae. *Memoirs, Torrey Bot. Club* **8**, 1–112.

Loeffler, J. E. and van Overbeek, J. (1964). Kinin activity in coconut milk. *In* "Régulateurs Naturels de la Croissance Végétale", pp. 77–82. Colloq. Internationaux du Centre National de la Recherche Scientifique, Paris, No. 123.

Lofland, H. B., Jr. (1950). *In vitro* culture of the cotton embryo. *Bot. Gaz.* **111**, 307–311.

Long, T. J. and Haber, A. H. (1965a). Opposite effects of galacturonic acid on initial versus subsequent elongation of seminal roots of germinating wheat. *Bot. Gaz.* **126**, 97–100.

Long, T. J. and Haber, A. H. (1965b). Growth of embryos on synthetic media after excision from gamma-irradiated (525 kr) and unirradiated wheat grains. *Radiation Bot.* **5**, 223–231.

Loo, S. W. and Wang, F. H. (1943). The culture of young conifer embryos *in vitro. Science* **98**, 544:

Lott, J. N. A., Larsen, P. L. and Darley, J. J. (1971). Protein bodies from the cotyledons of *Cucurbita maxima. Canad. Jour. Bot.* **49**, 1777–1782.

Luckwill, L. C. (1948). The hormone content of the seed in relation to endosperm development and fruit drop in the apple. *Jour. Hort. Sci.* **24**, 32–44.

Luckwill, L. C. (1953). Studies of fruit development in relation to plant hormones. I. Hormone production by the developing apple seed in relation to fruit drop. *Jour. Hort. Sci.* **28**, 14–24.

Lugo-Lugo, H. (1955a). Effects of nitrogen on the germination of *Vanilla planifolia* seeds. *Amer. Orch. Soc. Bull.* **24**, 309–312.

Lugo-Lugo, H. (1955b). The effect of nitrogen on the germination of *Vanilla planifolia*. *Amer. Jour. Bot.* **42**, 679–684.   ·

Maciejewska, U., Ryc, M., Maleszewski, S. and Lewak, S. (1974). Embryonal dormancy and the development of photosynthetic activity in apple seedlings. *Physiol. Végét.* **12**, 115–122.

Magoon, M. L. and Khanna, K. R. (1963). Haploids. *Caryologia* **16**, 191–235.

Magrou, J., Mariat, F. and Rose, H. (1949). Sur la nutrition azotée des Orchidées. *Compt. Rend. Acad. Sci. Paris* **229**, 685–688.

Maheshwari, N. (1958). *In vitro* culture of excised ovules of *Papaver somniferum*. *Science* **127**, 342.

Maheshwari, N. and Lal, M. (1958). *In vitro* culture of ovaries of *Iberis amara* L. *Nature* **181**, 631–632.

Maheshwari, N. and Lal, M. (1961a). *In vitro* culture of ovaries of *Iberis amara* L. *Phytomorphology* **11**, 17–23.

Maheshwari, N. and Lal, M. (1961b). *In vitro* culture of excised ovules of *Papaver somniferum* L. *Phytomorphology* **11**, 307–314.

Maheshwari, P. (1950). "An Introduction to the Embryology of Angiosperms". McGraw-Hill, New York.

Maheshwari, P. (1959). Test-tube fruits and seeds. *Jour. Indian Bot. Soc.* **38**, 161–170.

Maheshwari, P. (1964). Embryology in relation to taxonomy. *In* "Vistas in Botany" (W. B. Turrill, ed.), Vol. 4, pp. 55–97. Pergamon Press, Oxford.

Maheshwari, P. and Baldev, B. (1961). Artificial production of buds from the embryos of *Cuscuta reflexa*. *Nature* **191**, 197–198.

Maheshwari, P. and Baldev, B. (1962). *In vitro* induction of adventive buds from embryos of *Cuscuta reflexa* Roxb. *In* "Plant Embryology—A Symposium", pp. 129–138. Council of Scientific and Industrial Research, New Delhi.

Maheshwari, P. and Johri, B. M. (1950). Development of embryo sac, embryo and endosperm in *Helixanthera ligustrina* (Wall.) Dans. *Nature* **165**, 978–979.

Maheshwari, P. and Kapil, R. N. (1966). Some Indian contributions to the embryology of angiosperms. *Phytomorphology* **16**, 239–291.

Maheshwari, P. and Rangaswamy, N. S. (1958). Polyembryony and *in vitro* culture of embryos of *Citrus* and *Mangifera*. *Indian Jour. Hort.* **15**, 275–282.

Maheshwari, P. and Rangaswamy, N. S. (1965). Embryology in relation to physiology and genetics. *In* "Advances in Botanical Research" (R. D. Preston, ed.), Vol. 2, pp. 219–321. Academic Press, New York and London.

Maheshwari, P. and Sachar, R. C. (1963). Polyembryony. *In* "Recent Advances in the Embryology of Angiosperms" (P. Maheshwari, ed.), pp. 265–296. International Society of Plant Morphologists, Delhi.

Maheshwari, P. and Singh, B. (1952). Embryology of *Macrosolen cochinchinensis*. *Bot. Gaz.* **114**, 20–32.

Maheshwari, P. and Singh, H. (1967). The female gametophyte of gymnosperms. *Biol. Rev.* **42**, 88–130.

MAHESHWARI, P., JOHRI, B. M. and DIXIT, S. N. (1957). The floral morphology and embryology of the Loranthoideae (Loranthaceae). *Jour. Madras Univ.* **B27**, 121–136.

MAHESHWARI, S. C. and GUPTA, G. R. P. (1965). Production of adventitious embryoids *in vitro* from stem callus of *Foeniculum vulgare*. *Planta* **67**, 384–386.

MAHESHWARI, S. C. and JOHRI, M. M. (1963). Respiration in the developing ovules of poppy (*Papaver somniferum* L.). *Naturwiss.* **50**, 718–719.

MAHESHWARI, S. C., JOHRI, M. M. and BHALLA, P. R. (1964). Acidic auxins in the maturing ovules of cotton (*Gossypium hirsutum* var. Indore 2). *Indian Jour. Exptl Biol.* **2**, 198–202.

MALHOTRA, S. S. and SPENCER, M. (1970). Changes in the respiratory, enzymatic, and swelling and contraction properties of mitochondria from cotyledons of *Phaseolus vulgaris* L. during germination. *Plant Physiol.* **46**, 40–44.

MALIGA, P., MÁRTON, L. and SZ-BREZNOVITS, A. (1973a). 5-Bromodeoxyuridine-resistant cell lines from haploid tobacco. *Plant Sci. Lett.* **1**, 119–121.

MALIGA, P., SZ-BREZNOVITS, A. and MÁRTON, L. (1973b). Streptomycin-resistant plants from callus culture of haploid tobacco. *Nature New Biol.* **244**, 29–30.

MALY, R. (1958). Die Mutabilität der Plastiden von *Antirrhinum majus* L. Sippe 50. *Zeitschr. Vererbungslehre* **89**, 692–696.

MARCUS, A. (1969). Seed germination and the capacity for protein synthesis. *Symp. Soc. Exptl Biol.* **23**, 143–160.

MARCUS, A. and FEELEY, J. (1964). Activation of protein synthesis in the imbibition phase of seed germination. *Proc. Natl Acad. Sci. U.S.A.* **51**, 1075–1079.

MARCUS, A. and FEELEY, J. (1965). Protein synthesis in imbibed seeds. II. Polysome formation during imbibition. *Jour. Biol. Chem.* **240**, 1675–1680.

MARCUS, A. and FEELEY, J. (1966). Ribosome activation and polysome formation *in vitro*: Requirement for ATP. *Proc. Natl Acad. Sci. U.S.A.* **56**, 1770–1777.

MARCUS, A., FEELEY, J. and VOLCANI, T. (1966). Protein synthesis in imbibed seeds. III. Kinetics of amino acid incorporation, ribosome activation, and polysome formation. *Plant Physiol.* **41**, 1167–1172.

MARIAT, F. (1949). Action de l'acide nicotinique sur la germination et le développement des embryons de *Cattleya*. *Compt. Rend. Acad. Sci. Paris* **229**, 1355–1357.

MARIAT, F. (1952). Recherches sur la physiologie des embryons d'Orchidées. *Rev. Gén. Bot.* **59**, 324–377.

MARINOS, N. G. (1970). Embryogenesis of the pea (*Pisum sativum*). I. The cytological environment of the developing embryo. *Protoplasma* **70**, 261–279.

MARRÈ, E. (1967). Ribosome and enzyme changes during maturation and germination of the castor bean seed. *In* "Current Topics in Developmental Biology" (A. A. Moscona and A. Monroy, eds), Vol. 2, pp. 75–105. Academic Press, New York and London.

MARRÈ, E., COCUCCI, S. and STURANI, E. (1965). On the development of the ribosomal system in the endosperm of germinating castor bean seeds. *Plant Physiol.* **40**, 1162–1170.

MARTIN, G. C., MASON, M. I. R. and FORDE, H. I. (1969). Changes in endogenous growth substances in the embryos of *Juglans regia* during stratification. *Jour. Amer. Soc. Hort. Sci.* **94**, 13–17.

MASAND, P. and KAPIL, R. N. (1966). Nutrition of the embryo sac and embryo—A morphological approach. *Phytomorphology* **16**, 158–175.

MASCARENHAS, J. P. (1971). RNA and protein synthesis during pollen development and tube growth. *In* "Pollen: Development and Physiology" (J. Heslop-Harrison, ed.), pp. 201–222. Butterworths, London.

MATSUBARA, S. (1962). Studies on a growth promoting substance, "embryo factor", necessary for the culture of young embryos of *Datura tatula in vitro*. *Bot. Mag. Tokyo* **75**, 10–18.

MATSUBARA, S. (1964a). Abnormal organ formation from the hypocotyl of cultured young embryos in *Pharbitis nil* (preliminary report). *Sci. Rep. Kyoto Pref. Univ. (Nat. Sci., Liv. Sci. & Welf. Sci.) Ser. A* **15**, 15–16.

MATSUBARA, S. (1964b). Effect of nitrogen compounds on the growth of isolated young embryos of *Datura*. *Bot. Mag. Tokyo* **77**, 253–259.

MATSUBARA, S. (1964c). Effect of *Lupinus* growth factor on the *in vitro* growth of embryos of various plants and carrot root tissue. *Bot. Mag. Tokyo* **77**, 403–411.

MATSUBARA, S. and KOSHIMIZU, K. (1966). Factors with cytokinin activity in young *Lupinus* seeds and their partial purification. *Bot. Mag. Tokyo* **79**, 389–396.

MATSUBARA, S. and NAKAHIRA, R. (1965a). Some factors affecting the growth of young embryo *in vitro*. *Sci. Rep. Kyoto Pref. Univ. (Nat. Sci., Liv. Sci. & Welf. Sci.) Ser. A* **16**, 1–6.

MATSUBARA, S. NAKAHIRA, R. (1965b). Contents of embryo factor in developing seeds of *Datura* and *Lupinus*. *Sci. Rep. Kyoto Pref. Univ. (Nat. Sci., Liv. Sci. & Welf. Sci.) Ser. A* **16**, 7–10.

MATSUBARA, S. and NAKAHIRA, R. (1966). *In vitro* formation of adventitious organs by *Pharbitis* embryos. *Nature* **211**, 1208–1210.

MATSUBARA, S. and OGAWA, J. (1963). Chromatographic separation of a growth factor for embryo and of gibberellin-like substances from the diffusate of *Lupinus* seed. *Bot. Mag. Tokyo* **76**, 440–445.

MATSUBARA, S., KOSHIMIZU, K. and NAKAHIRA, R. (1966). Growth-promoting factors in young *Lupinus* seeds. *Sci. Rep. Kyoto Pref. Univ. (Nat. Sci., Liv. Sci. & Welf. Sci.) Ser. A* **17**, 5–11.

MATTINGLY, C. F. and COLLINS, G. B. (1974). The use of anther-derived haploids in *Nicotiana*. III. Isolation of nullisomics from monosomic lines. *Chromosoma* **46**, 29–36.

MAUGINI, E. and FIORDI, A. C. (1970). Passaggio di materiale dalle cellule del tappeto alla cellula centrale dell'archegonio e al proembrione di *Ginkgo biloba* L. *Caryologia* **23**, 415–440.

MAUNEY, J. R. (1961). The culture *in vitro* of immature cotton embryos. *Bot. Gaz.* **122**, 205–209.

MAUNEY, J. R., CHAPPELL, J. and WARD, B. J. (1967). Effects of malic acid salts on growth of young cotton embryos *in vitro*. *Bot. Gaz.* **128**, 198–200.

MAUNEY, J. R., HILLMAN, W. S., MILLER, C. O., SKOOG, F., CLAYTON, R. A. and STRONG, F. M. (1952). Bioassay, purification, and properties of a growth factor from coconut. *Physiol. Plantarum* **5**, 485–497.

MAYER, A. M. and POLJAKOFF-MAYBER, A. (1975). "The Germination of Seeds" (second edition). Pergamon Press, Oxford.

MCARTHUR, J. A. and BRIGGS, W. R. (1970). Phytochrome appearance and distribution in the embryonic axis and seedling of Alaska peas. *Planta* **91**, 146–154.

MCCANCE, R. A. and WIDDOWSON, E. M. (1940). The chemical composition of foods. *Med. Res. Council (Brit.) Spec. Report Ser.* No. 235, p. 150.

MCCARTHY, W. J., APP, A. A. and CROTTY, W. J. (1971). The effect of calcium on *in vitro* polyphenylalanine synthesis by rice ribosomes. *Biochim. Biophys. Acta* **246**, 132–140.

MCKEE, H. S., ROBERTSON, R. N. and LEE, J. B. (1955). Physiology of pea fruits. I. The developing fruit. *Austral. Jour. Biol. Sci.* **8**, 137–163.

MCLANE, S. R. and MURNEEK, A. E. (1952). The detection of syngamin, an indigenous plant hormone, by culture of immature corn embryos. *Univ. Missouri Agric. Sta. Res. Bull.*, 496.

MCLEAN, S. W. (1946). Interspecific crosses involving *Datura ceratocaula* obtained by embryo dissection. *Amer. Jour. Bot.* **33**, 630–638.

MCRARY, W. (1940). Nitrogen metabolism of the plant embryo. *Bot. Gaz.* **102**, 89–96.

MCWILLIAM, A. A., SMITH, S. M. and STREET, H. E. (1974). The origin and development of embryoids in suspension cultures of carrot (*Daucus carota*). *Ann. Bot. N.S.* **38**, 243–250.

MEEUSE, A. D. and OTT, E. C. J. (1962). The occurrence of chlorophyll in *Nelumbo* seeds. *Acta Bot. Neerl.* **11**, 228.

MEHRA, A. and MEHRA, P. N. (1972). Differentiation in callus cultures of *Mesembryanthemum floribundum*. *Phytomorphology* **22**, 171–176.

MEHRA, A. and MEHRA, P. N. (1974). Organogenesis and plantlet formation *in vitro* in almond. *Bot. Gaz.* **135**, 61–73.

MEHRA, P. N. and MEHRA, A. (1971). Morphogenetic studies in *Pterotheca falconeri*. *Phytomorphology* **21**, 174–191.

MEIER, H. (1958). On the structure of cell walls and cell wall mannans from ivory nuts and from dates. *Biochim. Biophys. Acta* **28**, 229–240.

MEISTER, A. (1965). "The Biochemistry of Amino Acids" (second edition), Vol. I. Academic Press, New York and London.

MELCHERS, G. (1971). Haploide Pflanzen aus Pollen. *Umschau* **7**, 223–229.

MELCHERS, G. (1972). Haploid higher plants for plant breeding. *Zeitschr. Pflanzenzüchtg* **67**, 19–32.

MELCHERS, G. and BERGMANN, L. (1959). Untersuchungen an Kulturen von haploiden Geweben von *Antirrhinum majus*. *Ber. Deut. Bot. Ges.* **71**, 459–473.

MELCHERS, G. and LABIB, G. (1970). Die Bedeutung haploider höherer

Pflanzen für Physiologie und Pflanzenzüchtung. *Ber. Deut. Bot. Ges.* **83**, 129–150.

MELETTI, P. (1960). Allevamento su farina di embrioni isolati di graminaceae. *Genet. Agraria* **10**, 366–372.

MELETTI, P. and D'AMATO, F. (1961). The embryo transplantation technique in the study of embryo-endosperm relations in irradiated seeds. *In* "Effects of Ionizing Radiations on Seeds". Proceedings of the Symposium on the Effects of Ionizing Radiations on Seeds and their Significance for Crop Improvement, pp. 47–55. International Atomic Energy Agency, Vienna.

MELETTI, P., FLORIS, C. and D'AMATO, F. (1968). The mutagenic effect of the irradiated endosperm in water-soaked seeds of *durum* wheat. *Mutation Res.* **6**, 169–172.

MENKE, W. and FRICKE, B. (1964). Beobachtungen über die Entwicklung der Archegonien von *Dryopteris filixmas*. *Zeitschr. Naturforsch.* **19b**, 520–524.

MERICLE, L. W., EUNUS, A. M. and MERICLE, R. P. (1955). Effects of maleic hydrazide on embryologic development. I. *Avena sativa. Bot. Gaz.* **117**, 142–147.

MERICLE, L. W. and MERICLE, R. P. (1957). Irradiation of developing plant embryos. I. Effects of external irradiation (X-rays) on barley embryogeny, germination, and subsequent seedling development. *Amer. Jour. Bot.* **44**, 747–756.

MERICLE, L. W. and MERICLE, R. P. (1961). Radiosensitivity of the developing plant embryo. *In* "Fundamental Aspects of Radiosensitivity". *Brookhaven Symp. Biol.* **14**, 262–286.

MERICLE, L. W. and MERICLE, R. P. (1962). Mutation induction by proembryo irradiation. *Radiation Bot.* **1**, 195–202.

MERICLE, L. W. and MERICLE, R. P. (1969). Cytological consequences of proembryo irradiation. *Radiation Bot.* **9**, 269–282.

MERRY, J. (1942). Studies in the embryo of *Hordeum sativum*—II. The growth of the embryo in culture. *Bull. Torrey Bot. Club* **69**, 360–372.

MESTRE, J.-C., BA, L.-T. and GUIGNARD, J.-L. (1972). Perte du pouvoir embryogène d'une souche de cal de carotte à la suite d'un bref passage en culture sur un milieu défini. Recherche de la cause de ce phénomène: Rôle du 2,4 D. *Compt. Rend. Acad. Sci. Paris* **274**, 54–57.

MESTRE, J.-C., BA, L.-T. and GUIGNARD, J.-L. (1973). L'induction expérimentale de la perte du pouvoir embryogène d'une souche de cal de carotte. *Soc. Bot. Fr. Mémoires, Colloq. Morphologie*, 215–222.

MEYER, J. R. (1945a). Metodo papa proteção das orquideas quando netiradas dos recipients de cultura para adatação ao meio exterior. *O'Biologico* **11**, 48–51.

MEYER, J. R. (1945b). Ação de una heteroauxina sobre o crescimento de "seedlings" de orquideas. *O'Biologico* **11**, 151–153.

MEYER, J. R. (1945c). The use of tomato juice in the preparation of a medium for the germination of orchid seeds. *Amer. Orch. Soc. Bull.* **14**, 99–101.

MEYER, J. R. and PELLOUX, A. (1948). Ação estimulante de dois hormonios de crescimento vegetal sobre germinação de sementes e desenvolvimento de seedling de orquideas em meio assimbiotico. *O'Biologico* **14**, 151–161.

References

MEYER-TEUTER, H. and REINERT, J. (1973). Correlation between rate of cell division and loss of embryogenesis in longterm tissue cultures. *Protoplasma* **78**, 273–283.

MIAH, A. J. and BRUNORI, A. (1970). Chromosomal aberrations induced in meristematic regions of the embryos in X-irradiated dry seeds of *Vicia faba*. *Radiation Bot.* **10**, 87–94.

MICHAELIS, P. (1925). Zur Cytologie und Embryoentwicklung von *Epilobium*. *Ber. Deut. Bot. Ges.* **43**, 61–67.

MIETTINEN, J. K. and WARIS, H. (1958). A chemical study of the neomorphosis induced by glycine in *Oenanthe aquatica*. *Physiol. Plantarum* **11**, 193–199.

MIFLIN, B. J. (1969a). A technique for the sterile culture of germinating barley embryos. *Jour. Exptl Bot.* **20**, 808–809.

MIFLIN, B. J. (1969b). The inhibitory effects of various amino acids on the growth of barley seedlings. *Jour. Exptl Bot.* **20**, 810–819.

MIKULSKA, E. and RODKIEWICZ, B. (1967a). Ultrastructure of the maturing embryo sac of *Lilium regale*. *Acta Soc. Bot. Polon.* **36**, 555–566.

MIKULSKA, E. and RODKIEWICZ, B. (1967b). Fine structure of four-nucleate stages and the central cell of *Lilium regale* embryo sac. *Flora* **157**, 365–372.

MILLER, C. O. (1961). A kinetin-like compound in maize. *Proc. Natl Acad. Sci. U.S.A.* **47**, 170–174.

MILLER, E. V. (1958). The accumulation of carbohydrates by seeds and fruits. *In* "Handbuch der Pflanzenphysiologie" (W. Ruhland, ed.), Vol. 6, pp. 871–880. Springer-Verlag, Berlin.

MILLER, H. A. and WETMORE, R. H. (1945). Studies in the developmental anatomy of *Phlox drummondii* Hook. I. The embryo. *Amer. Jour. Bot.* **32**, 588–599.

MILLERD, A. (1975). Biochemistry of legume seed proteins. *Ann. Rev. Plant Physiol.* **26**, 53–72.

MILLERD, A. and WHITFELD, P. R. (1973). Deoxyribonucleic acid and ribonucleic acid synthesis during the cell expansion phase of cotyledon development in *Vicia faba* L. *Plant Physiol.* **51**, 1005–1010.

MILLERD, A., SIMON, M. and STERN, H. (1971). Legumin synthesis in developing cotyledons of *Vicia faba* L. *Plant Physiol.* **48**, 419–425.

MIRÀMON, A. (1957). The requirement of biotin for the synthesis of fats from acetate by developing embryos of flax, *Linum usitatissimum*. *Bios* **28**, 181–186.

MITCHELL, R. C. and VILLIERS, T. A. (1972). Polysome formation in light-controlled dormancy. *Plant Physiol.* **50**, 671–674.

MITRA, G. C. and CHATURVEDI, H. C. (1970). Fruiting plants from *in vitro* grown leaf tissue of *Rauvolfia serpentina* Benth. *Curr. Sci.* **39**, 128–129.

MITRA, G. C. and CHATURVEDI, H. C. (1972). Embryoids and complete plants from unpollinated ovaries and from ovules of *in vitro*-grown emasculated flower buds from *Citrus* spp. *Bull. Torrey Bot. Club* **99**, 184–189.

MITRA, G. C. and KAUL, K. N. (1964). *In vitro* culture of root and stem callus of *Rauvolfia serpentina* Benth. for reserpine. *Indian Jour. Exptl Biol.* **2**, 49–51.

MITRA, J. and DATTA, C. (1951a). Role of ascorbic acid on the growth of excised embryos of jute. *Sci. & Cult.* **16**, 428–430.

MITRA, J. and DATTA, C. (1951b). Nutritional malformations in excised embryos of jute. *Sci. & Cult.* **16**, 531.

MITSUDA, H., MURAKAMI, K., KUSANO, T. and YASUMOTO, K. (1969). Fine structure of protein bodies isolated from rice endosperm. *Arch. Biochem. Biophys.* **130**, 678–680.

MITSUDA, H., YASUMOTO, K., MURAKAMI, K., KUSANO, T. and KISHIDA, H. (1967). Studies on the proteinaceous subcellular particles in rice endosperm: electron-microscopy and isolation. *Agric. Biol. Chem.* **31**, 293–300.

MIZUSHIMA, U., MURAKAMI, K. and HOZYO, Y. (1955). Influence of anomalous environmental conditions upon embryogenesis in the rice plant, *Oryza sativa* L. I. Histochemical observation of normal embryogenesis. *Tohoku Jour. Agric. Res.* **6**, 1–19.

MIZUSHIMA, U., MURAKAMI, K. and HOZYO, Y. (1956). Influence of anamalous environmental conditions upon embryogenesis in the rice plant, *Oryza sativa* L. II. Abnormal development induced by chemical substances. *Tohoku Jour. Agric. Res.* **6**, 179–206.

MODRZEJEWSKI, R., GUZOWSKA, I. and ZENKTELER, M. (1970). Regeneracja fragmentów dojrałego zarodka *Cuscuta lupuliformis* Krock. w hodowli *in vitro*. *Poznan. Towar. Przy. Nauk, Prace Komisji Biologicz.* **33**, 39–53.

MOGENSEN, H. L. (1972). Fine structure and composition of the egg apparatus before and after fertilization in *Quercus gambelii*: The functional ovule. *Amer. Jour. Bot.* **59**, 931–941.

MOGENSEN, H. L. (1973). Some histochemical, ultrastructural, and nutritional aspects of the ovule of *Quercus gambelii*. *Amer. Jour. Bot.* **60**, 48–54.

MOHAN RAM, H. Y. and KAMINI, I. (1964). Embryology and fruit development in *Withania somnifera* Dunal. *Phytomorphology* **14**, 574–587.

MOHAN RAM, H. Y. and SATSANGI, A. (1963). Induction of cell division in the mature endosperm of *Ricinus communis* during germination. *Curr. Sci.* **32**, 28–29.

MOHAN RAM, H. Y. and WADHI, M. (1964). Endosperm in Acanthaceae. *Phytomorphology* **14**, 388–413.

MOHAN RAM, H. Y. and WADHI, M. (1965). Culture of excised leaves and leaf explants of *Kalanchoe pinnata* Pers. *In* "Tissue Culture" (C. V. Ramakrishnan, ed.), pp. 274–282. Dr. W. Junk, The Hague.

MOIR, W. W. G. (1957). Orchids and gibberellin compounds. *Bull. Pacific Orch. Soc. Hawaii* **14**, 92–96.

MOLLIARD, M. (1921). Sur le développement des plantules fragmentées. *Compt. Rend. Soc. Biol.* **84**, 770–772.

MONIN, J. (1959). Un aspect de l'action de la gibbérelline sur la dormance embryonnaire: Cas des embryons d'*Evonymus europaeus*. *84ᵉ Congres des Soc. Savantes*, 473–480.

MONIN, J. (1964). Étude sur la dormance des graines de *Fraxinus pennsylvanica* var. *subintergerrima* (Vahl) Fern. *Compt. Rend. Séan. Soc. Biol.* **158**, 2283–2286.

MONIN, J. (1966). Modalités d'élimination de la dormance des graines d'*Evonymus europaeus* L. au cours de leur stratification. *Compt. Rend. Séan. Soc. Biol.* **160**, 2262–2264.

MONIN, J. (1967a). Étude sur l'evolution de la teneur en gibbérellines des embryos d'*Evonymus europaeus* L. au cours de la stratification des graines. *Bull. Soc. Fr. Physiol. Végét.* **13**, 171–177.

MONIN, J. (1967b). Action de divers agents sur la dormance des embryons d'*Evonymus europaeus* L. isolés ou entourés de leur albumen. *Compt. Rend. Séan. Soc. Biol.* **161**, 576–579.

MONIN, J. (1967c). Action de différents agents de levée de dormance au cours de la stratification des graines d'*Evonymus europaeus* L. *Compt. Rend. Acad. Sci. Paris* **264**, 300–302.

MONIN, J. (1967d). Isolement d'un inhibiteur chez les embryons dormants d'*Evonymus europaeus* L. *Compt. Rend. Acad. Sci. Paris* **264**, 1997–1999.

MONIN, J. (1967e). Étude d'un inhibiteur existant chez les embryons dormants d'*Evonymus europaeus* L. *Compt. Rend. Acad. Sci. Paris* **264**, 2367–2370.

MONIN, J. (1967f). Action de l'acide *trans*-para-coumarique sur la dormance des embryons d'*Evonymus europaeus* L. *Compt. Rend. Acad. Sci. Paris* **264**, 2999–3001.

MONIN, J. (1967g). Recherches d'auxines chez les embryons d'*Evonymus europaeus* L. *Compt. Rend. Acad. Sci. Paris* **265**, 329–332.

MONIN, J. (1968). Étude de la teneur en substances de croissance des embryons d'*Evonymus europaeus* L. en rapport avec leur dormance. *Bull. Soc. Fr. Physiol. Végét.* **14**, 25–29.

MONNIER, M. (1968a). Comparison du développement des embryons immatures de *Capsella bursa-pastoris*, *in vitro* et *in situ*. *Bull. Soc. Bot. Fr.* **115**, 15–29.

MONNIER, M. (1968b). Action de l'acide 2,4-dichlorophénoxyacétique sur l'ovaire fécondé et sur l'embryon du *Nicotiana tabacum*. *Rev. Gén. Bot.* **75**, 319–326.

MONNIER, M. (1970). Croissance et survie des embryons de *Capsella bursa-pastoris* cultivés *in vitro* dans diverses solutions minérales. *Rev. Gén. Bot.* **77**, 73–83.

MONNIER, M. (1973). Croissance et développement des embryons globulaires de *Capsella bursa-pastoris* cultivés *in vitro* dans un milieu à base d'une nouvelle solution minérale. *Soc. Bot. Fr. Mémoires, Colloq. Morphologie*, 179–194.

MORGAN, D. T., JR., and RAPPLEYE, R. D. (1951). Polyembryony in maize and lily (*Lilium regale*) following X-irradiation of the pollen. *Jour. Hered.* **42**, 90–93.

MORO, M. S., POMERANZ, Y. and SHELLENBERGER, J. A. (1963). The effects of gibberellic acid on alpha-amylase in wheat endosperms. *Phyton* **20**, 59–64.

MORRISON, J. W., HANNAH, A. E., LOISELLE, R. and SYMKO, S. (1959). Cytogenetic studies in the genus *Hordeum*. II. Interspecific and intergeneric crosses. *Canad. Jour. Plant Sci.* **39**, 375–383.

MORY, Y. Y., CHEN, D. and SARID, S. (1972). Onset of deoxyribonucleic acid synthesis in germinating wheat embryos. *Plant Physiol.* **49**, 20–23.

MORY, Y. Y., CHEN, D. and SARID, S. (1975). *De novo* biosynthesis of deoxyribonucleic acid polymerase during wheat embryo germination. *Plant Physiol.* **55**, 437–442.

MOURAS, A. and LUTZ, A. (1973). Sur les variations du comportement organogène des cultures de tissus de carotte sauvage produisant des pseudo-embryons. *Compt. Rend. Acad. Sci. Paris* **277**, 1151–1153.

MUKHERJEE, R. K., BHANJA, A. and SIRCAR, S. M. (1966). Growth substances separated from the fruits of *Cassia fistula*. *Physiol. Plantarum* **19**, 448–458.

MUKHERJI, D. K. (1951). Embryo culture as an aid to seed testing. *Proc. Natl Inst. Sci. India* **17B**, 253–259.

MÜNTZING, A. (1930). Über Chromosomenvermehrung in *Galeopsis*-Kreuzungen und ihre phylogenetische Bedeutung. *Hereditas* **14**, 153–172.

MURASHIGE, T. (1974). Plant propagation through tissue cultures. *Ann. Rev. Plant Physiol.* **25**, 135–166.

MURGAI, P. (1959). *In vitro* culture of the inflorescences, flowers and ovaries of an apomict, *Aerva tomentosa* Forsk. *Nature* **184**, 72–73.

MURNEEK, A. E. (1954). The embryo and endosperm in relation to fruit development, with special reference to the apple, *Malus sylvestris*. *Proc. Amer. Soc. Hort. Sci.* **64**, 573–582.

MUZIK, T. J. (1954). Development of fruit, seed, embryo, and seedling of *Hevea brasiliensis*. *Amer. Jour. Bot.* **41**, 39–43.

MUZIK, T. J. (1956). Studies on the development of the embryo and seed of *Hevea brasiliensis* in culture. *Lloydia* **19**, 86–91.

NAG, K. K. and JOHRI, B. M. (1969). Organogenesis and chromosomal constitution in embryo callus of *Nuytsia floribunda*. *Phytomorphology* **19**, 405–408.

NAG, K. K. and JOHRI, B. M. (1971). Morphogenic studies on endosperm of some parasitic angiosperms. *Phytomorphology* **21**, 202–218.

NAG, K. K. and STREET, H. E. (1973). Carrot embryogenesis from frozen cultured cells. *Nature* **245**, 270–272.

NAGL, W. (1962a). Über Endopolyploidie, Restitutionskernbildung und Kernstrukturen im Suspensor von Angiospermen und einer Gymnosperme. *Öster. Bot. Zeitschr.* **109**, 431–494.

NAGL, W. (1962b). 4096–Ploidie und "Riesenchromosomen" im Suspensor von *Phaseolus coccineus*. *Naturwiss.* **49**, 261–262.

NAGL, W. (1970). Temperature-dependent functional structures in the polytene chromosomes of *Phaseolus*, with special reference to the nucleolus organizers. *Jour. Cell Sci.* **6**, 87–107.

NAGL, W. (1973). The angiosperm suspensor and the mammalian trophoblast: Organs with similar cell structure and function? *Soc. Bot. Fr. Mémoires, Colloq. Morphologie*, 289–302.

NAKAJIMA, T. (1962). Physiological studies of seed development, especially embryonic growth and endosperm development. *Bull. Univ. Osaka Pref. Ser. B* **13**, 13–48.

NAKAJIMA, T. and MORISHIMA, H. (1958). Studies on embryo culture in plants. II. Embryo culture of interspecific hybrids in *Oryza*. *Jap. Jour. Breed.* **8**, 105–110.

NAKAJIMA, T. and YAMAGUCHI, T. (1967). On the embryogenesis observed in tissue culture of carrot, *Daucus carota* L. *Bull. Univ. Osaka Pref. Ser. B* **19**, 43–49.

NAKAMURA, A. and ITAGAKI, R. (1973). Anther culture in *Nicotiana* and the characteristics of the haploid plants. *Jap. Jour. Breed.* **23**, 71–78.

NAKATA, K. (1971). Competition among pollen grains for haploid tobacco plant formation by anther culture. I. Analysis with leaf color character. *Jap. Jour. Breed.* **21**, 29–34.

NAKATA, K. and KURIHARA, T. (1972). Competition among pollen grains for haploid tobacco plant formation by anther culture. II. Analysis with resistance to tobacco mosaic virus (TMV) and wildfire diseases, leaf color, and leafbase shape characters. *Jap. Jour. Breed.* **22**, 92–98.

NAKATA, K. and TANAKA, M. (1968). Differentiation of embryoids from developing germ cells in anther culture of tobacco. *Jap. Jour. Genet.* **43**, 65–71.

NAKAZAWA, S. (1950). Origin of polarity in the eggs⋅ of *Sargassum confusum*. *Sci. Rep. Tohoku Univ. 4th Ser. (Biol.)* **18**, 424–433.

NAKAZAWA, S. (1951). Invalid stratification of the egg polarity in *Coccophora* and *Sargassum*. *Sci. Rep. Tohoku Univ. 4th Ser. (Biol.)* **19**, 73–78.

NARAYANA, R. (1954). Contribution to the embryology of *Dendrophthoe* Mart. *Phytomorphology* **4**, 173–179.

NARAYANASWAMI, S. (1956). Plant endosperm and its culture *in vitro*. *Sci. & Cult.* **22**, 132–136.

NARAYANASWAMI, S. (1959a). Experimental studies on growth of excised grass embryos *in vitro*—I. Overgrowth of the scutellum of *Pennisetum* embryos. *Phytomorphology* **9**, 358–367.

NARAYANASWAMI, S. (1959b). Experimental studies on growth of excised grass embryos *in vitro*. II. Effect of maleic hydrazide on embryo growth. *Bull. Torrey Bot. Club* **86**, 248–258.

NARAYANASWAMI, S. (1962). Morphological variations in growth and differentiation of embryos *in vitro*. *In* "Proceedings of the Summer School of Botany" (P. Maheshwari, B. M. Johri and I. K. Vasil, eds), pp. 231–239. Ministry of Scientific Research and Cultural Affairs, New Delhi.

NARAYANASWAMI, S. (1963). Studies on growth of excised grass embryos in culture. *In* "Plant Tissue and Organ Culture—A Symposium" (P. Maheshwari and N. S. Rangaswamy, eds), pp. 302–313. International Society of Plant Morphologists, Delhi.

NARAYANASWAMI, S. and NORSTOG, K. (1964). Plant embryo culture. *Bot. Rev.* **30**, 587–628.

NARAYANASWAMI, S. and RANGASWAMY, N. S. (1959). Plant embryo culture. *Memoirs, Indian Bot. Soc.* **2**, 27–29.

NARAYANASWAMY, S. and CHANDY, L. P. (1971). *In vitro* induction of haploid, diploid and triploid androgenic embryoids and plantlets in *Datura metel* L. *Ann. Bot. N.S.* **35**, 535–542.

NARAYANASWAMY, S. and GEORGE, L. (*née* CHANDY) (1972). Morphogenesis of belladonna (*Atropa belladonna* L.) plantlets from pollen in culture. *Indian Jour. Exptl Biol.* **10**, 382–384.

NASON, A. (1950). The distribution and biosynthesis of niacin in germinating corn. *Amer. Jour. Bot.* **37**, 612–623.

NATARAJA, K. (1971a). Formation of shoot buds and callus in seed cultures of *Euphorbia. In* "Morphogenesis in Plant Cell, Tissue and Organ Cultures— International Symposium", pp. 66–67 (abstract). University of Delhi.

NATARAJA, K. (1971b). Morphogenic variations in callus cultures derived from floral buds and anthers of some members of Ranunculaceae. *Phytomorphology* **21**, 290–296.

NATARAJA, K. and KONAR, R. N. (1970). Induction of embryoids in reproductive and vegetative tissues of *Ranunculus sceleratus* L. *in vitro. Acta Bot. Neerl.* **19**, 707–716.

NATARAJAN, A. T. and SWAMINATHAN, M. S. (1958). Indirect effects of radiation and chromosome breakage. *Indian Jour. Genet. Plant Breed.* **18**, 220–223.

NAYLOR, J. M. (1966). Dormancy studies in seed of *Avena fatua.* 5. On the response of aleurone cells to gibberellic acid. *Canad. Jour. Bot.* **44**, 19–32.

NAYLOR, J. M. and SIMPSON, G. M. (1961a). Dormancy studies in seed of *Avena fatua.* 2. A gibberellin-sensitive inhibitory mechanism in the embryo. *Canad. Jour. Bot.* **39**, 281–295.

NAYLOR, J. M. and SIMPSON, G. M. (1961b). Bioassay of gibberellic acid using excised embryos of *Avena fatua* L. *Nature* **192**, 679–680.

NEAL, G. E. and BEEVERS, H. (1960). Pyruvate utilization in castor-bean endosperm and other tissues. *Biochem. Jour.* **74**, 409–416.

NÉTIEN, G. and RAYNAUD, J. (1972). Formation d'embryons dans la culture *in vitro* de tissus de *Conium maculatum* L. *Bull. Mensl Soc. Linné. Lyon* **41**, 49–51.

NÉTIEN, G., BEAUCHÊNE, G. and MENTZER, C. (1951). Influence du "lait de mais" sur la croissance des tissus de carotte *in vitro. Compt. Rend. Acad. Sci. Paris* **233**, 92–93.

NETOLITZKY, F. (1926). Anatomie der Angiospermen-Samen. *In* "Handbuch der Pflanzenanatomie" (K. Linsbauer, ed.), Vol. 10. Borntraeger, Berlin.

NEUMANN, K.-H. and DEGARCIA, H. (1974). Uber den Einfluss der Saccharosekonzentration auf die Entwicklung von Embryonen aus Zellsuspensionen von *Daucus carota* L. *Zeitschr. Pflanzenphysiol.* **74**, 85–90.

NEWCOMB, M. and CLELAND, R. E. (1946). Aseptic cultivation of excised plant embryos. *Science* **104**, 329–330.

NEWCOMB, W. (1973a). The development of the embryo sac of sunflower *Helianthus annuus* before fertilization. *Canad. Jour. Bot.* **51**, 863–878.

NEWCOMB, W. (1973b). The development of the embryo sac of sunflower *Helianthus annuus* after fertilization. *Canad. Jour. Bot.* **51**, 879–890.

NEWCOMB, W. and FOWKE, L. C. (1973). The fine structure of the change from the free nuclear to cellular condition in the endosperm of chickweed *Stellaria media. Bot. Gaz.* **134**, 236–241.

NEWCOMB, W. and FOWKE, L. C. (1974). *Stellaria media* embryogenesis: The development and ultrastructure of the suspensor. *Canad. Jour. Bot.* **52**, 607–614.

NEWCOMB, W. and STEEVES, T. A. (1971). *Helianthus annuus* embryogenesis: Embryo sac wall projections before and after fertilization. *Bot. Gaz.* **132**, 367–371.

NEWCOMB, W. and WETHERELL, D. F. (1970). The effects of 2,4,6-trichlorophenoxyacetic acid on embryogenesis in wild carrot tissue cultures. *Bot. Gaz.* **131**, 242–245.

NICHOLLS, P. B. (1967). The isolation of indole-3-acetyl-2-0-myo-inositol from *Zea mays. Planta* **72**, 258–264.

NICKELL, L. G. (1951). Embryo culture of weeping crabapple. *Proc. Amer. Soc. Hort. Sci.* **57**, 401–405.

NIIMOTO, D. H. and SAGAWA, Y. (1961). Ovule development in *Dendrobium. Amer. Orch. Soc. Bull.* **30**, 813–819.

NIIZEKI, H. and OONO, K. (1968). Induction of haploid rice plant from anther culture. *Proc. Japan Acad.* **44**, 554–557.

NIIZEKI, H. and OONO, K. (1971). Rice plants obtained by anther culture. *In* "Les Cultures de Tissus de Plantes", pp. 251–257. Colloq. Internationaux du Centre National de la Recherche Scientifique, Paris, No. 193.

NIIZEKI, M. (1973a). Studies on plant cell and tissue culture. I. Production of haploid plants of tobacco by anther culture. *Jour. Fac. Agric. Hokkaido Univ.* **57**, 161–178.

NIIZEKI, M. (1973b). Studies on plant cell and tissue culture. II. Effect of different kinds of media on the variation of chromosome numbers in tobacco callus. *Jour. Fac. Agric. Hokkaido Univ.* **57**, 179–191.

NIIZEKI, M. (1974a). Studies on plant cell and tissue culture. IV. Effect of para-fluorophenylalanine on haploid and diploid cells of tobacco plant *in vitro. Jour. Fac. Agric. Hokkaido Univ.* **57**, 349–356.

NIIZEKI, M. (1974b). Studies on plant cell and tissue culture. V. Effect of different kinds of media on the variation of chromosome numbers in tobacco callus and regenerated plant. *Jour. Fac. Agric. Hokkaido Univ.* **57**, 357–367.

NIIZEKI, M. and GRANT, W. F. (1971). Callus, plantlet formation, and polyploidy from cultured anthers of *Lotus* and *Nicotiana. Canad. Jour. Bot.* **49**, 2041–2051.

NIIZEKI, M. and KITA, F. (1974). Studies on plant cell and tissue culture. III. *In vitro* induction of callus from anther culture of forage crops. *Jour. Fac. Agric. Hokkaido Univ.* **57**, 293–300.

NILES, J. J. (1951). Hybridization methods with paddy. *Trop. Agricult Ceylon* **107**, 25–29.

NISHI, S., KAWATA, J. and TODA, M. (1959). On the breeding of interspecific hybrids between two genomes, "c" and "a", of *Brassica* through the application of embryo culture techniques. *Jap. Jour. Breed.* **8**, 215–222.

NISHI, T. and MITSUOKA, S. (1969). Occurrence of various ploidy plants from anther and ovary culture of rice plant. *Jap. Jour. Genet.* **44**, 341–346.

NISHIYAMA, I. and UEMATSU, S. (1967). Radiobiological studies in plants— XII. Embryogenesis following X-irradiation of pollen in *Lycopersicum pimpinellifolium. Radiation Bot.* **7**, 481–489.

NITSCH, C. (1974). La culture de pollen isolé sur milieu synthétique. *Compt. Rend. Acad. Sci. Paris* **278**, 1031–1034.

NITSCH, C. and NORREEL, B. (1973). Effet d'un choc thermique sur le pouvoir embryogène du pollen de *Datura innoxia* cultivé dans l'anthère ou isolé de l'anthère. *Compt. Rend. Acad. Sci. Paris* **275**, 303–306.

Nitsch, J. P. (1949). Culture of fruits *in vitro*. *Science* **110**, 499.

Nitsch, J. P. (1951). Growth and development *in vitro* of excised ovaries. *Amer. Jour. Bot.* **38**, 566–577.

Nitsch, J. P. (1953). The physiology of fruit growth. *Ann. Rev. Plant Physiol.* **4**, 199–236.

Nitsch, J. P. (1958). Présence de gibbérellines dans l'albumen immature du pommier. *Bull. Soc. Bot. Fr.* **105**, 479–482.

Nitsch, J. P. (1963). The *in vitro* culture of flowers and fruits. *In* "Plant Tissue and Organ Culture—A Symposium" (P. Maheshwari and N. S. Rangaswamy, eds), pp. 198–204. International Society of Plant Morphologists, Delhi.

Nitsch, J. P. (1969). Experimental androgenesis in *Nicotiana*. *Phytomorphology* **19**, 389–404.

Nitsch, J. P. (1970). Embryogénèse expérimentale. I. Production d'embryons à partir de grains de pollen. *Bull. Soc. Bot. Fr. Mémoires* **117**, 19–29.

Nitsch, J. P. (1971). The production of haploid embryos from pollen grains. *In* "Pollen: Development and Physiology" (J. Heslop-Harrison, ed.), pp. 234–236 (abstract). Butterworths, London.

Nitsch, J. P. (1972). Haploid plants from pollen. *Zeitschr. Pflanzenzüchtg* **67**, 3–18.

Nitsch, J. P. and Nitsch, C. (1969). Haploid plants from pollen grains. *Science* **163**, 85–87.

Nitsch, J. P. and Nitsch, C. (1970). Obtention de plantes haploïdes à partir de pollen. *Bull. Soc. Bot. Fr.* **117**, 339–359.

Nitsch, J. P., Nitsch, C. and Hamon, S. (1968). Réalisation expérimentale de l'"androgenèse" chez divers *Nicotiana*. *Compt. Rend. Séanc. Soc. Biol.* **162**, 369–372.

Nitsch, J. P., Nitsch, C. and Hamon, S. (1969a). Production de *Nicotiana* diploïdes à partir de cals haploïdes cultivés *in vitro*. *Compt. Rend. Acad. Sci. Paris* **269**, 1275–1278.

Nitsch, J. P., Nitsch, C. and Péreau-Leroy, P. (1969b). Obtention de mutants à partir de *Nicotiana* haploïdes issus de grains de pollen. *Compt. Rend. Acad. Sci. Paris* **269**, 1650–1652.

Nitsch, J. P., Pratt, C., Nitsch, C. and Shaulis, N. J. (1960). Natural growth substances in Concord and Concord seedless grapes in relation to berry development. *Amer. Jour. Bot.* **47**, 566–576.

Nitzsche, W. (1970). Herstellung haploider Pflanzen aus *Festuca-Lolium*-Bastarden. *Naturwiss.* **57**, 199–200.

Noggle, G. R. and Wynd, F. L. (1943). Effects of vitamins on germination and growth of orchids. *Bot. Gaz.* **104**, 455–459.

Norreel, B. (1970). Étude cytologique de l'androgénèse expérimentale chez *Nicotiana tabacum* et *Datura innoxia*. *Bull. Soc. Bot. Fr.* **117**, 461–478.

Norreel, B. (1972). Étude comparative de la répartition des acides ribonucléiques au cours de l'embryogenèse zygotique et de l'embryogenèse androgénétique chez le *Nicotiana tabacum* L. *Compt. Rend. Acad. Sci. Paris* **275**, 1219–1222.

Norreel, B. (1973a). Cultures de tissus végétaux et embryogenèse non zygotique. *Soc. Bot. Fr. Mémoires, Colloq. Morphologie*, 71–98.

Norreel, B. (1973b). Étude comparative de l'évolution de la surface cellulaire et du nombre de cellules au cours des embryogenèses androgenétique et zygotique. *Compt. Rend. Acad. Sci. Paris* **276**, 2657–2660.

Norreel, B. and Nitsch, J. P. (1968). La formation d'"embryos végétatifs" chez *Daucus carota* L. *Bull. Soc. Bot. Fr.* **115**, 501–514.

Norreel, B. and Nitsch, J. P. (1970). Embryologie expérimentale. II. Production d'embryons à partir de tissus végétatifs. *Bull. Soc. Bot. Fr. Mémoires* **117**, 30–39.

Norreel, B. and Rao, P. S. (1974). Étude de l'action des rayons γ sur l'embryogenèse somatique chez le *Daucus carota* L. Effet sur la croissance des cals et l'initiation des embryons *in vitro*. *Compt. Rend. Acad. Sci. Paris* **279**, 751–754.

Norstog, K. J. (1955). Responses of the oat coleorhiza to various treatments in culture. *Ohio Jour. Sci.* **55**, 340–342.

Norstog, K. J. (1956a). The growth of barley embryos on coconut milk media. *Bull. Torrey Bot. Club* **83**, 27–29.

Norstog, K. J. (1956b). Growth of rye-grass endosperm *in vitro Bot. Gaz.* **117**, 253–259.

Norstog, K. (1961). The growth and differentiation of cultured barley embryos. *Amer. Jour. Bot.* **48**, 876–884.

Norstog, K. (1965a). Development of cultured barley embryos. I. Growth of 0·1–0·4-mm embryos. *Amer. Jour. Bot.* **52**, 538–546.

Norstog, K. (1965b). Induction of apogamy in megagametophytes of *Zamia integrifolia*. *Amer. Jour. Bot.* **52**, 993–999.

Norstog, K. (1967). Studies on the survival of very small barley embryos in culture. *Bull. Torrey Bot. Club* **94**, 223–229.

Norstog, K. (1969a). Morphology of coleoptile and scutellum in relation to tissue culture responses. *Phytomorphology* **19**, 235–241.

Norstog, K. (1969b). Coleorhiza responses to certain plant growth regulators in cultures of excised barley embryos. *Trans. Illinois State Acad. Sci.* **62**, 312–315.

Norstog, K. (1970). Induction of embryolike structures by kinetin in cultured barley embryos. *Develop. Biol.* **23**, 665–670.

Norstog, K. (1972a). Early development of the barley embryo: Fine structure. *Amer. Jour. Bot.* **59**, 123–132.

Norstog, K. (1972b). Factors relating to precocious germination in cultured barley embryos. *Phytomorphology* **22**, 134–139.

Norstog, K. (1973). New synthetic medium for the culture of premature barley embryos. *In Vitro* **8**, 307–308.

Norstog, K. (1974). Nucellus during early embryogeny in barley: Fine structure. *Bot. Gaz.* **135**, 97–103.

Norstog, K. and Klein, R. M. (1972). Development of cultured barley embryos. II. Precocious germination and dormancy. *Canad. Jour. Bot.* **50**, 1887–1894.

NORSTOG, K. and RHAMSTINE, E. (1967). Isolation and culture of haploid and diploid cycad tissues. *Phytomorphology* **17**, 374–381.

NORSTOG, K. and SMITH, J. (1963). Culture of small barley embryos on defined media. *Science* **142**, 1655–1656.

NÖTH, M. H. and ABEL, W. O. (1971). Zur Entwicklung haploider Pflanzen aus unreifen Mikrosporen verschiedener *Nicotiana*-Arten. *Zeitschr. Pflanzenzüchtg* **65**, 277–284.

NOUGARÈDE, A. (1963a). Premières observations sur l'infrastructure et sur l'évolution des cellules des jeunes ébauches foliaires embryonnaires du *Tropaeolum majus* L. : Cytologie de la déshydratation de maturation. *Compt. Rend. Acad. Sci. Paris* **257**, 1335–1338.

NOUGARÈDE, A. (1963b). Premières observations sur l'infrastructure et sur l'évolution des cellules des jeunes ébauches foliaires embryonnaires du *Tropaeolum majus* L. Cytologie de l'hydratation germinative et des premières étapes de la germination. *Compt. Rend. Acad. Sci. Paris* **257**, 1495–1497.

NOVÁK, F. J. (1974). Induction of haploid callus in anther cultures of *Capsicum* sp. *Zeitschr. Pflanzenzüchtg* **72**, 46–54.

NUCHOWICZ, A. (1965). Recherches sur la culture d'embryons et de fragments d'embryons d'*Arachis hypogaea* L. I. Essais orientatifs. *Agricultura, Louvain* **3**, 3–37.

NUTI RONCHI, V. and MARTINI, G. (1962). Germinabilità, sviluppo delle plantule e frequenza di aberrazioni cromosomiche in rapporto all'età del seme nel frumento. *Caryologia* **15**, 293–302.

NUTMAN, P. S. (1939). Studies in vernalisation of cereals. VI. The anatomical and cytological evidence for the formation of growth-promoting substances in the developing grain of rye. *Ann. Bot. N.S.* **3**, 731–757.

OAKS, A. (1965). The synthesis of leucine in maize embryos. *Biochim. Biophys. Acta* **111**, 79–89.

OAKS, A. and BEEVERS, H. (1964). The requirement for organic nitrogen in *Zea mays* embryos. *Plant Physiol.* **39**, 37–43.

OBENDORF, R. L. and MARCUS, A. (1974). Rapid increase in adenosine 5′-triphosphate during early wheat embryo germination. *Plant Physiol.* **53**, 779–781.

OGAWA, Y. (1963a). Changes in the content of gibberellin-like substances in ripening seed and pod of *Lupinus luteus*. *Plant Cell Physiol.* **4**, 85–94.

OGAWA, Y. (1963b). Gibberellin-like substances occurring in the seed of *Pharbitis nil* Chois and their change in contents during the seed development. *Plant Cell Physiol.* **4**, 217–225.

OHTA, Y. and FURUSATO, K. (1957). Embryoculture in *Citrus*. *Seiken Zihô, Rep. Kihara Inst. Biol. Res.* No. 8, 49–54.

OLNEY, H. O. and POLLOCK, B. M. (1960). Studies of rest period. II. Nitrogen and phosphorus changes in embryonic organs of after-ripening cherry seed. *Plant Physiol.* **35**, 970–975.

OOTA, Y. (1958). A study of the relationship between water uptake and respiration of isolated bean germ-axes. *Physiol. Plantarum* **11**, 710–721.

OOTA, Y. and OSAWA, S. (1954). Migration of "storage PNA" from cotyledon into growing organs of bean seed embryo. *Experientia* **10**, 254–256.

Oota, Y., Fujii, R. and Osawa, S. (1953). Changes in chemical constituents during the germination stage of a bean, *Vigna sesquipedalis*. *Jour. Biochem. Japan* **40**, 649–661.

Öpik, H. (1965). Respiration rate, mitochondrial activity and mitochondrial structure in the cotyledons of *Phaseolus vulgaris* L. during germination. *Jour. Exptl Bot.* **16**, 667–682.

Öpik, H. (1968). Development of cotyledon cell structure in ripening *Phaseolus vulgaris* seeds. *Jour. Exptl Bot.* **19**, 64–76.

Ory, R. L. and Henningsen, K. W. (1969). Enzymes associated with protein bodies isolated from ungerminated barley seeds. *Plant Physiol.* **44**, 1488–1498.

Ory, R. L., Yatsu, L. Y. and Kircher, H. W. (1968). Association of lipase activity with the spherosomes of *Ricinus communis*. *Arch. Biochem. Biophys.* **123**, 255–264.

Ouyang, T.-W., Hu, H., Chuang, C.-C. and Tseng, C.-C. (1973). Induction of pollen plants from anthers of *Triticum aestivum* L. cultured *in vitro*. *Scient. Sinica* **16**, 79–95.

Ozsan, M. and Cameron, J. W. (1963). Artificial culture of small *Citrus* embryos, and evidence against nucellar embryony in highly zygotic varieties. *Proc. Amer. Soc. Hort. Sci.* **82**, 210–216.

Padmanabhan, D. (1964). The embryology of *Avicennia officinalis*—II. Endosperm. *Phytomorphology* **14**, 442–451.

Padmanabhan, D. (1967). Effect of fusaric acid on *in vitro* culture of embryos of *Phaseolus vulgaris* L. *Curr. Sci.* **36**, 214–215.

Paleg, L. G. (1960). Physiological effects of gibberellic acid. I. On carbohydrate metabolism and amylase activity of barley endosperm. *Plant Physiol.* **35**, 293–299.

Paleg, L. G. and Hyde, B. (1964). Physiological effects of gibberellic acid. VII. Electron microscopy of barley aleurone cells. *Plant Physiol.* **39**, 673–680.

Pammenter, N. W., Adamson, J. H. and Berjak, P. (1974). Viability of stored seed: Extension by cathodic protection. *Science* **186**, 1123–1124.

Pandey, K. K. (1973). Theory and practice of induced androgenesis. *New Phytol.* **72**, 1129–1140.

Pant, D. D. and Nautiyal, D. D. (1962). Seed cuticles in some modern cycads. *Curr. Sci.* **31**, 75–76.

Paranjothy, K. and Raghavan, V. (1970). Effects of fluorinated pyrimidines on the growth of excised pea embryos. *Plant Cell Physiol.* **11**, 259–271.

Paris, D. and Duhamet, L. (1953). Action d'un mélange d'acides aminés et de vitamines sur la prolifération des cultures de tissus de crown-gall de scorsonère; comparison avec l'action du lait de coco. *Compt. Rend. Acad. Sci. Paris* **236**, 1690–1692.

Paris, D., Rietsema, J., Satina, S. and Blakeslee, A. F. (1953). Effect of amino acids, especially aspartic and glutamic acid and their amides, on the growth of *Datura stramonium* embryos *in vitro*. *Proc. Natl Acad. Sci. U.S.A.* **39**, 1205–1212.

PARK, W.-M. and CHEN, S. S. C. (1974). Patterns of food utilization by the germinating lettuce seeds. *Plant Physiol.* **53**, 64–66.

PATIL, J. A. (1966). The culture *in vitro* of immature embryos of okra (*Abelmoschus esculentus*). *Indian Jour. Plant Physiol.* **9**, 59–65.

PAUL, A. K. and MUKHERJI, S. (1973). Initiation of active metabolism in the early imbibitional phase of mungbean (*Phaseolus aureus*) seeds. *Biol. Plantarum* **15**, 398–404.

PAUL, K. B., PATEL, C. S. and BISWAS, P. K. (1973). Changes in endogenous growth regulators in loblolly pine seeds during the process of stratification and germination. *Physiol. Plantarum* **28**, 530–534.

PAULSON, R. E. and SRIVASTAVA, L. M. (1968). The fine structure of the embryo of *Lactuca sativa*. I. Dry embryo. *Canad. Jour. Bot.* **46**, 1437–1445.

PAVLISTA, A. D. and HABER, A. H. (1970). Embryo expansion without protrusion in lettuce seeds. *Plant Physiol.* **46**, 636–637.

PEARSON, H. H. W. (1929). "Gnetales". Cambridge University Press.

PECKET, R. C. and SELIM, A. R. A. A. (1965). Embryo-culture in *Lathyrus*. *Jour. Exptl Bot.* **16**, 325–328.

PELLETIER, G. (1973). Les conditions et les premiers stades de l'androgenèse *in vitro* chez *Nicotiana tabacum*. *Soc. Bot. Fr. Mémoires, Colloq. Morphologie*, 261–268.

PELLETIER, G. and DURRAN, V. (1972). Recherche de tissus nourriciers pour la réalisation de l'androgenèse *in vitro* chez *Nicotiana tabacum*. *Compt. Rend. Acad. Sci. Paris* **275**, 35–37.

PELLETIER, G. and ILAMI, M. (1972). Les facteurs de l'androgénèse *in vitro* chez *Nicotiana tabacum*. *Zeitschr. Pflanzenphysiol.* **68**, 97–114.

PELLETIER, G., RAQUIN, C. and SIMON, G. (1972). La culture *in vitro* d'anthères d'asperge (*Asparagus officinalis*). *Compt. Rend. Acad. Sci. Paris* **274**, 848–851.

PERNER, E. (1965). Electronenmikroskopische Untersuchungen an Zellen von Embryonen im Zustand völliger Samenruhe. I. Mitteilung. Die zelluläre Strukturordnung in der Radicula lufttrockner Samen von *Pisum sativum*. *Planta* **65**, 334–357.

PEROTTI, R. (1913). Contributions all'embryologia delle Dianthaceae. *Ann. Bot. Rome* **11**, 371–385.

PETERSON, D. M. and TORREY, J. G. (1968). Amino acid incorporation in developing *Fucus* eggs. *Plant Physiol.* **43**, 941–947.

PETRŮ, E. (1970). Development of embryoids in carrot root callus culture (*Daucus carota* L.). *Biol. Plantarum* **12**, 1–5.

PHATAK, V. G. and AMBEGAOKAR, K. B. (1963). Embryological studies in Acanthaceae. V. Development of embryosac and endosperm in *Blepharis maderaspatensis* (Linn.) Roth. *Proc. Indian Acad. Sci.* **B57**, 88–95.

PHINNEY, B. O., WEST, C. A., RITZEL, M. and NEELY, P. M. (1957). Evidence for "gibberellin-like" substances from flowering plants. *Proc. Natl Acad. Sci. U.S.A.* **43**, 398–404.

PICARD, E. (1973). Influence de modifications dans les corrélations internes sur le devenir du gamétophyte mâle de *Triticum aestivum* L. *in situ* et en culture *in vitro*. *Compt. Rend. Acad. Sci. Paris* **277**, 777–780.

PICARD, E. and DE BUYSER, J. (1973). Obtention de plantules haploïdes de *Triticum aestivum* L. à partir de culture d'anthères *in vitro*. *Compt. Rend. Acad. Sci. Paris* **277**, 1463–1466.

PIECZUR, E. A. (1952). Effect of tissue cultures of maize endosperm on the growth of excised maize embryos. *Nature* **170**, 241–242.

PILET, P.-E. (1961). Culture *in vitro* de tissus de carotte et organogenèse. *Ber. Schweiz. Bot. Ges.* **71**, 189–208.

PINFIELD, N. J. and STOBART, A. K. (1969). Gibberellin-stimulated nucleic acid metabolism in the cotyledons and embryonic axes of *Corylus avellana* (*L.*) seeds. *New Phytol.* **68**, 993–999.

PINFIELD, N. J. and STOBART, A. K. (1972). Hormonal regulation of germination and early seedling development in *Acer pseudoplatanus* (*L.*). *Planta* **104**, 134–145.

PINFIELD, N. J., DAVIES, H. V. and STOBART, A. K. (1974). Embryo dormancy in seeds of *Acer platanoides*. *Physiol. Plantarum* **32**, 268–272.

PINFIELD, N. J., MARTIN, M. H. and STOBART, A. K. (1972) The control of germination in *Stachys alpina* (*L*). *New Phytol.* **71**, 99–104.

PISSAREV, W. E. and VINOGRADOVA, N. M. (1944). Hybrids between wheat and *Elymus*. *Doklady Akad. Sci. SSSR* **45**, 129–132.

PODDUBNAYA-ARNOLDI, V. A. (1959). Study of fertilization and embryogenesis in certain angiosperms using living material. *Amer. Naturl.* **93**, 161–169.

PODDUBNAYA-ARNOLDI, V. A. (1960). Study of fertilization in the living material of some angiosperms. *Phytomorphology* **10**, 185–198.

PODDUBNAYA-ARNOLDI, V. A., ZINGER, N. V. and PETROVSKAYA-BARANOVA, T. P. (1964). A histochemical investigation of the ovules, embryo sacs and seeds in some angiosperms. *In* "Pollen Physiology and Fertilization" (H. F. Linskens, ed.), pp. 3–7. North-Holland, Amsterdam.

POIRIER-HAMON, S., RAO, P. S. and HARADA, H. (1974). Culture of mesophyll protoplasts and stem segments of *Antirrhinum majus* (snapdragon): Growth and organization of embryoids. *Jour. Exptl Bot.* **25**, 752–760.

POLLARD, J. K., SHANTZ, E. M. and STEWARD, F. C. (1961). Hexitols in coconut milk: Their role in nurture of dividing cells. *Plant Physiol.* **36**, 492–501.

POLLOCK, B. M. and OLNEY, H. O. (1959). Studies of the rest period. I. Growth, translocation, and respiratory changes in the embryonic organs of the after-ripening cherry seed. *Plant Physiol.* **34**, 131–142.

POLLOCK, E. G. and JENSEN, W. A. (1964). Cell development during early embryogenesis in *Capsella* and *Gossypium*. *Amer. Jour. Bot.* **51**, 915–921.

POLLOCK, E. G. and JENSEN, W. A. (1967). Ontogeny and cytochemistry of the chalazal proliferating cells of *Capsella bursa-pastoris* (L.) Medic. *New Phytol.* **66**, 413–417.

PONTOVICH, V. É. and SVESHNIKOVA, I. N. (1966). Formation of *Papaver somniferum* L. embryos in ovules cultured *in vitro*. *Fiziol. Rastenii* (English translation) **13**, 91–102.

POWELL, L. E. and PRATT, C. (1964). Kinins in the embryo and endosperm of *Prunus persica*. *Nature* **204**, 602–603.

PRABHUDESAI, V. R. and NARAYANASWAMY, S. (1973). Differentiation of cytokinin-induced shoot buds and embryoids on excised petioles of *Nicotiana tabacum*. *Phytomorphology* **23**, 133–137.

PRABHUDESAI, V. R. and NARAYANASWAMY, S. (1974). Organogenesis in tissue cultures of certain Asclepiads. *Zeitschr. Pflanzenphysiol.* **71**, 181–185.

PRADERA, E. S., FERNANDEZ, E. and CALDERIN, O. (1942). Coconut water. A clinical and experimental study. *Amer. Jour. Diseas. Child.* **64**, 977–995.

PRICE, C. E. and MURRAY, A. W. (1969). Purine metabolism in germinating wheat embryos. *Biochem. Jour.* **115**, 129–133.

PRITCHARD, H. N. (1964a). A cytochemical study of embryo sac development in *Stellaria media*. *Amer. Jour. Bot.* **51**, 371–378.

PRITCHARD, H. N. (1964b). A cytochemical study of embryo development in *Stellaria media*. *Amer. Jour. Bot.* **51**, 472–479.

PRITCHARD, H. N. and BERGSTRESSER, K. A. (1969). The cytochemistry of some enzyme activities in *Stellaria media* embryos. *Experientia* **25**, 1116–1117.

PROKOF'EV, A. A. and RODINOVA, M. A. (1966). Change of sunflower embryo mitochondria at various stages of seed formation. *Fiziol. Rastenii* (English translation) **13**, 723–726.

PUNDIR, N. S. (1972). Experimental embryology of *Gossypium arboreum* L. and *G. hirsutum* L. and their reciprocal crosses. *Bot. Gaz.* **133**, 7–26.

PURI (*née* MURGAI), P. (1963). Growth *in vitro* of parthenogenetic embryos of *Aerva tomentosa* Forsk. *In* "Plant Tissue and Organ Culture—A Symposium" (P. Maheshwari and N. S. Rangaswamy, eds), pp. 281–291. International Society of Plant Morphologists, Delhi.

PURVIS, O. N. (1940). Vernalization of fragments of embryo tissue. *Nature* **145**, 462.

PURVIS, O. N. (1944). Studies in the vernalisation of cereals. VIII. The role of carbohydrate and nitrogen supply in the vernalisation of excised embryos of 'Petkus' winter rye. *Ann. Bot. N.S.* **8**, 285–314.

PURVIS, O. N. (1947). Studies in the vernalisation of cereals. X. The effect of depletion of carbohydrates on the growth and vernalisation response of excised embryos. *Ann. Bot. N.S.* **11**, 269–283.

PURVIS, O. N. (1948). Studies in vernalisation. XI. The effect of date of sowing and of excising the embryo upon the responses of Petkus winter rye to different periods of vernalisation treatment. *Ann. Bot. N.S.* **12**, 183–206.

PURVIS, O. N. (1961). The physiological analysis of vernalisation. *In* "Encyclopedia of Plant Physiology" (W. Ruhland, ed.), Vol. 16, pp. 76–122. Springer-Verlag, Berlin.

PURVIS, O. N. and GREGORY, F. G. (1953). Accelerating effect of an extract of vernalized embryos of winter rye on flower initiation of unvernalized embryos. *Nature* **171**, 687–688.

QUATRANO, R. S. (1968). Rhizoid formation in *Fucus* zygotes: Dependence on protein and ribonucleic acid synthesis. *Science* **162**, 468–470.

QUEDNOW, K. G. (1930). Beiträge zur Frage der Aufnahme gelöster Kohlenstoffverbindungen durch Orchideen und andere Pflanzen. *Bot. Arch.* **30**, 51–108.

QUISUMBING, E. and JULIANO, J. B. (1927). Development of ovule and embryo sac of *Cocos nucifera*. *Bot. Gaz.* **84**, 279–293.

RAACKE, I. D. (1957). Protein synthesis in ripening pea seeds. 2. Development of embryos and seed coats. *Biochem. Jour.* **66**, 110–113.

RABÉCHAULT, H. (1962). Recherches sur la culture "*in vitro*" des embryons de palmier à huile (*Elaeis guineensis* Jacq.). I. Effets de l'acide β indolyl-acétique. *Oléagineux* **17**, 757–764.

RABÉCHAULT, H. (1967). Relations entre le comportement des embryons de palmier à huile (*Elaeis guineensis* Jacq.) en culture *in vitro* et la teneur en eau des graines. *Compt. Rend. Acad. Sci. Paris* **264**, 276–279.

RABÉCHAULT, H. and AHÉE, J. (1966). Recherches sur la culture "*in vitro*" des embryons de palmier à huile (*Elaeis guineensis* Jacq.). III. Effets de la grosseur et de l'age des graines. *Oléagineux* **21**, 729–734.

RABÉCHAULT, H. and CAS, G. (1973). Relations entre l'inhibition par la caféine de la croissance des embryons de caféiers et leur teneur en phénols totaux. *Compt. Rend. Acad. Sci. Paris* **277**, 2697–2700.

RABÉCHAULT, H., AHÉE, J. and GUÉNIN, G. (1968). Recherches sur la culture *in vitro* des embryons de palmier à huile (*Elaeis guineensis* Jacq.). IV. Effets de la teneur en eau des noix et de la durée de leur stockage. *Oléagineux* **23**, 233–237.

RABÉCHAULT, H., AHÉE, J. and GUÉNIN, G. (1969). Devéloppement *in vitro* des embryons de palmier à huile (*Elaeis guineensis* Jacq. var *dura* Becc.) extraits de graines dormantes ou non dormantes au cours de leur déshy-dratation naturelle. *Compt. Rend. Acad. Sci. Paris* **268**, 1728–1731.

RABÉCHAULT, H., AHÉE, J. and GUÉNIN, G. (1970). Colonies cellulaires et formes embryoïdes obtenues *in vitro* à partir des cultures d'embryons de palmier à huile (*Elaeis guineensis* Jacq. var. *dura* Becc.). *Compt. Rend. Acad. Sci. Paris* **270**, 3067–3070.

RADFORTH, N. W. and BONGA, J. M. (1960). Differentiation induced as season advances in the embryo-gametophyte complex of *Pinus nigra* var. *austriaca*, using indole acetic acid. *Nature* **185**, 332.

RADFORTH, N. W. and PEGORARO, L. C. (1955). Assessment of early dif-ferentiation in *Pinus* proembryos transplanted to *in vitro* conditions. *Trans. Roy. Soc. Canada* **49**, 69–82.

RADFORTH, W. N. (1936). The development *in vitro* of the proembryo of *Ginkgo*. *Trans. Roy. Canad. Inst.* **21**, 87–94.

RADLEY, M. (1958). The distribution of substances similar to gibberellic acid in higher plants. *Ann. Bot. N.S.* **22**, 297–307.

RADLEY, M. and DEAR, E. (1958). Occurrence of gibberellin-like substances in the coconut. *Nature* **182**, 1098.

RAGHAVAN, V. (1964a). Effects of certain organic nitrogen compounds on growth *in vitro* of seedlings of *Cattleya*. *Bot. Gaz.* **125**, 260–267.

RAGHAVAN, V. (1964b). Interaction of growth substances in growth and organ initiation in the embryos of *Capsella*. *Plant Physiol.* **39**, 816–821.

RAGHAVAN, V. (1965). Hormonal control of embryogenesis in *Capsella*. *In* "Proceedings of an International Conference on Plant Tissue Culture"

(P. R. White and A. R. Grove, eds), pp. 357–369. McCutchan Publishing Co., Berkeley.

RAGHAVAN, V. (1966). Nutrition, growth and morphogenesis of plant embryos. *Biol. Rev.* **41**, 1–58.

RAGHAVAN, V. (1967). Plant embryo culture. *In* "Methods in Developmental Biology" (F. H. Wilt and N. K. Wessells, eds), pp. 413–424. Thomas Y. Crowell Co., New York.

RAGHAVAN, V. (1975a). Nucleic acid synthesis during embryogenesis in vascular plants. *In* "Form, Structure and Function in Plants" (H. Y. Mohan Ram, J. J. Shah and C. K. Shah eds), pp. 94–101. Sarita Prakashan, Meerut, India.

RAGHAVAN, V. (1975b). Induction of haploid plants from anther cultures of henbane. *Zeitschr. Pflanzenphysiol.* **76**, 89–92.

RAGHAVAN, V. (1975c). Applied aspects of embryo culture. *In* "Applied and Fundamental Aspects of Plant Cell, Tissue and Organ Culture" (J. Reinert and Y. P. S. Bajaj, eds) (in press) Springer-Verlag, Berlin.

RAGHAVAN, V. (1975d). Unpublished observations.

RAGHAVAN, V. and BARUAH, H. K. (1958). Studies on the tannins of arecanut (*Areca catechu* L.). I. Development of the tannin cells. *Phyton* **10**, 35–42.

RAGHAVAN, V. and TORREY, J. G. (1963). Growth and morphogenesis of globular and older embryos of *Capsella* in culture. *Amer. Jour. Bot.* **50**, 540–551.

RAGHAVAN, V. and TORREY, J. G. (1964a). Inorganic nitrogen nutrition of the seedlings of the orchid, *Cattleya*. *Amer. Jour. Bot.* **51**, 264–274.

RAGHAVAN, V. and TORREY, J. G. (1964b). Effects of certain growth substances on the growth and morphogenesis of immature embryos of *Capsella* in culture. *Plant Physiol.* **39**, 691–699.

RAM, M. (1956). Floral morphology and embryology of *Trapa bispinosa* Roxb. with a discussion on the systematic position of the genus. *Phytomorphology* **6**, 312–323.

RAM, M. (1957). Morphological and embryological studies in the family Santalaceae—I. *Comandra umbellata* (L.) Nutt. *Phytomorphology* **7**, 24–35.

RAM, M. (1959a). Morphological and embryological studies in the family Santalaceae—II. *Exocarpus*, with a discussion of its systematic position. *Phytomorphology* **9**, 4–19.

RAM, M. (1959b). Morphological and embryological studies in the family Santalaceae—III. *Leptomeria* R. Br. *Phytomorphology* **9**, 20–33.

RAMAKRISHNAN, T., INDIRA, M. and SIRSI, M. (1957). A new growth-promoting factor for *Mycobacterium tuberculosis*. *Nature* **179**, 1356–1357.

RAMAN, K. and GREYSON, R. I. (1974). *In vitro* induction of embryoids in tissue cultures of *Nigella damascena*. *Canad. Jour. Bot.* **52**, 1988–1989.

RANDOLPH, L. F. (1936). Developmental morphology of the caryopsis in maize. *Jour. Agric. Res.* **53**, 881–916.

RANDOLPH, L. F. (1945). Embryo culture of *Iris* seed. *Bull. Amer. Iris Soc.* **97**, 33–45.

RANDOLPH, L. F. (1959). Advances in tissue and organ culture. *Memoirs, Indian Bot. Soc.* **2**, 1–6.

RANDOLPH, L. F. and COX, L. G. (1943). Factors influencing the germination of *Iris* seed and the relation of inhibiting substances to embryo dormancy. *Proc. Amer. Soc. Hort. Sci.* **43**, 284–300.

RANDOLPH, L. F. and KHAN, R. (1960). Growth response of excised mature embryos of *Iris* and wheat to different culture media. *Phytomorphology* **10**, 43–49.

RANDOLPH, M. L. and HABER, A. H. (1961). Production and decay of free radicals induced by X-irradiation of dry lettuce seeds. *In* "Effects of Ionizing Radiations on Seeds". Proceedings of the Symposium on the Effects of Ionizing Radiations on Seeds and their Significance for Crop Improvement, pp. 57–65. International Atomic Energy Agency, Vienna.

RANGAN, T. S. (1965). Morphogenesis of the embryo of *Cistanche tubulosa* Wight in vitro. *Phytomorphology* **15**, 180–182.

RANGAN, T. S. and RANGASWAMY, N. S. (1968). Morphogenic investigations of parasitic angiosperms. I. *Cistanche tubulosa* (Orobanchaceae). *Canad. Jour. Bot.* **46**, 263–266.

RANGAN, T. S. and RANGASWAMY, N. S. (1969). Morphogenic investigations on parasitic angiosperms. III. *Cassytha filiformis* (Lauraceae). *Phytomorphology* **19**, 292–300.

RANGAN, T. S., MURASHIGE, T. and BITTERS, W. P. (1968). *In vitro* initiation of nucellar embryos in monoembryonic *Citrus*. *HortSci.* **3**, 226–227.

RANGAN, T. S., MURASHIGE, T. and BITTERS, W. P. (1969). *In vitro* studies of zygotic and nucellar embryogenesis in *Citrus*. *Proc. First Internat. Citrus Symp.*, 225–229.

RANGASWAMY, N. S. (1958a). Culture of nucellar tissue of *Citrus* in vitro. *Experientia* **14**, 111.

RANGASWAMY, N. S. (1958b). *In vitro* culture of nucellus and embryos of *Citrus*. *In* "Modern Developments in Plant Physiology" (P. Maheshwari, ed.), pp. 104–105. University of Delhi.

RANGASWAMY, N. S. (1959). Morphogenetic response of *Citrus* ovules to growth adjuvants in culture. *Nature* **183**, 735–736.

RANGASWAMY, N. S. (1961). Experimental studies on female reproductive structures of *Citrus microcarpa* Bunge. *Phytomorphology* **11**, 109–127.

RANGASWAMY, N. S. (1963). Studies on culturing seeds of *Orobanche aegyptiaca* Pers. *In* "Plant Tissue and Organ Culture—A Symposium" (P. Maheshwari and N. S. Rangaswamy, eds), pp. 345–354. International Society of Plant Morphologists, Delhi.

RANGASWAMY, N. S. (1967). Morphogenesis of seed germination in angiosperms. *Phytomorphology* **17**, 477–487.

RANGASWAMY, N. S. and PROMILA (1972). Morphogenesis of the adult embryo of *Azadirachta indica* A. Juss. *Zeitschr. Pflanzenphysiol.* **67**, 377–379.

RANGASWAMY, N. S. and RANGAN, T. S. (1963). *In vitro* culture of embryos of *Cassytha filiformis* L. *Phytomorphology* **13**, 445–449.

RANGASWAMY, N. S. and RANGAN, T. S. (1966). Effects of seed germination-stimulants on the witchweed *Striga euphrasioides* (Vahl) Benth. *Nature* **210**, 440–441.

RANGASWAMY, N. S. and RANGAN, T. S. (1971). Morphogenic investigations on parasitic angiosperms. IV. Morphogenesis in decotylated embryos of *Cassytha filiformis* L. Lauraceae. *Bot. Gaz.* **132**, 113–119.

RANGASWAMY, N. S. and RAO, P. S. (1963). Experimental studies on *Santalum album* L. Establishment of tissue culture of endosperm. *Phytomorphology* **13**, 450–454.

RANGASWAMY, N. S. and SHIVANNA, K. R. (1967). Induction of gamete compatibility and seed formation in axenic cultures of a diploid self-incompatible species of *Petunia*. *Nature* **216**, 937–939.

RANGASWAMY, N. S. and SHIVANNA, K. R. (1971). Overcoming self-incompatibility in *Petunia axillaris*. II. Placental pollination *in vitro*. *Jour. Indian Bot. Soc. Silver Jubilee Vol.* 50A, 286–296.

RAO, P. S. (1965a). The *in vitro* fertilization and seed formation in *Nicotiana rustica* L. *Phyton* **22**, 165–167.

RAO, P. S. (1965b). *In vitro* induction of embryonal proliferation in *Santalum album* L. *Phytomorphology* **15**, 175–179.

RAO, P. S. and NARAYANASWAMI, S. (1972). Morphogenetic investigations in callus cultures of *Tylophora indica*. *Physiol. Plantarum* **27**, 271–276.

RAO, P. S. and RANGASWAMY, N. S. (1971). Morphogenic studies in tissue cultures of the parasite *Santalum album* L. *Biol. Plantarum* **13**, 200–206.

RAO, P. S., HANDRO, W. and HARADA, H. (1973a). Bud formation and embryo differentiation in *in vitro* cultures of *Petunia*. *Zeitschr. Pflanzenphysiol.* **69**, 87–90.

RAO, P. S., HANDRO, W. and HARADA, H. (1973b). Hormonal control of differentiation of shoots, roots and embryos in leaf and stem cultures of *Petunia inflata* and *Petunia hybrida*. *Physiol. Plantarum* **28**, 458–463.

RAO, P. S., NARAYANASWAMY, S. and BENJAMIN, B. D. (1970). Differentiation *ex ovulo* of embryos and plantlets in stem tissue cultures of *Tylophora indica*. *Physiol. Plantarum* **23**, 140–144.

RAO, V. S. and KHAN, A. A. (1975). Enhancement of polyribosome formation by gibberellic acid and 3′,5′-adenosine monophosphate in barley embryos. *Biochem. Biophys. Res. Comm.* **62**, 25–30.

RAPPAPORT, J. (1954). *In vitro* culture of plant embryos and factors controlling their growth. *Bot. Rev.* **20**, 201–225.

RAPPAPORT, J., SATINA, S. and BLAKESLEE, A. F. (1950a). Ovular tumors and inhibition of embryo growth in incompatible crosses of *Datura*. *Science* **111**, 276–277.

RAPPAPORT, J., SATINA, S. and BLAKESLEE, A. F. (1950b). Extracts of ovular tumors and their inhibition of embryo growth in *Datura*. *Amer. Jour. Bot.* **37**, 586–595.

RAQUIN, C. and PILET, V. (1972). Production de plantules à partir d'anthères de pétunias cultivées *in vitro*. *Compt. Rend. Acad. Sci. Paris* **274**, 1019–1022.

RASCH, E. and WOODARD, J. W. (1959). Basic proteins of plant nuclei during normal and pathological cell growth. *Jour. Biophys. Biochem. Cytol.* **6**, 263–276.

RASHID, A. and STREET, H. E. (1973). The development of haploid embryoids from anther cultures of *Atropa belladonna* L. *Planta* **113**, 263–270.

RASHID, A. and STREET, H. E. (1974a). Growth, embryogenic potential and stability of a haploid cell culture of *Atropa belladonna* L. *Plant Sci. Lett.* **2**, 89–94.

RASHID, A. and STREET, H. E. (1974b). Segmentations in microspores of *Nicotiana sylvestris* and *Nicotiana tabacum* which lead to embryoid formation in anther cultures. *Protoplasma* **80**, 323–334.

RATNAMBA, S. P. and CHOPRA, R. N. (1974). *In vitro* induction of embryoids from hypocotyls and cotyledons of *Anethum graveolens* seedlings. *Zeitschr. Pflanzenphysiol.* **73**, 452–455.

RAU, M. A. (1950). The suspensor haustoria of some species of *Crotalaria* Linn. *Ann. Bot. N.S.* **14**, 557–562.

RAU, M. A. (1956). Studies in growth *in vitro* of excised ovaries—I. Influence of colchicine on the embryo and endosperm in *Phlox drummondii* Hook. *Phytomorphology* **6**, 90–96.

RAZI, B. A. (1949). Embryological studies of two members of the Podostemaceae. *Bot. Gaz.* **111**, 211–218.

RAZMOLOGOV, V. P. (1973). Tissue culture from the generative cell of the pollen grain of *Cupressus* spp. *Bull. Torrey Bot. Club* **100**, 18–22.

RÉDEI, G. (1955). *Triticum durum abyssinicum* × *Secale cereale* hybridek elöállítása mesterséges embyro nevelés segítségevel. *Növénytermelés* **4**, 365–367.

RÉDEI, G. and RÉDEI, G. (1955a). Adatok a búza szemtermésének fejlödéséhez. *Növénytermelés* **4**, 133–140.

RÉDEI, G. and RÉDEI, G. (1955b). Rearing wheats from ovaries cultured *in vitro*. *Acta Bot. Acad. Sci. Hungar.* **2**, 183–186.

RÉDEI, G. and RÉDEI, G. (1955c). Developing wheat embryos excised from ovaries cultured *in vitro*. *Experientia* **11**, 387–388.

REDEMANN, C. T., WITTWER, S. H. and SELL, H. M. (1951). The fruit-setting factor from the ethanol extracts of immature corn kernels. *Arch. Biochem. Biophys.* **32**, 80–84.

REEVES, R. G. and BEASLEY, J. O. (1935). The development of the cotton embryo. *Jour. Agric. Res.* **51**, 935–944.

REID, J. S. G. and MEIER, H. (1972). The function of the aleurone layer during galactomannan mobilisation in germinating seeds of fenugreek (*Trigonella foenum-graecum* L.), crimson clover (*Trifolium incarnatum* L.) and lucerne (*Medicago sativa* L.): A correlative biochemical and ultrastructural study. *Planta* **106**, 44–60.

REINERT, J. (1958). Morphogenese und ihre Kontrolle an Gewebekulturen aus Carotten. *Naturwiss.* **45**, 344–345.

REINERT, J. (1959). Über die Kontrolle der Morphogenese und die Induktion von Adventivembryonen an Gewebekulturen aus Karotten. *Planta* **53**, 318–333.

REINERT, J. (1963). Experimental modification of organogenesis in plant tissue cultures. *In* "Plant Tissue and Organ Culture—A Symposium" (P. Maheshwari and N. S. Rangaswamy, eds), pp. 168–177. International Society of Plant Morphologists, Delhi.

REINERT, J. (1967). Some aspects of embryogenesis in somatic cells of *Daucus carota*. *Phytomorphology* **17**, 510–516.

REINERT, J. (1968). Factors of embryo formation in plant tissues cultivated *in vitro*. *In* "Les Cultures de Tissus de Plantes", pp. 33–40. Colloq. Nationaux du Centre National de la Recherche Scientifique, Paris, No. 920.

REINERT, J. and BACKS, D. (1968). Control of totipotency in plant cells growing *in vitro*. *Nature* **220**, 1340–1341.

REINERT, J. and TAZAWA, M. (1969). Wirkung von Stickstoffverbindungen und von Auxin auf die Embryogenese in Gewebekulturen. *Planta* **87**, 239–248.

REINERT, J., BACKS, D. and KROSING, M. (1966). Faktoren der Embryogenese in Gewebekulturen aus Kulturformen von Umbelliferen. *Planta* **68**, 375–378.

REINERT, J., BACKS-HÜSEMANN, D. and ZERBAN, H. (1971). Determination of embryo and root formation in tissue cultures from *Daucus carota*. *In* "Les Cultures de Tissus de Plantes", pp. 261–268. Colloq. Internationaux du Centre National de la Recherche Scientifique, Paris, No. 193.

REINERT, J., TAZAWA, M. and SEMENOFF, S. (1967). Nitrogen compounds as factors of embryogenesis *in vitro*. *Nature* **216**, 1215–1216.

REINHOLZ, E. (1959). Beeinflussung der Morphogenese embryonaler Organe durch ionisierende Strahlungen. I. Keimlingsanomalien durch Röntgenbestrahlung von *Arabidopsis thaliana*–Embryonen in verschiedenen Entwicklungsstadien. *Strahlentherapie* **109**, 537–553.

REJMAN, E. and BUCHOWICZ, J. (1971). The sequence of initiation of RNA, DNA and protein synthesis in the wheat grains during germination. *Phytochemistry* **10**, 2951–2957.

REJMAN, E. and BUCHOWICZ, J. (1973). RNA synthesis during the germination of wheat seed. *Phytochemistry* **12**, 271–276.

RENNER, O. (1914). Befruchtung und Embryobildung bei *Oenothera lamarckiana* und einigen verwandten Arten. *Flora* **107**, 115–150.

REST, J. A. and VAUGHAN, J. G. (1972). The development of protein and oil bodies in the seed of *Sinapis alba* L. *Planta* **105**, 245–262.

RICARDO, M. J., JR. and ALVAREZ, M. R. (1971). Ultrastructural changes associated with utilization of metabolite reserves and trichome differentiation in the protocorm of *Vanda*. *Amer. Jour. Bot.* **58**, 229–238.

RIETSEMA, J. and BLONDEL, B. (1959). Growth processes in the embryo and seed. *In* "Blakeslee: The Genus *Datura*" (A. G. Avery, S. Satina and J. Rietsema, eds), pp. 196–219. Ronald Press, New York.

RIETSEMA, J. and SATINA, S. (1959). Barriers to crossability: Post-fertilization. *In* "Blakeslee: The Genus *Datura*" (A. G. Avery, S. Satina and J. Rietsema, eds), pp. 245–262. Ronald Press, New York.

RIETSEMA, J., SATINA, S. and BLAKESLEE, A. F. (1953a). The effect of sucrose on the growth of *Datura stramonium* embryos *in vitro*. *Amer. Jour. Bot.* **40**, 538–545.

RIETSEMA, J., SATINA, S. and BLAKESLEE, A. F. (1953b). The effect of indole-3-acetic acid on *Datura* embryos. *Proc. Natl Acad. Sci. U.S.A.* **39**, 924–933.

RIETSEMA, J., SATINA, S. and BLAKESLEE, A. F. (1954). On the nature of the embryo inhibitor in ovular tumors of *Datura*. *Proc. Natl Acad. Sci. U.S.A.* **40**, 424–431.

RIETSEMA, J., BLONDEL, B., SATINA, S. and BLAKESLEE, A. F. (1955). Studies on ovule and embryo growth in *Datura*. I. A growth analysis. *Amer. Jour. Bot.* **42**, 449–455.

RIJVEN, A. H. G. C. (1952). *In vitro* studies on the embryo of *Capsella bursa-pastoris*. *Acta Bot. Neerl.* **1**, 157–200.

RIJVEN, A. H. G. C. (1955). Effects of glutamine, asparagine and other related compounds on the *in vitro* growth of embryos of *Capsella bursa-pastoris*. *Koninkl. Nederl. Akad. Wetensch. Proc.* **C58**, 368–376.

RIJVEN, A. H. G. C. (1956). Glutamine and asparagine as nitrogen sources for the growth of plant embryos *in vitro*: A comparative study of 12 species. *Austral. Jour. Biol. Sci.* **9**, 511–527.

RIJVEN, A. H. G. C. (1958). Effects of some inorganic nitrogenous substances on growth and nitrogen assimilation of young plant embryos *in vitro*. *Austral. Jour. Biol. Sci.* **11**, 142–154.

RIJVEN, A. H. G. C. (1960). On the utilization of γ-aminobutyric acid by wheat seedlings. *Austral. Jour. Biol. Sci.* **13**, 132–141.

RIJVEN, A. H. G. C. (1961). Regulation of glutamyl transferase level in germinating wheat embryos. *Biochim. Biophys. Acta* **52**, 213–215.

RIJVEN, A. H. G. C. and BANBURY, C. A. (1960). Role of the grain coat in wheat grain development. *Nature* **188**, 546–547.

RIVIÈRES, R. (1959). Sur la culture *in vitro* d'embryons isolés de Polypodiacées. *Compt. Rend. Acad. Sci. Paris* **248**, 1004–1007.

RIZZINI, C. T. (1973). Dormancy in seeds of *Anona crassiflora* Mart. *Jour. Exptl Bot.* **24**, 117–123.

ROBBINS, W. J. (1922). Cultivation of excised root tips and stem tips under sterile conditions. *Bot. Gaz.* **73**, 376–390.

ROBERTS, B. E. and OSBORNE, D. J. (1973). Protein synthesis and loss of viability in rye embryos. The lability of transferase enzymes during senescence. *Biochem. Jour.* **135**, 405–410.

ROBERTS, B. E., PAYNE, P. I. and OSBORNE, D. J. (1973). Protein synthesis and the viability of rye grains. Loss of activity of protein-synthesizing systems *in vitro* associated with a loss of viability. *Biochem. Jour.* **131**, 275–286.

ROBERTS, E. H. (1972). "Viability of Seeds". Chapman and Hall, London.

ROBINSON, K. R. and JAFFE, L. F. (1975). Polarizing fucoid eggs drive a calcium current through themselves. *Science* **187**, 70–72.

RODKIEWICZ, B. and MIKULSKA, E. (1967). The micropylar and antipodal cells of the *Lilium regale* embryo sac observed with the electron microscope. *Flora* **158**, 181–188.

ROGOZIŃSKA, J. (1960). Zmienność frakcji azotowych w izolowanych zarodkach *Lupinus mutabilis* Sweet. *Acta Bot. Soc. Polon.* **29**, 733–741.

ROLLIN, P. (1972). Seed germination. *In* "Phytochrome" (K. Mitrakos and

W. Shropshire, Jr., eds), pp. 229–254. Academic Press, New York and London.

ROMMEL, M. (1958). Eine vereinfachte Methode der Embryokultur bei Getreide. *Der Züchter* **28**, 149–151.

ROMMEL, M. (1960). The effect of embryo transplantation on seed set in some wheat-rye crosses. *Canad. Jour. Plant Sci.* **40**, 388–395.

RONDET, P. (1958). Répartition et signification des acides ribonucléiques au cours de l'embryogenèse chez *Lens culinaris* L. *Compt. Rend. Acad. Sci. Paris* **246**, 2396–2399.

RONDET, P. (1961). Répartition et signification des acides ribonucléiques au cours l'embryogenèse chez *Myosurus minimus* L. *Compt. Rend. Acad. Sci. Paris* **253**, 1725–1727.

RONDET, P. (1962). L'organogenèse au cours de l'embryogenèse chez l'*Alyssum maritimum* Lamk. *Compt. Rend. Acad. Sci. Paris* **255**, 2278–2280.

ROSS, J. D. and BRADBEER, J. W. (1968). Concentrations of gibberellin in chilled hazel seeds. *Nature* **220**, 85–86.

ROSS, J. D. and BRADBEER, J. W. (1971). Studies in seed dormancy. V. The content of endogenous gibberellins in seeds of *Corylus avellana* L. *Planta* **100**, 288–302.

ROST, T. L. (1972). The ultrastructure and physiology of protein bodies and lipids from hydrated dormant and nondormant embryos of *Setaria lutescens* (Gramineae). *Amer. Jour. Bot.* **59**, 607–616.

ROWSELL, E. V. and GOAD, L. J. (1962a). The constituent of wheat binding latent β-amylase. *Biochem. Jour.* **84**, 73P (abstract).

ROWSELL, E. V. and GOAD, L. J. (1962b). Latent β-amylase of wheat: Its mode of attachment to glutenin and its release. *Biochem. Jour.* **84**, 73P–74P (abstract).

ROY, S. C. (1972). Effect of mitomycin-C and colchicine on callus tissues and embryo cultured *in vitro*. *Indian Jour. Exptl Biol.* **10**, 244–246.

ROY CHOWDHURY, C. (1962). The embryogeny of Conifers: A review. *Phytomorphology* **12**, 313–338.

RUDNICKI, R. (1969). Studies on abscisic acid in apple seeds. *Planta* **86**, 63–68.

RUDNICKI, R., SIŃSKA, I. and LEWAK, S. (1972). The influence of abscisic acid on the gibberellin content in apple seeds during stratification. *Biol. Plantarum* **14**, 325–329.

RYCHTER, A. and LEWAK, S. (1971). Apple embryo peroxidases. *Phytochemistry* **10**, 2609–2613.

RYCZKOWSKI, M. (1960). Changes of the osmotic value during the development of the ovule. *Planta* **55**, 343–356.

RYCZKOWSKI, M. (1961). Changes in the specific gravity of the central vacuolar sap in developing ovules. *Bull. Acad. Polon. Sci. Ser. Sci. Biol.* **9**, 261–266.

RYCZKOWSKI, M. (1962). Changes in the concentration of sugars in developing ovules. *Acta Soc. Bot. Polon.* **31**, 53–65.

RYCZKOWSKI, M. (1965). Changes in osmotic value of the endosperm sap and differentiation of the egg cell in developing ovules of *Cycas revoluta* (Gymnospermae). *Bull. Acad. Polon. Sci. Ser. Sci. Biol.* **13**, 557–559.

RYCZKOWSKI, M. (1969). Changes in osmotic value of the central vacuole and endosperm sap during growth of the embryo and ovule. *Zeitschr. Pflanzenphysiol.* **61**, 422–429.

SABHARWAL, P. S. (1962). *In vitro* culture of nucelli and embryos of *Citrus aurantifolia* Swingle. *In* "Plant Embryology—A Symposium", pp. 239–243. Council of Scientific and Industrial Research, New Delhi.

SABHARWAL, P. S. (1963). *In vitro* culture of ovules, nucelli and embryos of *Citrus reticulata* Blanco var. Nagpuri. *In* "Plant Tissue and Organ Culture— A Symposium" (P. Maheshwari and N. S. Rangaswamy, eds), pp. 265–274. International Society of Plant Morphologists, Delhi.

SACCARDO, F. and DEVREUX, M. (1967). Caractéristiques morphologiques et cytogénétiques de quinze mutants de tabac. *Caryologia* **20**, 239–256.

SACCARDO, F. and DEVREUX, M. (1971). Effets des rayons gamma appliqués en début et fin de période de repos sur le zygote de tabac. *Radiation Bot.* **11**, 303–308.

SACHAR, R. C. (1955). The embryology of *Argemone mexicana* L.—A reinvestigation. *Phytomorphology* **5**, 200–218.

SACHAR, R. C. and BALDEV, B. (1958). *In vitro* growth of ovaries of *Linaria maroccana* Hook. *Curr. Sci.* **27**, 104–105.

SACHAR, R. C. and GUHA, S. (1962). *In vitro* growth of achenes of *Ranunculus sceleratus* L. *In* "Plant Embryology—A Symposium", pp. 244–253. Council of Scientific and Industrial Research, New Delhi.

SACHAR, R. C. and IYER, R. D. (1959). Effect of auxin, kinetin and gibberellin on the placental tissue of *Opuntia dillenii* Haw. cultured *in vitro*. *Phytomorphology* **9**, 1–3.

SACHAR, R. C. and KANTA, K. (1958). Influence of growth substances on artificially cultured ovaries of *Tropaeolum majus* L. *Phytomorphology* **8**, 202–218.

SACHAR, R. C. and KAPOOR, M. (1958). Influence of kinetin and gibberellic acid on the test tube seeds of *Cooperia pedunculata* Herb. *Naturwiss.* **45**, 552–553.

SACHAR, R. C. and KAPOOR, M. (1959). *In vitro* culture of ovules of *Zephyranthes*. *Phytomorphology* **9**, 147–156.

SACHAR, R. C. and MOHAN RAM, H. Y. (1958). The embryology of *Eschscholzia californica* Cham. *Phytomorphology* **8**, 114–124.

SACHER, J. A. (1956). Observations on pine embryos grown *in vitro*. *Bot. Gaz.* **117**, 206–214.

SACHET, M.-H. (1948). Fertilization in six incompatible species crosses of *Datura*. *Amer. Jour. Bot.* **35**, 302–309.

SACHS, J. (1859). Physiologische Untersuchungen über die Keimung der Schminkbohne (*Phaseolus multiflorus*). *Sitzungsber. Kaiser Akad. Wiss. Wien. Math.-Naturl. Cl.* **37**, 57–119.

SACHS, J. (1862). Zur Keimungsgeschichte der Gräser. *Bot. Zeit.* **20**, 145–151.

SACHS, J. (1887). "Lectures on the Physiology of Plants", English translation by H. Marshall Ward. Clarendon Press, Oxford.

SAMPSON, K. (1916). Note on a sporeling of *Phylloglossum* attached to a prothallus. *Ann. Bot.* **30**, 605–607.

SANDERS, M. E. (1948). Embryo development in four *Datura* species following self and hybrid pollinations. *Amer. Jour. Bot.* **35**, 525–532.

SANDERS, M. E. (1950). Development of self and hybrid *Datura* embryos in artificial culture. *Amer. Jour. Bot.* **37**, 6–15.

SANDERS, M. E. and BURKHOLDER, P. R. (1948). Influence of amino acids on growth of *Datura* embryos in culture. *Proc. Natl Acad. Sci. U.S.A.* **34**, 516–526.

SANDERS, M. E. and ZIEBUR, N. K. (1963). Artificial culture of embryos. *In* "Recent Advances in the Embryology of Angiosperms" (P. Maheshwari, ed.), pp. 297–325. International Society of Plant Morphologists, Delhi.

SANKHLA, N. and SANKHLA, D. (1967). Growth response of excised ovaries of *Reseda odorata* in sterile culture. *Biol. Plantarum* **9**, 61–63.

SANKHLA, N., SANKHLA, D. and CHATTERJI, U. N. (1967a). *In vitro* proliferation of "colored callus" from cotyledons of excised embryos of *Merremia dissecta*. *Zeitschr. Pflanzenphysiol.* **57**, 198–200.

SANKHLA, N., SANKHLA, D. and CHATTERJI, U. N. (1967b). *In vitro* induction of proliferation in female gametophytic tissue of *Ephedra foliata* Boiss. *Naturwiss.* **54**, 203.

SANKHLA, N., SANKHLA, D. and CHATTERJI, U. N. (1967c). Production of plantlets from callus derived from root-tip of excised embryos of *Ephedra foliata* Boiss. *Naturwiss.* **54**, 349.

SANSOME, E. R., SATINA, S. and BLAKESLEE, A. F. (1942). Disintegration of ovules in tetraploid–diploid and in incompatible species crosses in *Datura*. *Bull. Torrey Bot. Club* **69**, 405–420.

SANWAL, M. (1962). Morphology and embryology of *Gnetum gnemon* L. *Phytomorphology* **12**, 243–264.

SAPRE, A. B. (1963). Utilization of embryo culture technique in rice breeding. *In* "Plant Tissue and Organ Culture—A Symposium" (P. Maheshwari and N. S. Rangaswamy, eds), pp. 314–315. International Society of Plant Morphologists, Delhi.

SARIĆ, M. R. (1957). The radiosensitivity of seeds of different ontogenetic development. I. The effects of X-irradiation on oat seeds of different phases of ontogenetic development. *Radiation Res.* **6**, 167–172.

SARIĆ, M. R. (1958). The dependence of irradiation effects in seed on the biological properties of the seed. *In* "Isotopes in Agriculture". Proceedings, Second UN International Conference on the Peaceful Uses of Atomic Energy, Vol. 27, pp. 233–248. United Nations, Geneva.

SARIĆ, M. (1961). The effects of irradiation in relation to the biological traits of the seed irradiated. *In* "Effects of Ionizing Radiations on Seeds". Proceedings of the Symposium on the Effects of Ionizing Radiations on Seeds and their Significance for Crop Improvement, pp. 103–116. International Atomic Energy Agency, Vienna.

SASAKI, S. and BROWN, G. N. (1969). Changes in nucleic acid fractions of seed components of red pine (*Pinus resinosa* Ait.) during germination. *Plant Physiol.* **44**, 1729–1733.

SASAKI, S. and BROWN, G. N. (1971). Polysome formation in *Pinus resinosa* at initiation of seed germination. *Plant Cell Physiol.* **12**, 749–758.

SASAKI, S. and KOZLOWSKI, T. T. (1969). Utilization of seed reserves and currently produced photosynthates by embryonic tissues of pine seedlings. *Ann. Bot. N.S.* **33**, 473–482.

SATINA, S. and BLAKESLEE, A. F. (1935). Fertilization in the incompatible cross *Datura stramonium* × *D. metel*. *Bull. Torrey Bot. Club* **62**, 301–312.

SATINA, S. and RIETSEMA, J. (1959). Seed development. *In* "Blakeslee: The Genus *Datura*" (A. G. Avery, S. Satina and J. Rietsema, eds), pp. 181–195. Ronald Press, New York.

SATINA, S., RAPPAPORT, J. and BLAKESLEE, A. F. (1950). Ovular tumors connected with incompatible crosses in *Datura*. *Amer. Jour. Bot.* **37**, 576–586.

SATO, S. (1956a). On the reduction of Janus Green B by plant embryo slices and the mechanism of the specific staining of mitochondria. *Bot. Mag. Tokyo* **69**, 87–90.

SATO, S. (1956b). The distribution of succinic dehydrogenase and mitochondria in the embryos of *Phaseolus vulgaris*. *Bot. Mag. Tokyo* **69**, 137–141.

SATO, S. (1956c). Studies on the participation of succinic dehydrogenase in the reduction of tetrazolium salt by plant embryo homogenates. *Bot. Mag. Tokyo* **69**, 273–280.

SATO, S. (1962a). Studies on the reduction of tetrazolium salt by plant tissues. I. The reduction of TTC by plant tissue homogenate, with special reference to its relation to plant age and cyanide effect. *Cytologia* **27**, 97–105.

SATO, S. (1962b). Studies on the reduction of tetrazolium salt by plant tissues. II. Effect of plasmolysis on the reduction of TTC in plant cell. *Cytologia* **27**, 158–171.

SATO, T. (1974). Callus induction and organ differentiation in anther culture of poplars. *Jour. Jap. Forest Soc.* **56**, 55–62.

SATSANGI, A. and MOHAN RAM, H. Y. (1965). A continuously growing tissue culture from the mature endosperm of *Ricinus communis* L. *Phytomorphology* **15**, 26–30.

SAUTER, J. J. (1969). Cytochemische Untersuchung der Histone in Zellen mit unterschiedlicher RNS- und Protein-Synthese. *Zeitschr. Pflanzenphysiol.* **60**, 434–449.

SAWYER, M. L. (1925). Crossing *Iris pseudacorus* and *I. versicolor*. *Bot. Gaz.* **79**, 60–72.

SCHAFFNER, M. (1906). The embryology of the Shepherd's Purse. *Ohio Naturl.* **7**, 1–8.

SCHAFFSTEIN, G. (1938). Untersuchungen über die Avitaminose der Orchideenkeimlinge. *Jahrb. Wiss. Bot.* **86**, 720–752.

SCHARPÉ, A. and VAN PARIJS, R. (1971). Polytene nuclei in parenchyma cells of cotyledons, "*Pisum sativum*" L. in relation to ribosome and protein synthesis. *Arch. Intern. Physiol. Biochim.* **79**, 1042–1043.

SCHEIBE, J. and LANG, A. (1965). Lettuce seed germination: Evidence for a reversible light-induced increase in growth potential and for phytochrome mediation of the low temperature effect. *Plant Physiol.* **40**, 485–492.

SCHLOSSER-SZIGAT, G. (1962). Artbastardierung mit Hilfe der Embryokultur bei Steinklee (*Melilotus*). *Naturwiss.* **49**, 452–453.

SCHNARF, K. (1929). Embryologie der Angiospermen. *In* "Handbuch der Pflanzenanatomie" (K. Linsbauer, ed.), Vol. 10/2. Borntraeger, Berlin.

SCHNARF, K. (1931). "Vergleichende Embryologie der Angiospermen". Borntraeger, Berlin.

SCHNARRENBERGER, C., OESER, A. and TOLBERT, N. E. (1972). Isolation of protein bodies on sucrose gradients. *Planta* **104**, 185–194.

SCHNEPF, E. and NAGL, W. (1970). Über einige Strukturbesonderheiten der Suspensorzellen von *Phaseolus vulgaris*. *Protoplasma* **69**, 133–143.

SCHOOLER, A. B. (1959). Gibrel an aid to embryo culture media for *Hordeum vulgare*. *N. Dakota Acad. Sci.* **13**, 16–18.

SCHOOLER, A. B. (1960a). Wild barley hybrids. I. *Hordeum compressum* × *Hordeum pusillum*. *Jour. Hered.* **51**, 179–181.

SCHOOLER, A. B. (1960b). Wild barley hybrids. II. *Hordeum marinum* × *Hordeum compressum*. *Jour. Hered.* **51**, 243–246.

SCHOOLER, A. B. (1960c). The effect of gibrel and gibberellic acid (K salt) in embryo culture media for *Hordeum vulgare*. *Agron. Jour.* **52**, 411.

SCHOOLER, A. B. (1962). Technique of crossing wild and domestic barley species. *Bimonth. Bull. N. Dakota Agric. Exptl Sta.* **22** (7), 16–17.

SCHOU, L. (1951). On chlorophyll formation in the dark in excised embryos of *Pinus jeffreyi*. *Physiol. Plantarum* **4**, 617–620.

SCHROEDER, C. A. (1968). Adventive embryogenesis in fruit pericarp tissue *in vitro*. *Bot. Gaz.* **129**, 374–376.

SCHULZ, P. and JENSEN, W. A. (1969). *Capsella* embryogenesis: The suspensor and the basal cell. *Protoplasma* **67**, 139–163.

SCHULZ, P. and JENSEN, W. A. (1971). *Capsella* embryogenesis: The chalazal proliferating tissue. *Jour. Cell. Sci.* **8**, 201–227.

SCHULZ, P. and JENSEN, W. A. (1973). *Capsella* embryogenesis: The central cell. *Jour. Cell Sci.* **12**, 741–763.

SCHULZ, P. and JENSEN, W. A. (1974). *Capsella* embryogenesis: The development of the free nuclear endosperm. *Protoplasma* **80**, 183–205.

SCHULZ, R. and JENSEN, W. A. (1968a). *Capsella* embryogenesis: The early embryo. *Jour. Ultrastr. Res.* **22**, 376–392.

SCHULZ, R. and JENSEN, W. A. (1968b). *Capsella* embryogenesis: The synergids before and after fertilization. *Amer. Jour. Bot.* **55**, 541–552.

SCHULZ, R. and JENSEN, W. A. (1968c). *Capsella* embryogenesis: The egg, zygote and young embryo. *Amer. Jour. Bot.* **55**, 807–819.

SEHGAL, C. B. (1964). Artificial induction of polyembryony in *Foeniculum vulgare* Mill. *Curr. Sci.* **33**, 373–375.

SEHGAL, C. B. (1969). Experimental studies on maize endosperm. *Beitr. Biol. Pflanzen* **46**, 233–238.

SEHGAL, C. B. (1972). *In vitro* induction of polyembryony in *Ammi majus* L. *Curr. Sci.* **41**, 263–264.

SEN, B. and CHAKRAVARTI, S. C. (1947). Vernalization of excised mustard embryos. *Nature* **159**, 783–784.

SEN, B. and VERMA, G. (1959). Cultivation of mustard embryo and seedling fragments. *Memoirs, Indian Bot. Soc.* **2**, 36–39.

SEN, B. and VERMA, G. (1963). Studies on embryo fragments. *In* "Plant Tissue and Organ Culture—A Symposium" (P. Maheshwari and N. S. Rangaswamy, eds), pp. 326–331. International Society of Plant Morphologists, Delhi.

SEN, N. K. and MUKHOPADHYAY, I. (1961). Studies in embryo culture of some pulses. *Indian Agriculturist* **5**, 48–56.

SEN, S. and OSBORNE, D. J. (1974). Germination of rye embryo following hydration–dehydration treatments: Enhancement of protein and RNA synthesis and earlier induction of DNA replication. *Jour. Exptl Bot.* **25**, 1010–1019.

SETTERFIELD, G., STERN, H. and JOHNSTON, F. B. (1959). Fine structure in cells of pea and wheat embryos. *Canad. Jour. Bot.* **37**, 65–72.

SEUFERT, R. (1965). Die Strahlenempfindlichkeit verschiedener Entwicklungsstadien der Tetraploiden *Oenothera berteriana*. *Radiation Bot.* **5**, 153–169.

SHAFER, T. H. and KRIEBEL, H. B. (1974). Histochemistry of RNA during pollen tube growth and early embryogenesis in eastern white pine. *Canad. Jour. Bot.* **52**, 1519–1523.

SHANTZ, E. M. and STEWARD, F. C. (1952). Coconut milk factor: The growth promoting substances in coconut milk. *Jour. Amer. Chem. Soc.* **74**, 6133–6135.

SHANTZ, E. M. and STEWARD, F. C. (1955a). The general nature of some nitrogen free growth-promoting substances from *Aesculus* and *Cocos*. *Plant Physiol.* **30** (Suppl.), xxxv (abstract).

SHANTZ, E. M. and STEWARD, F. C. (1955b). The identification of compound A from coconut milk as 1,3-diphenylurea. *Jour. Amer. Chem. Soc.* **77**, 6351–6353.

SHANTZ, E. M. and STEWARD, F. C. (1964). Growth promoting substances from the environment of the embryo. II. The growth-stimulating complexes of coconut milk, corn and *Aesculus*. *In* "Régulateurs Naturels de la Croissance Végétale", pp. 59–75. Colloq. Internationaux du Centre National de la Recherche Scientifique, Paris, No. 123.

SHANTZ, E. M. and STEWARD, F. C. (1968). A growth substance from the vesicular embryo sac of *Aesculus*. *In* "Biochemistry and Physiology of Plant Growth Substances" (F. Wightman and G. Setterfield, eds), pp. 893–909. Runge Press, Ottawa.

SHARP, W. R., DOUGALL, D. K. and PADDOCK, E. F. (1971a). Haploid plantlets and callus from immature pollen grains of *Nicotiana* and *Lycopersicon*. *Bull. Torrey Bot. Club* **98**, 219–222.

SHARP, W. R., RASKIN, R. S. and SOMMER, H. E. (1971b). Haploidy in *Lilium*. *Phytomorphology* **21**, 334–337.

SHARP, W. R., RASKIN, R. S. and SOMMER, H. E. (1972). The use of nurse culture in the development of haploid clones in tomato. *Planta* **104**, 357–361.

SHARP, W. R., CALDAS, L. S., CROCOMO, O. J., MONACO, L. C. and CARVALHO, A. (1973). Production of *Coffea arabica* callus of three ploidy levels and subsequent morphogenesis. *Phyton* **31**, 67–74.

SHAW, M. and SRIVASTAVA, B. I. S. (1964). Purine-like substances from coconut endosperm and their effect on senescence in excised cereal leaves. *Plant Physiol.* **39**, 528–532.

SHIMAMURA, T. (1931). A note on the mitotic division in the proembryo of *Ginkgo*, with special reference to a chromatin elimination. *Bot. Mag. Tokyo* **45**, 525–530.

SHIMAMURA, T. (1935). Zur Cytologie des Befruchtungsvorganges bei *Cycas* und *Ginkgo* unter Benutzung der Feulgenschen Nuclealreaktion. *Cytologia* **6**, 465–473.

SHIMAMURA, T. (1956). Cytochemical studies on the fertilization and proembryo of *Pinus thunbergii*. *Bot. Mag. Tokyo* **69**, 524–529.

SHIVANNA, K. R. (1965). *In vitro* fertilization and seed formation in *Petunia violacea* Lindl. *Phytomorphology* **15**, 183–185.

SHUSTER, L. and GIFFORD, R. H. (1962). Changes in 3'-nucleotidase during the germination of wheat embryos. *Arch. Biochem. Biophys.* **96**, 534–540.

SIDDIQUI, S. A. (1964). *In vitro* culture of ovules of *Nicotiana tabacum* L. var. N.P. 31. *Naturwiss.* **51**, 517.

SIEGEL, S. M. (1952). Secretion of phosphorylase by red kidney bean embryos. *Bot. Gaz.* **114**, 139–141.

SIEGEL, S. M. (1953). Effects of exposures of seeds to various physical agents. II. Physiological and chemical aspects of heat injury in the red kidney bean embryo. *Bot. Gaz.* **114**, 297–312.

SIGEE, D. C. (1972a). Pattern of cytoplasmic DNA synthesis in somatic cells of *Pteridium aquilinum*. *Exptl Cell Res.* **73**, 481–486.

SIGEE, D. C. (1972b). The origin of cytoplasmic DNA in the mature egg cell of *Pteridium aquilinum*. *Protoplasma* **75**, 323–334.

SIGEE, D. and BELL, P. R. (1968). Deoxyribonucleic acid in the cytoplasm of the female reproductive cells of *Pteridium aquilinum*. *Exptl Cell Res.* **49**, 105–115.

SIGEE, D. C. and BELL, P. R. (1971). The cytoplasmic incorporation of tritiated thymidine during oogenesis in *Pteridium aquilinum*. *Jour. Cell Sci.* **8**, 467–487.

SIMMONDS, J. A. and DUMBROFF, E. B. (1974). High energy charge as a requirement for axis elongation in response to gibberellic acid and kinetin during stratification of *Acer saccharum* seeds. *Plant Physiol.* **53**, 91–95.

SIMOLA, L. K. (1971). Subcellular organization of developing embryos of *Bidens cernua*. *Physiol. Plantarum* **25**, 98–105.

SIMON, E. W. and MEANY, A. (1965). Utilization of reserves in germinating *Phaseolus* seeds. *Plant Physiol.* **40**, 1136–1139.

SIMPSON, G. M. (1965). Dormancy studies in seed of *Avena fatua*. 4. The role of gibberellin in embryo dormancy. *Canad. Jour. Bot.* **43**, 793–816.

SINGH, A. P. and MOGENSEN, H. L. (1975). Fine structure of the zygote and early embryo in *Quercus gambelii*. *Amer. Jour. Bot.* **62**, 105–115.

SINGH, B. (1952). A contribution to the floral morphology and embryology of *Dendrophthoe falcata* (L.f.) Ettingsh. *Jour. Linn. Soc.* **53**, 449–473.

SINGH, H. (1961). The life history and systematic position of *Cephalotaxus drupacea* Sieb. et Zucc. *Phytomorphology* **11**, 153–197.

SINGH, U. P. (1963). Raising nucellar seedlings of some Rutaceae *in vitro*. *In* "Plant Tissue and Organ Culture—A Symposium" (P. Maheshwari and N. S. Rangaswamy, eds), pp. 275–277. International Society of Plant Morphologists, Delhi.

SIŃSKA, I. and LEWAK, S. (1970). Apple seeds gibberellins. *Physiol. Végét.* **8**, 661–667.

SIRCAR, S. M. and DAS, T. M. (1951). Growth hormones of rice grains germinated at different temperatures. *Nature* **168**, 382–383.

SIRCAR, S. M. and LAHIRI, A. N. (1956). Studies on the physiology of rice. XII. Culture of excised embryos in relation to endosperm auxin and other growth factors. *Proc. Natl. Inst. Sci. India* **22B**, 212–225.

SIRCAR, S. M., DAS, T. M. and LAHIRI, A. N. (1955). Germination of rice embryo under water and its relation of growth to endosperm fractions. *Nature* **175**, 1046–1047.

SIWECKA, M. A. and SZARKOWSKI, J. W. (1974). Changes in distribution of ribosomes during germination of rye (*Secale cereale* L.) embryos. *Cytobios* **9**, 217–225.

SKENE, K. G. M. (1969). Stimulation of germination of immature bean embryos by gibberellic acid. *Planta* **87**, 188–192.

SKENE, K. G. M. (1970). The gibberellins of developing bean seeds. *Jour. Exptl Bot.* **21**, 236–246.

SKENE, K. G. M. and CARR, D. J. (1961). A quantitative study of the gibberellin content of seeds of *Phaseolus vulgaris* at different stages of their development. *Austral. Jour. Biol. Sci.* **14**, 13–25.

SKIRM, G. W. (1942). Embryo culturing as an aid to plant breeding. *Jour. Hered.* **33**, 211–215.

SKOVSTED, A. (1935). Cytological studies in cotton. III. A hybrid between *Gossypium davidsonii* Kell. and *G. sturtii* F. Muell. *Jour. Genet.* **30**, 397–405.

SMIRNOV, A. M. and PAVLOV, A. M. (1964). Cultivation of corn embryos without scutella and from immature seeds under sterile conditions. *Fiziol. Rastenii* (English translation) **11**, 347–351.

SMITH, C. G. (1974). The ultrastructural development of spherosomes and oil bodies in the developing embryo of *Crambe abyssinica*. *Planta* **119**, 125–142.

SMITH, C. W. (1965). Growth of excised embryo shoot apices of wheat *in vitro*. *Nature* **207**, 780–781.

SMITH, C. W. (1967). Growth of excised embryo shoot apices of wheat *in vitro*. *Ann. Bot. N.S.* **31**, 593–605.

SMITH, C. W. (1968). The effect of growth substances on growth of excised embryo shoot apices of wheat *in vitro*. *Ann. Bot. N.S.* **32**, 593–600.

SMITH, D. E. (1958). Effect of gibberellins on certain orchids. *Amer. Orch. Soc. Bull.* **27**, 742–747.

SMITH, D. L. (1971). Nuclear changes in the cotyledons of *Pisum arvense* L. during germination. *Ann. Bot. N.S.* **35**, 511–521.

SMITH, D. L. (1973). Nucleic acid, protein, and starch synthesis in developing cotyledons of *Pisum arvense* L. *Ann. Bot. N.S.* **37**, 795–804.

SMITH, F. (1932). Raising orchid seedlings asymbiotically under tropical conditions. *Gard. Chron.* **91**, 9–11.

SMITH, F. G. (1952). The mechanism of tetrazolim reaction in corn embryos. *Plant Physiol.* **27**, 445–456.

SMITH, H. H. (1974). Model systems for somatic cell plant genetics. *BioScience* **24**, 269–276.

SMITH, J. G. (1973). Embryo development in *Phaseolus vulgaris*. II. Analysis of selected inorganic ions, ammonia, organic acids, amino acids, and sugars in the endosperm liquid. *Plant Physiol.* **51**, 454–458.

SMITH, P. G. (1944). Embryo culture of a tomato species hybrid. *Proc. Amer. Soc. Hort. Sci.* **44**, 413–416.

SMITH, S. M. and STREET, H. E. (1974). The decline of embryogenic potential as callus and suspension cultures of carrot (*Daucus carota* L.) are serially subcultured. *Ann. Bot. N.S.* **38**, 223–241.

SMITHERS, A. G. and SUTCLIFFE, J. F. (1967). A method for the removal of mineral salts from coconut milk and a comparison of some of the growth promoting properties of whole and desalted coconut milk in carrot root tissue cultures. *Ann. Bot. N.S.* **31**, 333–350.

SMOLEŃSKA, G. and LEWAK, S. (1971). Gibberellins and the photosensitivity of isolated embryos from non-stratified apple seeds. *Planta* **99**, 144–153.

SMOLEŃSKA, G. and LEWAK, S. (1974). The role of lipases in the germination of dormant apple embryos. *Planta* **116**, 361–370.

SOBOLEV, A. M., BUTENKO, R. G. and SUVOROV, V. I. (1971). Tissue culture of castor endosperm as a model for studying biosynthesis of storage protein and phytin. *Fiziol. Rastenii* (English translation) **18**, 232–239.

SOLACOLU, T. and CONSTANTINESCO, D. (1936). Action de l'acide-β-indolylacétique sur le développement des plantules. *Compt. Rend. Acad. Sci. Paris* **203**, 437–440.

SOLOMON, B. (1950). Inhibiting effect of autoclaved malt preventing the *in vitro* growth of *Datura* embryos. *Amer. Jour. Bot.* **37**, 1–5.

SOMMER, H. E. and BROWN, C. L. (1974). Plantlet formation in pine tissue cultures. *Amer. Jour. Bot.* **61** (Suppl.), 11 (abstract).

SONDHEIMER, E., TZOU, D. S. and GALSON, E. C. (1968). Abscisic acid levels and seed dormancy. *Plant Physiol.* **43**, 1443–1447.

SOPORY, S. K. and MAHESHWARI, S. C. (1972). Production of haploid embryos by anther culture technique in *Datura innoxia*—A further study. *Phytomorphology* **22**, 87–90.

SOPORY, S. K. and MAHESHWARI, S. C. (1973). Similar effects of iron-chelating agents and cytokinins on the production of haploid embryos from the pollen grains of *Datura innoxia*. *Zeitschr. Pflanzenphysiol.* **69**, 97–99.

SOUÈGES, R. (1936). "La Différenciation. III. La Différenciation Organique." Hermann, Paris.

SOUÈGES, R. (1951). "Embryogénie et Classification". Hermann, Paris.

SOUÈGES, R. (1923). Embryogénie des Juncacées. Développement de l'embryon chez le *Luzula forsteri* DC. *Compt. Rend. Acad. Sci. Paris* **177**, 705–708.

SPOERL, E. (1948). Amino acids as sources of nitrogen for orchid embryos. *Amer. Jour. Bot.* **35**, 88–95.

SPOERL, E. and CURTIS, J. T. (1948). Studies on the nitrogen nutrition of orchid embryos. III. Amino acid nitrogen. *Amer. Orch. Soc. Bull.* **17**, 307–312.

SRIVASTAVA, B. I. S. (1963a). Ether-soluble and ether-insoluble auxins from immature corn kernels. *Plant Physiol.* **38**, 473–478.

SRIVASTAVA, B. I. S. (1963b). Investigation of purinelike compounds in immature maize kernels, germinating barley seeds, and yeast. *Arch. Biochem. Biophys.* **103**, 200–205.

SRIVASTAVA, L. M. and PAULSON, R. E. (1968). The fine structure of the embryo of *Lactuca sativa*. II. Changes during germination. *Canad. Jour. Bot.* **46**, 1447–1453.

SRIVASTAVA, P. S. (1971a). *In vitro* induction of triploid roots and shoots from mature endosperm of *Jatropha panduraefolia*. *Zeitschr. Pflanzenphysiol.* **66**, 93–96.

SRIVASTAVA, P. S. (1971b). *In vitro* growth requirements of mature endosperm of *Ricinus communis*. *Curr. Sci.* **40**, 337–339.

SRIVASTAVA, P. S. (1973). Formation of triploid 'plantlets' in endosperm cultures of *Putranjiva roxburghii*. *Zeitschr. Pflanzenphysiol.* **69**, 270–273.

STARITSKY, G. (1970). Embryoid formation in callus tissues of coffee. *Acta. Bot. Neerl.* **19**, 509–514.

STEBBINS, G. L. (1958). The inviability, weakness, and sterility of interspecific hybrids. *In* "Advances in Genetics" (M. Demerec, ed.), Vol. 9, pp. 147–215. Academic Press, New York and London.

STEHSEL, M. L. and WILDMAN, S. G. (1950). Interrelations between tryptophan, auxin and nicotinic acid during development of the corn kernel. *Amer. Jour. Bot.* **37**, 682–683 (abstract).

STEIN, O. L. and QUASTLER, H. (1963). The use of tritiated thymidine in the study of tissue activation during germination in *Zea mays*. *Amer. Jour. Bot.* **50**, 1006–1011.

STEINBAUER, G. P. (1937). Dormancy and germination of *Fraxinus* seeds. *Plant Physiol.* **12**, 813–824.

STERLING, C. (1949). Preliminary attempts in larch embryo culture. *Bot. Gaz.* **111**, 90–94.

STERNHEIMER, E. P. (1954). Method of culture and growth of maize endosperm *in vitro*. *Bull. Torrey Bot. Club* **81**, 111–113.

STEWARD, F. C. (1960). Carrots and coconuts—some investigations on growth. *Garden Jour.* **10**, 87–90, 94.

STEWARD, F. C. (1963a). The control of growth in plant cells. *Scient. Amer.* **209** (4), 104–113.

STEWARD, F. C. (1963b). Totipotency and variation in cultured cells: Some metabolic and morphogenetic manifestations. *In* "Plant Tissue and Organ Culture—A Symposium" (P. Maheshwari and N. S. Rangaswamy, eds), pp. 1–25. International Society of Plant Morphologists, Delhi.

STEWARD, F. C. (1963c). Carrots and coconuts: Some investigations on growth. *In* "Plant Tissue and Organ Culture—A Symposium" (P. Maheshwari and N. S. Rangaswamy, eds), pp. 178–197. International Society of Plant Morphologists, Delhi.

STEWARD, F. C. (1967). Totipotency of angiosperm cells. Its significance for morphology and embryology. *Phytomorphology* **17**, 499–507.

STEWARD, F. C. (1970a). Totipotency, variation and clonal development of cultured cells. *Endeavour* **29**, 117–124.

STEWARD, F. C. (1970b). From cultured cells to whole plants. The induction and control of their growth and morphogenesis. *Proc. Roy. Soc. Lond.* **B175**, 1–30.

STEWARD, F. C. and CAPLIN, S..M. (1952). Investigations on growth and metabolism of plant cells. IV. Evidence on the role of the coconut-milk factor in development. *Ann. Bot. N.S.* **16**, 491–504.

STEWARD, F. C. and MAPES, M. O. (1963). The totipotency of cultured carrot cells: Evidence and interpretations from successive cycles of growth from phloem cells. *Jour. Indian Bot. Soc.*, Maheshwari Comm. Vol. 42A, 237–247.

STEWARD, F. C. and MAPES, M. O. (1971a). Morphogenesis in aseptic cell cultures of *Cymbidium*. *Bot. Gaz.* **132**, 65–70.

STEWARD, F. C. and MAPES, M. O. (1971b). Morphogenesis and plant propagation in aseptic cultures of *Asparagus*. *Bot. Gaz.* **132**, 70–79.

STEWARD, F. C. and SHANTZ, E. M. (1959). Biochemistry and morphogenesis: Knowledge derived from plant tissue cultures. *In* "Biochemistry of Morphogenesis". Proceedings of the Fourth International Congress of Biochemistry, Vienna (W. G. Nickerson, ed.), Vol. 6, pp. 223–236. Pergamon Press, New York.

STEWARD, F. C. and STREET, H. E. (1947). The nitrogenous constituents of plants. *Ann. Rev. Biochem.* **16**, 471–502.

STEWARD, F. C., AMMIRATO, P. V. and MAPES, M. O. (1970). Growth and development of totipotent cells. Some problems, procedures and perspectives. *Ann. Bot. N.S.* **34**, 761–787.

STEWARD, F. C., KENT, A. E. and MAPES, M. O. (1966). The culture of free plant cells and its significance for embryology and morphogenesis. *In* "Current Topics in Developmental Biology" (A. A. Moscona and A. Monroy, eds), Vol. 1, pp. 113–154. Academic Press, New York and London.

STEWARD, F. C., MAPES, M. O. and AMMIRATO, P. V. (1969). Growth and morphogenesis in tissue and free cell cultures. *In* "Plant Physiology—A Treatise" (F. C. Steward, ed.), Vol. 5B, 329–376. Academic Press, New York and London.

STEWARD, F. C., MAPES, M. O. and MEARS, K. (1958a). Growth and organized development of cultured cells. II. Organization in cultures grown from freely suspended cells. *Amer. Jour. Bot.* **45**, 705–708.

STEWARD, F. C., MAPES, M. O. and SMITH, J. (1958b). Growth and organized development of cultured cells. I. Growth and division of freely suspended cells. *Amer. Jour. Bot.* **45**, 693–703.

STEWARD, F. C. with BLAKELEY, L. M., KENT, A. E. and MAPES, M. O. (1964a). Growth and organization in free cell cultures. *In* "Meristems and Differentiation". *Brookhaven Symp. Biol.* **16**, 73–88.

STEWARD, F. C. with MAPES, M. O., KENT, A. E. and HOLSTEN, R. D. (1964b). Growth and development of cultured plant cells. *Science* **143**, 20–27.

STEWARD, F. C. with SHANTZ, E. M., POLLARD, J. K., MAPES, M. O. and MITRA, J. (1961). Growth induction in explanted cells and tissues: Metabolic and morphogenetic manifestations. *In* "Synthesis of Molecular and Cellular Structure" (D. Rudnick, ed.), pp. 193–246. Ronald Press, New York.

STINGL, G. (1907). Experimentelle Studie über die Ernährung von pflanzlichen Embryonen. *Flora* **97**, 308–331.

STODDART, J. L., THOMAS, H. and ROBERTSON, A. (1973). Protein synthesis patterns in barley embryos during germination. *Planta* **112**, 309–321.

STOKES, P. (1952a). A new technique for the sterile culture of plant embryos *in vitro*. *Nature* **170**, 242.

STOKES, P. (1952b). A physiological study of embryo development in *Heracleum sphondylium* L. I. The effect of temperature on embryo development. *Ann. Bot. N.S.* **16**, 441–447.

STOKES, P. (1952c). A physiological study of embryo development in *Heracleum sphondylium* L. II. The effect of temperature on after-ripening. *Ann. Bot. N.S.* **16**, 571–576.

STOKES, P. (1953a). A physiological study of embryo development in *Heracleum sphondylium* L. III. The effect of temperature on metabolism. *Ann. Bot. N.S.* **17**, 157–173.

STOKES, P. (1953b). The stimulation of growth by low temperature in embryos of *Heracleum sphondylium* L. *Jour. Exptl Bot.* **4**, 222–234.

STOKES, P. (1965). Temperature and seed dormancy. *In* "Encyclopedia of Plant Physiology" (W. Ruhland, ed.), Vol. 15/2, pp. 746–803. Springer-Verlag, Berlin.

STONE, E. C. and DUFFIELD, J. W. (1950). Hybrids of sugar pine embryo culture. *Jour. Forest.* **48**, 200–201.

STOPES, M. C. and FUJII, K. (1906). The nutritive relations of the surrounding tissues to the archegonia in gymnosperms. *Beih. Bot. Centralbl.* **20**, 1–24.

STOUTAMIRE, W. P. (1963). Terrestrial orchid seedlings. *Austral. Plants* **2**, 119–122.

STOUTAMIRE, W. P. (1964). Seeds and seedlings of native orchids. *Michigan Botanist* **3**, 107–119.

STOWE, B. B. and THIMANN, K. V. (1953). Indolepyruvic acid in maize. *Nature* **172**, 764.

STRAUS, J. (1954). Maize endosperm tissue grown *in vitro*. II. Morphology and cytology. *Amer. Jour. Bot.* **41**, 833–839.

STRAUS, J. (1958). Spontaneous changes in corn endosperm tissue cultures. *Science* **128**, 537–538.

STRAUS, J. (1959). Anthocyanin synthesis in corn endosperm tissue cultures. II. Identity of the pigments and general factors. *Plant Physiol.* **34**, 536–541.

STRAUS, J. (1960). Maize endosperm tissue grown *in vitro*. III. Development of a synthetic medium. *Amer. Jour. Bot.* **47**, 641–647.

STRAUS, J. and LaRUE, C. D. (1954). Maize endosperm tissue grown *in vitro*. I. Culture requirements. *Amer. Jour. Bot.* **41**, 687–694.

STREET, H. E. (1966). The nutrition and metabolism of plant tissue and organ cultures. *In* "Cells and Tissues in Culture" (E. N. Willmer, ed.), Vol. 3, pp. 533–629. Academic Press, New York and London.

STREET, H. E. (1969). Growth in organized and unorganized systems. *In* "Plant Physiology—A Treatise" (F. C. Steward, ed.), Vol. 5B, pp. 3–225. Academic Press, New York and London.

STREET, H. E. and WITHERS, L. A. (1974). The anatomy of embryogenesis in culture. *In* "Tissue Culture and Plant Science 1974" (H. E. Street, ed.), pp. 71–100. Academic Press, New York and London.

SUBRAMANIAN, V. (1974). Protein synthesis during early germination. *Curr. Sci.* **43**, 297–299.

SUBRAMANYAM, K. (1949). An embryological study of *Lobelia pyramidalis* Wall., with special reference to the mechanism of nutrition of the embryo in the family Lobeliaceae. *New Phytol.* **48**, 365–373.

SUBRAMANYAM, K. (1950). An embryological study of *Levenhookia dubia* Sond. in Lehm. *Proc. Natl Inst. Sci. India* **16**, 245–253.

SUBRAMANYAM, K. (1953). The nutritional mechanism of embryo-sac and embryo in the families Campanulaceae, Lobeliaceae and Stylidiaceae. *Jour. Mysore Univ.* **13**, 355–358.

SUBRAMANYAM, K. (1960a). Nutritional mechanism of the seed. 1. Nutritional mechanism of the embryo sac. *Jour. Madras Univ.* **B30**, 29–44.

SUBRAMANYAM, K. (1960b). Nutritional mechanism of the seed. 2. Nutritional mechanism of the embryo. *Jour. Madras Univ.* **B30**, 45–56.

SULBHA, K. and SWAMINATHAN, M. S. (1959). Effect of grafting on fruit-set and embryo development in crosses between *Corchorus olitorius* and *C. capsularis*. *Curr. Sci.* **28**, 460–461.

SUN, C.-S., WANG, C.-C. and CHU, C.-C. (1974a). Cell division and differentiation of pollen grains in *Triticale* anthers cultured *in vitro*. *Scient. Sinica* **17**, 47–55.

SUN, C.-S., WANG, C.-C. and CHU, C.-C. (1974b). The ultrastructure of plastids in the albino pollen-plants of rice. *Scient. Sinica* **17**, 793–802.

SUN, C.-S., WANG, C.-C. and CHU, Z.-C. (1973). Cytological studies on the androgenesis of *Triticale*. *Acta Bot. Sinica* **15**, 163–173.

SUN, M. H. and ULLSTRUP, A. J. (1971). *In vitro* growth of corn endosperm. *Bull. Torrey Bot. Club* **98**, 251–258.

SUNDERLAND, N. (1970). Pollen plants and their significance. *New Scientist* **47**, 142–144.

SUNDERLAND, N. (1971). Anther culture: A progress report. *Sci. Prog. Oxford* **59**, 527–549.

SUNDERLAND, N. (1973). Pollen and anther culture. *In* "Plant Tissue and Cell Culture" (H. E. Street, ed.), pp. 205–239. University of California Press, Berkeley.

SUNDERLAND, N. and WELLS, B. (1968). Plastid structure and development in green callus tissues of *Oxalis dispar*. *Ann. Bot. N.S.* **32**, 327–346.

SUNDERLAND, N. and WICKS, F. M. (1969). Cultivation of haploid plants from tobacco pollen. *Nature* **224**, 1227–1229.

SUNDERLAND, N. and WICKS, F. M. (1971). Embryoid formation in pollen grains of *Nicotiana tabacum*. *Jour. Exptl Bot.* **22**, 213–226.

SUNDERLAND, N., COLLINS, G. B. and DUNWELL, J. M. (1974). The role of nuclear fusion in pollen embryogenesis of *Datura innoxia* Mill. *Planta* **117**, 227–241.

SUSSEX, I. M. (1972). Somatic embryos in long-term carrot tissue cultures: Histology, cytology, and development. *Phytomorphology* **22**, 50–59.

SUSSEX, I. M. and FREI, K. A. (1968). Embryoid development in long-term tissue cultures of carrot. *Phytomorphology* **18**, 339–349.

SUSSEX, I. M. and STEEVES, T. A. (1958). Experiments on the control of fertility of fern leaves in sterile culture. *Bot. Gaz.* **119**, 203–208.

SUSSEX, I., CLUTTER, M., WALBOT, V. and BRADY, T. (1973). Biosynthetic activity of the suspensor of *Phaseolus coccineus*. *Caryologia* **25** (Suppl.), 261–272.

SWAIN, R. R. and DEKKER, E. E. (1969). Seed germination studies. III. Properties of a cell-free amino acid incorporating system from pea cotyledons; possible origin of cotyledonary α-amylase. *Plant Physiol.* **44**, 319–325.

SWAMINATHAN, M. S., CHOPRA, V. L. and BHASKARAN, S. (1962). Cytological aberrations observed in barley embryos cultured in irradiated potato mash. *Radiation Res.* **16**, 182–188.

SWAMY, B. G. L. (1942). Female gametophyte and embryogeny in *Cymbidium bicolor* Lindl. *Proc. Indian Acad. Sci.* **B15**, 194–201.

SWAMY, B. G. L. (1943). Gametogenesis and embryogeny in *Eulophea epidendraea* Fischer. *Proc. Natl Inst. Sci. India* **9**, 59–65.

SWAMY, B. G. L. (1946). Embryology of *Habenaria*. *Proc. Natl Inst. Sci. India* **12**, 413–426.

SWAMY, B. G. L. (1949). Embryological studies in the Orchidaceae. II. Embryogeny. *Amer. Midland Naturl.* **41**, 202–232.

SWAMY, B. G. L. (1960). Contributions to the embryology of *Cansjera rheedii*. *Phytomorphology* **10**, 397–409.

SWAMY, B. G. L. and GANAPATHY, P. M. (1957). A new type of endosperm haustorium in *Nothapodytes foetida*. *Phytomorphology* **7**, 331–336.

SWAMY, B. G. L. and PARAMESWARAN, N. (1963). The helobial endosperm. *Biol. Rev.* **38**, 1–50.

SWAMY, B. G. L. and RAO, J. D. (1963). The endosperm of *Opilia amentacea* Roxb. *Phytomorphology* **13**, 423–428.

TAKAO, A. (1959). A cytochemical study on the proteid vacuoles in the egg of *Pinus thunbergii* Parl. *Bot. Mag. Tokyo* **72**, 289–297.

TAKAO, A. (1960). Histochemical studies on the embryogenesis in *Pinus thunbergii* Parl. *Bot. Mag. Tokyo* **37**, 379–388.

TAKAO, A. (1962). Histochemical studies on the formation of some leguminous seeds. *Jap. Jour. Bot.* **18**, 55–72.

TAKEUCHI, M. (1956). Embryogenesis in *Phaseolus vulgaris*. *Jour. Fac. Sci. Univ. Tokyo Sec. III. Bot.* **6**, 439–452.

TAMAOKI, T. and ULLSTRUP, A. J. (1958). Cultivation *in vitro* of excised endosperm and meristem tissues of corn. *Bull. Torrey Bot. Club* **85**, 260–272.

TAMMES, P. M. L. (1959). Nutrients in the giant embryosac-vacuole of the coconut. *Acta Bot. Neerl.* **8**, 493–496.

TANAKA, M. and NAKATA, K. (1969). Tobacco plants obtained by anther culture and the experiment to get diploid seeds from haploids. *Jap. Jour. Genet.* **44**, 47–54.

TANIFUJI, S., ASAMIZU, T. and SAKAGUCHI, K. (1969). DNA-like RNA synthesized in pea embryos at very early stage of germination. *Bot. Mag. Tokyo* **82**, 56–68.

TAO, K. L. and KHAN, A. A. (1974). Increases in activites of aminoacyl-tRNA synthetases during cold-treatment of dormant pear embryo. *Biochem. Biophys. Res. Comm.* **59**, 764–770.

TAUSSIG, H. B. (1962). A study of the German outbreak of *Phocomelia*. *Jour. Amer. Med. Assoc.* **180**, 1106–1114.

TAVENER, R. J. A. and LAIDMAN, D. L. (1969). Induction of lipase activity in the starchy endosperm of germinating wheat grains. *Biochem. Jour.* **113**, 32P (abstract).

TAYLOR, J. W. (1957). Growth of non-stratified peach embryos. *Proc. Amer. Soc. Hort. Sci.* **69**, 148–151.

TAYLOR, R. L. (1967). The foliar embryos of *Malaxis paludosa*. *Canad. Jour. Bot.* **45**, 1553–1556.

TAZAWA, M. and REINERT, J. (1969). Extracellular and intracellular chemical environments in relation to embryogenesis *in vitro*. *Protoplasma* **68**, 157–173.

TEPPER, H. B., HOLLIS, C. A., GALSON, E. C. and SONDHEIMER, E. (1967). Germination of excised *Fraxinus ornus* embryos with and without phleomycin. *Plant Physiol.* **42**, 1493–1496.

TERAOKA, H. (1967). Proteins of wheat embryos in the period of vernalization. *Plant Cell Physiol.* **8**, 87–95.

THÉVENOT, C. and CÔME, D. (1971a). Germination des embryons de pommier (*Pirus malus* L.) dormants amputés d'une partie plus ou moins importante de leurs cotylédons. *Compt. Rend. Acad. Sci. Paris* **272**, 1240–1243.

THÉVENOT, C. and CÔME, D. (1971b). Influence de la température et du mode d'imbibition sur la germination des embryons de pommier (*Pirus malus* L.) non dormants. *Compt. Rend. Acad. Sci. Paris* **273**, 2515–2517.

THÉVENOT, C. and CÔME, D. (1973a). Mise en évidence de l'influence des cotylédons sur la levée de dormance et la germination de l'axe de l'embryon de pommier (*Pirus malus* L.). *Physiol. Végét.* **11**, 161–169.

THÉVENOT, C. and CÔME, D. (1973b). Influence de la présentation de l'eau et de l'oxygène sur la germination des embryons de pommier (*Pirus malus* L.). *Compt. Rend. Acad. Sci. Paris* **277**, 401–404.

THOMAS, A. P. W. (1901). Preliminary account of the prothallium of *Phylloglossum*. *Proc. Roy. Soc. Lond.* **69**, 285–291.

THOMAS, E. and STREET, H. E. (1970). Organogenesis in cell suspension cultures of *Atropa belladonna* L. and *Atropa belladonna* cultivar *lutea* Döll. *Ann. Bot. N.S.* **34**, 657–669.

THOMAS, E. and STREET, H. E. (1972). Factors influencing morphogenesis in excised roots and suspension cultures of *Atropa belladonna*. *Ann. Bot. N.S.* **36**, 239–247.

THOMAS, E. and WENZEL, G. (1975). Embryogenesis from microspores of rye. *Naturwiss.* **62**, 40–41.

THOMAS, E., KONAR, R. N. and STREET, H. E. (1972). The fine structure of the embryogenic callus of *Ranunculus sceleratus* L. *Jour. Cell Sci.* **11**, 95–109.

THOMAS, H., WEBB, D. P. and WAREING, P. F. (1973). Seed dormancy in *Acer*: Maturation in relation to dormancy in *Acer pseudoplatanus* L. *Jour. Exptl Bot.* **24**, 958–967.

THOMAS, K. J. (1963). Effect of colchicine on embryos of *Chlorophytum laxum* grown *in vitro*. In "Plant Tissue and Organ Culture—A Symposium" (P. Maheshwari and N. S. Rangaswamy, eds), pp. 316–319. International Society of Plant Morphologists, Delhi.

THOMAS, M.-J. (1970a). Premières recherches sur les besoins nutritifs des embryons isolés du *Pinus silvestris* L. Embryons indifférenciés et en voie de différenciation. *Compt. Rend. Acad. Sci. Paris* **270**, 1120–1123.

THOMAS, M.-J. (1970b). Premières recherches sur les besoins nutritifs des embryons isolés due *Pinus silvestris* L. Embryons différenciés. *Compt. Rend. Acad. Sci. Paris* **270**, 2648–2651.

THOMAS, M.-J. (1972). Comportement des embryons de trois espèces de pins (*Pinus mugo* Turra, *Pinus silvestris* L. et *Pinus nigra* Arn.) isolés au moment de leur clivage et cultivés *in vitro*, en présence de cultures-nourrices. *Compt. Rend. Acad. Sci. Paris* **274**, 2655–2658.

THOMAS, M.-J. (1973a). Étude comparée des développements *in situ* et *in vitro* des embryons de pins: Influence des cultures-nourrices sur les embryons isolés durant les premières phases de leur développement. *Soc. Bot. Fr. Mémoires, Colloq. Morphologie*, 147–178.

THOMAS, M.-J. (1973b). Étude cytologique de l'embryogenèse du *Pinus silvestris* L.—I. Du proembryon cellulaire au clivage embryonnaire. *Rev. Cytol. Biol. Végét.* **36**, 165–252.

THOMAS, M.-J. and CHESNOY, L. (1969). Observations relatives aux mitochondries Feulgen positives de la zone périnucléaire de l'oosphère du *Pseudotsuga menziesii* (Mirb.) Franco. *Rev. Cytol. Biol. Végét.* **32**, 165–182.

THOMPSON, W. P. and JOHNSTON, D. (1945). The cause of incompatibility between barley and rye. *Canad. Jour. Res.* **C23**, 1–15.

THOMSON, R. B. (1934). Modification of the form of the haustorium in *Marsilia* on development in culture fluid. *Trans. Roy. Canad. Inst.* **20**, 69–72.

TILLETT, S. S. (1966). Cucurbit embryo culture. *Turtox News* **44** (1), 2–7.

TISSAOUI, T. and CÔME, D. (1973). Levée de dormance de l'embryon de pommier (*Pirus malus* L.) en l'absence d'oxygène et de froid. *Planta* **111**, 315–322.

TOBIN, E. M. and BRIGGS, W. R. (1969). Phytochrome in embryos of *Pinus palustris*. *Plant Physiol.* **44**, 148–150.

TONOLO, A. (1961). Colture artificiali di *Claviceps purpurea* (Fr.) Tul. I. Infezione e produzione di sclerozi su piante di cereali cresciute in ambiente sterile. *Rendicon. Isti. Super. Sanita* **24**, 452.

TOOLE, E. H., HENDRICKS, S. B., BORTHWICK, H. A. and TOOLE, V. K. (1956). Physiology of seed germination. *Ann. Rev. Plant Physiol.* **7**, 299–324.

TOROSIAN, C. D. (1971). Ultrastructural study of endosperm haustorial cells of *Lobelia dunnii* Greene (Campanulaceae, Lobelioideae). *Amer. Jour. Bot.* **58**, 456–457 (abstract).

TORREY, J. G. (1973). Plant embryo. *In* "Tissue Culture. Methods and Applications" (P. F. Kruse, Jr. and M. K. Patterson, Jr., eds), pp. 166–170. Academic Press, New York and London.

TOURTE, Y. (1968). Observations sur le comportement du noyau, des plastes et des mitochondries au cours de la maturation de l'oosphère du *Pteridium aquilinum* L. *Compt. Rend. Acad. Sci. Paris* **266**, 2324–2326.

TOURTE, Y. (1970). Nature, origine et évolution d'enclaves cytoplasmiques particulières au cours de l'oogénèse chez le *Pteridium aquilinum* (L.). *Rev. Cytol. Biol. Végét.* **33**, 311–324.

TREUB, M. (1884). Études sur les Lycopodiacees: Le prothalli de *Lycopodium cernuum* L. *Ann. Jard. Bot. Buitenzorg* **4**, 107–135.

TRIONE, S. O., DE HUNAU, R. C. and PONT LEZICA, R. F. (1971a). Estudios sobre crecimiento y morfogénesis en plantas. II. Algunas caracteristicas y propiedades del jugo nucelar-endospérmico de especies del genero *Prunus*. *Phyton* **28**, 39–50.

TRIONE, S. O., PONT LEZICA, R. F. and DE HUNAU, R. C. (1971b). Estudios sobre crecimiento y morfogénesis en plantas. I. Efectos promotores del crecimiento del jugo nucelar-endospérmico de almendra y durazno. *Phyton* **28**, 27–37.

TRUSCOTT, F. H. (1966). Some aspects of morphogenesis in *Cuscuta gronovii*. *Amer. Jour. Bot.* **53**, 739–750.

TUKEY, H. B. (1933a). Artificial culture of sweet cherry embryos. *Jour. Hered.* **24**, 7–12.

TUKEY, H. B. (1933b). Growth of the peach embryo in relation to growth of fruit and season of ripening. *Proc. Amer. Soc. Hort. Sci.* **30**, 209–218.

TUKEY, H. B. (1933c). Embryo abortion in early-ripening varieties of *Prunus avium*. *Bot. Gaz.* **94**, 433–468.

TUKEY, H. B. (1934a). Growth of the embryo, seed, and pericarp of the sour cherry (*Prunus cerasus*) in relation to season of fruit ripening. *Proc. Amer. Soc. Hort. Sci.* **31**, 125–144.

TUKEY, H. B. (1934b). Artificial culture methods for isolated embryos of deciduous fruits. *Proc. Amer. Soc. Hort. Sci.* **32**, 313–322.

TUKEY, H. B. (1938). Growth patterns of plants developed from immature embryos in artificial culture. *Bot. Gaz.* **99**, 630–665.

TUKEY, H. B. (1944). The excised-embryo method of testing the germinability of fruit seed with particular reference to peach seed. *Proc. Amer. Soc. Hort. Sci.* **45**, 211–219.

TUKEY, H. B. and CARLSON, R. F. (1945a). Morphological changes in peach seedlings induced by after-ripening treatments of the seed. *Proc. Amer. Soc. Hort. Sci.* **46**, 203–024 (abstract).

TUKEY, H. B. and CARLSON, R. F. (1945b). Morphological changes in peach seedlings following after-ripening treatments of the seeds. *Bot. Gaz.* **106**, 431–440.

TUKEY, H. B. and LEE, F. A. (1937). Embryo abortion in the peach in relation to chemical composition and season of fruit ripening. *Bot. Gaz.* **98**, 586–597.

TUKEY, H. B. and YOUNG, J. O. (1942). Gross morphology and histology of developing fruit of the apple. *Bot. Gaz.* **104**, 3–25.

TULECKE, W. (1953). A tissue derived from the pollen of *Ginkgo biloba*. *Science* **117**, 599–600.

TULECKE, W. (1959). The pollen cultures of C. D. LaRue: A tissue from the pollen of *Taxus*. *Bull. Torrey Bot. Club* **86**, 283–289.

TULECKE, W. (1960). Arginine-requiring strains of tissue obtained from *Ginkgo* pollen. *Plant Physiol.* **35**, 19–24.

TULECKE, W. (1964). A haploid tissue culture from the female gametophyte of *Ginkgo biloba*. *Nature* **203**, 94–95.

TULECKE, W. (1965). Haploidy versus diploidy in the reproduction of cell type. *In* "Reproduction: Molecular, Subcellular, and Cellular" (M. Locke, ed.), pp. 217–241. Academic Press, New York and London.

TULECKE, W. (1967). Studies on tissue cultures derived from *Ginkgo biloba* L. *Phytomorphology* **17**, 381–386.

TULECKE, W. and SEHGAL, N. (1963). Cell proliferation from the pollen of *Torreya nucifera*. *Contrib. Boyce Thompson Inst.* **22**, 153–163.

TULECKE, W., WEINSTEIN, L. H., RUTNER, A. and LAURENCOT, H. J., JR. (1961). The biochemical composition of coconut water (coconut milk) as related to its use in plant tissue culture. *Contrib. Boyce Thompson Inst.* **21**, 115–128.

TZOU, D.-S., GALSON, E. C. and SONDHEIMER, E. (1973). The metabolism of hormones during seed germination and release from dormancy. III. The effects and metabolism of zeatin in dormant and nondormant ash embryos. *Plant Physiol.* **51**, 894–897.

ULRICH, R. (1942). Observations sur la croissance de quelques fruits. *Rev. Sci.* **80**, 24–30.

ULRICH, R. (1957). Observations biométriques sur la croissance des fruits de lunaire (*Lunaria biennis*) Moench. *Bull. Soc. Bot. Fr.* **84**, 645–654.

USHA, S. V. (1965). *In vitro* pollination in *Antirrhinum majus* L. *Curr. Sci.* **34**, 511–513.

UTTAMAN, P. (1949a). A preliminary investigation into the viability of immature embryos of corn under conditions of cold storage at freezing point. *Curr. Sci.* **18**, 52–53.

UTTAMAN, P. (1949b). Culturing of pro-embryos of normal diploid corn (maize) aged 3 to 7 days. *Curr. Sci.* **18**, 215–216.

UTTAMAN, P. (1949c). The effect of cocoanut water on the growth of immature embryos of corn (maize). *Curr. Sci.* **18**, 251–252.

UTTAMAN, P. (1949d). Embryo culture to obtain F_1 plants of incompatible crosses in corn (maize). *Curr. Sci.* **18**, 297–299.

UTTAMAN, P. (1949e). A study in contrast of the effects of cocoanut water on the growth of immature embryos of corn (maize) when applied before and after germination of the embryo. *Curr. Sci.* **18**, 343–344.

VACIN, E. F. and WENT, F. W. (1949). Use of tomato juice in the asymbiotic germination of orchid seeds. *Bot. Gaz.* **111**, 175–183.

VALLADE, J. (1970). Développement embryonnaire chez un *Petunia hybrida* hort. *Compt. Rend. Acad. Sci. Paris* **270**, 1893–1896.

VALLADE, J. (1973). La formation des cotylédons et l'orientation du plan cotylédonaire chez le *Petunia hybrida* hort. *Soc. Bot. Fr. Mémoires*, *Colloq. Morphologie*, 355–366.

VALLADE, J. and BUGNON, F. (1974). Etapes du développement proembryonnaire et processus morphogénétiques fondamentaux. *Compt. Rend. Acad. Sci. Paris* **278**, 747–750.

VALLADE, J. and CORNU, A. (1973). Étude cytophotométrique des stades zygotiques chez le pétunia. *Compt. Rend. Acad. Sci. Paris* **276**, 2793–2796.

VANDENBELT, J. M. (1945). Nutritive value of coconut. *Nature* **156**, 174–175.

VAN DER EB, A. A. and NIEUWDORP, P. J. (1967). Electron microscopic structure of the aleuron cells of barley during germination. *Acta Bot. Neerl.* **15**, 690–699.

VAN DER PLUIJM, J. E. (1964). An electron microscopic investigation of the filiform apparatus in the embryo sac of *Torenia fournieri*. *In* "Pollen Physiology and Fertilization" (H. F. Linskens, ed.), pp. 8–16. North-Holland, Amsterdam.

VAN DE WALLE, C. and BERNIER, G. (1969). The onset of cellular synthetic activity in roots of germinating corn. *Exptl Cell Res.* **55**, 378–384.

VAN ONCKELEN, H. A., VERBEEK, R. and KHAN, A. A. (1974). Relationship of ribonucleic acid metabolism in embryo and aleurone to β-amylase synthesis in barley. *Plant Physiol.* **53**, 562–568.

VAN OVERBEEK, J. (1942). Hormonal control of embryo and seedlings. *Cold Spring Harb. Symp. Quant. Biol.* **10**, 126–133.

VAN OVERBEEK, J. (1941). Hormonal control of embryo and seedling. *Cold* in coconut milk essential for growth and development of very young *Datura* embryos. *Science* **94**, 350–351.

VAN OVERBEEK, J., CONKLIN, M. E. and BLAKESLEE, A. F. (1941b). Chemical stimulation of ovule development and its possible relation to parthenogenesis. *Amer. Jour. Bot.* **28**, 647–656.

VAN OVERBEEK, J., CONKLIN, M. E. and BLAKESLEE, A. F. (1942). Cultivation *in vitro* of small *Datura* embryos. *Amer. Jour. Bot.* **29**, 472–477.

VAN OVERBEEK, J., SIU, R. and HAAGEN-SMIT, A. J. (1944). Factors affecting the growth of *Datura* embryos *in vitro*. *Amer. Jour. Bot.* **31**, 219–224.

VAN STADEN, J. (1973). Changes in endogenous cytokinins of lettuce seed during germination. *Physiol. Plantarum* **28**, 222–227.

VAN STADEN, J. (1974). Cell division-inducing substances in the liquid endosperm of *Hyphaene natalensis*. *Jour. S. Afr. Bot.* **40**, 189–192.

VAN STADEN, J. and BROWN, N. A. C. (1972). Characterization of germination inhibitors in seed extracts of four South African species of Proteaceae. *Jour. S. Afr. Bot.* **38**, 135–150.

VAN STADEN, J. and DREWES, S. E. (1974). Identification of cell division inducing compounds from coconut milk. *Physiol. Plantarum* **32**, 347–352.

VAN STADEN, J. and WAREING, P. F. (1972). The effect of light on endogenous cytokinin levels in seeds of *Rumex obtusifolius*. *Planta* **104**, 126–133.

VAN STADEN, J., BROWN, N. A. C. and BUTTON, J. (1972a). The effects of applied hormones on germination of excised embryos of *Protea compacta* R. Br. *on vitro*. *Jour. S. Afr. Bot.* **38**, 211–214.

VAN STADEN, J., WEBB, D. P. and WAREING, P. F. (1972b). The effect of stratification on endogenous cytokinin levels in seeds of *Acer saccharum*. *Planta* **104**, 110–114.

VAN TIEGHEM, P. (1873). Recherches physiologiques sur la germination. *Ann. Sci. Nat. (V. Bot.)* **17**, 205–224.

VAN WENT, J. L. (1970a). The ultrastructure of the synergids of *Petunia*. *Acta Bot. Neerl.* **19**, 121–132.

VAN WENT, J. L. (1970b). The ultrastructure of the egg and central cell of *Petunia*. *Acta Bot. Neerl.* **19**, 313–322.

VAN WENT, J. L. (1970c). The ultrastructure of the fertilized embryo sac of *Petunia*. *Acta Bot. Neerl.* **19**, 468–480.

VAN WENT, J. L. and LINSKENS, H. F. (1967). Die Entwicklung des sogenannten "Fadenapparates" in Embryosack von *Petunia hybrida*. *Der Züchter* **37**, 51–56.

VARNER, J. E. and CHANDRA, G. R. (1964). Hormonal control of enzyme synthesis in barley endosperm. *Proc. Natl Acad. Sci. U.S.A.* **52**, 100–106.

VARNER, J. E. and SCHIDLOVSKY, G. (1963). Intracellular distribution of proteins in pea cotyledons. *Plant Physiol.* **38**, 139–144.

VARNER, J. E., BALCE, L. V. and HUANG, R. C. (1963). Senescence of cotyledons of germinating peas. Influence of axis tissue. *Plant Physiol.* **38**, 89–92.

VASIL, I. K. and HILDEBRANDT, A. C. (1966a). Variations of morphogenetic behavior in plant tissue cultures. I. *Cichorium endivia*. *Amer. Jour. Bot.* **53**, 860–869.

VASIL, I. K. and HILDEBRANDT, A. C. (1966b). Variations of morphogenetic behavior in plant tissue cultures. II. *Petroselinum hortense*. *Amer. Jour. Bot.* **53**, 869–874.

VASIL, I. K., HILDEBRANDT, A. C. and RIKER, A. J. (1964). Endive plantlets from freely suspended cells and cell groups grown *in vitro*. *Science* **146**, 76–77.

VASIL, V. (1959). Morphology and embryology of *Gnetum ula* Brongn. *Phytomorphology* **9**, 167–215.

VASIL, V. (1963). *In vitro* culture of embryos of *Gnetum ula* Brongn. In "Plant Tissue and Organ Culture—A Symposium" (P. Maheshwari and N. S. Rangaswamy, eds), pp. 278–280. International Society of Plant Morphologists, Delhi.

VASSILEVA-DRYANOVSKA, O. A. (1966a). Development of embryo and endosperm produced after irradiation of pollen in *Tradescantia*. *Hereditas* **55**, 129–148.

VASSILEVA-DRYANOVSKA, O. A. (1966b). Induced chromosomal aberrations and other abnormalities in the embryo sac following pollen irradiation in *Melandrium rubrum*. *Hereditas* **55**, 149–159.

VASSILEVA-DRYANOVSKA, O. A. (1966c). The induction of haploid embryos and tetraploid endosperm nuclei with irradiated pollen in *Lilium*. *Hereditas* **55**, 160–165.

VAZART, B. (1958). Différenciation des cellules sexuelles et fécondation chez les Phanérogames. *Protoplasmatologia* **7**, 3a, 1–158.

VAZART, B. (1971). Infrastructure de microspores de *Nicotiana tabacum* L. susceptibles de se développer en embryoïdes après excision et mise en culture des anthères. *Compt. Rend. Acad. Sci. Paris* **272**, 549–552.

VAZART, B. (1973a). Formation d'embryoïdes à partir de microspores de tabac: Évolution de l'infrastructure des cellules au cours de la première semaine de culture des anthères. *Soc. Bot. Fr. Mémoires, Colloq. Morphologie*, 243–260.

VAZART, B. (1973b). Ultrastructure des microspores de tabac dans les anthères embryogènes. *Caryologia* **25** (Suppl.), 303–314.

VAZART, B. and VAZART, J. (1965a). Infrastructure de l'ovule de lin, *Linum usitatissimum* L. L'assise jaquette ou endothélium. *Compt. Rend. Acad. Sci. Paris* **261**, 2927–2930.

VAZART, B. and VAZART, J. (1965b). Infrastructure de l'ovule de lin, *Linum usitatissimum* L. Les cellules du sac embryonnaire. *Compt. Rend. Acad. Sci. Paris* **261**, 3447–3450.

VAZART, B. and VAZART, J. (1966). Infrastructure du sac embryonnaire du lin (*Linum usitatissimum* L.). *Rev. Cytol. Biol. Végét.* **29**, 251–266.

VAZART, J. (1956). Etude cytologique de la reproduction sexuée chez quelques Ptéridophytes. *Rev. Cytol. Biol. Végét.* **17**, 263–432.

VAZART, J. (1969). Organisation et ultrastructure du sac embryonnaire du lin (*Linum usitatissimum* L.). *Rev. Cytol. Biol. Végét.* **32**, 227–240.

VECHER, A. S. and MATOSHKO, I. V. (1965). Accumulation of nucleic and other forms of phosphorus and total and protein nitrogen in maturing lupine seeds. *Biochemistry* (English translation) **30**, 808–814.

VEDEL, F. and D'AOUST, M. J. (1970). Polyacrylamide gel analysis of high molecular weight ribonucleic acid from etiolated and green cucumber cotyledons. *Plant Physiol.* **46**, 81–85.

VEEN, H. (1961). The effect of gibberellic acid on the embryo growth of *Capsella bursa-pastoris*. *Koninkl. Nederl. Akad. Wetensch. Proc.* **C64**, 79–85.

VEEN, H. (1962). Preliminary report on effects of kinetin on embryonic growth *in vitro* of *Capsella* embryos. *Acta Bot. Neerl.* **11**, 228–229.

VEEN, H. (1963). The effect of various growth-regulators on embryos of *Capsella bursa-pastoris* growing *in vitro*. *Acta Bot. Neerl.* **12**, 129–171.

VENKATESWARLU, J. and RAO, G. R. (1958). A contribution to the life-history of *Rubia cordifolia* Linn. *Jour. Indian Bot. Soc.* **37**, 442–454.

VERMYLEN-GUILLAUME, M. (1969). Quelques aspects de la structure d'embryons de carotte (*Daucus carota* L.) obtenus *in vitro*. *Bull. Soc. Roy. Bot. Belgique* **102**, 181–195.

VIJAYARAGHAVAN, M. R. and RATNAPARKHI, S. (1972). Some aspects of embryology of *Alectra thomsoni*. *Phytomorphology* **22**, 1–8.

VIJAYARAGHAVAN, M. R., JENSEN, W. A. and ASHTON, M. E. (1972). Synergids of *Aquilegia formosa*—Their histochemistry and ultrastructure. *Phytomorphology* **22**, 144–159.

VILLIERS, T. A. (1967a). Cytolysomes in long-dormant plant embryo cells. *Nature* **214**, 1356–1357.

VILLIERS, T. A. (1967b). Crystalloid structure in the microbodies of plant embryo cells. *Life Sci.* **6**, 2151–2156.

VILLIERS, T. A. (1968a). Intranuclear crystals in plant embryo cells. *Planta* **78**, 11–16.

VILLIERS, T. A. (1968b). An autoradiographic study of the effect of the plant hormone abscisic acid on nucleic acid and protein metabolism. *Planta* **82**, 342–354.

VILLIERS, T. A. (1971). Cytological studies in dormancy. I. Embryo maturation during dormancy in *Fraxinus excelsior*. *New Phytol.* **70**, 751–760.

VILLIERS, T. A. (1972a). Cytological studies in dormancy. II. Pathological ageing changes during prolonged dormancy and recovery upon dormancy release. *New Phytol.* **71**, 145–152.

VILLIERS, T. A. (1972b). Cytological studies in dormancy. III. Changes during low-temperature dormancy release. *New Phytol.* **71**, 153–160.

VILLIERS, T. A. (1972c). Seed dormancy. *In* "Seed Biology" (T. T. Kozlowski, ed.), Vol. 2, pp. 219–281. Academic Press, New York and London.

VILLIERS, T. A. and WAREING, P. F. (1960). Interaction of growth inhibitor and a natural germination stimulator in the dormancy of *Fraxinus excelsior* L. *Nature* **185**, 112–114.

VILLIERS, T. A. and WAREING, P. F. (1964). Dormancy in fruits of *Fraxinus excelsior* L. *Jour. Exptl Bot.* **15**, 359–367.

VILLIERS, T. A. and WAREING, P. F. (1965a). The possible role of low temperature in breaking the dormancy of seeds of *Fraxinus excelsior* L. *Jour. Exptl Bot.* **16**, 519–531.

VILLIERS, T. A. and WAREING, P. F. (1965b). The growth-substance content of dormant fruits of *Fraxinus excelsior* L. *Jour. Exptl Bot.* **16**, 533–544.

VISSER, T. (1956). The growth of apple seedlings as affected by after-ripening, seed maturity and light. *Koninkl. Nederl. Akad. Wetenschap. Proc.* **C59**, 325–334.

VOLD, B. S. and SYPHERD, P. S. (1968a). Modification in transfer RNA during the differentiation of wheat seedlings. *Proc. Natl Acad. Sci. U.S.A.* **59**, 453–458.

VOLD, B. S. and SYPHERD, P. S. (1968b). Changes in soluble RNA and ribonuclease activity during germination of wheat. *Plant Physiol.* **43**, 1221–1226.

VON VEH, R. (1936). Eine neue Methode der Anzucht von Sämlingen, unabhängig von Ruheperioden und Jahreszeit (bei Äpfeln, Birnen, Quitten, Pflaumen, Kirschen). *Der Züchter* **8**, 145–151.

VON VEH, R. (1939). Über Entwicklungsbereitschaft und Wüchsigkeit der Embryonen von Apfel, Pfirsich u.a. *Der Züchter* **11**, 249–255.

VOSS, H. (1939). Nachweis des inaktiven Wuchsstoffes, eines wuchsstoffantagonisten und deren wachstums-regulatorische Bedeutung. *Planta* **30**, 252–285.

VOZZO, J. A. (1973). Anatomic observations of dormant, stratified, and germinated *Quercus nigra* embryos. *Phytomorphology* **23**, 245–255.

VYSKOT, B. and NOVÁK, F. J. (1974). Experimental androgenesis *in vitro* in *Nicotiana clevelandii* Gray and *N. sanderae* hort. *Theoret. Appl. Genet.* **44**, 138–140.

WADHI, M. and MOHAN RAM, H. Y. (1964). Morphogenesis in the leaf callus of *Kalanchoe pinnata* Pers. *Phyton* **21**, 143–147.

WAGNER, G. and HESS, D. (1972). *In vitro*-Befruchtungen bei *Petunia hybrida*. *Zeitschr. Pflanzenphysiol*. **69**, 262–269.

WAGNER, G. and HESS, D. (1974). Haploide, diploide und triploide Pflanzen von *Petunia hybrida* aus Pollenkörnern. *Zeitschr. Pflanzenphysiol*. **73**, 273–276.

WALBOT, V. (1971). RNA metabolism during embryo development and germination of *Phaseolus vulgaris*. *Develop. Biol*. **26**, 369–379.

WALBOT, V. (1972). Rate of RNA synthesis and tRNA end-labeling during early development of *Phaseolus*. *Planta* **108**, 161–171.

WALBOT, V. (1973a). Effect of actinomycin D on growth and RNA synthesis during germination of *Phaseolus vulgaris*. *Caryologia* **25**, (Suppl.), 273–278.

WALBOT, V. (1973b). RNA metabolism in developing cotyledons of *Phaseolus vulgaris*. *New Phytol*. **72**, 479–483.

WALBOT, V., CLUTTER, M. and SUSSEX, I. M. (1972). Reproductive development and embryogeny in *Phaseolus*. *Phytomorphology* **22**, 59–68.

WALBOT, V., PECK, H. D. and DURE, L. S., III (1974). Characterization of a ribonucleoprotein particle believed to contain messenger RNA. *Federation Proc*. **34**, 1552 (abstract).

WALBOT, V., BRADY, T., CLUTTER, M. and SUSSEX, I. (1972). Macromolecular synthesis during plant embryogeny: Rates of RNA synthesis in *Phaseolus coccineus* embryos and suspensors. *Develop. Biol*. **29**, 104–111.

WALL, J. R. (1954). Interspecific hybrids of *Cucurbita* obtained by embryo culture. *Proc. Amer. Soc. Hort. Sci*. **63**, 427–430.

WALTON, D. C. (1966). Germination of *Phaseolus vulgaris*. I. Resumption of axis growth. *Plant Physiol*. **41**, 298–302.

WALTON, D. C. and SOOFI, G. S. (1969). Germination of *Phaseolus vulgaris*. III. The role of nucleic acid and protein synthesis in the initiation of axis elongation. *Plant Cell Physiol*. **10**, 307–315.

WALTON, D. C., SOOFI, G. S. and SONDHEIMER, E. (1970). The effects of abscisic acid on growth and nucleic acid synthesis in excised embryonic bean axes. *Plant Physiol*. **45**, 37–40.

WANG, C.-C., CHU, C.-C., SUN, C.-S., WU, S.-H., YIN, K.-C. and HSÜ, C. (1973). The androgenesis in wheat (*Triticum aestivum*) anthers cultured *in vitro*. *Scient. Sinica* **16**, 218–225.

WANG, C.-C., SUN, C.-S. and CHU, Z.-C. (1974). On the conditions for the induction of rice pollen plantlets and certain factors affecting the frequency of induction. *Acta Bot. Sinica* **16**, 43–54.

WANG, Y.-Y., SUN, C.-S., WANG, C.-C. and CHIEN, N.-F. (1973). The induction of the pollen plantlets of *Triticale* and *Capsicum annuum* from anther culture. *Scient. Sinica* **16**, 147–151.

WARD, M. and WETMORE, R. H. (1954). Experimental control of development in the embryo of the fern, *Phlebodium aureum*. *Amer. Jour. Bot*. **41**, 428–434.

WARDLAW, C. W. (1949). Experiments on organogenesis in ferns. *Growth* **13** (Suppl.), 93–131.

WARDLAW, C. W. (1955). "Embryogenesis in Plants". Methuen, London.

WARDLAW, C. W. (1965a). General physiological problems of embryogenesis in plants. *In* "Encyclopedia of Plant Physiology" (W. Ruhland, ed.), Vol. 15/1, pp. 424–442. Springer-Verlag, Berlin.

WARDLAW, C. W. (1965b). Physiology of embryonic development in cormophytes. *In* "Encyclopedia of Plant Physiology" (W. Ruhland, ed.), Vol. 15/1, pp. 844–965. Springer-Verlag, Berlin.

WARDLAW, C. W. (1965c). "Organization and Evolution in Plants". Longmans, London.

WARDLAW, C. W. (1968a). "Essays on Form in Plants". Manchester University Press.

WARDLAW, C. W. (1968b). "Morphogenesis in Plants. A Contemporary Study". Methuen, London.

WARMKE, H., RIVERA-PEREZ, E. and FERRER-MONGE, J. A. (1946). The culture of sugar cane embryos *in vitro*. *Inst. Trop. Agric. Univ. Puerto Rico 4th Ann Rep.*, 22–23.

WATANABE, A., NITTA, T. and SHIROYA, T. (1973). RNA synthesis in germinating red bean seeds. *Plant Cell Physiol.* **14**, 29–38.

WATANABE, Y. (1974). Meiotic chromosome behaviours of áuto-pentaploid rice plant derived from anther culture. *Cytologia* **39**, 283–288.

WATERS, L. C. and DURE, L. S., III (1966). Ribonucleic acid synthesis in germinating cotton seeds. *Jour. Mol. Biol.* **19**, 1–27.

WEAVER, J. B., JR. (1957). Embryological studies following interspecific crosses in *Gossypium*. I. *G. hirsutum × G. arboreum*. *Amer. Jour. Bot.* **44**, 209–214.

WEAVER, J. B., JR. (1958). Embryological studies following interspecific crosses in *Gossypium*. II. *G. arboreum × G. hirsutum*. *Amer. Jour. Bot.* **45**, 10–16.

WEBB, D. P. and WAREING, P. F. (1972). Seed dormancy in *Acer*: Endogenous germination inhibitors and dormancy in *Acer pseudoplatanus* L. *Planta* **104**, 115–125.

WEBSTER, G. T. (1955). Interspecific hybridization of *Melilotus alba × M. officinalis* using embryo culture. *Agron. Jour.* **47**, 138–142.

WEEKS, D. P. and MARCUS, A. (1971). Preformed messenger of quiescent wheat embryos. *Biochim. Biophys. Acta* **232**, 671–684.

WENZEL, G. and THOMAS, E. (1974). Observations on the growth in culture of anthers of *Secale cereale*. *Zeitschr. Pflanzenzüchtg* **72**, 89–94.

WERCKMEISTER, P. (1934). Über die künstliche Aufzucht von Embryonen aus *Iris*-Bastardsamen. *Gartenbauwiss.* **8**, 606–607.

WERCKMEISTER, P. (1936). Über Herstellung und künstliche Aufzucht von Bastarden der Gattung *Iris*. *Gartenbauwiss.* **10**, 500–520.

WERCKMEISTER, P. (1952). Embryokulturversuche zur Frage des Keimversuches von Irissamen aus der Sektion Regelia Foster. *Ber. Deut. Bot. Ges.* **65**, 321–325.

WERCKMEISTER, P. (1962). Studien über die künstliche Aufzucht von Embryonen ausserhalb des Samens in der Gattung *Iris* L. *In* "Advances in Horticultural Science and their Applications" (J.-C. Garnaud, ed.), Vol. 2, pp. 351–359. Pergamon Press, Oxford.

WETHERELL, D. F. and HALPERIN, W. (1963). Embryos derived from callus tissue cultures of the wild carrot. *Nature* **200**, 1336–1337.

WETMORE, R. H. and STEEVES, T. A. (1971). Morphological introduction to growth and development. *In* "Plant Physiology—A Treatise" (F. C. Steward, ed.), Vol. 6A, pp. 3–166. Academic Press, New York and London.

WHEELER, C. T. and BOULTER, D. (1967). Nucleic acids of developing seeds of *Vicia faba*. *Jour. Exptl Bot.* **18**, 229–240.

WHITAKER, D. M. (1940). Physical factors of growth. *Growth* **4** (Suppl.), 75–90.

WHITE, N. H. (1950). The significance of endosperm. *Austral. Jour. Sci.* **13**, 7–9.

WHITE, P. R. (1932). Plant tissue cultures. A preliminary report of results obtained in the culturing of certain plant meristems. *Arch. Exptl. Zellforsch.* **12**, 602–620.

WHITE, P. R. (1943). "A Handbook of Plant Tissue Culture". J. Cattell Press, Lancaster, Pa.

WHITTIER, D. P. and STEEVES, T. A. (1960). The induction of apogamy in the bracken fern. *Canad. Jour. Bot.* **38**, 925–930.

WILMAR, C. and HELLENDOORN, M. (1968). Growth and morphogenesis of *Asparagus* cells cultured *in vitro*. *Nature* **217**, 369–370.

WILSON, K. S. and CUTTER, V. M., JR. (1952). The distribution of acid phosphatases during development of the fruit of *Cocos nucifera*. *Amer. Jour. Bot.* **39**, 57–58.

WILSON, K. S. and CUTTER, V. M., JR. (1953). Acid phosphatase distribution patterns in the developing embryo of *Cocos nucifera*. *Jour. Elisha Mitchell Sci. Soc.* **69**, 170–174.

WILSON, S. B. and BONNER, W. D., JR. (1971). Studies of electron transport in dry and imbibed peanut embryos. *Plant Physiol.* **48**, 340–344.

WINKLER, A. J. and WILLIAMS, W. O. (1935). Effect of seed development on the growth of grapes. *Proc. Amer. Soc. Hort. Sci.* **33**, 430–434.

WITHNER, C. L. (1942). Nutrition experiments with orchid seedlings. *Amer. Orch. Soc. Bull.* **11**, 112–114.

WITHNER, C. L. (1943). Ovule culture: A new method for starting orchid seedlings. *Amer. Orch. Soc. Bull.* **11**, 261–263.

WITHNER, C. L. (1951). Effects of plant hormones and other compounds on the growth of orchids. *Amer. Orch. Soc. Bull.* **20**, 276–278.

WITHNER, C. L. (1953). Germination of "Cyps." *Orch. Jour.* **2**, 473–477.

WITHNER, C. L. (1955). Ovule culture and growth of *Vanilla* seedlings. *Amer. Orch. Soc. Bull.* **24**, 380–392.

WITHNER, C. L. (1959). Orchid physiology. *In* "The Orchids" (C. L. Withner, ed.), pp. 315–360. Ronald Press, New York.

WOCHOK, Z. S. (1973a). Microtubules and multivesicular bodies in cultured tissues of wild carrot: Changes during transition from the undifferentiated to the embryonic condition. *Cytobios* **7**, 87–95.

WOCHOK, Z. S. (1973b). DNA synthesis during development of adventive embryos of wild carrot. *Biol. Plantarum* **15**, 107–111.

Wochok, Z. S. and Burleson, B. (1974). Isoperosidase activity and induction in cultured tissues of wild carrot : A comparison of proembryos and embryos. *Physiol. Plantarum* **31**, 73–75.

Wochok, Z. S. and Wetherell, D. F. (1971). Suppression of organized growth in cultured wild carrot tissue by 2-chloroethylphosphonic acid. *Plant Cell Physiol.* **12**, 771–774.

Wollgiehn, R. (1960). Untersuchungen über den Zusammenhang zwischen Nucleinsäure- und Eiweisstoffwechsel in reifenden Samen. *Flora* **148**, 479–483.

Woo, S.-C. and Su, H.-Y. (1975). Doubled haploid rice from *indica* and *japonica* hybrids through anther culture. *Bot. Bull. Acad. Sinica* **16**, 19–24.

Woo, S.-C. and Tung, I.-J. (1972). Induction of rice plants from hybrid anthers of *indica* and *japonica* cross. *Bot. Bull. Acad. Sinica* **13**, 67–69.

Wood, A. and Bradbeer, J. W. (1967). Studies in seed dormancy. II. The nucleic acid metabolism of the cotyledons of *Corylus avellana* L. seeds. *New Phytol.* **66**, 17–26.

Woodard, J. W. (1956). DNA in gametogenesis and embryogeny in *Tradescantia*. *Jour. Biophys. Biochem. Cytol.* **2,** 765–777.

Woodard, J. W. (1958). Intracellular amounts of nucleic acids and protein during pollen grain growth in *Tradescantia*. *Jour. Biophys. Biochem. Cytol.* **4**, 383–390.

Woodcock, C. L. F. and Bell, P. R. (1968a) The distribution of deoxyribonucleic acid in the female gametophyte of *Myosurus minimus*. *Histochemie* **12**, 289–301.

Woodcock, C. L. F. and Bell, P. R. (1968b). Features of the ultrastructure of the female gametophyte of *Myosurus minimus*. *Jour. Ultrastr. Res.* **22**, 546–563.

Woodroof, J. G. and Woodroof, N. C. (1927). The development of the pecan nut (*Hicoria pecan*) from flower to maturity. *Jour. Agric. Res.* **34**, 1049–1063.

Wright, J. E. and Srb, A. M. (1950). Inhibition of growth in maize embryos by canavanine and its reversal. *Bot. Gaz.* **112**, 52–57.

Wright, S. T. C. (1956). Studies of fruit development in relation to plant hormones. III. Auxins in relation to fruit morphogenesis and fruit drop in the black currant *Ribes nigrum*. *Jour. Hort. Sci.* **31**, 196–211.

Wynd, F. L. (1933). Sources of carbohydrate for germination and growth of orchid seedlings. *Ann. Missouri Bot. Gard.* **20**, 569–581.

Yadav, S. P. and Das, H. K. (1974). Discontinuous incorporation of amino acids in embryo proteins of wheat during germination. *Develop. Biol.* **36**, 183–186.

Yamada, T., Nakagawa, H. and Sinotô, Y. (1967). Studies on the differentiation in cultured cells. I. Embryogenesis in three strains of *Solanum* callus. *Bot. Mag. Tokyo* **80**, 68–74.

Yamada, T., Shôji, T. and Sinotô, Y. (1963). Formation of calli and free cells in the tissue culture of *Tradescantia reflexa*. *Bot. Mag. Tokyo* **76**, 332–339.

Yamamoto, N., Sasaki, S., Asakawa, S. and Hasegawa, M. (1974). Polysome formation induced by light in *Pinus thunbergii* seed embryos. *Plant Cell Physiol.* **15**, 1143–1146.

YANUSHKEVICH, S. I. (1963). On the possibility of the modification of the ionizing radiation genetical effect in barley. *In* "Genetics Today", Proceedings of the XI International Congress of Genetics (S. J. Geerts, ed.), Vol. 1, p. 92 (abstract). Pergamon Press, Oxford.

YATES, R. C. and CURTIS, J. T. (1949). The effect of sucrose and other factors on the shoot–root ratio of orchid seedlings. *Amer. Jour. Bot.* **36**, 390–396.

YATSU, L. Y. (1965). The ultrastructure of cotyledonary tissue from *Gossypium hirsutum* L. seeds. *Jour. Cell Biol.* **25**, 193–199.

YATSU, L. and ALTSCHUL, A. M. (1963). Lipid-protein particles: Isolation from seeds of *Gossypium hirsutum*. *Science* **142**, 1062–1064.

YOMO, H. and VARNER, J. E. (1973). Control of formation of amylases and proteases in the cotyledons of germinating peas. *Plant Physiol.* **51**, 708–713.

YOMO, J. (1958). Barley malt. Sterilization of barley seeds and the formation of amylase by separated embryos and endosperms. *Hakko Kyokaishi* **16**, 444–448 (*Chem. Abstr.* 1960, **54**, col. 1669).

YOMO, J. (1960). Studies on the amylase-activating substance. V. Purification of the amylase-activating substance from the barley green malt. *Hakko Kyokaishi* **18**, 603–606 (*Chem. Abstr.* 1961, **55**, col. 26145).

YOO, B. Y. (1970) Ultrastructural changes in cells of pea embryo radicles during germination. *Jour. Cell Biol.* **45**, 158–171.

YOO, B. and JENSEN, W. A. (1966). Changes in nucleic acid content and distribution during cotton embryogenesis. *Exptl Cell Res.* **42**, 447–459.

YOSHII, Y. (1925). Über die Reifungsvorgänge des *Pharbitis*-samens mit besonderer Rücksicht auf die Keimungsfühigkeit des unreifen Samens. *Jour. Fac. Sci. Univ. Tokyo Sec. III. Bot.* **1**, 1–139.

YOUNG, J. L. and VARNER, J. E. (1959). Enzyme synthesis in the cotyledons of germinating seeds. *Arch. Biochem. Biophys.* **84**, 71–78.

YOUNG, R. S. (1952). Growth and development of the blueberry fruit (*Vaccinium corymbosum* L. and *V. angustifolium* Ait.). *Proc. Amer. Soc. Hort. Sci.* **59**, 167–172.

ZAGAJA, S. W. (1961). Preliminary results of investigations on the effect of low temperatures on the development of seedlings from immature embryos of sweet and sour cherries. *Bull. Acad. Polon. Sci. Ser. Biol.* **9**, 113–115.

ZAGAJA, S. W. (1962). After-ripening requirements of immature fruit tree embryos. *Hort. Res.* **2**, 19–34.

ZAGAJA, S. W., HOUGH, L. F. and BAILEY, C. H. (1960). The responses of immature peach embryos to low temperature treatments. *Proc. Amer. Soc. Hort. Sci.* **75**, 171–180.

ZDRUIKOVSKAIA-RIKHTER, A. I. and BABASIUK, M. S. (1974). Pollination and fertilization of seed buds in the culture *in vitro*. *Doklady Akad. Sci. SSSR* **218**, 1482–1484.

ZENCHENKO, V. A. (1954). The influence of the geographical latitude on the activity of peroxidase in barley embryos. *Doklady Akad. Nauk SSSR* **98**, 439–442.

ZENCHENKO, V. A. (1964). Cytochrome oxidase and peroxidase activity in the embryos of seeds of Northern origin. *Fiziol. Rastenii* (English translation) **11**, 228–230.

ZENCHENKO, V. A. (1965). Nonenzymatic oxidation in embryos of wheat seeds of Northern origin. *Fiziol. Rastenii* (English translation) **12**, 205–209.

ZENK, M. H. (1974). Haploids in physiological and biochemical research. *In* "Haploids in Higher Plants. Advances and Potential" (K. J. Kasha, ed.), pp. 339–353. University of Guelph.

ZENKTELER, M. (1965). Test tube fertilization in *Dianthus caryophyllus* Linn. *Naturwiss.* **52**, 645–646.

ZENKTELER, M. (1967). Test-tube fertilization of ovules in *Melandrium album* Mill. with pollen grains of several species of the Caryophyllaceae family. *Experientia* **23**, 775–776.

ZENKTELER, M. (1970). Test-tube fertilization of ovules in *Melandrium album* Mill. with pollen grains of *Datura stramonium* L. *Experientia* **26**, 661–662.

ZENKTELER, M. (1971). *In vitro* production of haploid plants from pollen grains of *Atropa belladonna* L. *Experientia* **27**, 1087.

ZENKTELER, M. (1972). Development of embryos and seedlings from pollen grains in *Lycium halimifolium* Mill. in the *in vitro* culture. *Biol. Plantarum* **14**, 420–422.

ZENKTELER, M. (1973). *In vitro* development of embryos and seedlings from pollen grains of *Solanum dulcamara*. *Zeitschr. Pflanzenphysiol.* **69**, 189–192.

ZENKTELER, M. A. and GUZOWSKA, I. (1970). Cytological studies on the regenerating mature female gametophyte of *Taxus baccata* L. and mature endosperm of *Tilia platyphyllos* Scop. in *in vitro* culture. *Acta Soc. Bot. Polon.* **39**, 161–173.

ZENKTELER, M., HILDEBRANDT, A. C. and COOPER, D. C. (1961). Growth *in vitro* of mature and immature carrot embryos. *Phyton* **17**, 125–128.

ZIEBUR, N. K. (1951). Factors influencing the growth of plant embryos. *In* "Plant Growth Substances" (F. Skoog, ed.), pp. 253–261. University of Wisconsin Press.

ZIEBUR, N. K. and BRINK, R. A. (1951). The stimulative effect of *Hordeum* endosperms on the growth of immature plant embryos *in vitro*. *Amer. Jour. Bot.* **38**, 253–256.

ZIEBUR, N. K., BRINK, R. A., GRAF, L. H. and STAHMANN, M. A. (1950). The effect of casein hydrolysate on the growth *in vitro* of immature *Hordeum* embryos. *Amer. Jour. Bot.* **37**, 144–148.

ZINGER, N. V. and PODDUBNAYA-ARNOLDI, V. A. (1966). Application of histochemical techniques to the study of embryonic processes in certain orchids. *Phytomorphology* **16**, 111–124.

ZWAR, J. A., KEFFORD, N. P., BOTTOMLEY, W. and BRUCE, M. I. (1963). A comparison of plant cell division inducers from coconut milk and apple fruitlets. *Nature* **200**, 679–680.

Author Index

Italicized numbers refer to References section.

Index to Plant Names

Numbers refer to pages in the text on which reference to the plant is made, either under its common name or under its scientific name. A plant designated by its common name in the text is listed in the index under its scientific name. A few plants for which scientific names are not used in the text are listed under their common names. Italicized numbers designate pages on which the plant is cited in a table or in a figure.

Subject Index